BIOTECHNOLOGY FOR DEGRADATION OF TOXIC CHEMICALS IN HAZARDOUS WASTES

BIOTECHNOLOGY FOR DEGRADATION OF TOXIC CHEMICALS IN HAZARDOUS WASTES

Edited by

R.J. Scholze, Jr., E.D. Smith, J.T. Bandy

U.S. Army Construction Engineering Research Laboratory
Champaign, IL

Y.C. Wu

Department of Civil and Environmental Engineering
New Jersey Institute of Technology
Newark, NJ

J.V. Basilico

U.S. Environmental Protection Agency
Washington, DC

NOYES DATA CORPORATION
Park Ridge, New Jersey, U.S.A.

Copyright © 1988 by Noyes Data Corporation
Library of Congress Catalog Card Number 88-1554
ISBN: 0-8155-1148-5
Printed in the United States

Published in the United States of America by
Noyes Data Corporation
Mill Road, Park Ridge, New Jersey 07656

10 9 8 7 6 5 4 3 2 1

Library of Congress Cataloging-in-Publication Data

Biotechnology for degradation of toxic chemicals in hazardous wastes /
    edited by R.J. Scholze, Jr. ... [et al].
        p. cm.
        Papers presented at the International Conference of Innovative
    Biological Treatment of Toxic Wastewaters, organized by the
    Consortium for Biological Waste Treatment Research and Technology,
    and others, held in Arlington, Va., June 1986.
        Bibliography: p.
        Includes index.
        ISBN 0-8155-1148-5
        1. Hazardous wastes--Congresses. 2. Sewage--Purification-
    -Biological treatment--Congresses. 3. Biotechnology--Congresses.
    I. Scholze, R.J. (Richard J.) II. International Conference of
    Innovative Biological Treatment of Toxic Wastewaters (1986 :
    Arlington, Va.)
    TD811.5.B56 1988   88-1554
    628.3'51--dc19   CIP

Foreword

The state-of-the-art of biotechnology for degradation of toxic chemicals in hazardous wastes is discussed in this book. It is based on a conference held in Arlington, Virginia in June of 1986 to assess the applicability of using biotechnology for the treatment of hazardous/toxic wastewaters.

Full-scale application of biotechnology for the treatment of municipal and industrial wastewaters has been practiced for many years. However, whether this technology can be employed for detoxification and destruction of hazardous chemicals in aqueous and solid media is not yet fully understood. Removal of toxic and refractory organics in wastewater, groundwater, and leachate may be more efficient as a result of combining biological treatment with other treatment technologies such as chemical and physical methods. Development of standard techniques for biotoxicity detection and toxicity reduction evaluation is essential and extremely important to both technical determination and decisions on the future policy for hazardous waste management. The papers presented in the book describe current research in biotechnology for degradation of toxic chemicals in hazardous wastes.

Thirty-seven of the fifty-five papers presented at the conference are included in the book, along with the results of a research needs workshop held concurrently.

The information in the book is from *International Conference on Innovative Biological Treatment of Toxic Wastewaters,* edited by R.J. Scholze, Jr., Y.C. Wu, E.D. Smith, J.T. Bandy, and J.V. Basilico for the U.S. Army Construction Engineering Research Laboratory; the U.S. Environmental Protection Agency; the National Science Foundation; the Naval Civil Engineering Laboratory; and the Consortium for Biological Waste Treatment Research and Technology, consisting of the University of Pittsburgh, the New Jersey Institute of Technology, the University of Delaware, and Vanderbilt University; April 1987.

The table of contents is organized in such a way as to serve as a subject index and provides easy access to the information contained in the book.

> Advanced composition and production methods developed by Noyes Data Corporation are employed to bring this durably bound book to you in a minimum of time. Special techniques are used to close the gap between "manuscript" and "completed book." In order to keep the price of the book to a reasonable level, it has been partially reproduced by photo-offset directly from the original report and the cost saving passed on to the reader. Due to this method of publishing, certain portions of the book may be less legible than desired.

ACKNOWLEDGEMENTS

The Conference Organizing Committee would like to acknowledge the invaluable contributions of the individuals listed below:

We thank the keynote speakers, Norbert Jaworski of USEPA, and Suellen W. Pirages of the National Solid Wastes Management Association, and Richard E. Speece of Drexel University. We wish to thank G.A. Anderson of the University of Newcastle Upon Tyne, Stephen W. Maloney of USA-CERL, John Liskowitz of New Jersey Institute of Technology, W.W. Eckenfelder, Jr., of Vanderbilt University, Yutaka Terashima of Kyoto University, John Bandy of USA-CERL, James Basilico of USEPA, Ronald Unterman of General Electric Company, Michael S. Switzenbaum of the University of Massachusetts, Roy Miller of U.S. Army for presiding at the technical sessions.

The Organizing Committee sincerely appreciates the contributions of Edward H. Bryan of NSF and A.F. Gaudy, Jr., of the University of Delaware in chairing the Research Needs Workshop and those of Robert W. Peters of Purdue University, Conley Hansen of Utah State University, Olli H. Tuovinen of Ohio State University, Ronald Unterman of General Electric Company, Michael S. Switzenbaum of the University of Massachusetts, James Basilico of USEPA, Dolloff Bishop of USEPA, Morgan E. Kommer of Aluminum Company of America, and D.B. Chan of the Naval Civil Engineering Research Laboratory in participating in the panel discussion.

Preface

The Consortium for Biological Waste Treatment Research and Technology which consists of the University of Delaware, the Vanderbilt University, the University of Pittsburgh and the new member, New Jersey Institute of Technology, has sponsored the bi-annual fixed film technology conference since 1980. This year the conference will focus on the subject of biological treatment of toxic wastewaters and the treatment technologies which will be covered and discussed include not only the fixed-film processes but also the other alternatives such as activated sludge or a combination of biological treatment with physical or chemical treatment. The main purpose of this conference is to understand the state-of-the-art of biological treatment technology for toxic wastewater management.

As usual, the conference is organized by the Consortium for Biological Waste Treatment Research and Technology in cooperation with the Army Construction Engineering Research Laboratory, the U.S. Naval Civil Engineering Research Laboratory, the U.S. Environmental Protection Agency, and the U.S. National Science Foundation. However, the Center for Research in Hazardous and Toxic Substances (formed by the New Jersey Institute of Technology, the Rutgers University, the Princeton University, the Stevens Institute of Technology, and the University of Medicine and Dentistry of New Jersey) has also actively participated in organizing the conference. The Members of the Conference Organizing Committee are: Dr. A.F. Gaudy, Jr. of the University of Delaware, Drs. Ed D. Smith and John Bandy and Mr. Richard Scholze of Army Construction Engineering Research Laboratory, Dr. Ed D. Bryan of National Science Foundation, Dr. D.B. Chan of Naval Civil Engineering Research Laboratory, Prof. W.W. Eckenfelder, Jr. of Vanderbilt University, Dr. Roy Miller of U.S. Army, and Mr. James V. Basilico, Thomas Pheiffer, Kent Dostal and Dr. Dolloff F. Bishop of Environmental Protection Agency.

The keynote addresses for the plenary session were given by: Dr. Norbert Jaworski of Environmental Protection Agency, Dr. Suellen W. Pirages of National Solid Waste Management Association, and Dr. Richard E. Speece of Drexel University. On behalf of the Organizing Committee, we wish to thank them for accepting our invitation. Finally, I personally would like to thank all of the conference participants and assistants for their support and interest in this conference.

Department of Civil and Environmental Engineering Y.C. Wu
New Jersey Institute of Technology
Newark, New Jersey 07102

NOTICE

These proceedings have been reviewed by the U.S. Army Construction Engineering Research Laboratory (USA-CERL), the University of Pittsburgh, the National Science Foundation (NSF) and the U.S. Environmental Protection Agency (USEPA) and approved for publication. Approval does not signify that the contents necessarily reflect the views and policies of USA-CERL, the University of Pittsburgh, the NSF, or the USEPA; nor does mention of trade names or commercial products constitute endorsement or recommendation for use.

The material presented herein is published as submitted by the authors. No attempt was made by the conference co-sponsors to edit, reformat, or alter the material provided except where necessary for production requirements or where obvious errors were detected. Any statements or views here presented are totally those of the authors and are neither condoned or disputed by the conference co-sponsors. On this basis the Publisher assumes no responsibility nor liability for errors or any consequences arising from the use of the information contained herein.

Final determination of the suitability of any information, procedure, or product for use contemplated by any user, and the manner of that use, is the sole responsibility of the user. The book is intended for informational purposes only. Expert advice should be obtained at all times when implementation is being considered, particularly where toxic chemicals, or other hazardous materials or processes are encountered.

Contents and Subject Index

KEYNOTE ADDRESS—BIOLOGICAL TREATMENT OF TOXICS IN WASTEWATER: THE PROBLEMS AND OPPORTUNITIES 1
Dolloff F. Bishop and Norbert A. Jaworski
  Introduction. 1
  The Problems 3
  EPA Research Approach. 8
    Treatability Research. 9
    Toxicity Detection and Reduction 12
    Innovative Approaches to Enhance Toxics Control—The
      Opportunities. 12
  Integrated System Engineering. 18
  Conclusions 20
  References. 20

KEYNOTE ADDRESS—HAZARDOUS WASTE MANAGEMENT: BIOLOGICAL TREATMENT 25
Suellen W. Pirages
  Role of Biological Treatment Technologies. 26
  Commercial Development. 27
    Degradation 27
    Concentration. 28
    Diverse Target Constituents. 28
    Consistency 29
    Relatively Low Cost. 30
  Stringent Treatment Requirements. 30
  Impact on Development of Biological Technologies 32
  Conclusion. 34

KEYNOTE ADDRESS—TOXICITY ASSAYS AND MOLECULAR STRUCTURE TOXICITY 36
R.E. Speece, N. Nirmalakhandan and Peter C. Jurs
    Introduction 36
    Toxicity Assays 37
    QSAR Background 40
    QSAR 43
    Log P Modelling 44
    Molecular Connectivity 48
    Chemometrics 54
    Drexel Correlations 54
    Summary 64
    References 64

COMPETITIVE KINETIC MODEL OF SUSPENDED-GROWTH INHIBITED BIOLOGICAL SYSTEMS 67
Pablo B. Sáez
    Introduction 67
    Mass Balances 68
    Kinetics Relationships 69
        Inhibitor Utilization Rate 69
        Substrate Utilization Rate 70
        Active Bacterial Growth Rate 71
    Competitive Kinetic Model 72
        General Model 72
        Limiting Cases 73
    Effect of the Parameters of the Model 74
    Summary and Conclusions 79
    References 80

TOXICITY OF NICKEL IN METHANE FERMENTATION SYSTEMS: FATE AND EFFECT ON PROCESS KINETICS 82
Sanjoy K. Bhattacharya and Gene F. Parkin
    Introduction 82
    Materials and Methods 83
    Kinetics of Acetate and Propionate Utilization 84
    Toxicity Kinetics 85
    Results and Discussion 86
        Slug Addition 88
        Continuous Addition 88
        Fate of Nickel 93
    Application of Uncompetitive-Inhibition-Coefficient Model 97
        Acclimation 97
        Using Measured Nickel in the Model 97
        Continuous Addition 97
    Conclusions 98
    References 100

HEAVY METAL REMOVAL BY AQUATIC MACROPHYTES IN A TEMPERATE CLIMATE AQUATIC TREATMENT SYSTEM 102
Paul L. Bishop and Jan DeWaters
- Introduction. 102
- Experimental Methods 104
- Results and Discussion 108
- Conclusions 117
- References 118

REMOVAL OF HEAVY METAL BY RECYCLING OF WASTE SLUDGE IN THE ACTIVATED SLUDGE PROCESS. 120
Kuei-lang Tsai and Pak-shing Cheung
- Introduction. 121
- Experimentals. 122
- Results and Discussion 123
  - Wastewater with Single Heavy Metal Ions. 123
    - Wastewater with Low Concentration of Cu. 123
    - Wastewater with High Concentration of Cu 123
  - Combined Wastewater with Different Heavy Metal Ions but Cu Dominant. 130
  - Combined Wastewater with Different Heavy Metal Ions but Cr and Fe Dominant. 130
- Conclusions 135
- References 136

THE EFFECT OF INORGANIC CATIONS ON BIOLOGICAL FIXED-FILM SYSTEMS. 137
Robert W. Peters
- Introduction. 137
- Objectives 141
- Results and Discussion 141
  - Trickling Filters 142
  - Anaerobic Filter 146
  - Rotating Biological Contactors (RBC's) 150
- Summary. 153
- Nomenclature. 153
- References 154

PHYSICAL-CHEMICAL AND ANAEROBIC FIXED FILM TREATMENT OF LANDFILL LEACHATE. 159
Dhandapani Thirumurthi, Shahid M. Rana and Thomas P. Austin
- Introduction. 159
- Current Treatment. 162
- Literature Review 162
- Objectives 164
- Experimental Procedure 165
  - Physical-Chemical Treatment. 165

Contents and Subject Index

    Biological Treatment 165
    Procedure 165
  Results 168
  Discussion of Results 168
    Physical-Chemical Treatment 168
    Biological Treatment 184
    Biomass Estimation 187
    Substrate (Removal) Utilization Rate 189
    Comparison with Previous Studies 189
  Conclusions 192
  References 194

TREATMENT OF LEACHATE FROM A HAZARDOUS WASTE LANDFILL SITE USING A TWO-STAGE ANAEROBIC FILTER 196
Y.C. Wu, O.J. Hao and K.C. Ou

  Introduction 196
  Materials and Methods 197
    Leachate 197
    System Description 197
    System Start-Up 199
    Analytical Methods 202
  Results and Discussion 204
    Start Up and Acclimation 204
    Influent Leachate Characteristics 205
    Waste Stabilization 208
    Kinetics 214
  Summary and Conclusions 215
  References 217

EFFECTS OF EXTENDED IDLE PERIODS ON HAZARDOUS WASTE BIOTREATMENT 221
A. Scott Weber, Mark R. Matsumoto, John G. Goeddertz and Alan J. Rabideau

  Introduction 221
  Experimental Approach 223
  Analytical Tests 224
  Results and Discussion 225
    Effect of Extended Idle Time Between Bioreactor Operation 225
    Effect of Temperature on Bioreactor Performance 228
    Effects of Powdered Activated Carbon (PAC) Addition to the
      Bioreactors 231
    Comparison of Effluent Quality Measures 239
  Summary 239
  References 242

A TECHNIQUE TO DETERMINE INHIBITION IN THE ACTIVATED SLUDGE PROCESS USING A FED-BATCH REACTOR 244
Andrew T. Watkin and W. Wesley Eckenfelder, Jr.

Introduction...244
Fed-Batch Reactor Methodology 245
    Fed-Batch Reactor Testing Protocol 245
    Theoretical Substrate and Oxygen Utilization Response in a
    Fed-Batch Reactor 247
Analysis of Experimental Results 249
    Differences Between Inhibition Constants of Various Sludges 252
Discussion 257
References..258

A COMPARISON OF THE MICROBIAL RESPONSE OF MIXED LIQUORS FROM DIFFERENT TREATMENT PLANTS TO INDUSTRIAL ORGANIC CHEMICALS 260
G. Lewandowski, D. Adamowitz, P. Boyle, L. Gneiding, K. Kim, N. McMullen, D. Pak and S. Salerno
    Background 261
    Procedures..261
    Microbial Characterization 263
    Results and Discussion 263
    References..272

SBR TREATMENT OF HAZARDOUS WASTEWATER—FULL-SCALE RESULTS 275
Kenneth L. Norcross III, Robert L. Irvine and Philip A. Herzbrun
    Introduction.......................................276
    Background 276
    Description of Facility 278
    Oxygen Transfer Theory.............................279
    Oxygen Transfer Evaluation Procedure 281
    Oxygen Transfer Discussion..........................281
    Process Results 283
    Summary...288
    References..291

THE USE OF PURE CULTURES AS A MEANS OF UNDERSTANDING THE PERFORMANCE OF MIXED CULTURES IN BIODEGRADATION OF PHENOLICS..292
G. Lewandowski, B. Baltzis and C. Peter Varuntanya
    Background 292
    Procedures..293
    Phenol-Degrading Species 295
    Pure-Culture Experiments...........................300
    Mixed-Culture Experiments and Kinetics 308
    Conclusions 308
    References..314

A COMPARTMENTALIZED ONE SLUDGE BIOREACTOR FOR SIMULTANEOUS REMOVAL OF PHENOL, THIOCYANATE, AND AMMONIA........316
Jeffrey H. Greenfield and Ronald D. Neufeld
- Introduction........316
- Literature Review........317
- Methods and Materials........319
  - Experimental System........319
  - Operation of Bioreactors........320
  - Experimental Methods........321
- Results and Discussion........321
  - One Compartment Bioreactor........322
  - Two Compartment Bioreactor........322
  - Three Compartment Bioreactor........325
  - Four Compartment Bioreactor........325
- Conclusions........326
- References........327

HIGH-RATE BIOLOGICAL PROCESS FOR TREATMENT OF PHENOLIC WASTES........329
Alan F. Rozich, Richard J. Colvin, and Anthony F. Gaudy, Jr.
- Introduction........329
- Background........330
- Materials and Methods........337
  - Bench-Scale Apparatus........337
  - Analyses........339
- Process Description........339
  - Bench Scale Results........339
  - Potential Field Applications........343
- Summary and Conclusions........347
- References........349
- List of Symbols........351
- Appendix........353

BIOTECHNOLOGY FOR THE TREATMENT OF HAZARDOUS WASTE CONTAMINATED SOILS AND RESIDUES........359
Mark E. Singley, Andrew J. Higgins, Vijay S. Rajput, Sumith Pilapitiya, Reba Mukherjee, and Ven Mercade
- Introduction........359
- Microcalorimeter........361
- References........373

BIOLOGICAL DEGRADATION OF POLYCHLORINATED BIPHENYLS........376
Ronald Unterman, Michael J. Brennan, Ronald E. Brooks, and Carl Johnson
- Introduction........376
- Biodegradation of Soil-Bound PCBs........377

Modeling a Mock Biodegradation Process 382
Summary. 385
References 386

**PROCESS DEVELOPMENT AND TREATMENT PLANT STARTUP
FOR AN EXPLOSIVES INDUSTRY WASTEWATER** 387
David M. Potter
   Introduction. 387
   Aerobic Treatability Study 390
   Two-Stage Anaerobic/Aerobic Treatability Study 398
   Comparison of Studies 410
   Full-Scale Start Up 414
   Summary. 418
   References 419

**UTILIZATION OF NITRITE OXIDATION INHIBITION TO
IMPROVE THE NITROGEN ELIMINATION PROCESS** 420
S. Suthersan and J.J. Ganczarczyk
   Introduction. 420
   Materials and Methods 422
   Results and Discussion 424
      Technical Concept of the Modification 424
      Definition of Parameters. 426
      Adaptation to Increasing FA Concentrations 426
      Deadaptation by Reverting to Favorable Conditions. 429
      Role of pH. 430
      Influence of Exposure Time 430
      Inhibitor: Biomass Ratio 434
      Effects on *Nitrosomonas* Due to the Inhibition Imposed on
         Nitrobacter 436
   Summary and Conclusions 438
   References 440

**ANAEROBIC TREATMENT OF MOLASSE SUGAR CANE
STILLAGE WITH HIGH MINERALS** 443
Michel Henry, Emmanuel Michelot, and Jean Pierre Jover
   Introduction. 444
   Material and Methods. 444
      Reactor. 444
      Analytical Methods 444
      Influents 444
   Results and Discussion 445
      Seeding and Start-Up of Reactor 446
      Transition to Molasse Cane Stillage. 446
      Influence of Salty Level on the System Stability 446
      Recovery of the System and Progression of Mineral Load 446
      Final Results 447
   Conclusions 448

POTENTIAL FOR ANAEROBIC TREATMENT OF HIGH SULFUR WASTEWATER IN A UNIQUE UPFLOW–FIXED FILM–SUSPENDED GROWTH REACTOR 449
 L. Syd Love
  Introduction. 449
  Sulfide Production in Anaerobic Degradation 450
  Sulfide Toxicity 451
  One Solution to Sulfide Toxicity 452
  An Alternative Solution to Sulfide Toxicity 452
  Advantages of the Sydlo Reactor 453
  The Sydlo Reactor—How It Operates 455
  Conclusion. 458
  References. 459
  Appendix A: HIPERION—Economical H_2S Removal by Direct
   Conversion to Elemental Sulfur. 460
   Process Description 460
   Advantage of the HIPERION Sulfur Removal Process. 462

ANAEROBIC DIGESTION OF CHEMICAL INDUSTRY WASTEWATERS CONTAINING TOXIC COMPOUNDS BY DOWNFLOW FIXED FILM TECHNOLOGY 463
 Michel Henry, Yves Thelier, and Jean Pierre Jover
  Introduction. 464
  Material and Methods. 464
   Reactor. 464
   Influents 465
   Analytical Methods 466
   Frequency of Sampling. 466
  Results 466
   Start-Up Phase—Loading Rate Increase—Influent NR 1 466
   Toxicity of CA—Effect of CA Shock During Start-Up Phase. 466
   Recovery of the System 467
   Performances Obtained with Influent NR 1 467
   Transition to Influent NR 2. 468
  Discussion. 470
  Conclusion. 471
  References. 471

TREATMENT OF PROCESS WASTEWATER FROM PETROCHEMICAL PLANT USING A ROTATING BIOLOGICAL CONTACTOR—A CASE STUDY 472
 Warren C. Davis, Jr. and Tom M. Pankratz
  General 472
  Introduction. 473
  Product (System) Description 476
  Pilot Test Chronology 477
  Test Methods 478
  Results and Discussion 478

Start-Up—November. 478
Flowrate—December 478
Disc Rotation—January. 479
PAC Addition—February. 479
Pre-aeration—March and April 480
Conclusions 482
References. 482

LAND TREATMENT OF NITROGUANIDINE WASTEWATER 483
Richard T. Williams, A. Ronald MacGillivray and David E. Renard
Introduction. 484
Materials and Methods 485
Results 486
Summary. 486
References. 487

COMBINED FIXED BIOLOGICAL FILM MEDIA AND EVAPORATIVE COOLING MEDIA TO SOLIDIFY HAZARDOUS WASTES FOR ENCAPSULATION AND EFFICIENT DISPOSAL 489
Sheldon F. Roe, Jr.
Introduction. 489
Discussion 490
Existing Technology 490
Conventional Evaporation. 490
Utility Cooling Tower Evaporation. 491
Description of Evaporative Cooling with Structured Packing . . . 491
C.L.E.A.R.S. Process 494
VOC Stripping 494
Biotechnology 494
Process Description 496
A Process Example 498
Evaporation—Biotechnology Combination 500
Evaporation—Anaerobic 500
Anaerobic—Evaporation 500
Aerobic—Evaporation. 501
Evaporation—Aerobic. 501
Ambient Temperature Evaporation. 501
Recommendations. 501
References. 502

FATE OF COD IN AN ANAEROBIC SYSTEM TREATING HIGH SULPHATE BEARING WASTEWATER 504
G.K. Anderson, J.A. Sanderson, C.B. Saw and T. Donnelly
Introduction. 504
Role of Sulphate Reducing Bacteria in Anaerobic Digestion. 505
Source of Sulphate Bearing Effluent in Edible Oil Refining 509
Description of Pilot Plants. 509
Results and Discussion 513

    Start-Up and Development of the SRB-Dominated System 513
    Inhibition of Sulphate Reducing Bacteria. 514
    Gross COD Removal Efficiency 514
    Relative Contribution of Carbonaceous COD and Sulphides to
      the Treated Effluent COD 518
    Removal of Carbonaceous COD 520
    COD:Sulphate Utilization Ratio. 520
    Downstream Treatment 526
Conclusions 528
References 529

THE FATE OF 4,6-DINITRO-o-CRESOL IN MUNICIPAL ACTIVATED SLUDGE SYSTEMS 532
Henryk Melcer and Wayne K. Bedford

Introduction. 532
Methods and Materials 533
Results and Discussion 536
    Conventional Parameters. 536
    Fate of 4,6-Dinitro-o-Cresol. 538
Conclusions 541
References 541

PILOT-SCALE ANAEROBIC BIOMASS ACCLIMATION STUDIES WITH A COAL LIQUEFACTION WASTEWATER 542
David N. Young, Eric B. Vale and Eric R. Hall

Introduction. 542
Materials and Methods 544
Experimental Design 547
Results and Discussion 548
    Wastewater Characteristics 548
    Co-Substrate Addition 550
    Low Severity Solvent Extraction Pretreatment 551
    Anhybrid Reactor Acclimation to Increasing Organic Loading
      Rate. 555
    GAC Fluidized Bed Reactor Acclimation to Increasing
      Organic Loading Rates. 561
    Comparative Performance of Anhybrid and GAC Fluidized
      Bed Reactors 568
Summary and Conclusions 569
References 571

ANOXIC/OXIC ACTIVATED SLUDGE TREATMENT OF CYANOGENS AND AMMONIA IN THE PRESENCE OF PHENOLS 573
Deanna J. Richards and Wen K. Shieh

Introduction. 573
Material and Methods. 574
    Experimental Units 574
    Experimental Design 577

Analytical Procedure 577
Results and Discussion 577
   TOC Removal. 577
   Cyanide and Thiocyanate Removal 578
   Formation and Removal of Ammonia-Nitrogen 579
Conclusion 581
References 582

PARTITIONING OF TOXIC ORGANIC COMPOUNDS ON MUNICIPAL WASTEWATER TREATMENT PLANT SOLIDS 584
Richard A. Dobbs, Michael Jelus and Kuang-Ye Cheng

Introduction 584
Materials and Methods 586
   Reagents 586
   Analytical Methods 586
   Wastewater Solids Preparation 587
Experimental Procedure 588
   Experimental Protocol Development 588
      Preliminary Studies 588
   Sorption or Partitioning of Selected Toxic Organics 593
Discussion of Results 593
Summary and Conclusions 598
References 599

PATAPSCO WASTEWATER TREATMENT PLANT TOXICITY REDUCTION EVALUATION 601
John A. Botts, Jonathan W. Braswell, Elizabeth C. Sullivan, William Goodfellow, Burton D. Sklar and Dolloff F. Bishop

Introduction 601
   Background 602
      Wastewater Sources 602
      Wastewater Treatment Facilities 602
   Key Project Tasks 604
Relationship of Plant Performance to Effluent Toxicity 605
   Influent Parameters versus Effluent Toxicity 605
   Operational Parameters versus Effluent Toxicity 606
   BOD and COD Removals versus Effluent Toxicity 607
   Performance During Toxic Events 608
Summary 611
Initial Study Results 611
   Wastewater Toxicity Characterization 611
Future Project Tasks 616
   Characterization of Plant Performance 616
   Identification of Specific Toxics 616
   Wastewater Treatability 619
Project Benefits 620
References 621

MICROTOX ASSESSMENT OF ANAEROBIC BACTERIAL TOXICITY ... 622
Doris S. Atkinson and Michael S. Switzenbaum
- Introduction ... 622
- Background ... 623
    - Anaerobic Toxicity Testing ... 623
    - Microtox Testing ... 624
- Methods ... 625
    - Literature Study ... 625
    - Anaerobic Toxicity Assays (ATA's) ... 626
    - Microtox Toxicity Assays ... 626
- Results ... 627
    - Literature Review ... 627
    - Laboratory Results ... 632
    - Combined Results ... 632
- Discussion ... 632
- Summary and Conclusions ... 636
- References ... 638

RESPIRATION-BASED EVALUATION OF NITRIFICATION INHIBITION USING ENRICHED *NITROSOMONAS* CULTURES ... 642
James E. Alleman
- Introduction ... 642
- Background ... 642
- Methods and Materials ... 644
    - Enriched Culture Development and Maintenance ... 644
    - Batch Bioassay Procedure ... 644
    - Respirometric Analysis ... 645
    - Ammonium-Nitrogen Oxidation Rate Analysis ... 645
- Results and Discussions ... 645
- Summary ... 649
- References ... 649

ASSESSMENT OF THE DEGREE OF TREATMENT REQUIRED FOR TOXIC WASTEWATER EFFLUENTS ... 651
G. Fred Lee and R. Anne Jones
- Introduction ... 651
- **Chemical Concentration-Based Effluent Limitations** ... 652
    - Ammonia Discharge Limits ... 654
    - Heavy Metal Discharge Limits ... 657
    - Chlorine Discharge Limits ... 660
- **Toxicity-Based Effluent Limitations** ... 661
    - "Quick and Easy" Toxicity Tests ... 663
- **Recommended Approach** ... 664
    - Part A—Chemical-Specific ... 665
    - Part B—Toxicity Testing ... 666
    - Variable Effluent Discharge Limits ... 668
    - Effluent Toxicity vs. Public Health ... 669
- **Hazardous Waste Treatment Facilities** ... 670

Toxicity of Sediment-Associated Contaminants............... 671
Conclusions 672
References... 673

RESEARCH NEEDS WORKSHOP 677
Edward H. Bryan and A.F. Gaudy, Jr., Chairman and Co-Chairman
    Discussion of Paper by Robert W. Peters 677
    Discussion of Paper Presented by Olli H. Tuovinen (Co-Authored
      by Conley Hansen) 677
    Preliminary Panel Comments............................ 678
      Remarks by Morgan Kommer 678
      Remarks by Dolloff F. Bishop......................... 678
      Remarks by D.B. Chan.............................. 680
      Remarks by Michael Switzenbaum 681
      Remarks by Ronald Unterman......................... 682
      Remarks by Anthony F. Gaudy 683
      Comments by Edward H. Bryan 683
    Panel Discussion 684
      Comments by C.P.L. Grady........................... 684
      Comments by Henryk Melcer......................... 688
    Closing Remarks by the Session Chairman 691

**INTRODUCTION TO THE CONSORTIUM FOR BIOLOGICAL
WASTEWATER TREATMENT RESEARCH AND TECHNOLOGY** 692
Anthony F. Gaudy, Jr.
    Introduction....................................... 692
    Background 693
    Current Activities 696

KEYNOTE ADDRESS—
BIOLOGICAL TREATMENT OF TOXICS IN WASTEWATER: THE PROBLEMS AND OPPORTUNITIES

Dolloff F. Bishop
Wastewater Research Division
Water Engineering Research Laboratory
U.S. Environmental Protection Agency
Cincinnati, Ohio

Norbert A. Jaworski
Office of Research and Development
U.S. Environmental Protection Agency
Washington, D.C.

INTRODUCTION

In 1976, with the consent decree on priority pollutants (1), the regulatory emphasis in the Clean Water Act for water pollution control in the United States shifted from regulation based upon conventional pollutants to regulations based upon specific toxics and conventional pollutants. Since that time, the Environmental Protection Agency's (EPA's) regulatory policy for control of toxics in wastewaters has evolved from effluent limitations, guidelines and standards for the primary industrial point sources for selected specific toxics to, in 1985, a National "Policy for the Development of Water Quality-Based Permit Limitations for Toxic Pollutants" (2). The water quality-based permit approach uses sensitive bioassays to indicate toxicity to the ecosystem of wastewater effluents. The approach also employs increasingly sensitive specific chemical monitoring to detect the discharge of specific toxics in wastewaters.

By itself, the water quality-based permit approach, with its broad and evolving detection capabilities, creates a challenge to the wastewater treatment industry to develop effective and economical techniques to control toxics in wastewaters. Parallel regulatory efforts on toxic and hazardous chemicals under the Toxic Substances Control Act, the Safe Drinking Water Act and the Resources Conservation and Recovery Act further strengthen the need for innovated wastewater treatment technology to manage toxic wastes in wastewaters.

For the Agency's regulatory policies on toxic and hazardous wastes to be effective, the Office of Research and Development must identify or develop practical and cost-effective approaches for control of these wastes. These approaches can be conveniently divided into the following general areas:

- waste reduction through industrial in-plant process modifications, recovery and reuse of toxics,
- biological treatment (aerobic and anaerobic) including advanced bioengineering concepts,
- physical-chemical treatment, and
- waste containment and immobilization.

A brief perspective on these areas as applied to toxics in wastewater is appropriate. The use of industrial in-plant process modification and toxics recovery and reuse is a desirable approach for minimizing toxic problems in wastewaters. Recovery and reuse will occur when the recovery cost is less than the costs of the treatment alternatives. Indeed, in the future with the increasing detection sensitivity for toxics and toxicity and improved risk assessment definition, the water quality-based permit approach can only encourage waste reduction efforts through in-plant recovery and reuse. For some wastes, recovery and reuse may be the only effective mechanism for environmental control.

Control of toxics through improved treatment, however, will remain an important tool for pragmatic management of toxics in wastewaters. As a treatment generality, EPA's experience has revealed that biological treatment usually is technically more effective and less costly than physical-chemical treatment for control of organic pollutants in wastewaters, especially those waters with complex mixtures of wastes. In some cases, a combination of biological and physical-chemical treatment may be the optimum treatment combination. Additionally, waste containment and immobilization techniques, such as landfills and other storage systems, offer at best solutions for only certain specialized toxics or hazardous wastes and do not provide satisfactory control for the bulk of the toxic and hazardous wastes in wastewaters. Thus, the regulatory challenge of consistently effective and economical treatment of toxics and hazardous wastes in wastewater hinges strongly on innovative biological treatment.

This Conference, thus, directly responds to the Agency's broad regulatory challenge. The individual papers that will be given here describe the direction and progress of research on the use of innovative biological treatment for control of toxics in wastewaters. In this address, we will describe the approach and scope of the Office of Research and Development's program for improving toxics control in wastewater treatment.

THE PROBLEMS

Early EPA research on biological treatment of toxics was designed to assess the removals of toxics in efficiently operated and acclimated treatment systems. The studies evaluated the process performance using steady or consistent influent concentrations of mixtures of representative priority pollutants under acclimated operating conditions. Typical spiked concentrations of the individual toxic organic ranged from 50 µg/L to 200 µg/L in pilot-scale municipal wastewater treatment systems (3,4,5). In a limited industrial field study (6) without spiked toxics, the influent toxics concentrations were consistently present and usually ranged from less than 1 milligram to a few milligrams per liter of individual toxic.

The results revealed typically good removals, often more than 95 percent (Table I) for semivolatile priority organics (3) and often exceeding 98 percent (Table II) for the volatile priority organics (4). The removal of indigenous trace metals (Table III) in these studies, however, varied ranging from about 20 to 90 percent, depending upon the individual metals (5), with pass-through concentrations of possible environmental concern. Required "categorical" pretreatment of the metals industry's discharges should often lead to acceptable metals pass-through and content in the municipal plant's sludges from conventional biological treatment.

Table I. Distribution of Semivolatile Organics in the Spiked Treatment System

	Final Effluent Conc. (μg/L)	Overall % Rem.	% in Primary Sludge	% in Waste Act. Sludge	% in Final Eff.	Total Mass Recovered (%)
PESTICIDES/PCBs						
Arochlor 1254	< 2.9	98	51	60	2	113
Heptachlor	< 2.3	93	33	35	7	75
Lindane	25.8	45	12	8	55	75
Toxaphene	< 2.9	98	31	27	2	60
PHENOLS						
2,4-dimethylphenol	< 0.9	99	0	0.5	1.1	1.6
Phenol	< 13.5	95	4.4	0.8	5	10.2
Pentachlorophenol	< 6.3	19	26	5	81	112
PHTHALATES						
Bis(2-ethylhexyl)phthalate	11.3	79	63	39	21	123
Butylbenzylphthalate	< 1.3	96	119	7	4	130
Diethylphthalate	< 1.2	97	7	9	3	19
Dimethylphthalate	< 0.8	98	0.3	2	2	4.3
Di-n-butylphthalate	< 2.7	94	39	11	6	56
Di-n-octylphthalate	< 4.8	83	91	43	17	151
POLYNUCLEAR AROMATIC HYDROCARBONS						
Acenaphthene	< 1.2	97	41	4	3	48
Anthracene	< 0.9	97	67	5	3	75
Benz(a)anthracene	< 0.6	98	67	18	2	87
Chrysene	< 1.2	97	75	13	3	91
Fluoranthene	< 1.9	94	84	13	6	103
Fluorene	< 0.7	98	51	3	2	56
Napththalene	< 0.7	99	22	0.5	1	23.5
Phenanthrene	< 1.1	97	59	1	3	63
Pyrene	< 2.0	94	107	7	6	120

Table II. Mean Concentrations Observed in the Experimental Treatment Sequence

	Inf. (µg/L)	Primary Eff. (µg/L)	Primary Sludge (µg/L)	RAS (µg/L)	Act. Sludge Eff. (µg/L)	Aer. Basin Off-gas (ng/L)	% Rem. by Primary Clarifier	% Rem. by Treatment Sequence
Methylene Chloride	118	89	<143	<21	<4.0	-	25	>97
1,1-Dichloroethene	79	34	<40	<1	<0.2	425	57	>99
Chloroform	137	143	<208	<7	3.6	838	0	>97
Carbon Tetrachloride	60	32	<14	<1	<0.2	322	47	>99
1,2-Dichloropropane	309	295	<461	<1	<6.0	1628	5	>98
Trichloroethylene	107	68	389	<1	<1.5	609	36	>99
1,1,2-Trichloroethane	133	155	<219	31	28.0	663	0	80
Dibromochloromethane	58	49	<10	<2	<7.0	1291	16	>88
Benzene	73	61	121	<1	<0.2	225	16	>99
1,1,1-Trichloroethane	132	96	<220	<1	<0.3	1022	27	>99
Bromodichloromethane	89	64	<25	<1	<0.2	1155	28	>99
Chlorobenzene	197	163	953	<5	<1.3	149	17	>99
Tetrachloroethylene and Tetrachloroethane	252	263	2033	25	16.0	1227	0	94
Toluene	255	198	974	<2	<0.6	672	22	>99
Ethylbenzene	82	69	766	<1	<0.2	243	16	>99

In these past studies, selected individual organics were poorly removed. Lindane, as an example (3), substantially passed through the aerobic treatment process. Furthermore, for some priority organics, even with efficient removals, residual pass-through concentrations of the organics could pose possible environmental concern (3). Finally, an important observation was made during these studies. The steady or consistent influent concentrations at the moderate levels encountered or added produced no measurable inhibitory or interference effects on the biological systems.

Table III. Metals Removal in Control Treatment Systems

Metal	Inf. (mg/L)	Prim. Eff. (mg/L)	Removal by Primary Clarifier (%)	Activated Sludge Eff. (mg/L)	Removal by Activated Sludge (%)	Total Rem. (%)
Ag[a]	8.0	9.0	0.0	5.0	44.0	37.5
As[a]	20.6	16.0	22.3	16.7	0.0	18.9
Cd[a]	20.9	17.0	18.7	7.9	53.5	62.2
Cr	0.63	0.51	19.0	0.34	33.3	46.0
Cu	0.80	0.57	28.8	0.16	71.9	80.0
Fe	4.29	2.44	43.1	1.01	58.6	76.5
Hg[a,b]	2.0	2.0	-	2.0	-	-
Mn	0.65	0.55	15.4	0.40	27.3	38.5
Ni	0.45	0.27	40.0	0.18	33.3	60.0
Pb	0.88	0.58	34.1	0.11	81.0	87.5
Zn	1.24	1.28	0.0	0.46	64.1	62.9

[a] Micrograms/liter.
[b] The Hg concentration is near the detection limit for the metal and removals cannot properly be calculated.

The completed EPA research has also evaluated the removal of spiked representative toxics in alternative treatment systems (7), as possible technologies for use on marine wastewater discharges, where the conventional pollutants (BOD, solids, and nutrients) do not significantly impact the water quality. The alternative pilot-scale systems included primary sedimentation, primary sedimentation plus filtration, chemical treatment in primary sedimentation, high rate trickling filters, facultative lagoons, and aerobic lagoons. These systems were compared with a primary sedimentatation-activated sludge system. The overall results revealed that conventional primary sedimentation-activated sludge (mean cell retention time ~ 7 days) provided the best overall control of toxics. The lagoons provided lower but fair control of the toxics. The remaining systems permitted significantly increased pass-through of toxics.

The EPA supported treatability studies (8,9) to assess the removal mechanisms in biological treatment. The fate of the toxic organics during treatment revealed (Tables IV and V) that acclimated operation at influent concentrations ranging from about 700 µg/L to 20 mg/L, even with relatively recalcitrant pentachlorophenol, produced very efficient removals of the individual organics, usually greater than 99 percent for acclimated operation. These studies with acclimated operation, however, revealed that the competitive removal mechanisms in the treatment systems, although minimizing pass-through of the organics in the wastewater effluent, could partially transfer the toxics into the wastewater sludges or, for volatile organics, into the air. Unacclimated operation increased the importance of the abiotic removal mechanisms, as well as the pass-through of the specific toxics.

Agency surveys such as the 40-City Survey (10) of municipal treatment plants and the EPA studies of industrial treatment plants (11) revealed variable influent concentrations and removal efficiencies of the individual priority pollutants in biological treatment plants in the field. The greater performance variability and poorer treatment efficiencies in the field were more noticeable in the municipal than in the industrial plants.

Completed and ongoing toxicity reduction studies also reveal the pass-through of residual toxicity. A completed study (12) revealed (Table VI) the variable presence and substantive pass-through of mutagenic materials in municipal wastewaters. An ongoing study (13) shows substantive but

processes of microorganisms may place limitations on the treatability of individual compounds in conventionally operated biological treatment. While diverse microbial populations occurring in the treatment of complex wastewaters are adaptive and thus can successfully degrade a wide range of organic compounds, the multiple organic nutrients in complex wastewaters, have variable thermodynamic-free energies available to support the metabolic activity of microorganisms. Indeed, individual organics (16) exhibit different effective energy for metabolic activity in different microorganisms and different degradation rates.

Recalcitrant organics--those with low available free energy, in the presence of sources of organics with more available free energy--may not be efficiently removed in the treatment available in the plant. The available data suggests that only a very small fraction of the toxics will be strongly recalcitrant to biodegradation producing substantive pass-through. More importantly, the acclimation of a much higher fraction of the toxic organics by the diverse microbial population in the process can be ineffective because of intermittent and highly variable amounts of the specific organic entering the industrial or municipal treatment system.

The abiotic competing removal mechanisms of volatilization and partitioning to the solids, although reducing water pollution, also cause other media pollution concerns. Transfer of volatile organics into the air from sewers, head works and the aerobic biological processes, even when substantive biodegradation occurs, cannot be fully prevented in conventionally operated wastewater treatment. Toxic compounds, especially the metals, partition to the sludges thus complicating the solids disposal processes on the land. Clearly, innovative operational approaches will be needed to minimize the limitations of conventionally operated wastewater treatment and to ensure, for control of toxics, more effective use of the substantial resources already invested in state-of-the-art wastewater treatment technology.

EPA RESEARCH APPROACH

With both problems and the treatment potential of biological treatment, an EPA approach is evolving to enhance the treatment potential and to minimize the problems. The toxics control program for wastewater treatment in the Agency's Water Engineering Research Laboratory in Cincinnati is following a three path approach in research. These paths are:

- treatability research on representative specific toxics,
- toxicity detection and reduction in treatment and pretreatment, and
- innovative concepts to enhance control of toxics in biological treatment.

Treatability Research

The treatability research aims at evaluating the equilibrium and kinetics relations of the principal removal mechanism in conventional primary-secondary (biological) treatment. The important mechanisms are:

- partitioning (sorption) on solids and biomass,
- volatilization (air stripping and surface desorption), and
- biodegradation (aerobic and anaerobic)

The goals of the research include developing appropriate bench-scale treatability protocols for experimentally determining the fate of specific toxics in and their impact on the conventional primary-secondary treatment system; determining kinetic and equilibrium data on representative toxics for the principal treatment mechanism; and, using the experimental data, developing predictive models for estimating the fate of toxics in conventional treatment from the chemical properties and structural activity relations of the individual toxics.

Substantive progress using first generation protocols has already been achieved. A paper (17) presented at this Conference describes an excellent correlation between the partitioning coefficient of organics from the water to the solids and biomass in wastewaters and the octanol/water partition coefficient of the individual toxics. The work also describes the kinetic rates for the partitioning of the toxics.

In volatilization, an early predictive model (18) developed for the EPA was only successful in predicting stripping from "clean water." The poor predictive performance, when the model was applied to biological treatment systems, was ascribed to solids and to biological activity

should also increase the effectiveness of toxics control by conventionally operated biological wastewater treatment systems. The problems of intermittent toxics pass-through and intermedia pollutant transfer, especially of volatiles to air, however, indicate a need for innovative approaches in biological treatment of wastewaters.

Table VII. Elements of a Toxicity Reduction Evaluation

- Evaluating in the Treatment Plant the Causes of Toxicity Pass-Through.
    - A. Plant Operations Deficiency.
    - B. Presences of Refractory Toxicity.
- Tracing Toxicity to Sources.
    - A. Bioassay Monitoring in Source Treatability Studies.
    - B. Alternative Monitoring Techniques.
- Identifying Toxicity Components.
    - A. Specific Chemical Identification.
    - B. Bioaccumulative Potential of Specific Toxics.
- Evaluating of Control Alternatives.
    - A. Industrial Process Modifications.
    - B. Pretreatment Alternatives
    - C. Central Treatment Plant Control of Toxicity

Studies in microbiology suggest that operational approaches can be used to increase degradation of recalcitrant compounds. A kinetics study in multiple carbon-source-limited growth (23) revealed that, under carbon-limited growth conditions, biogenic multiple carbon sources are used simultaneously and not sequentially by steady-state cultures of a marine bacteria. Additionally, the presence of multiple substrates produced substantially lower residual steady-state concentrations of the individual biodegradable substrate. Recently completed studies further reveal that the presence of multiple biogenic compounds under carbon-limited growth conditions also increases the growth rate and reduces the steady-state concentration of the recalcitrant xenobiotic organics in acclimated bacterial cultures; in one case (24), pentachlorophenol; in a second case (25), 2-chlorophenol.

in the biological process. Current volatilization studies (19) at the Water Engineering Research Laboratory have revealed that after partitioning equilibrium has been established, the stripping rate proceeds independently of the solids in the water (Figure 1). The solids do not affect the stripping rate as long as biodegradation is inhibited or does not occur. Further work is needed to establish the effects of surface active materials on the stripping process.

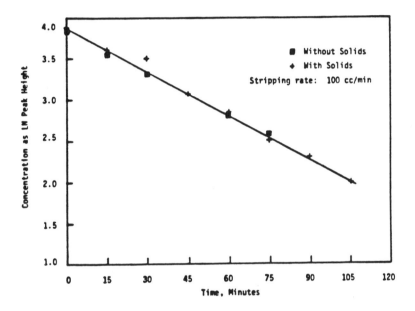

Figure 1. Impact of solids on stripping of chlorobenzene.

Initial progress on the biodegradation mechanism has also been achieved using an advanced electrolytic respirometric approach (20) to indicate relative acclimation requirements, intrinsic growth rate kinetics and kinetic inhibition rates for a few compounds. The technique (Figure 2) reduces the manpower required to develop the biodegradation data and, within two years, should provide a modest but sophisticated data base on representative toxics.

Toxicity Detection and Reduction

The Agency goals for research on toxicity detection and reduction in treatment and pretreatment are to:

- develop procedures for industrial and municipal wastewater treatment plants to isolate and identify the causes of effluent toxicity, and
- develop procedures to remove this toxicity either at the source or by enhanced treatment at the central plant.

The research approach features the conducting of case history studies called toxicity reduction evaluations (TREs) on plants known to pass-through toxics or toxicity. In these research TREs, the Agency is testing candidate alternative procedures for monitoring the treatment plants, for tracing the toxics or toxicity to their sources, and for identifying the specific toxics. Near real-time and inexpensive in-plant monitoring tools, which are alternatives to the classic bioassays for toxicity and conventional analytical methods for specific toxics, are also being evaluated. These monitoring alternatives include MICROTOX toxicity (21), respirometry (oxygen uptake methods) and adenosine triphosphate measurements.

The elements of a TRE at a treatment plant, after a problem of toxicity or toxics pass-through has been established, are shown in the Table VII. The results of the initial TRE studies (13,14,22), conducted at industrial and municipal wastewater treatment plants with good treatment of conventional pollutants but with toxicity pass-through problems, will be used to prepare Toxicity Reduction Evaluation Protocols for use by the wastewater treatment field.

Innovative Approaches to Enhance Toxics Control—The Opportunities

The above research paths are developing tools to increase regulatory effectiveness for toxics, especially through identifying and characterizing toxics problems in conventional wastewater treatment plants. Knowledge from treatability studies including the pretreatment studies

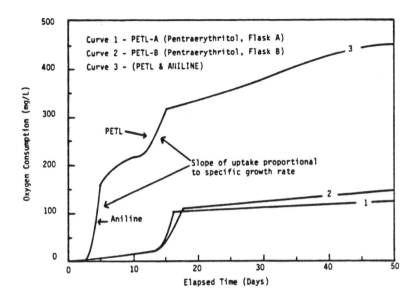

Figure 2. Biochemical oxygen uptake curve.

Successful but limited predictive models have already been developed (8) where appropriate experimental kinetic and equilibrium data have been established for the principal removal mechanisms. The current progress on the abiotic mechanisms and the relatively inexpensive development of the modest but sophisticated biodegradation data base suggest probable future success in developing tools for predicting the fate of toxics in conventional wastewater treatment by using the chemical properties and structural activity relations of the specific toxic compounds.

This research path also includes specific treatability studies on selected specific toxics for industrial treatment and pretreatment. When biological processes or conditions detrimental to biological treatment prevent successful control through biological methods, appropriate physical-chemical treatment options are being evaluated for these specific problem toxics but are beyond the scope of today's presentation.

These studies indicate that operational and microbiological mechanisms of metabolism to increase treatment efficiency of toxic organics are favored by the availability and/or control of:

- sufficient nitrogen, phosphorus, and trace nutrients,
- Appropriate dilution to prevent toxic inhibition,
- Co-metabolites, and
- Diverse microbial populations for improved adaptability.

Such conditions in wastewaters are usually best found in central municipal and industrial wastewaters. THE CHALLENGE THEN IS TO USE THE CENTRAL FACILITIES IN INNOVATIVE OPERATIONAL APPROACHES TO BETTER CONTROL THE TOXIC COMPOUNDS.

The Water Engineering Research Laboratory in Cincinnati, Ohio, has identified a series of innovative operational approaches to meet the challenge. Some of these approaches are being evaluated in ongoing research and TRE work. Others will undergo proof of concept testing as resources become available. The operational approaches are:

- controlled discharges of toxics into treatment,
- additives for improved toxic removal (clays, binding proteins, carbon, enzymes, etc.),
- anaerobic-aerobic treatment combinations,
- alternative operation of selected central plants as detoxification facilities, and
- bioengineered organisms.

The ongoing TRE studies are evaluating the effects of controlled discharges of toxics into the central treatment system. Simple storage and controlled releases should reduce the slugs of toxics currently reaching central treatment systems and thus minimize toxic pass-through as well as reduce chances for interference effects. The Laboratory is also supporting and has supported work on the use of additives in the central treatment system. Completed work includes assessment of powdered activated carbon as an additive in the biological process (8). Ongoing work involves developing and testing protocols with the National Sanitation Foundation (26) for the evaluation of bioaugmentation products. The testing protocols are being used to evaluate treatability effectiveness of existing bioaugmentation products, including microorganisms and their enzyme products, but can be applied to advanced products from bioengineering research.

The Water Engineering Research Laboratory also has a modest bioengineering research effort, cooperatively conducted with the Agency's Hazardous Waste Engineering Research Laboratory in Cincinnati, Ohio. In the toxics area, two projects are already showing promise. Recently, investigators (27) at Battelle-Columbus Laboratories isolated a pure culture of an anaerobe (called DCB-1) able to dehalogenate 3-chlorobenzoate (3-CBA). The gene coding for the anaerobic dehalogenating ability was found on the chromosome rather than on a plasmid. They next created a gene library of DCB-1 and cloned the entire chromosome into Escherichia (E.) coli. From this they were able to isolate a number of E. coli clones that contained pieces of the DCB-1 chromosome.

They are now testing those clones for expression of the 3-CBA activity. If E. coli is able to express the activity, it will be a major step forward, since it opens up a whole host of possibilities. For example, E. coli will be able to dehalogenate 3-CBA under facultative rather than strictly anaerobic conditions. It will be able to mobilize the activity into other species. Also, since it is hardier and able to grow much faster than anaerobes, it might be possible to obtain huge quantities of the enzyme cheaply and easily for possible use in the field instead of live organisms. The advantage of this lies in the fact that at present live cells of DCB-1 are not able to dehalogenate 3-CBA under sulfate-reducing conditions. The immobilized enzyme would.

Another factor is that the present anaerobe DCB-1 can only dehalogenate 3-CBA; it cannot grow on the product of the dehalogenation, benzoate. If the gene for 3-CBA dehalogenation were given to an organism able to grow on benzoate, then that organism could convert 3-CBA all the way to CO_2 and water.

In a second project (28) investigating the removal of the heavy metal cadmium from wastewater, researchers at the University of Washington, Seattle, have recently found that E. coli makes a cadmium binding peptide that is secreted outside the cell. In its native state, it is a high molecular weight aggregate of the peptide and cadmium bound tightly together. When the aggregate is reduced, the actual peptide is a low molecular weight molecule able to bind very tightly to cadmium. The investigators are trying to purify the peptide and deduce its chemical structure. Once that is done, it will then be immobilized onto a surface for the purpose of defining the binding kinetics to cadmium in aqueous solution.

From this work, the investigators have hypothesized a mechanism for the biological removal of cadmium in the environment. They suggest that the organism secretes the binding protein outside the cell, where it complexes with the toxic cadmium and prevents entry of the metal into the cell. The work can lead to innovative separation techniques for the control of cadmium in wastewaters.

Anaerobic-aerobic treatment combinations will be evaluated for improved toxics control in various alternative approaches. The TRE at the Patapsco Plant in Baltimore, Maryland, will assess the effect of sequential facultative anaerobic-aerobic operation on toxicity pass-through when the plant is converted to biological phosphorus removal. An anaerobic operating phase in a single sludge system will be used at Baltimore for biological phosphorus uptake. Indeed, bioengineered facultative dehalogenation developed in facultative organisms from the bioengineering research area may have a future role in the control of chlorinated organics in such operations. A second approach involves the use of separate stage anaerobic treatment, either as a pretreatment step at the toxics source or as a separate stage anaerobic process at the central treatment plant. The anaerobic treatment possibilities lead to the concept of using separate innovative operations at the central plant as a detoxification facility.

The substantial public and private investment in central conventional wastewater treatment systems can potentially be used to improve toxics control, not only in wastewater, but for toxics wastes in general. The use of selected central treatment plants as detoxification facilities requires unconventional operation of the wastewater collection and treatment systems. The potential for use as a detoxification center can be simply illustrated by describing several possible operational examples.

In the first example, the central industrial or municipal treatment plant may have capacity to successfully treat additional amounts of toxic wastes. These wastes are simply fed at the central plant into the wastewater at rates the plant can handle.

The next example involves the use of the central plant for treating the high COD leachates from landfills currently polluting groundwater. The leachates, usually from municipal landfills may or may not contain toxic components. Completed work (29) has evaluated the conventional treatment of these leachates by the central plant using sewers as the transport vehicle. Although technically feasible, such an approach produces substantive energy demands for aeration at the central plant and requires efficient anaerobic digestion of the difficult-to-digest aerobic biomass produced from the leachate.

An unconventional operational approach is to transport the high strength waste or leachate by truck directly to the central plant's digesters. Leachates, with substantive fatty acid contents from the landfill's incomplete anaerobic activity, are ideally suited for direct anaerobic digestion, using the diverse biomass and excess nutrients of the central treatment plant. The excess gas produced as recoverable energy may even exceed the energy requirements for the leachate transport. Such an approach also can be applied to anaerobically degradable toxics wastes currently handled in hazardous waste landfills.

Another example involves improved control of volatile toxic organics. The current operational approaches in central industrial or municipal wastewater treatment plants do not fully prevent volatilization transfer of at least a portion of the volatile organics into the air media. Even good biodegradation of the moderately volatile organics (Henry's Law Constant $\sim 10^{-3}$ m^3 atm/mole) such as benzene and toluene will, under acclimated conditions, only minimize the volatilization transfer (8). Compounds with Henry's Law Constant greater than those of the moderately volatile organics will exhibit increased volatilization transfer as their Henry's Law Constants increase, even in acclimated biological operation. Current techniques for controlling this problem include the use of steam distillation at the discharge source of these solvents to prevent their release into the atmosphere.

An unconventional operational alternative, however, could involve truck transport of wastewaters, at least those with reasonably high concentrations of volatiles, to the central plant for direct treatment in the plant's anaerobic digestion processes, or in specially established aerobic bioreactors, where control of the gas phase can be cost-effectively managed. Such reactors could include aerobic digesters using pure oxygen with oxygen recycle.

The use of special bioreactors to separately control recalcitrant toxic waste independently transported to the industrial or municipal wastewater treatment plant permits establishing a wide range of biological detoxification capabilities at these plants. Carbon-limited operation, with suitable co-metabolites, diverse biomass, nutrients, and appropriate dilution can be tailored to cost-effectively maximize the biodegradation of various recalcitrant toxic wastes. Aerobic-anaerobic treatment configurations can be readily established with back-up sources of organisms, co-metabolites and nutrients available from the central plant. To protect the bioreactors, rapid aerobic and anaerobic inhibition tests would be incorporated in the operational approach to determine appropriate feed rates to the reactors from toxic waste storage tanks. Indeed, a candidate aerobic inhibition test (30,31) using a batch-fed reactor is being described at this Conference.

As bioengineered organisms are developed, these organisms can be incorporated and controlled in the special reactors. The discharge from the special reactors could then be additionally treated through the treatment plant's main process system as back-up protection. Waste sources identified through TREs as being poorly treatable by conventional sewer transport and biological treatment would then have several alternative control options available including industrial inplant modifications, pretreatment at the discharge point and separate transport to a biological detoxification site at a central wastewater treatment plant.

INTEGRATED SYSTEM ENGINEERING

As a futuristic view, we envision an integrated systems engineering approach evolving--one that will take full advantage of technical, economic, and administrative capabilities of central water and wastewater treatment to control both toxics and traditional pollutants. The approach would include elements to prevent toxics transfer to the air, land, or groundwater. From the perspective of this Conference, the integrated management approach will focus on regulatory control of surface water quality using water quality-based permits.

The individual technology components in the integrated system (Figure 3) and in-stream water quality models for toxics are basically available. The linkage of the water supply quality and wastewater treatment effluent characteristics through the receiving water should soon be documented by Agency supported research to characterize surface water quality (32) and to develop surface water quality models (33) for assessing the impacts of effluent discharges and pollutant spills on the water quality of the lower Mississippi River.

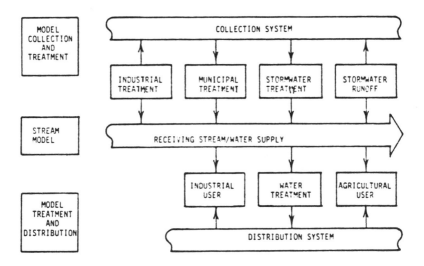

Figure 3. Components of the systems engineering concept.

The primary emphasis in the Mississippi study is on the movement of organics within the river and the effect discharges, under varying hydrological conditions, can have on concentrations of residual toxics at the raw waste supply intakes. As the Agency develops increasingly stringent maximum contaminant limits (MCL) for toxics in water supplies, the water utilities will be tempted to recover their increasing costs of treatment for drinking water from the upstream dischargers of the toxics. The linkage between wastewater effluent quality and water supply quality will strengthen the integrated systems engineering concept.

CONCLUSIONS

The toxicity impact on receiving water ecosystems, bioaccumulative uptake of toxics into the food chain, and effects of toxic discharges in wastewater effluents on the water quality of drinking water are strong factors driving the demand for improved technology for control of toxics. Ongoing EPA research should soon provide the scientific tools to support regulatory action by defining the environmental fate and effects and the health risk of toxics, by identifying the presence of toxics and toxicity in water and wastewater, and by tracing the toxics or toxicity to their sources. Effective toxics risk management through risk reduction, however, will require successful improvement of control technology.

The EPA is thus addressing approaches to improve control technology for managing toxics in wastewater. The costs for satisfactory control of the toxics will drive the EPA and the researchers in control technology to search for increasingly efficient and innovative technology. Systems engineering techniques will evolve to provide the tools for selection of technically effective and least costly combinations of control technology in industrial and municipal wastewater treatment and pretreatment and water treatment to meet the water quality and food chain requirements.

We at the EPA and you researchers in the field must meet the challenge for that technology innovation. This Conference addresses the critical core of the control technology--innovation in biological wastewater treatment. Best wishes in your endeavors this week.

REFERENCES

1. Natural Resources Defense Council (NRDC) et al. vs Train, 8 ERC 2120 (DDC 1976).
2. Technical Support Document for Water Quality-Based Toxics Control, Office of Water, U.S. Environmental Protection Agency, Washington, DC, September 1985.
3. Petrasek, A.C., et al., "Fate of Toxic Organic Compounds in Wastewater Treatment Plants," JWPCF, Vol. 55, No. 10, Oct., 1983, pp. 1286-1296.

4. Petrasek, A.C., Austern, B.M., and Neiheisel, T.W., "Removal and Partitioning of Volatile Organic Priority Pollutants in Wastewater Treatment," Proceedings of the Ninth U.S.-Japan Conference on Sewage Treatment Technology, EPA-600/9-85/014, Water Engineering Research Laboratory, U.S. Environmental Protection Agency, Cincinnati, Ohio, May, 1985, pp. 559-594.
5. Petrasek, A.C., and Kugelman, I.J., Metals Removals and Partitioning in Conventional Wastewater Treatment Plants," JWPCF, Vol. 55, No. 9, Sept., 1983, pp. 1183-1190.
6. Alsop, B.M., et al., Fate of Specific Organics in an Industrial Biological Wastewater Treatment Plant, Draft Report, Water Engineering Research Laboratory, U.S. Environmental Protection Agency, Cincinnati, Ohio, June 1984.
7. Hannah, S.A., et al., "Comparative Removal of Toxic Pollutants by Six Wastewater Treatment Processes," JWPCF, Vol. 58, No. 1, Jan., 1986, pp. 27-34.
8. Weber, W.J., and Jones, B.E., "Toxic Substance Removal in Activated Sludge and PAC Treatment Systems," EPA-600/2-86-045, Water Engineering Research Laboratory, U.S. Environmental Protection Agency, Cincinnati, Ohio, Apr., 1986.
9. Kirsch, E.J., Wukasch, E.J. Grady, C.P.L., "Fate of Eight Organic Priority Pollutants," Draft Report, EPA Cooperative Agreement No. CR-8076380-10, Water Engineering Research Laboratory, U.S. Environmental Protection Agency, Cincinnati, Ohio, 1985.
10. "Fate of Priority Toxic Pollutants in Publicly-Owned Treatment Works, EPA-440/1-82-303, Vol. I and II, Office of Water Regulation and Standards, U.S. Environmental Protection Agency, Washington, D.C., Sept., 1982.
11. Treatability Manual, Vol. III, Technology for Control/Removal of Pollutants, Office of Research and Development, U.S. Environmental Protection Agency, Washington, D.C., Revised Jan., 1983, pp. III.3.2.1-11 and 12.
12. Meier, J.R., and Bishop, D.F., "Evaluation of Conventional Treatment Processes for Removals of Mutagenic Activity from Municipal Wastewaters," JWPCF, Vol. 57, No. 10, Oct., 1985, pp. 999-1005.

13. Brasswell, J.W., et al., "Patapsco Wastewater Treatment Plant Toxicity Reduction Evaluation," Presented at the International Conference on Innovative Biological Treatment of Toxic Wastewaters, Arlington, Va., June 24-26, 1986.
14. Neiheisel, T.W., and Horning, W.B., "Influent Toxicity Assessment and Treatability in Municipal Wastewaters," Presented at the International Conference on Innovative Biological Treatment of Toxic Wastewaters, Arlington, VA, June 24-26, 1986.
15. Interferences at Publicly-Owned Treatment Works, Contract No. 68-03-1821, Water Engineering Research Laboratory, U.S. Environmental Protection Agency, Cincinnati, OH, 1986.
16. Stanier, R.Y., Adelberg, E.A., and Ingraham, J.L., The Microbial World, Fourth Ed., Prentice-Hall, Inc., Englewood Cliffs, NJ, 1970.
17. Dobbs, R.A., Jelus, M., and Cheng, K.Y., "Partitioning of Toxic Organic Compounds in Municipal Wastewater Treatment Plant Solids," Presented at the International Conference on Innovative Biological Treatment of Toxic Wastewaters, Arlington, VA, June 24-26, 1986.
18. Roberts, P.V., et al., Volatilization of Organic Pollutants in Wastewater Treatment--Model Studies," EPA-600/2-84-047, Municipal Environmental Research Laboratory, U.S. Environmental Protection Agency, Cincinnati, OH, Feb., 1984.
19. Rao, V.V., "Removal of Volatile Organics by Air Stripping," M.S. Thesis, Department of Civil and Environmental Engineering, University of Cincinnati, Cincinnati, OH, Feb., 1986.
20. Tabak, H.H., Lewis, R.F., and Oshima, A., "Electrolytic Respirometry Biodegradation Studies, CEC/OECD Ring Test of Respiration Method for Determination of Biodegradability," Municipal Environmental Research Laboratory, U.S. Environmental Protection Agency, Cincinnati, OH, Aug., 1984.
21. Beckman MICROTOX System Operations Manual (Beckman Instructions 015-555879), Beckman Instruments, Inc., Microbics Operations, Carlsbad, California, 1982.
22. Clark, R., et al., "Implementation of a Toxicity Reduction Evaluation at a Multi-Purpose Specialty Chemical Plant," Presented at the International Conference on Innovative Biological Treatment of Toxic Wastewaters, Arlington, VA, June 24-26, 1986.

23. Law, A.T., and Button, D.K., Multiple Carbon-Source-Limited Growth Kinetics of a Marine Coryneform Bacterium," Journal of Bacteriology, Vol. 129, No. 1, Jan., 1977, pp. 115-123.
24. Brown, E.J., et al., Pentachlorophenol Degradation: A Pure Culture and an Epilithic Microbial Consortium," Institute of Northern Engineering, University of Alaska-Fairbanks, 1986, Report in Review.
25. Cordone, L., "Effects of Feed Strength on the Kinetics of 2-Chlorophenol Removal from Multi-Component Media by Bacteria in Continuous Culture," Special Report, College of Engineering, Clemson University, May, 1985.
26. Bellen, G., and Anderson, M., "Bioaugmentation Product Test Methods Study and Standards Program," Contract No. S-11884-01-0, Water Engineering Research Laboratory, U.S. Environmental Protection Agency, Cincinnati, OH, 1984.
27. Pierce, G.E., "Determination of Enhancement of Anaerobic Dehalogenation and Degradation of Chlorinated Organics in Aqueous Solution," Cooperative Agreement No. CR-811120, Hazardous Waste Engineering Research Laboratory, U.S. Environmental Protection Agency, Cincinnati, OH, 1985.
28. Furlong, C.E., "Genetic Engineering Applied to Waste Stream Treatment," Cooperative Agreement No. CR-811948, Water Engineering Research Laboratory, U.S. Environmental Protection Agency, Cincinnati, OH, 1984.
29. Schuk, W.W. and James, S.C., "Treatment of Sanitary Landfill Leachate in Municipally-Owned Plants, Presented at the Wastewater Pretreatment and Toxicity Control Institute, March 13-14, 1986, University of Wisconsin-Milwaukee/Extension, Milwaukee, WI, 1986.
30. Grady, C.P.L., Loven, L.E., and Philbrook, D.M., "Assessment of Biodegradation Potential for Toxic Organic Pollutants," Presented at the 50th Annual Conference of the Water Pollution Control Federation, October 8. 1985.
31. Watkins, A.A. and Eckenfelder, W.W., "A Technique to Determine Inhibition in the Activated Sludge Process Using a Batch-Fed Reactor," Presented at the International Conference on Innovative Biological Treatment of Toxic Wastewaters, Arlington, VA, June 24-26, 1986.

32. Grayman, W.M., "Characterization of the Water Quality of the Lower Mississippi," EPA-600/2-85/043, Water Engineering Research Laboratory, U.S. Environmental Protection Agency, Cincinnati, OH, April 1985.
33. "Surface Water Quality Screening Model, A Case Study for Water Utility Management," " Sub-project of Cooperative Agreement with AWWA Research Foundation, CR-811335, Water Engineering Research Laboratory, U.S. Environmental Protection Agency, Cincinnati, OH, Aug. 20, 1984- Feb. 19, 1986.

KEYNOTE ADDRESS—
HAZARDOUS WASTE MANAGEMENT:
BIOLOGICAL TREATMENT

Suellen W. Pirages

National Solid Wastes Management Association
Washington, D.C.

Proper management of hazardous waste is a top priority of government officials, industrial managers, and private citizens. Since 1976, considerable effort has been expended toward this goal by

- identifying the origin of hazardous wastes,
- development of appropriate technologies for management of these wastes, and
- implementation of regulatory schemes that provide greater protection of public health and the environment.

Substantial progress has been made. We know that hazardous wastes are generated by the production of a variety of goods and services. Nearly every activity common to the current North American lifestyle results in the generation of some amount of wastes that could be hazardous if improperly managed. The production of medicines, household furniture and appliances, transportation vehicles, agricultural tools, and textiles are just a few examples of items whose manufacture generates hazardous wastes.

A waste service industry has emerged in response to the national concern for management of hazardous waste. The current members in the commercial service industry became involved in hazardous waste management activities only in response to the 1976 Resource Conservation and Recovery Act (RCRA). This new industry has made the commitment to manage wastes using state-of-the-art facilities and technologies that enhance protection to public health and the environment. At both the Federal and state level, regulatory programs are being developed that provide controls for all aspects of hazardous waste management. There exists a system to track the movement of hazardous wastes from point of origin to final disposal. Standards for proper transport of hazardous wastes have been developed. Minimum technological requirements for all management activities are being developed.

Although progress has been made, there is always room for improvement. New developments in facility design and treatment technologies must be encouraged. One of the areas that holds promise for better management capability is that of biological treatment technology, both for toxic liquids and solids. This paper discusses the role of biological treatment in the management scheme for hazardous wastes, the potential for commercial application, new stringent treatment standards, and the impact that mandates of the 1984 RCRA Amendments may have on the future application of biological treatment alternatives.

ROLE OF BIOLOGICAL TREATMENT TECHNOLOGIES

Application of microbial degradation and removal of undesirable constituents in industrial and municipal wastes is not a new concept. It is a commonly used process for general wastewater treatment activities and has been for many years. As the awareness of chemical contamination of the environment, much research on biological degradation of toxic chemicals has occurred. The topics to be presented during this conference attest to the fact that biological treatment of toxic constituents is possible. Among the range of treatment technologies, biological degradation ranks among the most effective. Its management application is enhanced by the potential to apply biological treatment in sequence with other chemical and thermal processes.

Another area for incorporating biological technologies in hazardous waste management activities is the recovery of reusable materials. Metals recovery is a very important area for biological applications. Because these inorganic elements cannot not be destroyed, an important goal is to recover and recycle metals, to the maximum extent possible. Using microbial-based technologies to recover inorganics may become an increasingly important area for further development as the US Environmental Protection Agency (US/EPA) establishes regulations about treatment of hazardous wastes. (This issue is discussed below in more detail.)

Finally, biological treatment has a very important application in the Superfund program. On-site cleanup of contaminated sites is one goal of the remedial program. Thus, in-situ treatment of contaminated soil and groundwater using biological processes is receiving increased attention as a cost-effective remedial action.

COMMERCIAL DEVELOPMENT

As the waste service industry attempts to respond to the management needs of the manufacturing sector of North America, biological treatment alternatives are receiving greater attention. In both Canada and the United States biological degradation of constituents in hazardous industrial wastewaters is now commercially available. There is room for more growth, however.

Such growth in availability will depend on the ability of this technology to meet specific criteria. If met, the desirability of commercial applications of biological technologies will be enhanced. Let me briefly outline these criteria.

1. Degradation. The preferred end-result for biological treatment is to have the constituents of concern completely

destroyed. Very often this technology simply removes the toxic component from one phase (i.e., liquid) to another (i.e., sludge or solid). When treating nonhazardous wastewaters, this is not particularly critical.

However, in hazardous waste management, the primary goal is to be able to delist treatment residues. By delisting a residue, it can be disposed in nonhazardous landfills at a much reduced cost. Within the current regulatory program, delisting is only possible, if the hazardous constituent of concern can no longer be detected in the treatment residue.

2. <u>Concentration</u>. Dilute hazardous wastes can pose a problem for cost-effective management. Many chemical and thermal treatment processes are only cost-effective on concentrated waste constituents. Biological treatment processes that concentrate these mixtures of dilute toxic constituents can be an important component in a sequential management strategy. For example, biological treatment may be used to concentrate organic constituents, followed by thermal treatment of the biological residue.

3. <u>Diverse</u> <u>Target</u> <u>Constituents</u>. The commercial industry serves a wide range of industrial clients. A major aspect of our services is to provide effective management of recalcitrant compounds and those waste streams that are mixtures of many hazardous constituents. Sometimes these mixtures include different organics chemicals; at other times, the wastes are mixtures of organic and inorganic elements. This latter situation is very difficult to treat at the present time.

In an ideal world, micro-organisms used in commercial applications should be capable of degrading mixtures of organics, not just single constituents. The efficiency of degradation should be relatively the same for all compounds found in the waste. In addition, it would be preferred that the micro-organisms could do "joint tasks". For example, in the same treatment batch, it would be desirable to degrade the organics present in a waste stream, while at the same time concentrating the inorganics for further treatment or recovery.

4. <u>Consistency</u>. Whether the treatment process yields a residue that is capable of delisting or one that will be treated further using another type of treatment process, consistency in the composition of the residue is essential. Variability in degradation efficiency among batches will prove costly to a waste management firm.

   Current regulations require continual monitoring of delisted treatment residue to assure that hazardous constituents are not present. If the end product of the biological process cannot be predicted reliably, then monitoring costs will be extensive, and biological treatment may not be cost-effective for commercial application.

   Also, many solidification treatment processes that could be used in sequence with biological concentration of hazardous constituents, often are sensitive to changes in constituent concentrations. For example, the effectiveness of a process may be constrained by maximum allowable concentrations of organics (i.e., less than 20%). If the biological process is too variable, effective solidification may not be possible. Again the impact for the management firm is unanticipated increased costs.

5. <u>Relatively low cost</u>. Currently, commercial waste management practices are dictated by client demands. These demands generally focus on reliable and <u>least costly</u> services. A generator will contract to have a waste treated using the least costly method available within regulatory constraints. Because of this fact of life, it has been difficult to reduce the use of landfills. Generators have not been willing to pay the high cost associated with alternative options. If biological treatment cannot compete with other processes (e.g., chemical and thermal applications) in effectiveness and cost, it will not gain a strong position in the hierarchy of commercial management alternatives.

STRINGENT TREATMENT REQUIREMENTS

In 1984, Congress passed the Hazardous and Solid Waste Amendments (HSWA). The purpose of these amendments was to reduce the nation's assumed dependence on land disposal of hazardous wastes. The intent of Congress was to force treatment of <u>all</u> hazardous wastes. Thus the mandate to the US/EPA was to determine which wastes should be restricted from land disposal unless treated.

Congress instructed the Agency to identify appropriate treatment alternatives and to set up a petition process for evaluating waivers to the treatment requirement. Congress indicated that if a generator could show that there would be no migration from a land disposal facility for "as long as the waste remained hazardous", then treatment prior to disposal may not be necessary.

Originally, the Agency proposed to set treatment standards using health-based criteria. Safe concentration levels for hazardous constituents would be identified and any waste that did not exceed those concentrations could be placed into land disposal facilities. A major uproar resulted among congressional leaders, however, about this proposal. Congress, it seems, wanted treatment for treatment sake. That is, regardless of the degree of hazard posed by a particular waste, the waste must be treated before placement in the land.

In response to Congress, the Agency is now considering a proposal establish performance standards for treatment of all listed hazardous wastes. This would be accomplished by identifying for a particular listed waste category the Best Demonstrated Available Technology (BDAT). The highest achievable performance level for that specific BDAT process become the designated performance standard for treatment of that waste.

Let me illustrate with an example. Listed waste F001 is halogenated and nonhalogenated solvents from nonspecific sources. The BDAT for F001 wastes is incineration. Current destruction efficiency requirements within the regulatory program for incineration is 99.9999% Thus, the performance level for treatment of F001 wastes presumably might be 99.9999%. Any chemical, biological, or combination thereof, treatment process that can achieve that level of degradation also could be used to treat F001 wastes. If that level of degradation level is not feasible using chemical and biological treatment, then by regulation only incineration can be used.

The Agency has provided some flexibility in the system by allowing waivers for specific wastes. If a particular waste within the category of F001 cannot be treated to the prescribed performance level due to special peculiarities, the generator of that specific waste can petition for a waiver. In granting the waiver, the Agency would identify the next BDAT for that waste and set the performance standard accordingly.

In addition, the Agency is maintaining a strict position on the "no migration" provision of the legislation. Any petitions to allow untreated wastes in land disposal facilities will have to prove either that no migration will occur, or that the wastes will be degraded prior to any potential migration.

IMPACT ON DEVELOPMENT OF BIOLOGICAL TECHNOLOGIES

This policy offers both benefits and disadvantages to commercial application of biological treatment in the RCRA and Superfund programs. There is some speculation that the Agency may establish two BDAT performance standards: one for concentrated wastes and one for dilute wastes. The first would continue to be incineration for most organic constituents and some type of solidification technology for inorganics. However, the latter may be based on biological treatment. If that is the case, it provides tremendous opportunities for further research and commercial development of a wide range of microbial applications.

Additional research will be needed to verify actual destruction efficiencies of the wide range of hazardous constituents found in industrial hazardous wastes. It will also require some real world testing and demonstration studies. Although much "real world" application has occurred with contaminated soils and groundwater at Superfund sites, RCRA applications represent different operating conditions and, in most instances, a more diverse range of contaminants than normally found at Superfund sites.

If biological treatment can be developed for RCRA wastes, the commercial opportunities should be significant. This will be particularly true, if the treatment residue can be shown to be nonhazardous and therefore the waste residue delisted.

What are the disadvantages? Given the complexity of the waste represented in single waste category (e.g., represented as F001), it will be difficult for the Agency to establish a "dilute" concentration level by which biological treatment performances can be evaluated. It is not impossible, but the data currently available may not be adequate for the task. The timeframe within which Agency staff must work may not be sufficient to develop the necessary data. Should that scenario turn out to be reality, biological treatment may be locked out of commercial applications. It might be possible for individual generators to gain specific waste waivers that would incorporate biological treatment. However, the nation would lose an opportunity to develop this valuable technology to its widest potential.

One might be able to develop a sequence of technologies that in combination could meet the BDAT performance standard. For example, some combination of biological, chemical and or thermal processes with an end-product residue that has reached the performance standard destruction levels is conceivable. However, preliminary discussions with the Agency suggests that such refinements of the decision-making process have not yet been recognized as essential.

Another disadvantage of this treatment policy is the adverse impact it may have on land treatment of industrial wastes. The Agency has acknowledged that it may be impossible to substantiate that "no migration" of hazardous constituents will occur at land treatment sites. If it is impossible to verify "no migration", the Agency may be forced, by the congressional mandate, to prohibit land treatment as a management alternative for RCRA wastes.

Also, application of the "no migration"

standard could preclude the use of _in-situ_ biological treatment at Superfund sites. There is much concern in Congress that Superfund activities do not now include standards developed under other environmental legislation. The proposed reauthorization of Superfund legislation has language that will require the Agency to incorporate all such standards where applicable. Because of the "no migration" mandate, it could be very difficult for the Agency to justify use of _in-situ_ biological treatment.

CONCLUSION

The commercial waste service industry recognizes the potential application of biological treatment for industrial hazardous wastes. Facilities employing this technique currently exist in Canada and the United States. Greater application and commercial development is possible, particularly if certain operational/performance criteria are met. These include

- complete destruction of hazardous constituents,
- the ability to concentrate target constituents for further treatment or for resource recovery,
- application of the technology to a diverse range of hazardous constituents found in mixtures, rather than in single constituent waste streams,
- consistency among treatment batches to reduce the need for expensive monitoring of the residue, and o the ability to be competitive financially with other treatment technologies that are currently available.

Advocates of biological technology should be concerned about and watchful of the developing Agency policy on RCRA treatment standards. Depending on the direction of this policy, commercial development of biological treatment technology could be affected adversely . It will be necessary for experts in this area to work with the Agency in identifying the needed data and in validating "real world" applications of these data.

If the Congressional mandate to treat all hazardous wastes includes the broadest possible range of management options, it will be to the benefit of all. However, a concerted effort is required to suure that policies, which optimize treatment opportunities, leave the door open for new research and development. Only with such policies can we be assured of providing maximum protection of public health and the environment.

KEYNOTE ADDRESS—
TOXICITY ASSAYS AND MOLECULAR STRUCTURE TOXICITY

R.E. Speece and N. Nirmalakhandan
Drexel University

Peter C. Jurs
Pennsylvania State University

INTRODUCTION

Many years ago Hartmann and Laubenberger (1968) concluded that because of their complexities, toxicity measurements and the evaluation and interpretation of the results are some of the most difficult tasks the sanitary biologist has to deal with. Then, while having lunch with Gerry Rohlich one day about 13 years ago, he made the observation "For every complicated problem, there is a simple, clear solution * * * * * * * that doesn't work !"

When one considers the complexity of a biological system's response to toxic chemicals, it would appear to be overly optimistic to attempt to formulate a rational quantitative - structure activity relationship (QSAR) based upon molecular structure or gross properties. Furthermore, when one considers the different biochemical features of the multitude of bacteria found in the environment, again it appears overly optimistic to attempt to base a toxicity assay on any given pure culture, or even a mixed culture for that matter. However, in view of the fact that there are over 60,000 chemicals in U.S. commerce and 200 to 1000 new chemicals are advanced into commercial production each year, some management tools are desperately needed to screen chemicals for toxicity. This is further reinforced by the relative expense involved in thorough evaluation of the ecotoxicological characteristics of chemicals before approval.

Thus, the need for "screening tools" is so pressing, that considerable value can still be derived from bacterial toxicity assays and QSAR methods which involve poorly understood mechanisms. Having accepted the premise that the purpose is as a "screening tool" and is being used to "flag" certain chemicals for further assay or to give early warning for operation of a treatment process, some very valuable insights can be gained from QSARs and the various microbial assays.

Jurs et al (1979) observed that the superior way to develop predictive capability is to understand, at the molecular level, the mechanisms that lead to the undesired cellular effects. Unfortunately, this information is not yet available for most classes of environmental chemicals. Thus, two choices are presented: study the mechanisms for a very few compounds to develop fundamental information for those few compounds, or use empirical methods to study larger data sets with correlative methods. The latter method comprises the SAR approach to chemical toxicity.

The concept that a relation exists between the structure of a molecule and its biological activity was formulated by Crum-Brown and Frazer in 1868. (Koch 1983). One of the greatest problems in QSAR studies is the translation of molecular structure into simple and unique numerical descriptors. These descriptors should be parameters of such fundamental properties of a molecule which determine possibly the biological activity and toxicity. Various parameters have been used in QSAR studies such as the Hansch constant, hydrophobic fragmental constant, partition coefficient in octanol/water, molvolume, surface activity, molar refraction, polarizability of electrons, Taft constant, Hammett factor, parachor and molecular connectivity indices.

QSARs do not indicate much about the kind of action of the compounds. However, when a high quality QSAR is found for a group of compounds, it might indicate that the action is similar (Konemann, 1981). Liu et al (1982) conclude that bacteria are now widely used in the assessment of chemical toxicity and have undergone sufficient evaluation and refinement to be considered reliable and reproducible.

TOXICITY ASSAYS

Toxicity assays can provide valuable insight into the possible impact of chemicals released into the environment and they can also be used advantageousluy to give early warning indications for treatment processes. A growing battery of standardized and innovative assays is described in the literature which can be used with varying levels of success. Among these are;

Microtox	Spirillum volutans
Algae	Act. Sl. Respiration
Protozoa	Act. Sl. Dehydrogenase
Daphnia	Nitrobacter
ISO	BOD
OECD	

Liu (1985) assayed six different pure cultures isolated from activated sludge with 12 environmental chemicals and found a chemicals' toxicity seems to vary dramatically from one species to another and this suggests there is little possibility of finding an ideal bacterial culture that is highly sensitive to all toxic compounds. For this reason, he concluded, a single pure culture study in the laboratory should not be liberally used or freely extrapolated to predict the ultimate consequence of new chemicals in the environment. He also observed that the lack of a general pattern between toxicants and microorganisms interaction clearly demonstrates the fallacy of those attempting to predict the environmental behavior of new chemicals based on a few sets of physical and chemical parameters. He proposes using a two species system for toxicity assay due to the extreme complexity and unpredictability of the biota-toxicant interactions.

Williamson and Johnson (1981) reported on a bioassay using freeze-dried Nitrobacteria as the test organism. This organism was intended to serve as a surrogate toxicity indicator for a fiberboard wastewater activated sludge system and they concluded that it met the criteria for a bioassay of industrial and municipal wastewater.

Klecka and Landi (1985) evaluated a wide variety of inorganic and organic chemicals to municipal activated sludge. They found the dose-response curve often appeared hyperbolic on a semi-log plot. This suggested to them that at low concentrations of the inhibitor, the receptor sites on the microorganisms are not saturated and thus the percent inhibition varies with chemical concentration. At high concentrations, the receptor sites eventually become saturated and thus the percent inhibition becomes independent of the chemical concentration. They conclude that although certain chemicals may be shown to be toxic to microorganisms in laboratory experiments, the reliability with which these results can be extrapolated to full-scale biological treatment systems is unclear. The response of a treatment plant to the presence of a toxicant may vary considerably from the laboratory due to differences in organic loading, biomass concentration, detention time, removal of the chemical by physical or biological mechanisms and as a result of biological adaptation. They recommend that in view of these factors, the toxicity screening test procedure should be regarded as a screening tool for identifying potential problems.

Bringmann and Kuhn (1980) studied the potential toxic action of water pollutants on model microorganisms characteristic of biological self-purification. The model bacteria was Pseudomonas putida, the model green algae was Scenedesmus quadricauda, and the model protozoa was Entosyphon sulcatrium. They assayed the toxicity threshold of 156 substances, both organic and inorganic, to the three model organisms. They defined toxicity threshold as the concentration at which the inhibitory action of a pollutant starts and will be present in that step of a dilution series of the pollutant having an extinction value at the end of the test period that is ≥ 3% below the mean value of extinction for non-toxic dilutions of the test cultures. These 156 substances were selected because they belong to different functional groups and are environmentally significant. They found from the results of their study that the damaging action of water pollutants was selectively directed against one group of model organisms, depending upon the different chemical nature and group classification of the pollutants involved.

From the number of 156 inorganic and organic pollutants tested by Bringann and Kuhn (1980), 23 exhibited a pronounced selective toxic action on the model bacteria, 47 on the model algae and 43 on the protozoa. They concluded that from the toxicity threshold ratios for these model organisms, a broader basis was available for assessing toxicity. Ecotoxicological tests of water pollutants using only one organism would give an incomplete and biased picture of the effects of the pollutants.

Dutka, et al (1983) compared several microbiological toxicity screening tests, e.g., Microtox, activated sludge respiration, activated sludge TTC-dehydrogenAse activity and Spirillum volutans. They found each procedure had its own toxicity sensitivity pattern. They broadly ranked the methods according to decreasing sensitivity as follows: Microtox, Spirillum, activated sludge respiration and activated sludge dehydrogenase activity. There were wide divergences between the test results. See Table I.

Table I Sensitivity of Four Acute Toxicity Screening Procedures to Various Chemicals (Dutka et al 1983)

Chemical	MTX-5	MTX-10	MTX-30	S. vol.	A. S. (.5 h)	Resp. (3 h)	A. S. TTC
3,5 DCP	3.2	3.0	2.9	5.0	38	22	80
CTMNH4Cl	1.35	0.98	0.86	1.45	5.8	5.5	18.5
NaLaurylSO4	3.19	2.1	1.8	4.15	188	135	48
Phenol	28	31.9	34.3	300	740	1000	1400
Cu	19.5	9.4	3.8	10	34	17	2.1
Hg	0.064	0.049	0.046	1	1.3	0.96	1.5
Zn	13.8	6.1	3.45	11.6	6.1	5.2	24

King (1984) assayed the toxicity of single chemicals and mixtures using mixed cultures. The bacterial assays used were: (1) Inhibition of respiration of activated sludge, (2) Inhibition of nitrification, (3) Microtox, (4) Inhibition of growth of sewage organisms and (5) Inhibitions of BOD5. The mixtures were added in proportion to their individual EC_{50} values and dilutions were made of this mixture to determine the new EC_{50} of the mixture. Then for comparison, the contribution of each chemical in the mixture was calculated, assuming that the chemicals did not interact and exerted the same effect as it would have exerted alone at that concentration, so that toxic effects were additive.

Table II shows the comparative summary of EC_{50} results of single compounds for the various microbial assays. The Microtox assay was very sensitive-in some cases giving EC50 more than 100 fold lower than the other methods. MBT and SLS were more toxic to nitrification than to respiration; while PCP had low toxicity to nitrifiers. Correlation coefficients for BOD_5 EC_{50} were not as good as those of the other methods.

Combinations of two and six chemicals showed little synergism or antagonisms to nitrifiers. This was generally true of all the assays. It is doubtful than any synergism or antagonism occurs, but Dutka and Kwan (1982) have reported large synergistic effects for mixtures of metals and detergents. Only MBT exhibited a significantly lower EC_{50} in the growth method than in the respiration method. It had been assumed that growth reflects viability and a wider range of microbial reactions of a microbial population and would be a more rigorous assay, but this did not prove to be so (King, 1984).

Table II Comparative Summary of EC-50 Results of Single Compounds Obtained by Different Methods (King 1984)

Chemical	ETAD	Nitrif.	Growth (6 h)	Growth (16 h)	BOD	MCTX
3,5 DCP	7	5	8	6	15	3.2
PCP	24	275	24	10	32	1.2
PCMX	36	12	41	41	50	8
DCP	17	27	12	4	7	0.055
PB	7	7	2	9	7	0.02
MBT	13	1	0.4	1.2	0.3	0.023
CuSO4.5 H2O	123	52	47	37	4.5	3.2
SLS	>500	24	352	545	>500	1.5

3,5 DCP = 3,5 Dichlorophenol
PCP = Pentachlorophenol
PCMX = p-Cl-m-Xylenol
DCP = 2,2 di-OH 5,5 di-Cl diphenyl Methane
PB = p Benzoquinone
MBT = Methylene bis Thiocyanate
SLS = Sodium Lauryl Sulfate

Variability is high in methods using mixed populations, even when taken from the same source. Brown, et al (1981) ISO (1984) and King and Painter (1986) recommended that EC_{50} results should be given as an order of magnitude to avoid undue emphasis on an individual value. Also EC_{50} data from screening tests, which are normally designed to be "fail-safe" may not always over-estimate effects in practice. For instance, streptomycin, known to be a growth inhibitor, did not affect nitrification in a batch test at 400 mg/l but in a fill and draw activated sludge plant, as little as 0.15 mg/l caused complete inhibition.

King (1984) concludes that the Microtox assay is very rapid and more precise than the other methods, but the EC50 values obtained are usually very much lower than those estimated from the other tests, and are therefore considered to be unsuitable for predictive purposes. However, the Microtox assay can be quite helpful for qualitative purposes of ranking the comparable toxicity of chemicals. The nitrifiers are generally considered to be more susceptible to inhibitory chemicals, but apart from specific inhibitors of nitrification, it is suggested they are no more susceptible than heterotrophs to inhibitory chemicals.

QSAR BACKGROUND

Lipnick (1985) has given a perspective on quantitative structure - activity relationships (QSAR) in ecotoxicology and observed that Meyer-Overton relationships had been demonstrated to correlate with increasing oil/water partition coefficients and increasing toxicity to various organisms of a large number of organic compounds. Meyer-Overton relationships have been demonstrated for numerous classes of organic compounds, such as hydrocarbons, alcohols, ethers, ketones, sulfuric, halogenated hydrocarbons, and other types of non reactive non-electrolytes (Albert 1979). This has become a very useful concept, but the mechanistic basis of Meyer-Overton action is still only poorly understood. Toxicity of this type is referred to as narcotic action, non-specific action, or physical toxicity and represents the adverse effect induced in organisms due to the presence of minimum molar concentrations of the toxicant in the biophase.

Lipnick (1985) indicated that QSAR relationships will be valid so long as the following conditions remain satisfied: (1) the experimental test duration leads to a pseudo steady state equilibrium betweeen the aquatic medium and the test organism, (2) the required toxic concentration is within experimental water solubility and (3) the chemical exhibits no more specific mechanism of action. For hydrophobic chemicals, steady state conditions will be achieved only within sufficiently long exposure durations.

Lipnick (1985) developed correlations between alcohols and other classes of narcotic chemicals. He noted the limitations of the narcotic toxicity QSAR's are: (1) solubility cutoff, (2) pharmacokinetic cutoff and (3) pro-electrophile mechanism. In a study of 55 alcohols containing no additional heteroatom functional groups, he observed that 5 compounds (Fig.1) were striking exceptions to the predictive model in that lethality and sickness were observed at concentrations ten to 100,000 times those predicted by Konemann's equation. Four of the compounds were primary or secondary allylic or propargylic alcohols and the fifth was a viconol diol. They proposed that the first four outliers had shown significantly greater toxicity than predicted because of a pro-electrophile mechanism involving their biotransformation to the corresponding á, ß-unsaturated aldehyde or ketone via the enzyme alcohol dehydrogenase. These aldehydes and ketones would be expected to exhibit highly electrophilic reactivity with endogenous nucleophilic macromolecules. It was hypothesised that some classes of chemicals such as these outliers inhibit only those enzymes which depend upon sulfhydryl groups for their activity, leading to covalent bond formation and inhibition of physiological function. This toxicity is clearly not narcotic. Lipnick (1985) proposed that it is reasonable to assume that a QSAR term of the reaction rate velocity, log k, of the predicted electrophile intermediate is required to model pro-electrophiles as a general toxicological class. An additional parameter the redox potential, E^*, is needed to quantitatively model the rate of biotransformation from pro-electrophile to electrophile.

$$\log (1/LC_{50}) = A \log P + B E^* + C \log k + D.$$

In a separate study Lipnick and Dunn (1983) analysed chemical toxicity and found acetonitrile and some esters appear to exhibit only physical toxicity. Other esters were only somewhat more toxic than predicted by narcosis. Those

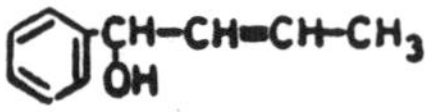

42   Biotechnology for Degradation of Toxic Chemicals

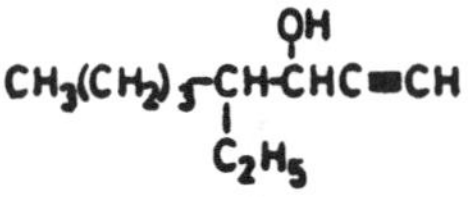

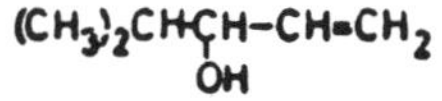

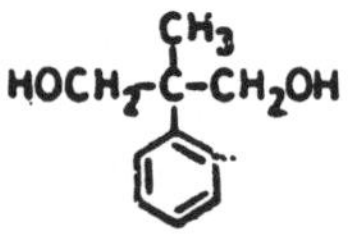

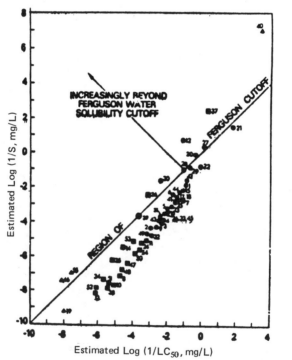

FIG 1. EXCEPTIONS TO PREDICTIVE MODEL
(LIPNICK 1985)

FIG 2. RELATIONSHIP BETWEEN WATER SOLUBILITY AND TOXICITY. "REGION OF FERGUSON CUTOFF" LINE REPRESENTS TOXICITY AT EXACT SATURATION IN AQUEOUS PHASE. CHEMICALS WITH LC50 ABOVE THIS LINE NOT EXPECTED TO SHOW TOXICITY UNDER LABORATORY CONDITIONS IN THE ABSENCE OF ORGANIC CARRIER SOLVENT. (R. L. LIPNICK et al)

chemicals showing the most extreme chemical toxicity were hydroquinone, acrolein, acrylonitrile and salicylaldehyde. These chemicals appeared to be more toxic because of their ability to act as electrophile. Hermans (1983) recently demonstrated that those non electrolytes that contain chemically reactive functional groups associated with electrophilic behavior are generally more toxic to fish than would be predicted by narcotic QSAR. He showed that the rate constants for covalent bond formation of such chemicals with the model nucleophile p-nitrobenzylpyridine correlates well with the acute fish toxicity of such chemicals. Lipnick and Dunn (1983) suggested a global electrophilicity parameter would be quite useful in correlating the toxicity of many of these chemicals.

Predicting the mode-of-action of chemicals remains a difficult task, and the errors associated with selecting the wrong structure-activity mechanism may be greater than the standard error of the estimate associated with any individual (Veith, et al 1983).

QSAR

For instance, structures that have a single functional group adjacent to the alcohol or carbonyl carbon may be more rapidly metabolized to a more toxic chemical resulting in computed toxicity much less than the actual toxicity. For example 2-chloroethanol had a computed narcotic toxicity of more than 1000 mg/l, but a measured LC_{50} of 37 mg/l. A likely metabolite is 2-chloroacetic acid, which is an enzyme inhibitor at levels below those that produce narcosis.

Lipnick (1985) observed that most practitioners employ the Hansch approach in their studies because of its mathematical simplicity and straightforward interpretation. In the Hansch approach, the effects of changes in substituents can be related through linear free energy relationships to a combination of hydrophobic, electronic and steric factors. The utility of the Hansch approach could be extended to more heterogenous data sets if more global molecular descriptors were available. He suggests that the dissociation constant together with log P can be used to model the biological transport of weak electrolytes. The dissociation constant can also be employed as a QSAR parameter in toxicity studies.

Chemical electrophilicity is a second global electronic parameter suggested by Lipnick (1985) that is likely to find increasing application in toxicology. This is confirmed by the work of Hermans (1983). Subsequently, the availability of a global steric parameter would provide the third property needed to extend the utility of the Hansch linear free energy relationship approach.

Lipnick (1985) concludes that QSAR methodologies potentially can provide the means to derive mathematical models to account for the transport, metabolism and receptor interaction of a series of toxicants. He goes on to say that a QSAR model will always be limited in its predictive ability by the extent to which biological mechanism and physicochemical property domains have been explored within those chemicals employed on a training set in its derivation.

LOG P MODELLING

Dearden (1985) concluded there is no more important parameter than lipophilicity in the quantitation of biological response as a multiplicity of published QSARs attests. Other factors, such as electronic and steric effects both intra-and intermolecular can affect the rate at which molecules arrive at a receptor site and the interaction with the receptor, but lipophilicity is generally dominant irrespective of the organism.

Hansch and Dunn (1972) derived QSAR equations in the following form:

$$\log (1/C) = A \log P + B$$

for 137 sets of data for narcotic chemicals tested on whole organisms, organs, cells, and pure cultures.

Leo (1985) reports many simple correlations between microbial toxicity and log P, such as the following:

$$\log (1/C) = 1.10 \log P + 0.21$$

for 50% inhibition of bacterial luminescence. However, he also notes that the biological activity bank contains many examples of enzyme inhibition where the overall log P does not help the correlation, but the value of the substituents in just one of the positions is clearly significant.

Hansch and Fujita (1964) analyzed the toxicity of phenols to Gram positive and Gram negative bacteria. They found the Hammett D was of little importance in determining the activity of phenol as measured by the phenol coefficient. However, they found good correlation with π term patterned after the Hammett D constant. They defined $\pi = \log [P_H/P_X]$ where P_H and P_X are the partition coefficient of the parent compound and derivative respectively. The equation for Gram positive bacteria was:

$$\log \text{Phenol Coeff.} = 0.951 \pi + 0.144 \qquad r^2 = 0.954$$

The equation for Gram Negative bacteria was:

$$\log \text{Phenol Coeff.} = -0.288 \pi^2 + 1.312 \pi + 0.139$$

The Gram negative organism was much less sensitive to the toxic action of phenols than the Gram positive organism.

The study of QSAR is an extremely complex endeavor. Attempts to predict toxicity from a single log P correlation can produce anomalous results. For example Liu and Thomson (1983) note that 1,2,3 trichlorobenzene and 1,3,5 trichlorobenzene have indentical log P values of 4.20. Yet the latter showed both stimulating and inhibitory effects while the former consistently stimulated microbial dehydrogenase activity. Also, para-dichlorobenzene was significantly less toxic than the ortho and meta isomers. They note that molecular geometry and electronic properties could be very important in eliciting a toxicant's biological acitivity.

Lipnick and Hood (1986) performed a QSAR study on the experimental results of Bringmann and Kuhn (1980). They developed baseline toxicity narcosis models from the data on monohydric saturated alcohols as follows:

Daphnia magna
$$\log(1/c) = 0.873 \log P - 5.022$$
$$n = 14 \quad r = 0.98 \quad s = 0.22$$

Pseudomonas putida
$$\log(1/c) = 0.738 \log P - 3.27$$
$$n = 14 \quad r = 0.93 \quad s = 0.26$$

Scenedesmus quadricauda
$$\log(1/c) = 0.897 \log P - 3.40$$
$$n = 13 \quad r = 0.96 \quad s = 0.28$$

Microcystis auruginosa
$$\log(1/c) = 0.909 \log P - 3.90$$
$$n = 14 \quad r = 0.95 \quad s = 0.33$$

Ueronema parduczi
$$\log(1/c) = 0.675 \log P - 3.166$$
$$n = 12 \quad r = 0.72 \quad s = 0.57$$

They then compared the toxicities of the remaining 98 organic non-electrolytes with predictions based upon these equations. They concluded that the increased activities of those compounds found to be more toxic than predicted were interpreted in terms of electrophile and pro-electrophile mechanism.

Vaishnav (1986) observed that while predicting the mode-of-action of chemicals remains a difficult task, it has been shown that gases, aliphatic and aromatic hydrocarbons, chlorinated hydrocarbons, alcohols, ethers, ketones, aldehydes, weak acids and bases, and some alphatic nitro compounds exhibit narcotic action in vertebrate animals. For a homologous series of ten straight chain alcohols, he was able to correlate the relationship between EC_{50} and log P by the following polynomial regression equation:

$$\log EC_{50} = -0.88 (\log P) + 0.07 (\log P)^2 - 0.23 \quad (r = 0.997)$$

Vaishnav (1986) then used this correlation equation to predict the EC_{50} values of 19 narcotic chemicals selected from several chemical classes with log P values ranging from -0.42 for methyl carbitol to 3.54 for 2-decanone. The ratios of predicted to experimental EC_{50} values of test chemicals ranged from 0.44 for naphthalene to 1.60 for benzyl alcohol with a mean value of 1.16 $\pm$ 0.35 as shown in Table III.

Figure 2 shows the relationship between water solubility and toxicity with the Ferguson cutoff indicated. Fig. 3 shows the frequency distribution of log P values for industrial chemicals. Solubility generally limits toxicity beyond log P = 5.5.

TABLE III
Selected Narcotic Industrial Chemicals, Logarithms of Their 1-Octanol/Water Partition Coefficients (Log P)[a], and Their Predicted[b] and Experimental[c] Microbial EC_{50} Values

Chemical	Log P	EC_{50} (mol/L) Predicted	EC_{50} (mol/L) Experimental	Ratio Predicted / Experimental EC_{50}
Methyl carbitol	−0.42	1.418	1.254	1.13
Acetone	−0.24	0.966	0.612	1.58
Isopropanol	0.11	0.472	0.514	0.92
2-Butanone	0.30	0.325	0.284	1.14
3-Methyl-2-butanone	0.62	0.178	0.144	1.24
Isobutanol	0.74	0.143	0.197	0.73
3-Pentanone	0.84	0.120	0.092	1.30
Cyclohexanone	0.95	0.099	0.071	1.39
Benzyl alcohol	1.11	0.075	0.047	1.60
2-Pentanol	1.19	0.066	0.053	1.25
Isoamyl alcohol	1.28	0.057	0.046	1.24
2-Hexanone	1.38	0.048	0.055	0.87
Cyclohexanol	1.42	0.045	0.031	1.45
4-Methyl-2-pentanol	1.60	0.034	0.024	1.42
Acetophenone	1.66	0.025	0.016	1.56
2-Octanone	2.46	0.010	0.013	0.77
2-Octanol	2.81	0.007	0.005	1.40
Naphthalene	3.29	0.004	0.009	0.44
2-Decanone	3.54	0.003	0.005	0.60

[a]Computed by fragment constant method of Hansch and Leo (1979).
[b]Concentration that would reduce maximum observed biodegradation rate by 50% as predicted from Log P value and its relationship:
Log EC_{50} (mol/L) = −0.88 (Log P) + 0.07 (Log P)^2 −0.23.
[c]Concentration that would reduce maximum observed biodegradation rate by 50% as manometrically determined employing respective chemical substrate and resting cells of an acclimated mixed microbial culture.

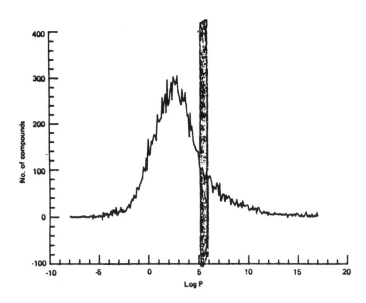

FIG 3. FREQUENCY DISTRIBUTION OF log P VALUES
FOR INDUSTRIAL CHEMICALS. (VEITH, DE FOE, AND KNUTH)

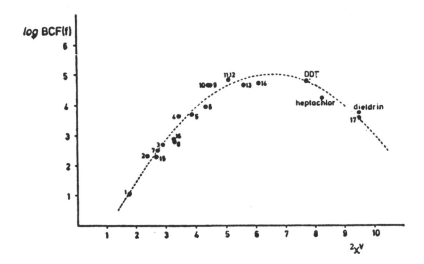

FIG 4. PLOT OF BCF(f) VS $^2\chi^v$ OF THE
STUDIED HALOCARBONS. (SABLJIC & PROTIC)
LOG (BCF) = $- 0.168 (\pm 0.013) (^2\chi^v)^2 + 2.222 (\pm 0.149) (^2\chi^v) - 2.323 (\pm 0.379)$
[n = 20 ; r = 0.971]

MOLECULAR CONNECTIVITY

Molecular connectivity is a method for the quantitation of molecular structure that encodes information about size, branching, cyclization, unsaturation and heteroatom content. The molecular connectivity method is based on molecular topology (number and type of atoms and chemical bonds). These are purely non-empirical data and their calculation is quite simple. A topological index is a numerical characteristic of a molecule based on a topological feature of the corresponding molecular graph. The method has the advantage of relative simplicity and flexibility. It can be used to represent molecular structure quantitatively at a number of levels of complexity. Each level provides some information uniquely related to the structure and through it to physical, chemical and biological characteristics. (Sabljic and Protic, 1982). For the calculation of these indices a numerical value is given to each atom in a molecule, except hydrogen, according to the number and type of its bonds, and to each bond, according to the values of the connected atoms. When the values of the atoms in a molecule are added, $^0X^v$ is obtained and $^1X^v$ is found when the values of the bonds are added (Konemann, 1981).

The structure of a molecule determines directly its physicochemical properties as well as its chemical and biochemical reactivity, even though such mechanisms may not be understood. This relation represents an important aspect of molecular toxicology. One of the greatest problems in QSAR studies is the translation of molecular structure into simple and unique numerical descriptions. The results of this approach will depend on the test organism species and the similarity of structure of the chemical compounds. Generally, it can be said that an increasing molecular connectivity index shows a stronger toxicity of the chemical compound. The objective limitation of such QSAR - studies consists in the enormous complexity of biological systems. The molecular connectivity index does not contain information about the reactivity of the functional groups or atoms in a molecule which are often responsible for specific toxic effects. On the contrary, it is a descriptor of the molecule as a whole. (Koch, 1982).

Kier and Hall (1977) reported that connectivity attempts to describe quantitatively these kinds (branching, heteroatoms, multiple bonding and cyclization features which are not readily quantitated in a form suitable for SAR of structural features at the same level of information as an atom count i.e., a numerical value which can be determined unambiguously for a given molecule and which is also transferrable from study to study. The chi (X) terms provide the information necessary for adequate structural description. (The ^1X index is a count of the bonds weighted by the degree of branching). It is now apparent that ^1X carries structural information: the number of atoms and the complexity of skeletal branching. For example, in chain lengthening X increases 0.5 with the addition of each methylene group. On the other hand X decreases with increasing complexity of branching.

Both the number of subgraphs and the weighting depends heavily on the structure. The number of ^3X_p subgraphs in o-xylene is 11 whereas it is 10 for the m and p isomers. The weighting is different however, in each isomer leading to unique values of ^3X_p: 2.540 for o-xylene, 2.199 for m-xylene and 2.305 for p-xylene. Thus ^3X_p reflects something of the difference in the pattern of branching in these three isomers.

There is contained in ^3X_p for the xylenes structural information which may play a significant role in a relationship to properties of xylene. It is then possible by a regression analysis search to isolate a small number of X terms which in concert, are highly related to property values. The molecular connectivity terms which appear in such significant correlations describe the dependence of the property on molecular structure. Hence, the ^nX_p terms are not the properties themselves but they describe numerically the significant structural characteristics to which the property is related.

Molecular connectivity correlates well with lipophilicity for a range of polar compounds including alcohols, ethers, ketones, carboxylic acids and amines:

$$\log P = 0.950\ ^1X - 1.48\ (r = 0.986\ \ s = 0.152)$$

Molecular connectivity is being increasingly applied to QSAR in environmental situations. For example, Sabljic and Protic found for a series of chlorinated arometic compounds in fish: (Fig. 4).

$$\log BCF = -0.171\ (^2X^v)^2 + 2.253\ (^2X^v) - 2.392$$

$$r = 0.970\ \ s = 0.297$$

Hall and Kier (1978) analyzed the toxicity data for 50 halogenated phenols to S aureus and found the ^1X index the best single variable for regression analysis as follows:

$$\log PC = 6.308\ (\pm 0.18) - 21.41\ (\pm 0.50)\ /\ ^1X$$

$$r = 0.975\ \ s = 0.20\ \ n = 49$$

This lead to the conclusion that there was no statistically significant difference in toxicity of the two halogens, chlorine and bromine. The presence of ^1X in the regression equation as a negative reciprocal, indicates that log PC decreases with decreasing ^1X value. Since ^1X decreases with increased chain branching, log PC also is predicted to decrease when alkyl side chains and branched rather than normal. Also the inverse relationship between log PC and ^1X indicates that successive increments in log PC decrease with the addition of each methylene group. The magnitude of ^1X decreases as the phenol ring becomes more highly substituted. Thus the equation predicts that for a given number of carbon atoms. Monoalkyl substitution leads to greater activity than polyalkyl substitutions.

Schultz, et al, (1982) correlated the antimicrobial activity of selected nitrogenous heterocyclic compounds with $^1\chi^v$ as follows;

$$\log(1/C) = 0.911\ ^1\chi_v - 2.969 \qquad r = 0.962 \quad s = 0.27$$

Finding this close relationship with $^1\chi_v$ makes it possible to describe some of the influences of molecular structure on toxicity. This indicates that (1) with increasing number of atoms, toxicity levels increase (2), successive methylation of a ring system increases toxicity (3) additional nitrogen atoms introduced into molecules of the same size decreases toxicity. Similar conclusions could be drawn from a Hansch analysis of the same data.

Hall and Kier (1978) used molecular connectivity to analyze antimicrobial activity of a phenyl propyl ether data set. Their results suggested the principal interactions were related to the para position of the phenyl ring and x/y portion of the ether side chain. Substitution leading to increased $^4\chi^v_{pc}$ values augmented toxicity. Bromine in the para-position and methyl in the meta-position increased toxicity, vic dihydroxy substitution decreased toxicity. In addition there appeared to be a whole molecule dispersion effect. Connectivity indices gave better correlations than Hansch π, D analysis.

Yoshioka, et al (1986) stated that one of the problems in the estimation of toxicity is which parameter should be employed. Log P is most popular as a descriptor of the membrane permeability and there are many reports on the realtion between biological activity and partition coefficient. They investigated molecular connectivity indices (X) because these can provide simple, flexible and direct structural description of organic molecules. They do not depend on experimentally determined physical properties and do not assume any mechanistic base. The found that $^3\chi_p$ gave a better correlation for the toxicity of 123 chemicals than did log P.

($r = 0.738$ for log P and $r = 0.829$ for $^3\chi_p$).

COMBINED LOG P AND MOLECULAR CONNECTIVITY MODELS

Ribo and Kaiser (1983) correlated Microtox, 30 min. EC-50 values with bacterial dehydrogenase activity (Bacillus sp.) IC-50 values and derived the following relationship for 20 chlorophenols:

IC 50 = -1.72 + 1.25 (30 min. EC 50)
n = 19, r^2 = 0.93

They were also able to correlate Microtox 30 min EC50 values with log P in the following relationship:

30 min EC50 = -0.679 + 0.678 log P
n = 20, r^2 = 0.79

With the exclusion of 2, 6 di- and 2,3,5,6 tetra-chlorophenol from the data set, the correlation coefficient increased significantly as follows.

30 min EC50 = -0.882 + 0.760 log P
n = 18, r^2 = 0.91

(It is shown below by Liu, et al (1982) that the presence of chloro substituents in the ortho-position to the phenolic OH group strongly decreases the phenol's toxicity.) Including an indicator variable (N) for the number of chlorines in the 2 and 6 position allows retention of the two chlorophenol isomers in the equation without loss of quality of fit. The new multiple linear regression equation is calcuated to be:

30 min EC50 = -0.82 + 0.805 log P -0.325 N
n = 20, r^2 = 0.87

This ortho effect has been explained in terms of hydrogen bonding and shielding of the OH group by the chloro substituents and is found for other biota as well.

Liu, et al (1982) determined IC50 concentrations for 24 chlorophenols. They noted the generalization that the toxicity of chlorophenols to microorganisms increases with the degree of chlorination in the phenol nucleus was not true for the ortho position. Penta-chlorophenol was less toxic than the lower chlorinated 2,3,4,5, TTP. Similarily, 2.6-dibromophenol (IC50 = 500 mg/l_ was less toxic than 2,4 dibromo phenol (IC 50 = 60 mg/l). The data were correlated with the following equation:

log (1/C) = 0.84 log P -2.54
n = 24, r^2 = 0.81

By including the Hammett's constant for ortho substitutions, the quality of the regression improved as follows:

log (1/C) = 0.98 log P -0.94 $\Sigma\nabla$ -2.68
n = 24, r^2 = 0.90

Boyd, et al (1981) compared log P and molecular connectivity in structure-activity analyses of some antimicrobial agents. They note that calculated log P values often differ markedly from experimentally determined values due to steric factors and the effects of inter - and intramolecular hydrogen bonding. However, the experimental determination of log P can be a rigorous and time consuming procedure. They attempt to use a readily calculable alternative to log P using molecular connectivity. They were able to demonstrate that antimicrobial data that had been correlated well with log P, could be correlated equally as well using molecular connectivity terms. For example, carboxylic acids toxicity was correlated

TABLE IV. IC_{50} of various chlorophenols and bromophenols to the culture TL81 expressed in mg L^{-1} and as logarithms of the inverse molar concentrations C [mol L^{-1}]

Chemicals	IC_{50}(ppm)	log $\frac{1}{C}$	Chemicals	IC_{50}(ppm)	log $\frac{1}{C}$
Phenol	2300	-1.39	2,3,4-TCP	13	1.18
2-MCP	700	-0.74	2,3,5-TCP	10	1.30
3-MCP	450	-0.54	2,3,6-TCP	190	0.02
4-MCP	400	-0.49	2,4,5-TCP	12	1.22
2,3-DCP	130	0.10	2,4,6-TCP	240	-0.08
2,4-DCP	75	0.34	3,4,5-TCP	5	1.60
2,5-DCP	85	-0.28	2,3,4,5-TTP	4	1.76
2,6-DCP	550	-0.53	2,3,5,6-TTP	54	0.63
3,4-DCP	52	0.50	PCP	9	1.47
3,5-DCP	25	0.81	2-MBP	550	-0.50
			3-MBP	380	-0.34
			4-MBP	400	-0.36
			2,4-DBP	60	0.62
			2,6-DBP	500	-0.30

as follows for log P:

$$\log (1/C) = 0.059 (\log P)^2 + 0.460 \log P + 3.754$$
$$r = 0.985 \quad s = 0.17 \quad n = 15$$

For the same correlation using molecular connectivity:

$$\log (1/C) = 7.062\ ^5X_p - 7.605\ ^6X_p + 1.320$$
$$r = 0.995 \quad s = 0.098 \quad n = 15$$

A set of 18 benzyl alcohols active against gram negative bacteria correlated as follows with log P and ∇;

$$\log (1/C) = 0.599 \log P + 0.421\ \nabla$$
$$r = 0.906 \quad s = 0.307 \quad n = 18$$

The best connectivity equation for the same set of data was;

$$\log (1/C) = 1.733\ ^5X_p + 2.037\ ^4X^v_p - 0.706\ ^4X_p + 2.363$$
$$r = 0.971 \quad s = 0.181 \quad n = 18$$

The fact that the ∇ term, highly significant in the above equation, was unnecessary using connectivity terms, suggests that connectivity may encode some measure of the electronic effects of substituents, probably via the valence connectivity term $^4\chi^v_p$ which takes into account the presence of hetero-atoms and the aromatic nature of the ring.

In the above cases, the use of higher order connectivity terms improved on the conventional correlations with physicochemical parameters. Boyd, et al (1981) hypothesize that since connectivity correlates well with many toxicity data sets, it may be a more fundamental representation of structure in structure-activity studies than log P, even though connectivity lacks the more easily interpreted physical significance of log P. Although improved methods are available for calculating log P, the connectivity terms themselves are absolute and as such possess no inherent error. Perhaps the main advantage of connectivity lies in the fact that the various terms are computer-calculated and thus conveniently available for subsequent multiple regression analysis. Since no experimental determinations of X are required, structure activity analysis involving connectivity is relatively quick and therefore useful, at least for an initial screening of data.

Niemi, et al (1985) state that one important area of research is the development of new variables or descriptors that quantitatively describe the structure of a chemical. They performed an exhaustive study on environmental chemicals using molecular connectivity indices (Kier and Hall, 1976). They analysed the biodegradability of 340 chemicals in the 5-day BOD test. They observed that eight principle components explained 93.5% of the variation in the original data. The first three principal components all convey generalized information on chlemical structure: (1) size, (2) degree of branchedness and (3) number of cycles. The remaining principal components are envisioned to convey both subtle and potentially important structural information. Murray, et al (1979) found a high correlation between log P and a form of the valence corrected molecular connectivity index when using distinct chemical groups such as esters and alcohols. However, Niemi, et al (1985) observed that with a much more diverse group of chemicals, principal components and polynomial regression of the principal components were relatively poor predictors of log P and neither was useful in their application. They identified several subgroups of a molecule that may be associated with persistence or degradability of a chemical.

With a group of 9 para-substituted phenols, the molar refractivity parameters (MR) correlated well with the Microtox 30 min EC50 values as follows:

30 min EC50 = 4.508 -0.451 (MR) $n = 9$, $r^2 = 0.76$

With these para substituted phenols log P correlations appear to form three different sets of compounds with each showing good correlation. All slopes were parallel. They speculated that the three lines may be somehow related
and that a physico-chemical or theoretical parameter may exist which will allow the grouping together of all three of these sets in a single equation.

CHEMOMETRICS

A computer software system called ADAPT has been developed by Jurs (1979). The first step is to identify, assemble, input, store and describe a data set of structures for chemicals that have been assayed for toxicity. The next step is to develop a computer-generated set of descriptors for each of the members of the data. Then pattern recognition methods are used as classifiers to discriminate toxicity levels. Next is to test the predictive ability of these discriminants for compounds of unknown toxicity. Finally, systematically reduce the molecular structure descriptors employed to the minimum required to retain discrimination.

The potential molecular structure descriptors are of three classes: (1) topological, (2) geometrical and (3) physicochemical. The descriptors currently available in ADAPT are:

>Fragment Descriptors
>>Counts of the number of atoms of each type
>>Number of bonds of each type
>>Molecular weight
>>Number of basis rings
>>Number of ring atoms
>
>Substructure Descriptors
>
>Environmental Descriptors
>>Code immediate surroundings of substructures
>
>Molecular Connectivity Descriptors
>>Indicates size, branchedness cyclicness
>
>Geometric Descriptors
>>Represent 3-D shape of molecules

DREXEL CORRELATIONS

In our laboratory we determined the 48 hr. IC_{50} concentration of 64 organic chemicals for an acetate enriched methanogenic culture. The culture had been maintained in our laboratory for many years and was maintained at a 50 day SRT and an acetate loading rate of 1 gm/l-d. Acetate was the only organic fed to the system along with the required inorganic salts. The VSS concentration was approximately 700 mg/l.

The assays were conducted using an anaerobic serum bottle technique. The cultures were spiked with a range of concentrations of each organic chemical and 10,000 mg/l $CaAc_2$ was injected as substrate. (The concentration was demonstrated to be non-inhibitory and resulted in gas production containing 25% CO_2 which facilitated pH control and the calcium tended to precipitate from solution as the acetate was metabolized.) The gas production was read for 48 hours. That concentration of the organic chemical which resulted in 48 hr. gas production which was 50% of the un-spiked control was taken as the 48 hr IC_{50}. The log of the relative toxicity number 50 (RTN_{50}) was used in the calculations as log RTN_{50} = log $1/IC_{50}$. Figure 5 shows the 2-D projection of the 3-D structure of these molecules arranged in the order of increasing toxicity. The ADAPT software system developed by Jurs was then used to analyze the results.

Molecules of the 64 chemicals were graphically constructed on an interactive terminal. The minimal bond energy and the resulting surface area and volume were calculated. Sixteen additional descriptors were calculated and a correlation matrix was prepared. From this matrix, after eliminating the interdependent descriptors, the following six were selected for further study:

Descriptor #	Name
24	# of CL
44	Kapa index
25	# of double bonds
21	# of O
28	Mol. wt.
35	Mol. Connectivity Index, ^3Xp

Stepwise multiple linear regression analysis relating log RTN_{50} to the above variables yielded the following results: [64 chemicals]

	Variable Entered	Equation	Multiple R
STEP 1.	24 = > # of CL	log RTN_{50} = 2.07 + 0.653('24')	0.675
STEP 2.	44 = > Kapa Index	log RTN_{50} = 1.52+0.73 ('24')+0.14 ('44')	0.725
STEP 3.	25 = ># of Dbl Bonds	log RTN_{50} = 1.30+0.76('24')+0.14('44')+0.36('25')	0.766
STEP 4.	21 => # OF O	log RTN_{50} = 1.59+0.70('24')+0.18('44')+0.56('25') -0.51('21')	0.812
STEP 5.	28 = > Mol. wt.	log RTN_{50} = 1.29+0.48('24')+0.09('44')+.60('25') -0.62('21')+0.007('28")	0.823
STEP 6.	35 => 3Xp	log RTN_{50} = 0.88+0.34('24')+0.09('44')+0.57('25') -0.68('21')+0.016('28')-0.54('35')	0.835

(See Fig.6)

56 Biotechnology for Degradation of Toxic Chemicals

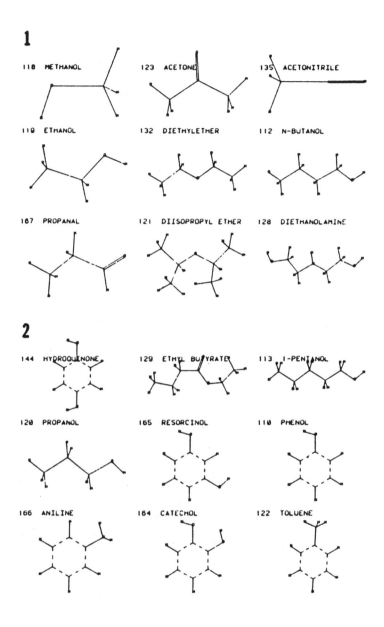

FIG 5. "MINIMAL BOND ENERGY" STRUCTURES OF 64 CHEMICALS.

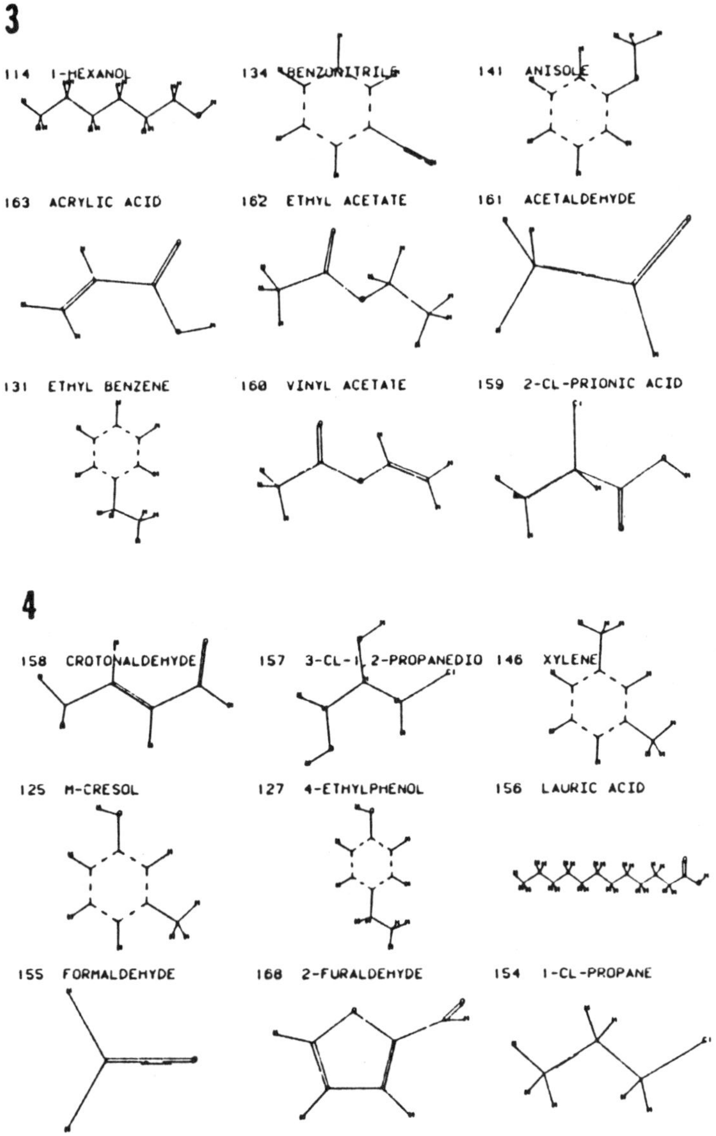

3

114 1-HEXANOL	134 BENZONITRILE	141 ANISOLE
163 ACRYLIC ACID	162 ETHYL ACETATE	161 ACETALDEHYDE
131 ETHYL BENZENE	160 VINYL ACETATE	159 2-CL-PRIONIC ACID

4

158 CROTONALDEHYDE	157 3-CL-1,2-PROPANEDIO	146 XYLENE
125 M-CRESOL	127 4-ETHYLPHENOL	156 LAURIC ACID
155 FORMALDEHYDE	168 2-FURALDEHYDE	154 1-CL-PROPANE

FIG 5. (continued)

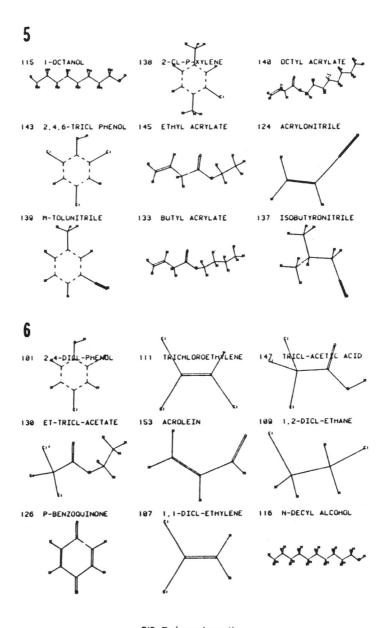

FIG 5. (continued)

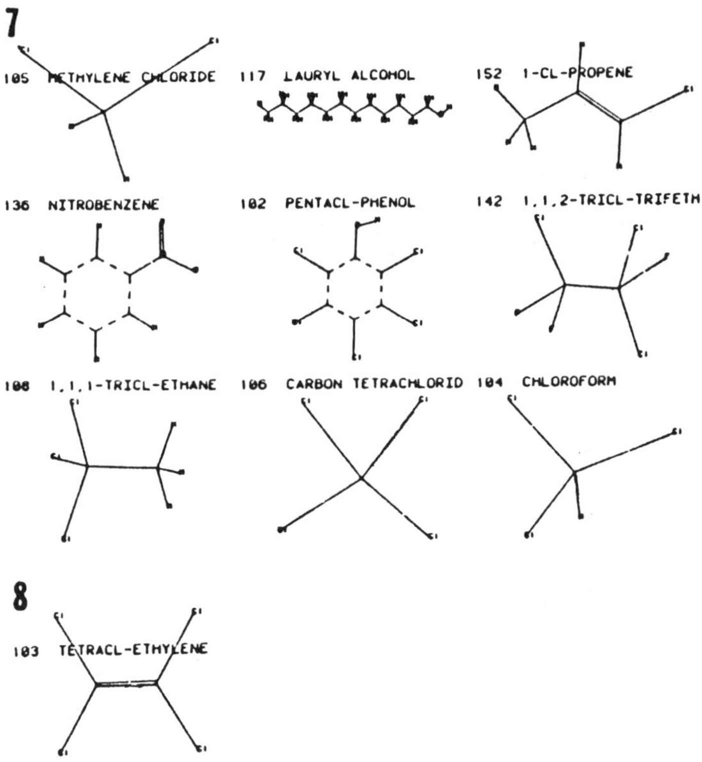

7

105 METHYLENE CHLORIDE   117 LAURYL ALCOHOL   152 1-CL-PROPENE

136 NITROBENZENE   102 PENTACL-PHENOL   142 1,1,2-TRICL-TRIFETH

108 1,1,1-TRICL-ETHANE   106 CARBON TETRACHLORID   104 CHLOROFORM

8

103 TETRACL-ETHYLENE

FIG 5. (continued)

When formaldehyde and carbon tetrachloride were deleted from the list, the R increased slightly to 0.838 with considerable reduction in standard error from 0.727 to 0.568.

When log P was added to the above list, the stepwise regression did not improve over 0.835 and log P did not contribute significantly enough to be included. ('F' to enter < 2.0)

Molar refractivity was estimated and was included along with the six variables selected earlier, for a stepwise multiple regression study; the same original equation resulted with molar refractivity being rejected ('F' to enter < 2.0).

A combination of five descriptors separated the 64 chemicals into the two groups depending on their respective log RTN_{50} values. When the "cut-off" point was set at log RTN_{50} = octanol, the following descriptors sorted the 64 chemicals with an overall accuracy of 88% (and with in accuracy of 92% for the toxic group):

Descriptor #	Name
24	# of CL
28	Mol. wt.
34	^2X_v
36	$^3X^v_c$
38	$^1X^v \cdot ^3X^v$

When log P was included, the efficiency dropped to 85%, Mol. Refractivity also did not improve the efficency over 86%.

The following 9 alcohols were regressed:

ID#	Name
1	Methanol
4	Ethanol
6	N-Butanol
12	1-Pentanol
13	Propanol
19	1-Hexanol
37	1-Octanol
54	N-Decyl Alcohol
56	Lauryl Alcohol

Correlation between log RTN_{50} and log P was:

$$\log RTN_{50} = 0.614 + 0.69 (\log P) \quad \{R=0.970 \text{ Std. error} = 0.644\}$$

Correlation between log RTN_{50} and $(\log P)^2$ was:

$$\log (RTN)_{50} = 0.919 + 0.137 (\log P)^2 \quad \{R = 0.906 \text{ Std. error} = 0.644\}$$

This compares with Vaishnav Equations for alcohols as follows:

$$\log EC50 = -0.88(\log P) + 0.07 (\log P)^2 - 0.23 \quad (r = 0.997)$$

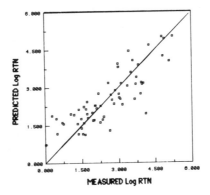

FIG 6. CORRELATION AMONG 64 COMPOUNDS.

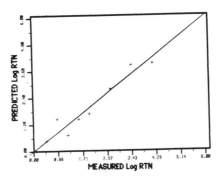

FIG 7. CORRELATION AMONG NINE ALCOHOLS.

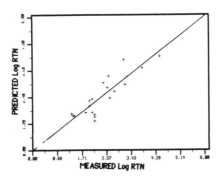

FIG 8. CORRELATION AMONG COMPOUNDS WITH RINGS.

The following 20 "ring" compounds were selected:

	ID#	Name
1.	10	Hydroquinone
2.	14	Resorcinol
3.	15	Phenol
4.	16	Aniline
5.	17	Catechol
6.	18	Toluene
7.	20	Benzonitrile
8.	21	Anisole
9.	25	Ethyl Benzene
10.	30	Xylene
11.	31	m-Cresol
12.	32	4-Ethyl Phenol
13.	35	2-Furaldehyde
14.	38	s-Cl-p-Xylene
15.	40	2,4,6 - Trichlorophenol
16.	43	m-Tolunitrile
17.	46	2,4-Dichlorophenol
18.	52	p-Benzoquinone
19.	58	Nitrobenzene
20.	59	Pentachlorophenol

A stepwise multiple regression analysis was performed using the original six descripters and the following additional ones: -

$$48 => (\# \text{ of Cl})^2$$
$$100 => \log P$$
$$101 => (\log P)^2$$
$$49 => (\# \text{ of Dbl Bonds})^2$$

Following results were obtained:

		Variable Entered	Multiple R
STEP 1.	48 ->	$(\# \text{ of Cl})^2$	0.599
STEP 2.	49 ->	$(\# \text{ of Dbl Bonds})^2$	0.651
STEP 3.	21 ->	# of O	0.694
STEP 4.	28 ->	Mol. wt.	0.757
(See Fig.8)			

In a separate study with 52 petrochemicals, (Chou, et al, 1978) the toxicities were clustered into toxic and non-toxic (4000 mg/l IC_{50}) groups with one descriptor $^3\chi^v_p$ as shown in Fig. 9. Only 4 chemicals were misclassified.

FIG 9. TOXICITY DENDROGRAM OF 52 PETROCHEMICALS.
(● - MISCLASSIFIED COMPOUNDS)

Analysis of the data of Tomlinson et al (1966) on chemicals which inhibited nitrification gave the following predictive equation;

$$\log RTN = 2.60 + 1.54 (\log P) - 0.42 (\log P)^2 + 0.21 (\# of S) - 0.32 (\# of O) + 0.33 (\# of N)$$
$$n = 32 \quad R = 0.807 \quad Std Err. = 0.66$$

SUMMARY

If the temptation to extrapolate QSAR models beyond the bounds in which they were derived can be resisted, then some very useful ecotoxicological predictive models can be used as "screening tools" to "flag" chemicals which may warrant more extensive testing. Likewise, if the ideocyncrasies of the various bacterial toxicity assays can be appropriately factored in, much qualitative information can be gleaned for various environmental purposes.

REFERENCES

Albert, A. (1979) Selective Toxicity Chapman and Hill.

Barffsnecht, C.F., Nichols, D. E, and Dunn, W.J. III (1975) J. Med. Chem. 18 p. 208.

Boyd, J.C., Millership, J.S. and Woolfson, A.D. (1982) "A Comparison of Log P and Molecular Connectivity in the Structure-Activity Analysis of Some Antomicrobial Agents" J. Pharm. Pharmacol. 34 p. 158-161.

Bringmann, G. and Kuhn, R. (1980) "Comparison of the Toxicity Thresholds of Water Pollutants to Bacteria, Algae, and Protozoa in the Cell Multiplication Inhibition Test" Water Research 14 p.231-241.

Brown, D., Hitz, H.R. and Schafer, L. (1981) "The Assessment of the Possible Inhibitory Effects of Dyestuffs on Aerobic Wastewater Bacteria: Experience with a Screening Test" Chemosphere 10 p. 245-261.

Chou, W.L., Speece, R.E., Siddiqi, R.H., and McKeon, K. (1978) "The Effect of Petrochemical Structure on Methane Fermentation Toxicity" Prog. Wat. Tech. 10 p. 545-558.

Crum-Brown A. and Frazer, T.R. (1868) Trans. Royal Soc. Edinburgh 25 p.151.

Dearden, J.C. (1985) "Partitioning and Lipophilicity in Quantitative Structure-Activity Relationship" Environmental Health Perspectives 61 p. 203-228.

Dutka, B.J. and Kwan, K.K. (1982) "Application of Four Bacterial Screening Procedures to Assess Changes in Toxicity of Chemicals in Mixtures" Environ. Poll. Series A 29 p. 125-134.

Dutka, B.J., Nyholm, N. and Petersen, J. (1983) "Comparison of Several Microbiological Toxicity Screening Tests" Water Res. 17 p. 1363-1368.

Hall, L.H. and Kier, L.B. (1978) "A Comparative Analysis of Molecular Connectivity, Hansch, Free-Wilson and Darc-Pelco Methods in the SAR of Halogenated Phenols" Eur. J. Med. Chem. - Chimica Therapeutica 13 p. 89-92.

Hall, L.H. and Kier, Lamont B. (1978) "Molecular Connectivity and Substructure Analysis" J. Pharm. Sciences 67 p. 1743-1747.

Hansch, C. and Dunn, W.J. III 1972 J. Pharm Sci 61 p. 1-19.

Hansch, Corwin and Fujito, Toshio (1964) "Rho, Sigma, Pi Analysis. A Method for the Correlation of Biological Activity and Chemical Structure" J. Am. Chem. Soc. 86 p. 1616-1626.

Hartmann L. and Laubenberger G. (1968) "Toxicity Measurements in Activated Sludge" Am. Soc. Civil Eng. (SA2 April) p. 247-256.

Hermans, J. (1983) "The Use of Quantitative Structure- Activity Relationships in Toxicity Studies with Aquatic Organisims. Correlation to Toxicity of Different Classes of Organic Chemicals with Poct. pKa and Chemical Reactivity " in Quantitative Approaches to Drug Design. J.C. Dearden, ed. Proceeding of Fourth European Symposium on Chemical Structure - Biological Activity: Quantitative Approaches. Elsevier p. 263-264.

ISO (1984) "Water Quality Activated Sludge Oxygen Consumption Inhibition Test". D15 8192, Int'l Org. for Std. Geneva.

Jurs, Peter C., Chou, J.T. and Yuan M. (1979) "Computer - Assisted Structure Activity Studies of Chemical Carcinogens. A Heterogenous Data Set." J. of Med. Chem 22 p. 476-483.

Kier, L.B. and Hall, L.H. (1976) Molecular Connectivity in Chemistry and Drug Research Academic Press.

Kier, L.B. and Hall, L.H. (19 77) "Structure - Activity Studies on Hallucinogenic Amphetamines Using Molecular Connectivity" J. Med. Chem 20 p. 1631-1644.

King, E.F. (1984) "A Comparative Study of Methods for Assessing the Toxicity to Bacteria of Single Chemicals and Mixtures" in Toxicity Screening Procedures Using Bacterial Systems ed. Liu. and Dutka, Dekker.

King, E.F. and Painter, H.A. (1986) "Inhibition of Respiration of Activated Sludge: Variability and Reproducibility of Results " Toxicity Assessment 1 p. 27-39.

Klecka, G.M. and Landi, L.P. (1985) "Evaluation of the OECD Activated Sludge, Respiration Inhibition Test" Chemosphere 14 p. 1239-1251.

Koch, R. (1982) "Molecular Connectivity and Acute Toxicity of Environmental Pollutants" Chemosphere 11 p. 925-931.

Koch, R. (1983) "Molecular Connectivity Index for Assessing Ecotoxicological Behavior of Organic Compounds" Toxicological and Environ. Chem. 6 p. 87-96.

Konemann, H. (1981) "QSAR in Fish Toxicity Studies-Part 1: Relationships for 50 Industrial Pollutants" Toxicology 19 p. 209-221.

Konemann, (1981) Toxicology 19 p. 223-228.

Lee, Albert (1985) "Parameter and Structure - Activity Data Base: Management for Maximum Utility" Environmental Health Perspectives 61 p. 275-285.

Lipnick, Robert L. (1985) "Validation and Extension of Fish Toxicity QSARS and Interspecies Comparsions for Certain Classes of Organic Chemicals" in QSAR in Toxicology and Xenobiochemistry ed. M.Tichy Vol 8 Pharmacochemistry Library Nauta and Rekker, Elsevier p. 38-52.

Lipnick, Robert L. and Dunn, W.J. III (1983) "A MLAB Study of Aquatic

Structure-Toxicity Relationships" in <u>Quantitative Approaches to Drug Design</u> ed. J. Dearden. vol .8. <u>Pharmacochemistry Library</u> Elsevier Press.

Lipnick, R.L. and Hood, Mark T. (1986) "Correlation of Chemical Structure and Toxicity of Industrial Organic Compounds to Daphnia, Algae, Bacteria and Protozoa" Abstract, <u>Soc. of Environ. Tox. and Chem.</u> Nov. 2, 1986 mtg.

Liu, D. (1985) "Effects of Bacterial Cultures on Microbial Toxicity Assessment" <u>Bull. Environm. Contamin. Toxicol.</u> <u>34</u> p. 330-339.

Liu, D. and Thomson K. (1983) "Toxicity Assessment of Chlorobenzenes Using Bacteria " <u>Bull. Environm. Contamin. Toxicol.</u> <u>31</u> p. 105-111.

Liu, D., Thomson, K., and Kaiser, K.L.E. (1982) <u>Bull. Environm. Contam. Toxicol.</u> <u>29</u> p. 130-136.

Murray W.J., Hall, L.H. and Kier, L.B. (1979) <u>J. Pharm. Sci. 64</u> p. 1978.

Newsome, L.D., Lipnick, Robert L. and Johnson. D.E. (1984) "Validation of Fish Toxicity QSAR's for Certain Non-Reactive Non-Electrolyte Organic Compounds" in <u>QSAR in Environmental Toxicology</u> ed. K.L.E. Kaiser Reidel Pub. p. 279-299.

Niemi, G.S., Regal, Ronald R., and Veith, Gilman D (1985) "Application of Molecular Connectivity Index and Multivariate Analysis in Environmental Chemisty" in <u>Environmental Applications of Chemometrics</u> ed. Joseph J. Breen and Philip E. Robinson ACS Symposium Series 292 p. 148-159.

Ribo, Juan, M. and Kaiser, Klaus (1983) "Effects of Selected Chemicals to Photoluminescent Bacteria and Their Correlations with Acute and Sublethal Effects on Other Organisms" <u>Chemosphere 12</u> p. 1421-1422.

Sabljic, A. and Protic, M. (1982) "Molecular Connectivity: A Novel Method for Prediction of Bioconcentration Factor for Hazardous Chemicals" <u>Chem-Biol. Interactions</u> <u>42</u> p. 301-310.

Schultz, T.W., Kier, L.B. and Hall, L.H. (1982) "Structure-Toxicity Relationships of Selected Nitrogenous Heterocyclic Compounds. III Relations Using Molecular Connectivity" <u>Bull. Environ. Contam. Toxicol.</u> <u>28</u> p. 373-378.

Tomlinson, T.G., Boon, A.G., and Trotman, C.N.A.(1966) "Inhibition of Nitrification in the Activated Sludge Process" <u>J. Applied Bact. 29</u>, p. 266-291.

Vaishnav, D.D. (1986) "Chemical Structure - Biodegradation Inhibition and Fish Acute Toxicity Relationships for Narcotic Industrial Chemicals" <u>Toxicity Assessment 1</u> p. 227-240.

Veith, G.D., Call D.J. and Brooke, L.T. (1983) "Structure-Toxicity Relationships for the Fathead Minnow Pimephales Promelas, Narcotic Industrial Chemicals" <u>Can J. Fish and Aquatic Sciences.</u> 40 p. 743-748.

Veith, Gilman D., DeFoe, David, and Knuth, Michael (1985) "Structure-Activity Relationships for Screening Organic Chemicals for Potential Ecotoxicity Effects" <u>Drug Metabolism Reviews 15</u> p. 1295-1303.

Williamson, K.S. and Johnson, D.G. (1981) "A Bacterial Bioassay for Assessment of Wastewater Toxicity" <u>Wat. Res. 15</u> p. 383-390.

Yoshioka , Y., Mizuno , T. ,Ose, Y. and Sata, T. (1986) "The Estimation for Toxicity of Chemicals on Fish by Physcio - Chemical Properties" <u>Chemosphere 15</u> p. .195-203.

COMPETITIVE KINETIC MODEL OF SUSPENDED-GROWTH INHIBITED BIOLOGICAL SYSTEMS

Pablo B. Sáez

*Department of Hydraulic Engineering
Catholic University of Chile
Santiago, Chile*

INTRODUCTION

The kinetic of suspended-growth biological processes used in wastewater treatment has continuously been studied during the last two decades. The behavior of that systems is better understood in the absence of toxic or inhibitory substances. The most used model for non-inhibited systems with one substrate is that proposed by Lawrence and McCarty (1), which is based on the assumption that the substrate utilization rate follows a Monod kinetics. Laterly, Sáez (2,3) generalized the model of Lawrence and McCarty in order to apply it to systems where the substrate utilization rate can be represented by zero-order, first-order, Monod and Contois kinetics.

Some researchers (4-6) have proposed kinetics models for non-inhibited systems with two substrates, in which one substrate, termed primary substrate, provides the cell's energy needs. The other substrate, called secondary substrate, can also be degraded by bacteria, which are, for all practical purposes, completely supported with the energy obtained from the oxidation of the primary substrate. These models have been tested against experimental data obtained on both biofilm (4,5) and suspended-growth (6) reactors.

The presence of inhibitory compounds in biological systems obviously decreases the rate at which biodegradable substances are stabilized. Rozich and Gaudy (7) developed a model to explain the behavior of systems with one inhibitory substrate, which used the Haldane function for relating growth rate to substrate concentration. This model has been used to analyze experimental results on aerobic degradation of phenol (8) and nitrification (9).

Different kinetics models have been proposed for biological systems in which two compounds, one substrate and one toxic, are present. The first group of models are based on that the kinetics coefficients of the Monod expression are functions of the toxic concentration (10-12). The other type of models involves competitive, uncompetitive, or noncompetitive reversible inhibition (13-15). However, the main limitation of these models is that they do not consider the possible degradation of the inhibitor compound, which may be, at least partially, metabolized. At this point, it should be mentioned that many organic substances belong to this category.

The purpose of this paper is to present a kinetic model of suspended-growth inhibited biological systems, based on the theory of competitive reversible inhibition. The model includes the effect of the presence of toxic compounds, which can be biodegradable or refractory. It can also be applied in situations involving primary and secondary substrates. Another goal is to help in the identification of research needs in kinetic modeling of inhibited systems.

MASS BALANCES

In order to formulate the model, a suspended-growth completely mixed reactor (CMR) was selected. Figure 1 shows a CMR with solids settling and recycle. The reactor has a volume V (L^3, where L is length), and is fed with a waste that has a substrate concentration So ($M_S L^{-3}$, where M_S is mass of substrate) and an inhibitor concentration Io ($M_I L^{-3}$, where M_I is mass of inhibitor). The rate of flow of the incoming waste is Q ($L^3 T^{-1}$, where T is time). It is supposed that the incoming active microorganisms concentration Xo ($M_X L^{-3}$, where M_X is mass of microorganisms) is negligible. The substrate, inhibitor, and active microorganisms concentrations within the reactor are the same that those in the reactor effluent, and are denoted as S ($M_S L^{-3}$), I ($M_I L^{-3}$), and X ($M_X L^{-3}$), respectively. The active microorganisms concentrations in the sedimentation basin overflow and underflow are referred to as Xe ($M_X L^{-3}$), and Xr ($M_X L^{-3}$), respectively. The flowrate of the recycle and waste sludge streams are q ($L^3 T^{-1}$), and W ($L^3 T^{-1}$), respectively. It is also assumed that the settler has no effect on the substrate and inhibitor concentrations.

At steady-state, the substrate and inhibitor mass balances around the reactor can be written as:

$$(-r_S) = (So - S) / HRT \qquad (1)$$

$$(-r_I) = (Io - I) / HRT \qquad (2)$$

where $(-r_S)$ = substrate utilization rate, $M_S L^{-3} T^{-1}$

$(-r_I)$ = inhibitor utilization rate, $M_I L^{-3} T^{-1}$
HRT = V/Q = hydraulic retention time, T

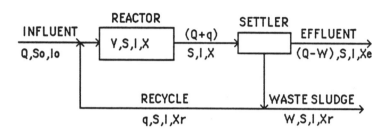

Figure 1. Schematic Diagram of a Completely Mixed Reactor with Solids Settling and Recycle.

If the total biological mass in the system is approximately equal to the microbial mass in the reactor, a materials balance for active microorganisms around the entire system, at steady-state, yields:

$$r_X = X / SRT \qquad (3)$$

where r_X = active bacterial growth rate, $M_X L^{-3} T^{-1}$
SRT = solids retention time, T

The solids retention time can be evaluated using:

$$SRT = X V / [(Q-W) Xe + W Xr] \qquad (4)$$

KINETICS RELATIONSHIPS

In order to obtain the effluent substrate and inhibitor concentrations, it is required to know the substrate and inhibitor utilization rates functions, as it can be observed from equations (1) and (2). It is also necessary to stipulate the active bacterial growth rate function since the utilization rates depend on the active microorganisms concentration.

Inhibitor Utilization Rate

It is postulated that the inhibitor utilization follows the general theory of enzymatic kinetic developed by Michaelis and Menten (13). This theory assumes that first the enzyme E' reversibly combines with the inhibitor I to form the enzyme-inhibitor complex E'I, and then the complex irreversibly decomposes to give free enzyme and product P':

$$E' + I \underset{}{\overset{K}{\longleftrightarrow}} E'I \qquad (5)$$

$$E'I \xrightarrow{k_I'} E' + P' \qquad (6)$$

where k_I' is the rate constant of the second reaction and K is the equilibrium constant of the first reaction, which is given by:

$$K = (E' \cdot I) / E'I \qquad (7)$$

It can be demonstrated, based on the reaction mechanism given by equations (5) and (6), that the inhibitor utilization rate is:

$$(-r_I) = k_I' \, I \, E_T' / (K + I) \qquad (8)$$

where E_T' represents the total enzyme concentration. In other words, E_T' includes both free and combined forms. If it is further supposed that the total enzyme concentration is proportional to the active microorganisms concentration, equation (8) takes the following final form, which is equivalent to that proposed by Monod:

$$(-r_I) = k_I \, I \, X / (K + I) \qquad (9)$$

where k_I = maximum specific inhibitor utilization rate, $M_I M_X^{-1} T^{-1}$
$\qquad\quad K$ = inhibitor saturation constant, $M_I L^{-3}$

Substrate Utilization Rate

The presence of the inhibitor affects the kinetic of the substrate utilization. Three main forms of reversible enzymatic inhibition have been proposed: competitive, uncompetitive, and noncompetitive. As an example, only competitive inhibition will be analyzed in this paper. For this type of inhibition, the following reaction mechanism has been postulated (13):

$$E + S \underset{}{\overset{K_S}{\longleftrightarrow}} ES \qquad (10)$$

$$E + I \underset{}{\overset{K_I}{\longleftrightarrow}} EI \qquad (11)$$

$$ES \xrightarrow{k'} E + P \qquad (12)$$

The enzyme E, different from the enzyme E' required to metabolize the inhibitor (see equations (5) and (6)), can reversibly combine with both the substrate S and the inhibitor I to form the enzyme-substrate ES and enzyme-inhibitor EI complexes, respectively. Finally, the complex ES irreversibly decomposes giving free enzyme and product P. The rate constant of reaction (12) is k', and K_S and K_I are the equilibrium constants of reactions (10) and (11), respectively, defined as:

$$K_S = (E \cdot S) / ES \qquad (13)$$

$$K_I = (E \cdot I) / EI \qquad (14)$$

Based upon the reaction mechanism given by equations (10), (11), and (12), the substrate utilization rate is obtained:

$$(-r_S) = k' \, S \, E_T \, / \, [K_S (1+I/K_I) + S] \qquad (15)$$

where E_T denotes the addition of the free and combined forms of enzyme E. If it is again supposed that E_T is proportional to the active cells concentration, the substrate utilization rate is given by:

$$(-r_S) = k \, S \, X \, / \, [K_S (1+I/K_I) + S] \qquad (16)$$

where
k = maximum specific substrate utilization rate, $M_S M_X^{-1} T^{-1}$
K_S = substrate saturation constant, $M_S L^{-3}$
K_I = inhibition coefficient, $M_I L^{-3}$

Active Bacterial Growth Rate

Biological cells obtain its energy requirements from the oxidation of different organic and/or inorganic molecules. This model assumes that substrate molecules are much more biodegradable than inhibitor molecules, or that substrate concentration is much greater than inhibitor concentration. Under such circumstances, bacterial growth is only due to the decomposition of the substrate, and the inhibitor degradation does not change the biological mass. Then, active bacterial growth rate is related to substrate utilization rate according to (16, 17):

$$r_X = Y (-r_S) - b X \qquad (17)$$

where Y = true yield coefficient, $M_X M_S^{-1}$
       b = decay coefficient, T^{-1}

COMPETITIVE KINETIC MODEL

General Model

Equations (1), (2), (3), (9), (16), and (17) completely describe the behavior of the system, and can be manipulated to yield expressions for effluent substrate and inhibitor concentrations as functions of the solids retention time:

$$[K_S (1 + I/K_I) + S] / S = k\,Y\,SRT / (1 + b\,SRT) \quad (18)$$

$$[(K + I)(Io - I)] / [I(So - S)] = k_I\,Y\,SRT / (1 + b\,SRT) \quad (19)$$

The competitive kinetic model of suspended-growth inhibited biological systems, given by equations (18) and (19), can be normalized using dimensionless substrate and inhibitor concentrations, S* and I*, respectively, and a dimensionless solids retention time parameter, SRT*:

$$S^* = S / K_S \quad (20)$$

$$I^* = I / K \quad (21)$$

$$SRT^* = k\,Y\,SRT / (1 + b\,SRT) \quad (22)$$

Substituting S*, I*, and SRT* in equations (18) and (19) yields:

$$[1 + K^* I^* + S^*] / S^* = SRT^* \quad (23)$$

$$[(1 + I^*)(Io^* - I^*)] / [I^* (So^* - S^*)] = KK^*\,SRT^* \quad (24)$$

where K* and KK* are dimensionless groups of kinetics parameters, defined by:

$$K^* = K / K_I \quad (25)$$

$$KK^* = (k_I / k)(K_S / K) \quad (26)$$

Equations (23) and (24) can be solved to give the effluent dimensionless substrate and inhibitor concentrations as functions of the dimensionless solids retention time:

$$S^* = [(B^{*2} - 4 A^* C^*)^{1/2} - B^*] / (2 A^*) \quad (27)$$

$$I^* = [S^* (SRT^* - 1) - 1] / K^* \quad (28)$$

where the dimensionless parameters A^*, B^*, and C^* depend on K^*, KK^*, So^*, Io^*, and SRT^*:

$$A^* = (SRT^* - 1) [SRT^* (1 - K^* KK^*) - 1] \quad (29)$$

$$B^* = K^* KK^* SRT^* [So^* (SRT^*-1)+1] - (SRT^*-1) (K^* Io^* - K^* + 2) \quad (30)$$

$$C^* = (1 - K^*) (1 + K^* Io^*) - K^* KK^* SRT^* So^* \quad (31)$$

The minimum dimensionless solids retention time, SRT^*_{MIN}, can be evaluated from equation (28) by setting I^* and S^* equal to Io^* and So^*, respectively:

$$SRT^*_{MIN} = (1 + K^* Io^* + So^*) / So^* \quad (32)$$

Finally, in order to equation (27) be well behaved yielding results with physical meaning, it is required that the following condition be satisfied:

$$KK^* K^* Io^* (1 + K^* Io^* + So^*) \leq (1 + Io^*) (1 + K^* Io^*) \quad (33)$$

Limiting Cases

The numerical values of the dimensionless groups of kinetics parameters K^* and KK^* have an important effect on the behavior of the system, and determine some limiting cases of the model. The condition $KK^* = 0$ represents the situation where the inhibitor is non-biodegradable at all. Substrate and inhibitor concentrations in the effluent can be found manipulating equations (23) and (24):

$$S^* = (1 + K^* Io^*) / (SRT^* - 1) \quad (34)$$

$$I^* = Io^* \quad (35)$$

The minimum dimensionless solids retention time is obtained from equation (32).

The system is completely inhibited when $K^* = \infty$. Under such circumstances, substrate and inhibitor concentrations do not change within the reactor, and their concentrations at all point in the reactor are the same than those in the incoming waste.

The case in which the inhibitor does not affect the substrate utilization rate is modeled by setting $K^* = 0$. In this type of system, the inhibitor is better called secondary substrate. The oxidation of the primary substrate completely supports bacteria, which can also metabolize the secondary substrate. The solution of equations (23) and (24) for this situation is:

$$S^* = (SRT^* - 1)^{-1} \tag{36}$$

$$I^* = [\,(D^{*2} + 4\,Io^*)^{1/2} - D^*\,] / 2 \tag{37}$$

where the dimensionless parameter D^* is given by:

$$D^* = KK^* \, SRT^* \, (So^* - S^*) - Io^* + 1 \tag{38}$$

The minimum dimensionless solids retention time can be evaluated using:

$$SRT^*_{MIN} = (1 + So^*) / So^* \tag{39}$$

EFFECT OF THE PARAMETERS OF THE MODEL

Figures 2-9 show the effect of the dimensionless kinetics parameters K^* and KK^* and the dimensionless incoming concentrations So^* and Io^* on the behavior of the system. It can be observed that K^*, So^*, and Io^* tremendously affect the minimum dimensionless solids retention time. On the other hand, that minimum time does not vary with changes in KK^*. Another interesting fact observed from all these figures is that the effect of all the dimensionless parameters of the model decreases as the dimensionless solids retention time becomes larger. The figures also stress the importance of the solids retention time in handling inhibited biological systems.

The effect of the inhibitor toxicity is shown in Figures 2 and 3. The bigger the K^*-value is the more toxic the inhibitor is since the quantity of enzyme E present in the form of enzyme-inhibitor complex EI increases. Then, the availability of enzyme E to form the enzyme-substrate complex ES, required to metabolize the substrate S, decreases, as it can be deduced from equations (10)-(14). This explains the decrease in substrate oxidation efficiency shown in Figure 2 when K^* increases. The oxidation of the inhibitor also decreases when K^* becomes larger, as illustrated in Figure 3. This is due to two reasons. First, the amount of active microorganisms, required to metabolize the inhibitor, decreases when K^* increases since the substrate utilization rate diminishes. Secondly, the bigger the K^*-value is the lesser the amount of enzyme-inhibitor complex E'I formed, which is required to oxidate the inhibitor, as it can be observed from equations (5)-(7).

Figures 4 and 5 illustrate the effect of the inhibitor biodegradability. It can be deduced from equation (9) that the bigger the KK^*-value is the more biodegradable the inhibitor is. It is important to point out that the inhibitor biodegradability has been neglected in most of the inhibited biological models, which gives a reduced substrate oxidation efficiency, as shown in Figure 4. For accurate prediction of the effluent substrate concentration, it is recommended to take into account the phenomenon just mentioned. Figure 5 shows the changes in the dimensionless effluent inhibitor concentration as KK^*

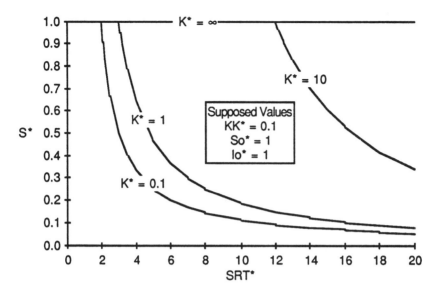

Figure 2. Effect of the Inhibitor Toxicity on the Dimensionless Effluent Substrate Concentration.

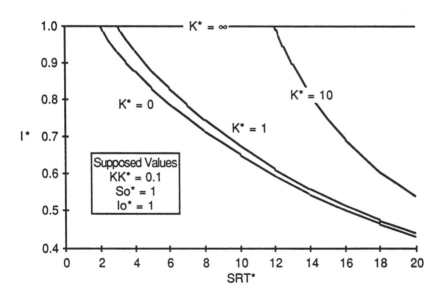

Figure 3. Effect of the Inhibitor Toxicity on the Dimensionless Effluent Inhibitor Concentration.

76 Biotechnology for Degradation of Toxic Chemicals

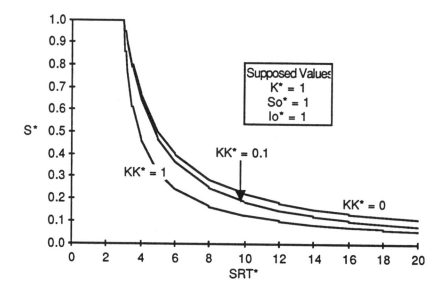

Figure 4. Effect of the Inhibitor Biodegradability on the Dimensionless Effluent Substrate Concentration.

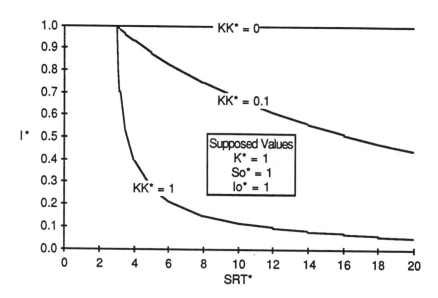

Figure 5. Effect of the Inhibitor Biodegradability on the Dimensionless Effluent Inhibitor Concentration.

varies.

Figures 6 and 7 illustrate the response of the model with changes in the incoming substrate concentration. It should be noted the logarithmic scale in Figure 6. An interesting observation is raised from Figure 6: the bigger the incoming substrate concentration is the lesser the effluent substrate concentration is, for all dimensionless solids retention time greater than SRT^*_{MIN}. This differs from the estimation of most models, which predict that the effluent substrate concentration is independent or directly proportional to the influent substrate concentration. It can be deduced from Figure 7 that the effluent inhibitor concentration decreases as So^* becomes larger. This can be explained by the fact that the substrate oxidation rate, and as a consequence the amount of active microorganisms, increases as the incoming substrate concentration becomes larger. Then, the inhibitor oxidation efficiency increases since there is a larger quantity of bacterial cells in the reactor.

The effect of the incoming inhibitor concentration is shown in Figures 8 and 9. It is observed that the effluent substrate and inhibitor concentrations become larger as the concentration of the inhibitor in the incoming waste increases, as expected.

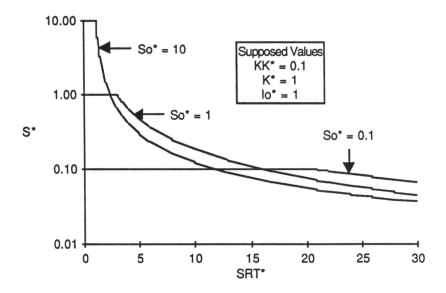

Figure 6. Effect of the Incoming Substrate Concentration on the Dimensionless Effluent Substrate Concentration.

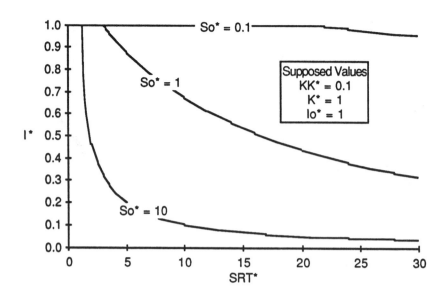

Figure 7. Effect of the Incoming Substrate Concentration on the Dimensionless Effluent Inhibitor Concentration.

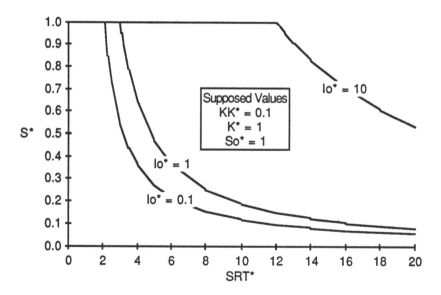

Figure 8. Effect of the Incoming Inhibitor Concentration on the Dimensionless Effluent Substrate Concentration.

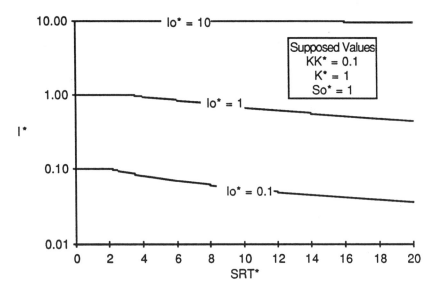

Figure 9. Effect of the Incoming Inhibitor Concentration on the Dimensionless Effluent Inhibitor Concentration.

SUMMARY AND CONCLUSIONS

A kinetic model of suspended-growth inhibited biological systems has been presented. The model is only based upon fundamental principles, which include mass balances and kinetics relationships. It has been supposed that the inhibitor may be biodegradable, following a Monod function. Another assumption is that the substrate utilization rate is given by the general theory of competitive reversible inhibition, and completely determine the cells growth rate. The model has been normalized to better understand the process. The effect of the inhibitor biodegradability and toxicity has been studied. The response of the model with changes in the influent substrate and inhibitor concentrations has also been discussed.

The main contribution of the model presented is that it helps in the identification of research needs in kinetic modeling of inhibited systems. First, it is clear that more experimental results are required. Second, the model has been structured to visualize its assumptions, which can be changed in order to find those that better represent the real world. For example, it is possible to construct models in which the inhibitor utilization rate follow an Haldane function, or the substrate utilization rate could be modeled by uncompetitive or noncompetitive reversible inhibition, or the active bacterial growth also depend on the inhibitor utilization rate, etc.

The model presented may be easily applied to batch or plug-flow suspended-growth reactors. The only changes are related to mass balances equations. Also, its principles could be used to model inhibited biofilm reactors.

ACKNOWLEDGEMENTS

This research has been supported in part by Research Grant 0126/85 from the National Comission of Scientific and Technological Research, Chile (CONICYT), and in part by Research Grant 016/86 from the Research Direction of Catholic University of Chile (DIUC).

REFERENCES

1. Lawrence, A.W., and McCarty, P.L., "Unified Basis for Biological Treatment Design and Operation," *Journal of the Sanitary Engineering Division*, ASCE, Vol. 96, No. SA3, 1970, pp.757-778.
2. Sáez, P.B., "Cinética de los Procesos Biológicos Usados en el Tratamiento de Aguas Residuales," presented at the November 6-11, 1983, X Interamerican Congress of Chemical Engineering, held at Santiago, Chile, Vol. II, pp.226-230.
3. Sáez, P.B., "Cinética de los Procesos Anaeróbicos," *Apuntes de Ingeniería*, Vol. 14, 1984, pp.91-125.
4. Namkung, E., Stratton, R.G., and Rittmann, B.E., "Predicting Removal of Trace Organic Compounds by Biofilms," *Journal Water Pollution Control Federation*, Vol. 55, 1983, pp.1366-1372.
5. Bouwer, E.J., and McCarty, P.L., "Utilization Rates of Trace Halogenated Organic Compounds in Acetate-Grown Biofilms," *Biotechnology and Bioengineering*, Vol. 27, 1985, pp.1564-1571.
6. García-Orozco, J.H., Fuentes, H.R., and Eckenfelder, W.W., "Modeling and Performance of the Activated Sludge-Powdered Carbon Process in the Presence of 4,6 Dinitro-o-Cresol," presented at the October 6-9, 1985, 58th Annual Conference of the Water Pollution Control Federation, held at Kansas City, USA.
7. Rozich, A.F., and Gaudy, A.F.,Jr., "Critical Point Analysis of Toxic Waste Treatment," *Journal of the Environmental Engineering Division*, ASCE, Vol. 110, No. 3, 1984, pp.562-572.
8. Colvin, R.J., and Rozich, A.F., "Phenol Growth Kinetics of Heterogeneous Populations in a Two-Stage Continuous Culture System," presented at the October 6-9, 1985, 58th Annual Conference of the Water Pollution Control Federation, held at Kansas City, USA.
9. Castens, D.J., and Rozich, A.F., "Analysis of Batch Nitrification Using Substrate Inhibition Kinetics," *Biotechnology and Bioengineering*, Vol. 28, 1986, pp.461-465.
10. Neufeld, R.D., and Valiknac, T., "Inhibition of Phenol Biodegradation by Thiocyanate," *Journal Water Pollution Control Federation*, Vol. 51, 1979, pp.2283-2291.
11. Neufeld, R.D., Greenfield, J.H., Hill, A.J., Rieder, C.B., and Adekoya, D.O., "Nitrification Inhibition Biokinetics," EPA-60012-83-111, 1983.
12. Greenfield, J.H., and Neufeld, R.D., "Inhibition of Nitrification by Coke

Plant Organics," presented at the October 6-9, 1985, 58th Annual Conference of the Water Pollution Control Federation, held at Kansas City, USA.
13. Lehninger, A.L., *Biochemistry*, Worth Publishers, Inc., New York, USA, 1976.
14. Parkin, G.F., and Speece, R.E., "Modeling Toxicity in Methane Fermentation Systems," *Journal of the Environmental Engineering Division*, ASCE, Vol. 108, 1982, pp.515.
15. Lewandowski, Z., Janta, K., and Mazierski, J., "Inhibition Coefficient Determination in Activated Sludge," *Water Resources*, Vol. 19, 1985, pp.671-674.
16. Heukelekian, H., Orford, H.E., and Manganelli, R., "Factors Affecting the Quantity of Sludge Production in the Activated Sludge Process," *Sewage and Industrial Wastes*, Vol. 23, 1951, pp.945-958.
17. Eckenfelder, W.W., and Weston, R.F., "Kinetics of Biological Oxidation," in McCabe, J., and Eckenfelder, W.W. (eds), *Biological Treatment of Sewage and Industrial Wastes*, Reinhold Publishing Corporation, Vol. 1, New York, USA, 1956.

TOXICITY OF NICKEL IN METHANE FERMENTATION SYSTEMS: FATE AND EFFECT ON PROCESS KINETICS

Sanjoy K. Bhattacharya and Gene F. Parkin

Department of Civil Engineering
Drexel University
Philadelphia, PA

INTRODUCTION

The applicability of anaerobic processes for treatment of industrial wastewaters and domestic sludges has been recognized for many years. However, anaerobic processes have yet to gain wide popularity. A major reason for the prevailing skepticism about the reliability of anaerobic waste treatment processes is the lack of quantitative information on the capability of such processes to handle potentially toxic wastes. It has long been realized that anaerobic systems are particularly vulnerable to high loadings of heavy metals. The most common single cause of stress in anaerobic digesters in England was reported to be heavy metal toxicity (1).

The toxicity of heavy metals depends upon the various chemical forms which the metal may assume under anaerobic conditions and at near-neutral pH. Mosey (2) stated that heavy metals only cause digestion failure when the concentration of their free ions exceeds a certain threshold concentration, which is directly related to the concentration of divalent sulfide ions present in the system. Ashley, et al. (3) studied the effect of nickel addition on various reactions occuring in anaerobic digesters fed primary sludge. They found that methane production was inhibited by soluble nickel levels above 1 mg/l and found many hydrolytic bacteria inhibited at levels above 12 mg/l. Some evidence has shown that anaerobic systems can adapt to heavy metals contained in an industrial wastewater by gradually increasing the heavy metal concentration in the feed (4,5). Gould and Genetelli (6) suggested that heavy metals are bound by weakly acidic functional groups. Ahring and Westermann (7) studied the effects of slug doses of Cd, Cu, and Ni on the thermophilic digestion of sewage sludge in semi-continuous systems; Ni was found to be completely inhibitory at 200 mg/l. However, Parkin, et al. (8) found from continuous addition experiments that 200 mg/l nickel could be tolerated by methanogenic bacteria with no decrease in process performance.

The fate of heavy metals in biological systems is of considerable importance. Hayes and Theis (9) attempted to determine the fate of various heavy metals using filtration and extraction procedures. They defined metals associated with the soluble phase, precipitated metals, and extracellular and intracellular metals. Results indicated a significant fraction was associated with the biomass. Callander and Barford (10,11) developed and applied a methodology for the quantitative assessment of the individual effects of precipitation and chelation of metal ions in an anaerobic digester. Nielson, et al. (12) showed that it is the free nickel that is sorbed by activated sludge.

From a review of the literature it appears that very little research has focused on kinetics of anaerobic systems exposed to nickel, or on the fate of nickel during anaerobic treatment. Little work has been done to quantify the effect of solids retention time (SRT). This research was undertaken to provide such information.

The objectives of the research were to study the kinetic effects of nickel on anaerobic utilization of acetate and propionate and to determine its fate in these systems. Nickel was added both as slug and continuous doses. With the shock of a slug addition, there should be minimal chance for acclimation to nickel by the bacteria. With continuous addition, however, the concentration of nickel can be increased gradually and the bacteria will have a chance to acclimate. In this manner, it should be possible to systematically address acclimation characteristics.

MATERIALS AND METHODS

Stock acetate and propionate enrichment cultures were developed and used in this research. Chemostats operated at 40, 25, and 15 days were used to determine base-line and toxicity kinetics. (With no biomass recycle, hydraulic retention time (HRT) equals SRT.) The chemostats were 2-L glass bottles containing 1.5-L of acetate or propionate enrichment culture. The cultures were continuously mixed with magnetic mixers and fed continuously with required nutrients and acetate or propionate using Manostat Cassette pumps (Manostat, New York, NY). The details of the experiments are reported elsewhere (13).

Chemostats were anaerobically seeded from stock cultures. Volatile acids were regularly monitored using a gas chromatograph (Shimadzu RIA with 1.7-m glass column, packed with 60/80 Carbopack C/0.3% Carbowax 20M/0.1% H_3PO_4 (Supelco, Inc., Bellefonte, PA)). Biomass was measured as volatile suspended solids (VSS) following Standard Methods (14). Attainment of steady state was arbitrarily defined as variation of VSS and acetate or propionate concentrations within 10 and 20 percent, respectively, for at least one week. Toxicity studies were started after steady state was reached. Slug doses were injected directly into the chemostats. For continuous addition studies, nickel was added to the respective feed bottles. Experiments were continued until new steady states were attained, unless a failure occurred. Failure was arbitrarily defined as accumulation of 3000 mg/l of volatile acids. At this point, the bicarbonate alkalinity was destroyed. Nickel was measured using a Perkin-Elmer Model 403 Atomic Absorption Spectrophotometer.

KINETICS OF ACETATE AND PROPIONATE UTILIZATION

Two basic equations relate biological growth and substrate utilization. The first describes the relationship between net rate of growth of bacteria and rate of substrate utilization (15,16):

$$dX_a/dt = Y(dS/dt) - bX_a \qquad (1)$$

where
- dX_a/dt = net growth rate of bacteria, mass/volume-time;
- dS/dt = rate of substrate utilization, mass/volume-time;
- Y = bacterial yield, mass bacteria/mass substrate;
- b = bacterial decay rate, day^{-1}.

The second equation relates the rate of substrate utilization to both the concentration of bacteria in the reactor and the concentration of substrate remaining (15,16,17):

$$dS/dt = kSX_a/(K_s+S) \qquad (2)$$

where
- k = maximum specific substrate utilization rate, mass substrate/mass bacteria-day;
- K_s = half-velocity coefficient, mass substrate/volume;
- S = substrate concentration, mass/volume.

Once the chemostats reached steady state, VSS and acetate and/or propionate were measured for a period of approximately one week. Using standard techniques (18), the baseline kinetic coefficients (Y, b, k, and K_s) were determined. Table I shows the values of the kinetic parameters for acetate and propionate utilization in anaerobic chemostats.

Table I. Baseline Values of Kinetic Coefficients

	Acetate Utilization	Propionate Utilization
k (mg/mg-day)	2.5	2.2
K_s (mg/l)	10	11
Y (mg/mg)	0.05	0.07
b (day^{-1})	0.01	0.016

TOXICITY KINETICS

The Uncompetitive-Inhibition-Coefficient model described by Parkin and Speece (19) was used in this research to quantify the effect of nickel. The equation for substrate utilization then becomes:

$$dS/dt = kSX_a/(K_s+S[1+T_x/K_I]) \qquad (3)$$

where
T_x = nickel concentration, mass/volume;
K_I = inhibition coefficient, mass/volume.

Table II lists the equations formulated to quantify the effects of slug and continuous addition of nickel in a chemostat using the Uncompetitive-Inhibition-Coefficient model. To solve the model equations, a value for K_I must be pre-determined in addition to the already-determined baseline kinetic parameters. The Fourth-Order Runge-Kutta method was used to numerically solve the model equations. An evaluation of the model is made in a later section by comparison with experimental data.

Table II. Equations used to Quantify Toxicity Kinetics

UNCOMPETITIVE INHIBITION IN A CHEMOSTAT

$$dS/dt = ([S_0-S]/\theta) - kSX_a/(K_s+[SA]) \qquad (4)$$

$$dX_a/dt = YkSX_a/(K_s+[SA]) - X_a(1/\theta+b) \qquad (5)$$

SLUG DOSE

$$A = 1+(T_x/K_I)[(\theta-1)/\theta]^t \qquad (6)$$

CONTINUOUS DOSE

$$A = 1+(T_x/K_I)[1-((\theta-1)/\theta)^t] \qquad (7)$$

Note:
θ = SRT and HRT
t = time after addition of nickel

Values of K_I can be determined if a chemostat reaches a new steady state after the addition of nickel. At steady state ($dS/dt = 0$; $dX_a/dt = 0$), the model equations reduce to simple algebraic equations which can then be solved to determine the effluent acetate or propionate concentration (19):

$$S = K_s(1+b\theta)/[\theta(Yk-b[1+T_x/K_I])-(1+T_x/K_I)] \tag{8}$$

or,

$$1/S = 1/S_{eff} - T_x/CK_I \tag{9}$$

where:

$$S_{eff} = K_s(1+b\theta)/[\theta(Yk-b[1+T_x/K_I])-1] \tag{10}$$

$$C = K_s(1+b\theta) \tag{11}$$

Since the decay coefficient, b, is very small, the effect of nickel on b can be neglected. Therefore:

$$S_{eff} \approx K_s(1+b\theta)/[\theta(Yk-b)-1] \tag{12}$$

Conceptual plots of $1/S$ versus T_x are shown in Figure 1. K_I can be calculated from the slopes of the plots. The term C varies with SRT causing variation in the slopes of the lines. Using the values of the baseline kinetics reported in Table 1 for the acetate enrichment culture, the above equations can be reduced to the following:

$$\text{SRT} = 15 \text{ days:} \quad S = 11.5/(0.725-T_x/K_I) \tag{13}$$
$$\text{SRT} = 25 \text{ days:} \quad S = 12.5/(1.875-T_x/K_I) \tag{14}$$
$$\text{SRT} = 40 \text{ days:} \quad S = 14.0/(3.6-T_x/K_I) \tag{15}$$

Use of these equations to predict system behavior will be demonstrated in a later section.

RESULTS AND DISCUSSION

Response to Nickel Toxicity

Nickel chloride ($NiCl_2 \cdot 6H_2O$, crystal, Fisher Certified) was used as the source of nickel.

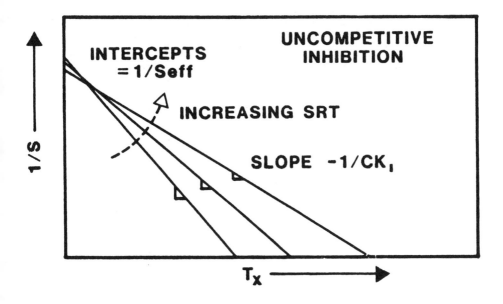

Figure 1. Effect of Uncompetitive Inhibition on Effluent Substrate Concentration

Slug Addition

When chemostats maintained for slug addition studies reached steady state, nickel chloride was injected directly into the chemostats. Samples for analysis were withdrawn frequently, especially during the first few days. Typical results for acetate enrichment cultures are shown in Figure 2. A slug dose of 100 mg/l caused failure in a 40-day-SRT system, but a 25-day system recovered from the intial shock that caused acetate concentrations to increase to 2000 mg/l. The same slug dose of 100 mg/l caused a lesser effect in the 15-day system (Figure 2). These observations show the potential advantage of a low-SRT/HRT system to withstand slug doses of nickel, since the toxicant washout is more rapid at lower HRTs. However, if the slug dose is great enough (e.g., 200 mg/l in Figure 2), the system will fail.

In propionate enrichment systems, both propionate and acetate were expected to accumulate after introduction of nickel. The purpose of these experiments was to determine whether the acetate utilizaers were more affected than the propionate utilizing organisms. A higher accumulation of acetate indicated that the acetate utilizers were more adversely affected than propionate utilizers (Figure 3). For some systems, the results were not as clear and additional calculations were necessary (13). In all cases, the acetate utilizers were more adversely affected than the propionate utilizers.

Continuous Addition

Different amounts of nickel chloride were added to the continuous feed for the chemostats. Typical results for acetate systems are shown in Figure 4. Nickel concentrations of up to 100 mg/l showed no significant effect on 40-day systems. The 25-day systems showed no major effect up to 70 mg/l of nickel. Chemostats maintained at a 15-day SRT could tolerate up to 50 mg/l, but a dose of 100 mg/l caused failure. These results indicate the advantage of higher SRTs for tolerating continuous nickel doses: higher nickel concentrations can be handles with longer SRTs.

Similar response patterns were observed with the propionate systems as indicated by the typical results shown in Figure 5. Acetate accumulation was again observed to be higher than propionate accumulation. This indicates that with both slug and continuous addition of nickel the acetate utilizing methanogens were always more adversely affected than the propionate utilizing bacteria.

The results of the effects of nickel on acetate and propionate cultures are summarized in Table III.

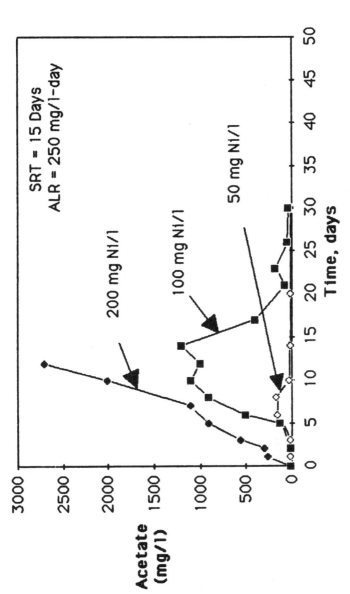

Figure 2 Effect of Slug Doses of Nickel on Acetate Utilization

90 Biotechnology for Degradation of Toxic Chemicals

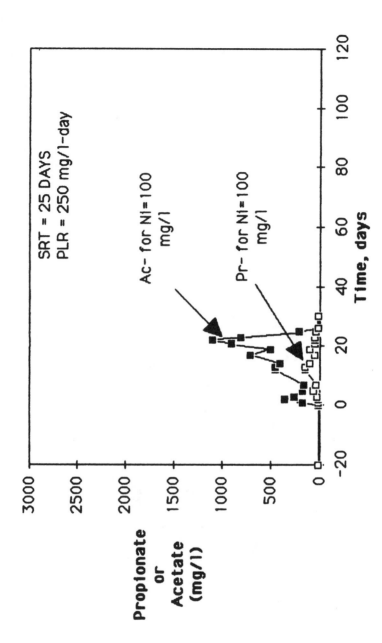

Figure 3 Effect of Slug Dose of Nickel on Propionate Utilization

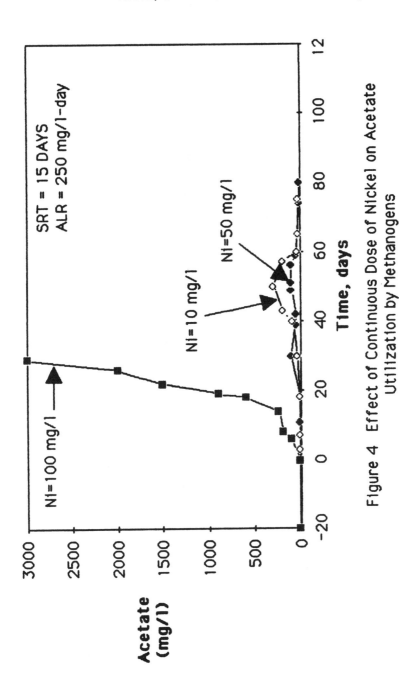

Figure 4 Effect of Continuous Dose of Nickel on Acetate Utilization by Methanogens

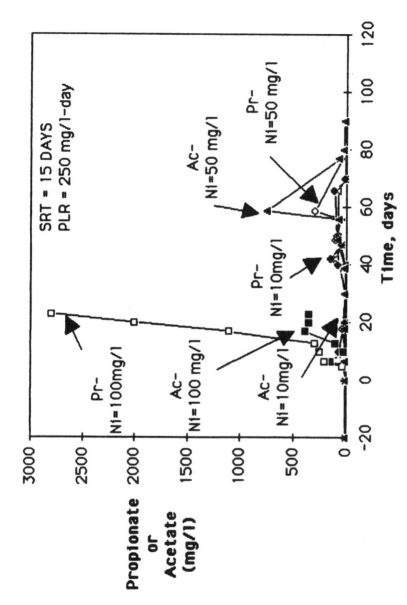

Figure 5 Effect of Continuous Dose of Nickel on Propionate Utilization

Table III. Effects of Nickel on Acetate and Propionate Utilization

SRT (days)	Nickel (mg/l)	Fate of System Acetate	Propionate
Slug Addition			
40	100	Failed	Failed
25	100	Recovered	Recovered
15	200	Failed	Failed
15	100	Recovered	Failed
15	70	*	Recovered
15	50	Small effect	*
Continuous Addition			
40	100	No effect	*
40	80	*	No effect
40	50	*	No effect
40	20	No effect	*
25	70	No effect	No effect
25	50	No effect	No effect
15	100	Failed	Failed
15	50	No effect	No effect
15	10	No effect	No effect

* - condition not tested

Fate of Nickel

Samples for nickel analysis were centrifuged and supernatants were analyzed to determine total soluble nickel (TSN). Centrifuged solid pellets of some samples were acidified with 1:1 HNO_3 and analyzed for nickel. The theoretical nickel was calculated from added nickel by taking dilution into account and was found to be within five percent of the sum of TSN and the nickel in the solid pellets; the mass balance for these samples was thus satisfactory. Variation of pH will significantly change TSN. However, the pH was essentially constant in all chemostats and thus should have little impact on TSN.

With slug doses it was observed that when TSN was about 6 mg/l or less, acetate or propionate utilization rates were not significantly affected. Figure 6

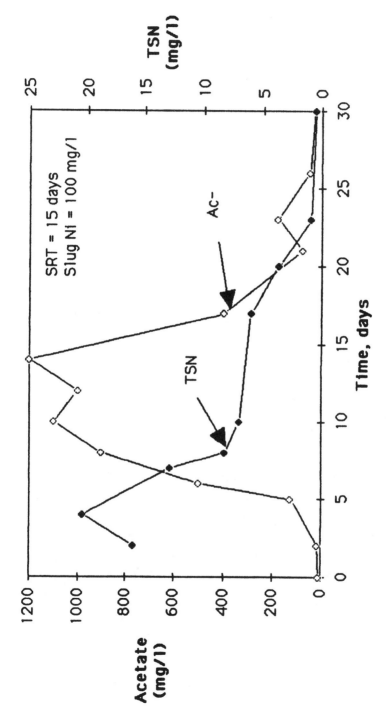

Figure 6 Measured Total Soluble Nickel after a Slug Dose

shows that in a 15-day system the accumulation of acetate started decreasing after TSN decreased to about 6 mg/l. Data in Table IV indicates that irrespective of the concentration of the slug dose, systems started recovering when measured TSN was between 5.4 and 6.2 mg/l.

Table IV. TSN Levels at Recovery from Slug Doses of Nickel

SRT (days)	Enrichment Culture	Dose (mg/l)	Measured TSN at Recovery Point (mg/l)	Day (Recovery Point)	Calculated Total Nickel* (mg/l)
15	Acetate	100	6.0	17	30.9
15	Propionate	70	5.4	10	35.1
15	Acetate	50	6.2	9	26.9

* - theoretical total nickel concentration expected after dilution

With continuous nickel addition, the tolerable TSN concentrations were very high, as shown in Table V. Tolerable TSN appears to increase with SRT. A 40-day system could tolerate up to 31.6 mg/l TSN. However, the 15-day systems failed with about the same TSN. This is strong evidence of the role of SRT; acclimation may be facilitated with long SRTs.

Table V. TSN Levels with Continuous Nickel Doses

SRT (days)	Enrichment Culture	Nickel Dose (mg/l)	Measured TSN* (mg/l)	Fate of System
40	Acetate	100	31.6	Survived
40	Propionate	80	16.0	Survived
40	Propionate	50	10.4	Survived
25	Propionate	70	18.6	Survived
25	Propionate	50	18.0	Survived
15	Acetate	100	30.7	Failed
15	Propionate	100	29.6	Failed
15	Acetate	50	22.4	Survived
15	Propionate	10	2.0	Survived

* - after 3xSRT

A combination of dialysis and ion exchange methods were used to estimate the concentrations of free nickel (Ni^{2+}) and high-MW nickel complexes. The ion exchange part of the experiment was done following the method of Cantwell, et al. (20). Dialysis bags (MW cut-off of 10,000; Spectrum, CA) were used to separate high-MW complexes (possibly protein-nickel or carbohydrate-nickel complexes). When the chemostat experiments were finished, 400-mL samples were taken for the dialysis experiments. Dowex 50W-X8, 20-50 mesh, resin (0.3 g) and 30 mL distilled water were placed in a dialysis bag which was placed in a glass bottle containing the sample and mixed with a magnetic stirrer. Dialysis-kinetics data indicated that equilibrium was reached after about 24 hours; however, the experiments were continued for three days. At the end of the experiment, samples from both inside and outside the bag were analyzed for nickel. The nickel associated with the resin was measured after elution with 1:1 HNO_3. Free nickel was determined following Cantwell, et al. (20); the distribution coefficient of nickel was determined as 1.2 l/g. The results of the dialysis-ion exchange experiments are shown in Table VI.

Table VI. Speciation of Nickel in Chemostats

System*	Measured TSN at Equilibrium		Nickel in Resin (mg/g)	Calculated Free Nickel (mg/l)	Fate of System
	Outside Bag (mg/l)	Inside Bag (mg/l)			
1	31.6	16.8	0.5	0.4	Recovered
2	30.7	30.2	7.8	6.5	Failed

* System 1 was a 40-day-SRT with a continuous dose of 100 mg/l (see Table V)
Table VI Cont'd

System 2 was a 15-day-SRT with a continuous dose of 100 mg/l (see TableV)

In System 1 (SRT = 40 days), from the equilibrium concentrations it appears that a significant amount (31.6 - 16.8 = 14.8 mg/l) of high-MW nickel complexes (>10,000 MW) was formed. However, in System 2 (SRT = 15 days), the equilibrium concentrations were almost equal on both sides of the dialysis bag, indicating that high-MW complexes were not formed. The difference in free nickel concentrations in the two systems with almost equal TSN indicates that in the 40-day systems the methanogens were perhaps able to acclimate to nickel by excreting extracellular polymeric compounds to keep the free nickel levels below 0.5 mg/l. Rittmann, et al. (21) have shown that anaerobic production of soluble polymers increases with increasing SRT for SRTs greater than about seven days.

APPLICATION OF UNCOMPETITIVE-INHIBITION-COEFFICIENT MODEL

In this section an evaluation of the model presented earlier is made by comparing the experimental data with predicted values. An estimation of K_I was first made from steady-state conditions using equation 8 (13). Model equations 4 and 5 were then solved numerically with a range of K_I values close to the estimated K_I to determine the best-fit value. The Fourth Order Runge-Kutta method was used for this numerical solution.

Acclimation

In the model equations, K_I is defined as a constant. For purposes of this paper, acclimation is defined as bacterial adaptation to nickel such that bacterial activity becomes "normal" (returns to non-toxic, steady state) in the toxic environment. Of course, it is also possible that a new steady state will be reached with higher acetate or propionate concentrations. This means that even if nickel concentration, T_x, does not change (except with dilution), K_I increases to such an extent that the ratio T_x/CK_I becomes very small (equation 9). Acclimation is possible with gradual exposure to nickel. Thus, a higher-SRT-system has an inherent advantage of facilitating acclimation to the same continuous nickel dose. The model predicts that a higher-SRT-system will be more tolerant than a lower-SRT-system. This tolerance is independent of acclimation because model equations do not take acclimation into account. Thus, a better potential for acclimation adds to the advantage of high-SRT-systems.

Using Measured Nickel in the Model

Ideally, free nickel concentration should be used as T_x in the model. This would facilitate system-independent model predictions. However, determination of free nickel is difficult. It was observed that after the "nickel demand" (by sulfide, etc.), total soluble nickel (TSN) is essentially proportional to the calculated nickel concentration (13). Thus, using TSN as T_x offers no advantage over using added nickel as T_x in the model. It should be noted that the "nickel demand" will be different for different wastewaters. For toxicity mechanism studies, both TSN and free nickel data are required.

Continuous Addition

The best-fit K_I values calculated from experimental data are shown in Table VII. Interestingly, values of K_I were not constant; K_I increased with increasing SRT. This indicates acclimation to higher nickel concentrations is possible with longer

SRTs.

Table VII. K_I Values for Continuous Addition of Nickel to Acetate Enrichment Cultures

SRT (days)	Nickel Dose (mg/l)	Best-Fit K (mg/l)
40	20-100	150
25	50-70	120
15	10-100	80

Figure 7 illustrates that at a 15-day SRT the predicted data (K_I = 80) compares well with experimental observations. Similar agreement was noted for other experimental conditions. In propionate systems, the K_I values were 10-20 percent higher, indicating higher resistance to nickel toxicity by propionate utilizing bacteria.

Slug Additions

The effect of slug doses of nickel could not be predicted well using the Uncompetitive-Inhibition-Coefficient model. A possible reason is that the "shock effect" of a slug dose could not be modeled with the existing equations. Further research is required to quantify the shock due to a slug addition.

CONCLUSIONS

The following conclusions are drawn from this study on the fate and kinetic effects of nickel on anaerobic acetate and propionate utilization.

1. Acetate-utilizing methanogens are more severely affected than propionate utilizers by nickel.

2. Massive slug doses of nickel immediately stop bacterial activity; smaller slug doses give lower-HRT systems a better chance to recover.

3. Response to nickel toxicity is similar to uncompetitive inhibition. The Uncompetitive-Inhibition-Coefficient model worked well for continuous addition of nickel. Best-fit K_I values varied between 80 and 150 mg/l

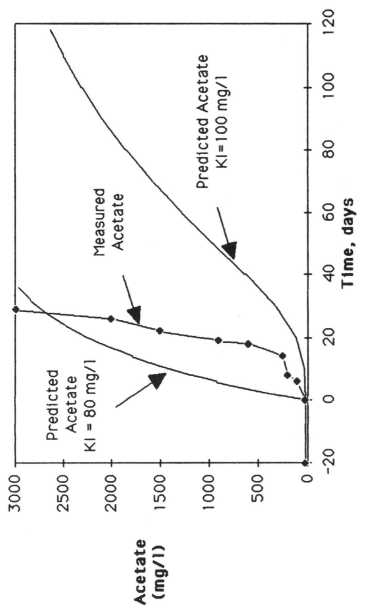

Figure 7 Comparison of Predicted and Observed Data with Continuous Additions of Nickel to Acetate Enrichment Cultures

depending on SRT. The observed increase in K_I with increasing SRT indicates acclimation.

4. The Uncompetitive-Inhibition-Coefficient model did not adequately describe the fate of systems exposed to slug doses of nickel.

5. Without acclimation to nickel, total soluble nickel concentrations above 6 mg/l caused failure; total soluble nickel concentrations as high as 31 mg/l could be tolerated with acclimation.

REFERENCES

1. British Notes on Water Pollution, (1971) "Inhibition in the Anaerobic Digestion Process for Sewage Sludge," No. 53, Water Research Centre, Stevenage Laboratory, Stevenage, Hertz, June.

2. Mosey, F.E. (1976) "Assessment of the Maximum Concentration of Heavy Metals in Crude Sewage which will not Inhibit the Anaerobic Digestion of Sludge", Water Pollution Control, 75, 1, 10.

3. Ashley, N.V., Davies, M., and Hurst, T.J. (1982) "The Effect of Increased Nickel Ion Concentrations on Microbial Populations in the Anaerobic Digestion of Sewage Sludge," Water Research, 16, 963.

4. Fannin, K. F., Conrad, J.R., Srivastava, V.J., Jerger, D.E., and Chynoweth, D.P. (1982) "Anaerobic Processes," Jour. Water Pollution Control Fed., 6, 612.

5. Kremer and Fradkin (1981) "Bacterial Adaptation to an Industrial Wastewater," In Symposium Papers: Energy from Biomass and Wastes V, D. L. Klass and J. W. Weatherly III (Eds.), Inst. of Gas Technol., Chicago, IL, 275.

6. Gould, M.S., and Genetelli, E.J. (1978b) "The Effect of Methylation and Hydrogen Ion Concentration on Heavy Metal Binding by Anaerobically Digested Sludges," Water Research, 12, 889.

7. Ahring, B.K. and Westermann, P. (1983) "Toxicity of Heavy Metals to Thermophilic Anaerobic Digestion," Eur. J. Appl. Microbiol. Biotechnol., 17, 365.

8. Parkin, G.F., Speece, R.E., Yang, C.H.J., and Kocher, W.M. (1983) "Response of Methane Fermentation Systems to Industrial Toxicants," Jour. Water Pollution Control Fed., 55, 44.

9. Hayes, T.D. and Theis, T.L. (1978) "The Distribution of Heavy Metals in Anaerobic Digestion," Jour. Water Pollution Control Fed., 50, 61.

10. Callander, I.J., and Barford, J.P., (1983a) "Precipitation, Chelation, and the Availability of Metals as Nutrients in Anaerobic Digestion, I. Methodology", Biotechnology and Bioengineering, XXV.

11. Callander, I.J. and Barford, J.P. (1983b) "Precipitation, Chelation, and the Availability of Metals as Nutrients in Anaerobic Digestion, II. Applications", Biotechnology and Bioengineering, XXV.

12. Nielson, J. S., Hrudey, S. E. and Cantwell, F. F. (1984) "Role of the Free Metal Ion Species in Soluble Nickel Removal by Activated Sludge," Environ. Sci. Technol., 18, 11.

13. Bhattacharya, S. K. (1986) "Toxic Substances in Methane Fermentation Systems," Ph.D. Dissertation, Drexel University, Philadelphia, PA.

14. Standard Methods for the Examination of Water and Wastewater, (1975) 14th ed., American Public Health Association, Washington, D.C.

15. Lawrence, A.W. and McCarty, P.L. (1970) "Unified Basis for Biological Treatment Design and Operation," Jour. Sanitary Eng. Div., ASCE, 96, 757.

16. van Uden, N. (1967) "Transport-Limited Growth in the Chemostat and its Competitive Inhibition; A Theoretical Treatment," Archiv. fur. Mikrobiologie, 58, 145.

17. Monod, J. (1949) "The Growth of Bacterial Cultures," Annual Reviews of Microbiology, 3, 371.

18. Metcalf and Eddy, Inc. (1979) Wastewater Engineering: Treatment, Disposal, Reuse, 2nd Edition, G. Tchbanoglous, Ed., McGraw-Hill, New York.

19. Parkin, G.F. and Speece, R.E. (1982) "Modeling Toxicity in Methane Fermentation Systems," Jour. Env. Eng. Div., ASCE, 108, 515.

20. Cantwell, F.F., Nielsen, J.S., and Hrudey, S.E. (1982) "Free Nickel Ion Concentration in Sewage by an Ion Exchange Column-Equilibration Method", Analytical Chemistry, 54, 9, 1498.

21. Rittmann, B. E., Bae, W., Namkung, E., and Lu, C.-J. (1986) "A Critical Evaluation of Microbial Product Formation in Biological Proceses, Water Science and Tech., in press.

HEAVY METAL REMOVAL BY AQUATIC MACROPHYTES IN A TEMPERATE CLIMATE AQUATIC TREATMENT SYSTEM

Paul L. Bishop
*Department of Civil Engineering
University of New Hampshire
Durham, New Hampshire*

Jan DeWaters
*Department of Environmental Sciences
and Engineering
University of North Carolina-Chapel Hill
Chapel Hill, North Carolina*

INTRODUCTION

In recent years, major efforts have been undertaken to develop wastewater treatment techniques which are effective at removing both organic and inorganic contaminants so that wastewater discharge and water quality standards can be met. This will become even more of a concern as the need for wastewater reuse increases. In many cases, tertiary or advanced waste treatment processes will be needed to meet these water quality goals. However, advanced wastewater treatment systems are generally energy-intensive, expensive and relatively ineffective for the removals of many trace contaminants.

Innovative wastewater treatment systems are needed to amend this situation. Indeed, the Clean Water Act, as amended in 1977 (PL 95-217), encourages the use of innovative and alternative technologies for the reclamation of wastewater. It further requires that energy conservation during treatment be achieved and that the treatment process be cost effective. The problem that currently exists is finding an equitable solution to implement advanced waste treatment systems in a cost effective manner. This is a particular problem in areas where the serviced population cannot finance and maintain such facilities.

One class of treatment system which has recently been proposed for advanced waste treatment in small communities is aquatic treatment. Aquatic treatment systems appear to be low cost, low energy demanding solutions to wastewater treatment (1). Aquatic treatment systems are ones in which wastewater is applied to natural or man-made wetlands (marshes, ponds, peat bogs, cypress domes, etc.) or to basins containing free-floating or submersed aquatic macrophytes. These systems not only remove organics contributing to biochemical oxygen demand (BOD) and total suspended solids (TSS) as in conventional secondary wastewater treatment systems, but also are very efficient at removing nutrients (nitrogen and phosphorus), heavy metals, and trace organics. They are also biomass generators which may generate plant material for anaerobic digestion and methane production (2).

In these systems treatment occurs through a combination of biological, physical and chemical processes which are similar to those occurring in conventional secondary treatment processes. Treatment occurs principally via sedimentation of suspended solids; biodegradation of organics by attached, pelagic and sediment bacteria; and metals and nutrient uptake by microorganisms and the aquatic plants.

The aquatic macrophytes play an essential role in aquatic treatment systems. The roots, stems and leaves of the plants in the water column provide a surface on which bacteria can grow and provide a media for filtration and adsorption of solids. They also provide oxygen to the bacteria through photosynthesis. The stems and leaves at or above the water surface help to attenuate sunlight and thus prevent growth of suspended algae, and reduce effects of wind on the water, thus minimizing roiling of sediments (3). In essence, aquatic treatment systems are similar to horizontal flow trickling filters in which macrophytes replace the trickling filter media as a support structure for bacteria. The plants play the additional role of nutrient and heavy metal assimilators (3,4).

Nearly all aquatic treatment systems have been established in tropical or sub-tropical environments and are based on the use of _Eichornia crassipes_ (water hyacinth), because of the difficulty in maintaining viable aquatic macrophytes under cold temperature conditions. Water hyacinths generally need temperatures in excess of 20°C for growth. Essentially all macrophytes tested to date for aquatic waste treatment are intolerant of cold temperatures. However, aquatic systems could be extremely useful in northern climates if suitable temperate climate plants could be discovered.

As a result, research was begun in 1982 at the University of New Hampshire to examine and assess the feasibility of using aquatic macrophytes and their associated bacteria in temperate regions as a combined secondary and tertiary wastewater treatment process. A number of temperate region submersed aquatic macrophytes were evaluated for potential use in these systems. Those showing promise were used in pilot-scale treatment units and tested over a two year period. The results indicate that an Elodea-based aquatic treatment system can operate on a seasonal basis at loading rates and hydraulic detention times similar to hyacinth-based systems while providing good secondary and tertiary treatment of domestic primary effluent and maintaining good plant productivity (5-10).

One phase of this research effort concerned the removal of heavy metals from the wastewater by the aquatic macrophytes. This paper discusses that phase of the research.

EXPERIMENTAL METHODS

Initially, a literature search was conducted to find species of submerged, temperate-climate macrophytes having traits which make them potentially applicable in an aquaculture wastewater treatment scheme. The plants were judged according to the following criteria: 1) the plant should be cold temperature tolerant; 2) the plant should be capable of surviving the high organic and nutrient loadings imposed by a wastewater environment; 3) it should have a high surface area:volume ratio in order that it's ability to act as a surface for microbial attachment and solids filtration/adsorption be enhanced; 4) it should be capable of vegetative reproduction; and 5) it should be available locally. The plants selected for further study were Elodea nuttallii, Myriophyllum heterophyllum, Ceratophyllum demersum, Cabomba caroliniana and Potamogeton amplifolius. Later, Lemna minor was also used in the pilot-scale studies.

The effects of light, temperature and wastewater strength on the productivity (g dry wt/m^2·day) of each of these four macrophytes and on their nutrient removal efficiencies was determined by use of batch experiments carried out in a greenhouse. In addition, the effect of mixing and aeration on productivity of E. nuttallii, M. heterophyllum and C. demersum was evaluated. The results of these studies have been published elsewhere (5-10). In summary, they indicated that E. nuttallii and M. heterophyllum had the potential to be used successfully in aquatic treatment systems. They could consistently provide better than secondary treatment on a year-round basis.

Pilot-scale aquatic macrophyte treatment studies were then carried out in an 8 m by 12 m (25 ft x 40 ft) greenhouse constructed at the Durham, NH Wastewater Treatment Facility. A furnace was installed in the greenhouse to maintain minimum winter air temperatures of 5°C. Fans were used to prevent excessive heating in the summer and for year round ventilation.

Twelve 120 liter reactors were constructed with surface areas of 1.0 m^2 each. Plastic screens were suspended horizontally in the tanks to keep the plant fragments from settling to the bottom of the reactors and to discourage the plants from rooting in sediment. Gentle mixing was provided by diffusing between 0.15 and 0.30 m^3/hr of air in each reactor. Tracer studies indicated that there was complete mixing in all tanks. Mixing was employed to enhance macrophyte productivity and simplify sampling procedures. Three groups of four reactors each had their own head tank. Various strengths of primary effluent (10 to 100 percent) could be produced in each of the head tanks by mixing primary effluent (from the primary clarifier located adjacent to the greenhouse) with tap water. Peristaltic pumps delivered the primary effluent mixtures to the reactors. Theoretical detention times were varied from 2.3 to 4.6 days. Control tanks containing no macrophytes were established for all the experimental runs. In addition, a set of tanks containing plastic aquaria plants were also monitored to distinguished between removal by aquatic plants and by the heterotrophic biofilms growing on the plants.

Four plants were evaluated at pilot-scale for their potential as temperate climate aquatic treatment macrophytes. Elodea nuttallii (Planch.) St. John, "waterweed"; Myriophyllum heterophyllum Michx., "water milfoil"; Ceratophyllum demersum L., "hornwort" or "coontail"; and Lemna minor, duckweed, were selected as candidates, based on our previous work or information derived from the literature. Plant specimens were collected from the Old Durham Reservoir,

Durham, NH and from Lake Winnipesaukee, Wolfboro, NH. The plants wee placed into the reactors at stocking densities of approximately 0.6 and 2.5 kg wet wt/m^2 surface area.

Seven experimental runs were conducted over the course of two years to examine the treatment efficiency and removal rates of various wastewater constituents by macrophyte-based reactors on a seasonal basis. The plants initially added to the reactors were maintained on a continuous basis for the entire two year period. Influent and effluent samples were taken from each reactor on an every other day basis over a seven day period (four times total) for each sampling run. The samples were analyzed for dissolved oxygen; pH; temperature; BOD_5; ammonia, nitrite and nitrate nitrogen; ortho and total phosphorus; and fecal and total coliforms.

Studies were also performed during the experimental runs to determine the heavy metal removal potential of temperate climate macrophyte aquatic treatment systems. Eight reactors were used, three containing approximately 1.5 kg of Elodea sp., two containing approximately 1.5 kg of Myriophyllum heterophyllum, one containing Lemna sp., one containing plastic aquatic plant material and the eighth containing no plant material (control). In the second run one of the M. heterophyllum tanks was mixed with E. nutallii. These studies were performed during August and December 1984 so that metal removal during summer and winter months could be compared. Operational parameters for these two studies are shown in Table 1. Water samples were taken every other day from each of the head tanks (reactor influent) and reactor effluents over a seven day period. On the fourth sampling day, samples were also taken of the plants, reactor sediments and plastic plant biofilms. Liquid samples were acidified with concentrated HNO_3 and stored at room temperature.

Plant and sediment samples were analyzed for their dry weight to wet weight ratio, and stored after drying for 72 h at 70°C. Reactor effluent samples were digested with concentrated nitric acid (5 ml/100 ml sample). The mixture was brought slowly to a boil and evaporated to a volume of 15-20 ml. An additional 5 ml HNO_3 was added, the dish was covered with a watchglass and the sample was heated to reflux. Heating and addition of HNO_3 was continued until digestion was complete (a clear, light colored solution). The digestate was filtered and brought back to a volume of 100 ml with Nanopure water. Blanks were handled in the same fashion, using Nanopure water as the blank.

Plant materials, head tank samples and sediment were

TABLE 1

OPERATIONAL PARAMETERS FOR HEAVY METAL REMOVAL STUDIES

Criteria	Typical Design Parameters		Operational Parameters	
	Primary Effluent	Secondary Effluent	Aug.'84 Run 6	Dec.'84 Run 7
Hydraulic retention time (d)	>50	>6	3.47	3.78
Hydraulic loading rate ($m^3 \cdot ha^{-1} \cdot d^{-1}$)	200	800	345.6	316.8
Maximum depth (m)	≤ 1.5	0.91	0.45	0.45
Minimum depth (m)	–	0.38	0.30	0.30
Organic loading rate (kg $kBOD_5 \cdot ha^{-1} \cdot d^{-1}$)	≤ 30	≤ 50	39.0–40.8	45.1–45.2
Nitrogen loading rate (kg $TN \cdot ha^{-1} \cdot d^{-1}$)	–	≤ 20	6.5–6.9	8.7–9.2
Ammonia loading rate (kg $NH_3-N \cdot ha^{-1} \cdot d^{-1}$)	–	–	6.30–6.34	9.09–9.26
Orthophosphate loading rate (kg $PO_4-P \cdot ha^{-1} \cdot d^{-1}$)	–	–	1.06–1.15	0.81–1.07
Temperature	>20	>20	22–26	10–16
Macrophyte stocking density (kg wet $wt \cdot m^{-2}$)	1–4	1–4	0.3–1.5	1.5
Harvesting	weekly–monthly	weekly–monthly	←——weekly–monthly——→	
Mixing	no	no	yes	yes

first dry ashed in a muffle furnace (600°C) and then digested with concentrated nitric acid as above.

Digested samples were analyzed for lead, cadmium, copper, nickel and zinc content using a Perkin-Elmer Model 2380 atomic absorption spectrophotometer. The method of standard additions was used for all analyses. Where metal concentrations were below detectable limits using flame atomic absorption spectrophotometry, they were analyzed using a Perkin-Elmer HGA-400 graphite furnace using the "Stabilized Temperature Platform Furnace" (STPF) technique (11).

RESULTS AND DISCUSSION

The results of research concerning treatment efficiency (BOD_5, nitrogen, phosphorus and coliform removal) and macrophyte productivity have been reported elsewhere (5-10). In general, they show that Elodea nuttallii-based aquatic treatment systems are feasible for combined secondary and tertiary treatment of domestic wastewater in temperate climates (see Table 2). The plants can grow year round, support an active biofilm and provide a high quality effluent with only a 2.5 to 4.5 day detention time. BOD_5 removals averaged 89.9 percent over the two year testing period, compared with 66.8 percent removal in control reactors containing no macrophytes. Effluent soluble BOd_5 concentrations were generally less than 5 mg/l in the E. nuttallii reactors. BOD_5 removal dropped off slightly during winter months, but was always greater than 75 percent. Myriophyllum heteophyllum also did well, even though it was not as productive. It appears that the biofilm growing on the plant surfaces is responsible for the BOD_5 removal, rather than the macrophyte. This was substantiated by the fact that reactors containing plastic plants on which biofilm grew were also effective at reducing BOd_5. BOd_5 removal in the Lemna minor-based reactors was poor, probably because of a lack of surface area for biofilm growth.

The macrophyte systems were very efficient at removing nitrogen from the wastewater. During summer months ammonia removal was in excess of 90 percent in the macrophyte-based reactors, while in the control reactors it was generally less than 20 percent. Removal efficiency dropped off during winter months, but was still generally greater than 50 percent.

Total nitrogen removal in the E. nuttallii based reactors averaged 64 percent in the summer and 32 percent in

TABLE 2

WASTEWATER TREATMENT EFFICIENCIES OF VARIOUS AQUATIC MACROPHYTE TREATMENT SYSTEMS

Date	Reactor Type[1]	Productivity $g/m^2 \cdot d$	Hydraulic Retention Time, d	Water Temp. °C	BOD_5	NH_3-N	Total N	Total P
3/9/84–	E	0.500	4.62	11	80.4	96.9	33.5	24.0
3/15/84	M	0.200	4.62		86.9	93.6	29.2	17.4
	C	–	4.62		80.1	7.2	0.0	9.1
4/17/84–	E	1.080	4.62	13	76.0	44.2	28.3	0.0
4/23/84	C	–	4.62		70.3	0.5	0.0	0.0
6/30/84–	E	1.665	3.47	23	91.6	85.4	64.6	31.4
7/6/84	M	–0.618	3.47		89.5	74.8	49.2	18.9
	C	–	3.47		83.6	21.6	0.0	18.3
7/25/84–	E	2.176	3.47	23	96.6	90.2	70.9	40.5
7/31/84	M	2.673	3.47		95.4	93.7	70.8	30.9
	P	–	3.47		65.5	96.0	0.0	0.0
	C	–	3.47		25.9	11.2	0.0	4.5
8/17/84–	E	3.248	3.47	21	97.4	95.2	63.6	44.5
8/23/84	M	–0.808	3.47		95.9	87.7	41.7	30.5
	L	1.901	3.47		83.0	45.5	28.7	43.2
	P	–	3.47		85.5	66.9	34.1	23.0
	C	–	3.47		72.0	65.3	3.1	15.4
11/27/84–	E	0.454	3.78	11	95.5	51.3	17.9	42.3
12/3/84	M	0.255	3.78		95.3	51.4	19.7	38.1
	L	0.020	3.78		47.8	12.0	0.9	37.5
	P	–	3.78		92.8	54.0	11.8	20.2
	C	–	3.78		72.3	11.3	2.0	42.0

[1] E = Elodea nuttallii
M = Myriophyllum heterophyllum
L = Lemna minor
P = Plastic plants
C = Control (no macrophyte)

the winter. Losses from control reactors were generally less than 3 percent. During periods when macrophytes were quite productive, plant uptake of nitrate accounted for 30 to 55 percent of the nitrogen removal from the system. Ammonia votalilization, denitrification and sedimentation accounted for the rest. During periods of low productivity, nitrogen removal by plant uptake accounted for only 7 to 30 percent of the nitrogen removed. Phosphorus removal was also greater in the macrophyte systems than in the control systems. E. nuttallii reactors removed approximately 35 percent of the influent phosphorus, as compared with 19 percent removal in the control reactors.

The principal topic of this research paper is the ability of these systems to remove heavy metals.

Table 3 provides data on metals removal from reactors containing E. nuttallii, M. heterophyllum, L. minor or plastic plants during a seven day period in August 1984. Table 4 provides comparable data for a seven day period in early December 1984. Table 5 is a summary which shows percent metal removals in each aquatic treatment system as a function of season. It should be noted that the reactor detention times were 3.5 to 3.8 days during these studies, so that reactor effluent data cannot be compared with reactor influent data for any particular day. Consequently, influent and effluent metal concentrations were averaged for each reactor in order to determine removal efficiencies. A statistical analysis of data showed that there was no significant difference between reactors containing like macrophytes; consequently, these data were combined.

It is readily apparent from Tables 3 and 4 that the Durham, N.H. wastewater contains very little cadmium or nickel (approximately 35 µg/l Cd and 70 µg/l Ni).

Very little nickel was removed by any of the aquatic treatment processes, with removal rates ranging from 0.0 to 14.3 percent and with an average removal of only 5.1 percent. Removal rates were slightly higher in the summer than in the winter, but not significantly so, indicating that plant uptake may not be an important removal mechanism for nickel. This is substantiated by the very low nickel concentrations found in the macrophyte tissues. These macrophyte concentrations, averaging only 11.7 µg/g in summer and 4.2 µg/g in winter, varied very little among different macrophytes (Lemna minor did the worst). The biofilms on the plastic plants contained 20 µg Ni/g in the summer and 13 µg/g in the winter, suggesting that much of the nickel measured in

TABLE 3

METAL REMOVAL BY VARIOUS AQUATIC MACROPHYTE TREATMENT SYSTEMS – August 17, 1984 – August 23, 1984

Reactor Number	Macrophyte Type	Macrophyte Productivity g/m^2·d	Sample Day	Concentration[1]				
				Cu	Pb	Cd	Ni	Zn
1	Elodea	4.00	1	16	120	38	68	204
			3	12	113	35	54	137
			5	14	105	36	38	158
			7	21	65	32	56	73
			Sediment	611	221	24	33	645
			Macrophyte	110	46	9	18	120
2	Elodea	3.10	1	16	145	22	58	168
			3	18	148	9	54	146
			5	17	91	24	42	124
			7	30	104	25	44	136
			Sediment	711	229	19	22	646
			Macrophyte	147	85	13	14	129
3	Elodea	2.65	1	17	130	34	66	76
			3	26	129	26	44	167
			5	21	119	33	52	122
			7	29	131	28	46	128
			Sediment	1029	225	21	16	629
			Macrophyte	93	44	7	14	133
4	Lemna	1.90	1	75	221	28	50	85
			3	144	210	35	64	237
			5	96	341	25	76	126
			7	82	150	33	76	191
			Sediment	559	125	50	18	590
			Macrophyte	47	20	8	7	29
5	Myriophyllum	0.02	1	25	128	32	48	179
			3	45	124	35	60	153
			5	10	110	26	58	165
			7	20	118	28	54	162
			Sediment	492	148	46	13	459
			Macrophyte	44	18	5	8	153

TABLE 3 (cont.)

Reactor Number	Macrophyte Type	Macrophyte Productivity $g/m^2 \cdot d$	Sample Day	Concentration[1]				
				Cu	Pb	Cd	Ni	Zn
6	Myriophyllum	-1.63	1	12	133	25	34	130
			3	14	110	21	72	184
			5	17	126	33	54	113
			7	23	139	36	58	107
			Sediment	511	130	28	33	416
			Macrophyte	63	41	9	9	164
7	Plastic Plants	–	1	68	149	28	48	95
			3	96	161	36	50	146
			5	75	151	29	48	133
			7	70	140	34	48	157
			Sediment	391	129	28	283	543
			Biofilm	752	91	8	20	425
8	Control	–	1	101	214	39	44	113
			3	108	192	33	56	224
			5	103	190	45	60	85
			7	132	154	36	40	119
			Sediment	540	102	43	8	443
Influent	–	–	1	207	455	37	44	184
			3	199	197	42	49	154
			5	270	158	35	69	173
			7	473	135	27	58	261

[1] Concentration units:

    Wastewater = µg/l
    Sediment = µg/g
    Macrophyte = µg/g
    Biofilm = µg/g

TABLE 4

METAL REMOVAL BY VARIOUS AQUATIC MACROPHYTE TREATMENT SYSTEMS –
November 27, 1984 – December 3, 1984

Reactor Number	Macrophyte Type	Macrophyte Productivity $g/m^2 \cdot d$	Sample Day	Concentration[1]				
				Cu	Pb	Cd	Ni	Zn
1	Elodea	0.48	1	27	193	34	84	150
			3	36	216	39	82	170
			5	27	187	36	84	192
			7	22	157	30	82	157
			Sediment	805	127	22	20	579
			Macrophyte	101	27	15	5	255
2	Elodea	0.46	1	24	112	35	112	156
			3	30	71	24	70	149
			5	41	146	36	80	152
			7	20	92	30	58	169
			Sediment	697	131	12	19	664
			Macrophyte	159	23	6	3	227
3	Elodea	0.33	1	33	115	39	85	168
			3	34	113	34	79	175
			5	35	185	38	78	135
			7	44	149	20	82	139
			Sediment	806	218	19	18	758
			Macrophyte	157	41	13	6	228
4	Lemna	0.02	1	166	138	30	86	208
			3	210	122	35	85	161
			5	182	144	40	81	152
			7	219	243	28	83	203
			Sediment	1014	216	14	15	507
			Macrophyte	40	22	4	3	98
5	Myriophyllum	0.26	1	20	152	33	88	182
			3	31	140	31	72	145
			5	64	197	25	69	155
			7	47	207	37	94	182
			Sediment	630	161	37	15	657
			Macrophyte	354	65	6	4	286

TABLE 4 (cont.)

Reactor Number	Macrophyte Type	Macrophyte Productivity $g/m^2 \cdot d$	Sample Day	Concentration[1]				
				Cu	Pb	Cd	Ni	Zn
6	Myriophyllum and Elodea	0.46	1	51	97	36	86	155
			3	20	142	38	82	163
			5	22	133	34	72	197
			7	30	128	35	77	163
			Sediment	489	148	21	20	497
			Macrophyte	239	45	9	4	222
7	Plastic Plants	–	1	25	120	29	92	211
			3	40	113	36	76	143
			5	58	143	28	71	168
			7	37	134	37	88	192
			Sediment	780	161	18	25	662
			Biofilm	647	101	8	13	249
8	Control	–	1	208	213	37	83	173
			3	234	125	35	84	154
			5	149	100	40	90	178
			7	194	116	35	78	172
			Sediment	610	11	1	4	400
Influent	–	–	1	406	215	38	85	274
			3	493	173	37	78	193
			5	343	133	33	92	166
			7	388	163	34	84	130

[1] Concentration units:

Wastewater = µg/l
Sediment = µg/g
Macrophyte = µg/g
Biofilm = µg/g

TABLE 5

PERCENT METAL REMOVAL BY VARIOUS AQUATIC MACROPHYTE TREATMENT SYSTEMS AS A FUNCTION OF SEASON

Macrophyte Type	Metal Removal, %									
	August					December				
	Cu	Pb	Cd	Ni	Zn	Cu	Pb	Cd	Ni	Zn
Elodea	93.1	50.6	20.0	7.4	29.2	92.4	15.4	5.7	5.1	13.4
Myriophyllum	92.8	47.7	14.3	2.2	22.7	90.1	-1.7	8.6	5.8	9.8
Myriophyllum/ Elodea	-	-	-	-	-	92.5	26.9	0.0	6.5	7.9
Lemna	65.5	2.5	12.9	0.0	16.0	52.3	5.4	2.9	1.2	1.6
Plastic Plants	73.1	37.6	11.4	14.3	31.2	90.2	25.4	5.7	3.6	3.0
Control	61.4	20.7	0.0	10.7	29.9	49.4	19.0	0.0	1.3	8.0

the macrophytes may have actually been associated with the biofilms on the macrophytes' surfaces. Sediment samples from each reactor also contained very little nickel, again suggesting that aquatic treatment systems may not be effective at removing nickel, at least when present in these low concentrations.

Cadmium operated in much the same fashion, except that removal rates were slightly higher, ranging from 11.4 to 20.0 percent in summer and 0.0 to 8.6 percent in winter. <u>Elodea nuttallii</u> was the best performer during the summer; removals during the winter were fairly comparable among macrophytes, with <u>L. minor</u> again doing poorly. Macrophyte cadmium concentrations were very low, averaging 8.5 µg/g in summer and 8.8 µg/g in winter. The plastic plant biofilms also contained 8.0 µg Cd/g dry weight. Cadmium sediment concentrations were also very low, averaging 32 µg/g in summer and 20 µg/g in winter.

Removal rates were higher for zinc, particularly in the summer. <u>E. nuttallii</u> reactors removed 29.2 percent of the influent zinc during the summer, while <u>M. heterophyllum</u> reactors removed 22.7 percent; <u>L. minor</u> reactors only removed 16.0 percent. Surprisingly, the control reactors, containing no macrophytes, did very well removing 29.9 percent of the influent zinc. This tends to indicate that zinc is probably removed principally by sedimentation rather than by macrophyte or biofilm uptake. This finding is substantiated by the high zinc concentrations in the sediments, ranging from 416 µg/g to 646 µg/g. The macrophyte tissues and plastic plant biofilms also contained relatively high concentrations of zinc, though (averaging 140 µg/g for the macrophytes and 425 µg/g for the plastic plant biofilms), suggesting that macrophyte or biofilm uptake may also be responsible for some removal. These two mechanisms are actually coupled, as much of the reactor sediment is composed of dead and decaying macrophyte material. Zinc removals were greatly reduced during winter months, again suggesting that bioaccumulation of zinc may be important. As with the other metals, removing of zinc by the <u>Lemna minor</u> reactors was poor, probably due to the lack of adequate vegetative surface area in the water column.

There was a substantial removal of lead in the submerged macrophyte-containing reactors. <u>Elodea nuttallii</u> removed over half of the influent lead during the summer, while <u>Myriophyllum heterophyllum</u> did almost as well. <u>Lemna minor</u> reactors did very poorly, removing essentially no lead from the wastewater. It even did considerably worse than the control reactor which contained no macrophytes (21 percent lead removal). The reactor containing plastic plants did much better than the control reactor, removing an average of 37.6 percent of the influent lead. This again suggests that a large proportion of the heavy metals removed in aquatic treatment systems are due to the epifloral biofilms. Plant uptake is very important, though. This can be seen by comparing summer and winter data for lead removal. The control reactors removed essentially the same percentage in summer as in winter, while the plastic plant reactors showed a moderate decline. Lead removal in the macrophyte reactors dropped substantially during the winter, probably due to the minimal productivities occurring then.

Copper removal was significant in the macrophyte reactors no matter what time of year it was. Better than 90 percent of the influent copper was removed from E. nuttallii and M. heterophyllum reactors in both summer and winter. This can be compared with the control reactors where 61 percent of the influent copper was removed in the summer and 49 percent in the winter. The majority of this removal was probably accomplished by the epifloral biofilms. The plastic plant reactors were also highly successful at removing copper, achieving the same removal rate (90 percent) in the macrophyte reactors during winter months when plant productivity was low. The copper concentration in the biofilms was very high, on the order of 760 µg/g.

CONCLUSIONS

These studies, performed under both warm and cold weather conditions using three different temperate climate aquatic macrophytes, suggest that the heavy metal removal efficiencies of various aquatic macrophyte treatment systems varies depending on the metal considered, but that the underlying removal mechanisms are probably the same. In essentially all cases, metal removals in the macrophyte-containing reactors were better than in the control reactors. Much of this additional removal is probably due to the epifloral biofilms growing on the macrophytes rather than to the macrophytes themselves. This is potentially very beneficial to the plants as it prevents high concentrations of toxic heavy metals from bioconcentrating in the macrophyte's tissues. Copper was almost entirely removed from the waste stream in aquatic treatment systems containing the temperate climate macrophytes _Elodea nuttallii_ and _Myriophyllum heterophyllum_; substantial quantities of lead and zinc were also removed. Removal rates for cadmium and nickel were much lower, but this may have been due to the low concentrations present in the raw wastewater.

ACKNOWLEDGMENTS

This research was supported by Grant Number CEE-8209851 from the National Science Foundation.

REFERENCES

1. Crites, R., "Economics of Aquatic Treatment Systems," _Aquaculture Systems for Wastewater Treatment: Seminar Proceedings and Engineering Evaluation_, EPA 430/9-80-006, 1979, pp. 475-485.

2. Benemann, J., "Energy from Wastewater Aquaculture Systems," _Aquaculture Systems for Wastewater Treatment: Seminar Proceedings and Engineering Evaluation_, EPA 430/9-80-006, 1979, pp. 441-458.

3. Stowell, R., Ludwig, R., Colt, J. and Tchobanoglous, G., "Concepts in Aquatic Treatment System Design," _Journal of the Environmental Engineering Division_, ASCE, Vol. 107, No. 5, 1980, pp. 919-941.

4. Pershe, E., "Combined Aquaculture Systems for Wastewater Treatment in Cold Climates: An Engineering Assessment," _Aquaculture Systems for Wastewater Treatment: An Engineering Assessment_, EPA 430/9-80-007, 1980, pp. 105-127.

5. White, M. and Bishop, P., "Evaluation of Five Species of Temperate Climate Aquatic Plants for Application in Aquaculture Treatment of Wastewater," _Environmental Engineering_, Proceedings of the 1984 National Environmental Engineering Conference, Los Angeles, CA, 1984, pp. 564-572.

6. Bishop, P., "Examination of a Combined Secondary and Tertiary Wastewater Treatment Process for Temperate Regions Using Aquatic Macrophytes and Their Epifloral

Biofilms," completion report submitted to the National Science Foundation, 1985.

7. Eighmy, T. and Bishop, P., "Preliminary Evaluation of Submerged Aquatic Macrophytes in a Pilot-Scale Aquatic Treatment System," Environmental Engineering, Proceedings of the 1985 National Conference on Environmental Engineering, Boston, MA, 1985, pp. 764-770.

8. Eighmy, T., Jahnke, L. and Bishop, P., "the Use of Elodea nuttallii in a Combined Secondary and Tertiary Aquatic Wastewater Treatment System," Proceedings, Conference on Research and Applications of Aquatic Plants for Water Treatment and Resource Recovery, Orlando, FL, 1986.

9. Kozak, P. and Bishop, P., "the Effect of Mixing and Aeration on the Productivity of Myriophyllum heterophyllum Michx. (Water Milfoil) During Aquatic Wastewater Treatment," Proceedings, Conference on Research and Applications of Aquatic Plants for Water Treatment and Resource Recovery, Orlando, FL, 1986.

10. Eighmy, T., "An Investigation of Aquatic Macrophyte Based Aquatic Wastewater Treatment Systems for Use in Temperate Climates," Ph.D. Dissertation, University of New Hampshire, Durham, NH, 1986.

11. Slavin, W., Carnrick, G., Manning, D. and Pruszkowska, E., "Recent Experiences with the Stabilized Temperature Platform Furnace and Zeeman Background Correction," Atomic Spectroscopy, Vol. 4, No. 3, May 1983, pp. 69-86.

REMOVAL OF HEAVY METAL BY RECYCLING OF WASTE SLUDGE IN THE ACTIVATED SLUDGE PROCESS

Kuei-lang Tsai

*Department of Hydraulic Engineering
Feng Chia University
Taichung, Taiwan
Republic of China*

Pak-shing Cheung

*Department of Environmental Science
Feng Chia University
Taichung, Taiwan,
Republic of China*

ABSTRACT

In the conventional activated sludge process, part of the settled sludge from the final clarifier will be returned to the aeration tank and the rest of them are being wasted. It has been known for a long time that the activated sludge did have very strong adsorption power. In this study the wasted sludge would be first recycled to the incoming sewage such that they could adsorb the heavy metals and part of the organic matters and then being removed in the primary settling tank. This could simultaneously protect the biological unit from excess imcoming toxic heavy metal ions and also reduce the organic load to the aeration tank. Thus considerable energy could be saved.

In this study, the jar tester was used to investigate the feasibility of this application. Results indicated that the activated sludge did have very big adsorption ability. As compared to plain sedimentation, with added activated sludge, the additional COD removal ranged from 11-52% and that for heavy metal varied from 15-32%.

INTRODUCTION

The conventional activated sludge process consists of 3 major parts (see fig.1), namely: Primary settling tank, aeration unit and final clarifier. Part of the settled sludge in the final clarifier would be returned to the aeration unit as working microorganism and the excess sludge, which amounts to about 0.8-1 kg.MLSS/kg.BOD removed, would be wasted.

The primary settling tank can normally remove about 30% of the incoming SS and BOD_5, i.e. the majority of the pollution load must be handled by the aeration unit.

Investigations in Germany[1][2] and North America[3][4] indicated that almost half of the organic matters would be removed after only half an hour of aeration, due to the adsorption power of the recycled activated sludge. But this amount of adsorbed materials have to be oxidized later in the aeration unit.

In this study the adsorptive capability of the activated sludge would be utilized in such a way that the wasted sludge would be suggested to be returned to the influent point, where

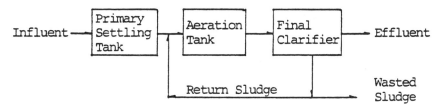

Conventional Activated Sludge System

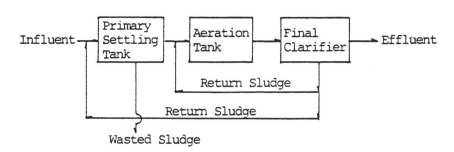

Modified Activated Sludge System

Figure 1   Conventional and Modified Activated Sludge System

the turbulence condition makes the thorough mixing of the two streams without any difficulty. Then the recycled sludge together with the adsorbed materials would be removed from the system in the primary settling tank. A comparison of this modified process with the conventional activated sludge unit is shown also in Fig.1. Since the adsorbed materials are being removed in the primary settling tank, the energy requirement in the aeration unit could be reduced and also with respect to the removal of toxic heavy metal ions, the biological system could also be protected without any additional financial burden.

EXPERIMENTALS

In order to verify the validity of the modification mentioned in the previous section. The whole study was divided into 3 parts, namely:

First part: Preliminary study using jar tester to determine the adsorption capability of the activated sludge.

Second Part: Pilot scale continuous activated sludge system to verify the amount of heavy metal ions that could be adsorbed by the recycled wasted sludge.

Third Part: The results obtained in the previous study would be tested in an actual activated sludge plant.

In this study only the first part as described above was carried out. All the equipments employed will be outlined as follows:

Bench scale activated sludge unit: This is a laboratory scale unit made of PVC plastic sheet with an aeration section of 15 ℓ and settling section of 9.5 ℓ. The activated sludge used in this study were produced batchwise by this unit for about 10 days.

Jar tester: This is a commerical Jar test with 6 stirrers in a series. The stirrers are paddle type driven by an electric motor, in which the speed of rotation could be adjusted.

Procedures of experiment: Wastewaters with heavy metal ions (eg. electroplating discharge) were collected, 500 ml of which were placed in a series of 6 beakers, in which different amount of activated sludge were introducted. Then the stirrers were set in rotation at a speed of 100 rpm for 10 minutes. After 30 minutes of sedimentation, the settled wastewater were analyzed for COD, heavy metal ions. This result would be compared with the same amount of wastewater under the same experimental conditions but without addition of activated sludge.

Chemical analysis: COD and heavy metal ions determination were carried out for each series of experiment according to the standard methods.

RESULTS AND DISCUSSION

Wastewater with Single Heavy Metal Ions

Wastewater with low concentration of Cu: The results are tabulated in Table 1 and represented in Fig.2. From Table 1, it can be been that a relative COD removal from 22.8% to 37.9% with an average removal of 30.1% could be obtained. While the corresponding figure for Cu are 0.8% to 50.1% with an average Cu removal of 32%.

From Fig.2 the data indicated that no clear sludge dose could be predicted both for COD and Cu removal. Although more sludge addition could bring about higher percentage removal, a satisfactory removal rate could also be obtained with relative lower volume of added sludge. Coupled with the unit mass removal, which is a decreasing function of the added sludge, it seemed that the amount of sludge added is not a deciding factor.

Wastewater with high concentration of Cu : The results are shown in Table 2 and illustrated in Fig 3. An average of 52.1 % of COD and 14.8% of Cu Removal rate could be achieved. Again the optimun sludge dose would not be predicted.

Wastewater with Ni ions : Table 3 summarized the results and these are shown in Fig. 4. The average COD removal amounted to 10.9% and for Ni removal 23.3%. From Fig 4, more sludge are shown to be more effective.

Table 1. Removal Efficiencies of Wastewater with Cu Ions (Low Concentration)

Sludge addition (mg) / Items		MLSS = 1590 mg/l			COD = 98.1 mg/l				Cu = 0.668 mg/l			
	Raw waste	0	10	30	60	90	120	150	180	210	240	AVE.
Sludge mass (mg)	–	0	15.9	48.0	95.6	143	191	239	287	334	382	–
COD(mg/l)	89.1	88.5	61.9	56.9	72.0	64.5	68.3	56.9	55.0	55.6	65.3	–
Unit Mass** removal	–	–	1.67	0.66	0.17	0.17	0.11	0.13	0.12	0.10	0.06	–
relative* percentage removal	–	–	30.1	35.7	18.6	27.1	22.8	35.7	37.9	37.2	26.2	30.1
Cu (mg/l)	0.668	0.665	0.660	0.390	0.414	0.516	0.443	0.389	0.408	0.332	0.517	–
Unit Mass** removal (10^{-3})	–	–	0.3	5.7	2.62	1.04	1.16	1.15	0.90	1.00	0.39	–
relative* percentage removal	–	–	0.8	41.4	37.7	22.4	33.4	41.5	38.6	50.1	22.3	32.0

Relative* percentage removal = $\dfrac{\text{settled COD or metal ions with sludge addition} - \text{settled COD or metal ions without sludge addition}}{\text{settled COD or metal ions without sludge addition}} \times 100\%$

Unit mass** removal = $\dfrac{\text{settled COD or metal ions without} - \text{settled COD or metal ions with sludge addition}}{\text{mg of added sludge}}$

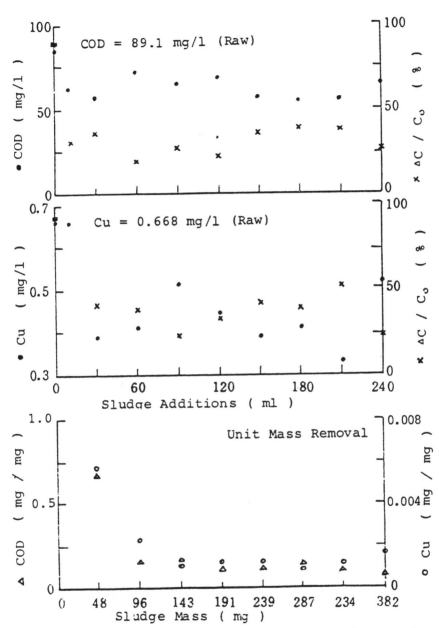

Figure 2 Removal Efficiencies and Unit Mass Removal of Wastewater with Cu Ions (low concentration)

Table 2. Removal Efficiencies of Wastewater with Cu Ions. (High Concentration)

MLSS = 2020 mg/l

Items \ Sludge addition (mg)	Raw waste	0	10	30	60	90	120	150	180	210	240	AVE.
Sludge mass (mg)	0	0	20	61	121	181	242	302	363	423	484	–
COD (mg/l)	170	158	75.0	73.3	68.2	75.8	91.0	72.0	69.5	69.5	85.9	–
Unit Mass** removal	–	–	4.11	1.40	0.74	0.45	0.28	0.28	0.24	0.21	0.15	–
relative* percentage removal	–	–	52.5	53.6	56.8	52.0	42.4	54.4	56.0	56.0	45.6	52.1
Cu (mg/l)	36.5	36.2	36.1	35.5	34.6	30.4	29.5	28.6	28.2	27.7	26.9	–
Unit Mass** removal	–	–	0.005	0.011	0.013	0.032	0.028	0.025	0.022	0.020	0.019	–
relative* percentage removal	–	–	0.276	1.93	4.42	16.0	18.5	21.0	22.1	23.5	25.7	14.8

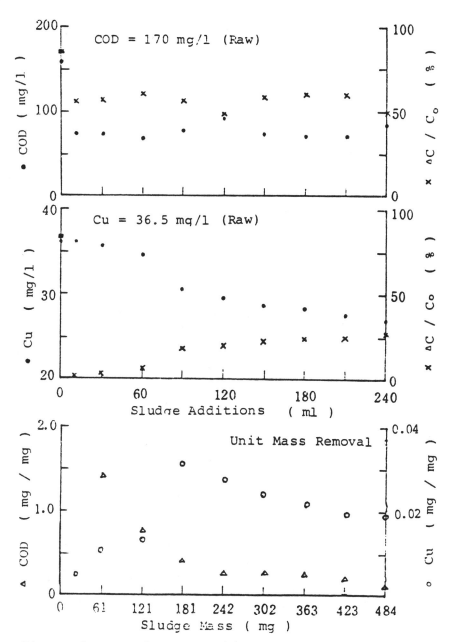

Figure 3 Removal Efficiencies and Unit Mass Removal of Wastewater with Cu Ions (high concentration)

Table 3. Removal Efficiencies of Wastewater with Ni Ions

Items	Sludge addition (ml)											
	MLSS = 1710 mg/l											
	Raw waste	0	10	30	60	90	120	150	180	210	240	AVE.
Sludge mass (mg)	0	0	17.1	53.3	103	154	205	257	308	359	411	–
COD (mg/l)	619	610	600	594	563	537	505	556	499	537	499	–
Unit Mass** removal	–	–	0.58	0.30	0.46	0.47	0.51	0.21	0.36	0.20	0.27	–
relative* percentage removal	–	–	1.64	2.62	7.70	12.0	17.2	8.85	18.2	12.0	18.2	10.9
Ni (mg/l)	20.6	19.8	19.7	16.9	16.7	16.7	16.3	14.5	11.7	12.3	11.9	–
Unit Mass** removal	–	–	0.006	0.054	0.030	0.020	0.017	0.021	0.026	0.021	0.019	–
relative* percentage removal	–	–	0.5	14.6	15.7	15.7	17.7	26.8	40.9	37.9	39.9	23.3

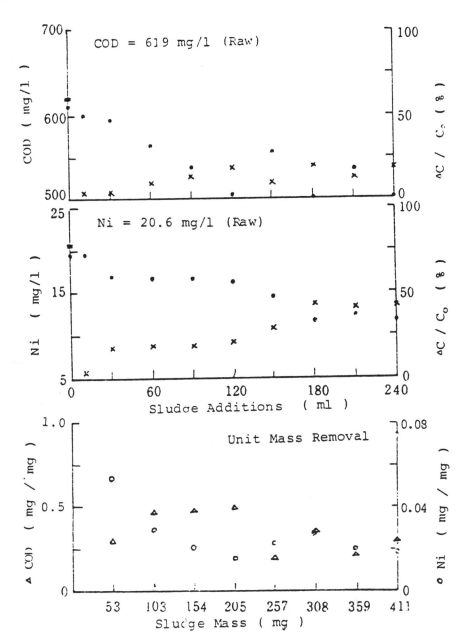

Figure 4   Removal Efficiencies and Unit Mass Removal of Wastewater with Ni Ions

Combined Wastewater with Different Heavy Metal Ions But Cu Dominant

Table 4 shows that the average COD removal amounted to 27.9% while Cu removal to 20.2%. Other metal ions (eg. Cr, Fe Mn and Zn), although very low in concentration with respect to Cu, did end up with a mixed picture with different added sludge. This might due to the sensitivity of the atomic absorption instrument used in the heavy metal determination. Similar observation is obtained in Fig. 5 as compared to the previous Figures.

Combined Wastewater with Different Heavy Metal Ions But Cr And Fe Dominant

Table 5 gives an average COD removal of 40%, Cr removal of 30.6% and Fe removal of 22.5%. Again in Fig.6, the removal efficiencies with respect to COD and Cr/Fe are not a clear function of added sludge.

Table 4. Removal Efficiencies of Combined Wastewater with Cu Ions Dominant

MLSS = 1770 mg/l

Items \ Sludge addition (mg)	Raw waste	0	10	30	60	90	120	150	180	210	240	AVE.
sludge mass (mg)	0	0	17.7	53.0	106	159	212	265	318	371	424	–
COD (mg/l)	400	378	361	400	306	211	278	256	267	245	256	–
Unit Mass** removal	–	–	0.96	–	0.68	1.05	0.47	0.46	0.35	0.36	0.29	–
relative* percentage removal	–	–	4.50	–	19.0	44.2	26.5	32.3	29.4	35.2	32.3	27.9
Cu (mg/l)	44.9	44.4	44.2	42.5	41.4	33.8	30.6	33.1	28.9	33.2	31.1	–
Unit Mass** removal	–	–	0.011	0.036	0.028	0.067	0.065	0.043	0.049	0.030	0.031	–
relative* percentage removal	–	–	0.450	4.28	6.76	23.9	31.1	25.5	34.9	25.2	30.0	20.2
Cr (mg/l)	0.096	0.041	0.190	0.121	0.149	0.111	0.139	0.134	0.133	0.167	0.088	–
Fe (mg/l)	0.821	0.726	0.235	0.781	0.607	–	0.841	1.021	0.866	0.575	0.676	–
Mn (mg/l)	0.041	0.029	0.032	0.008	0.027	0.035	0.055	0.086	0.041	0.047	0.039	–
Zn (mg/l)	0.480	0.988	1.203	0.231	0.42	0.777	0.478	0.530	2.041	0.409	1.31	–

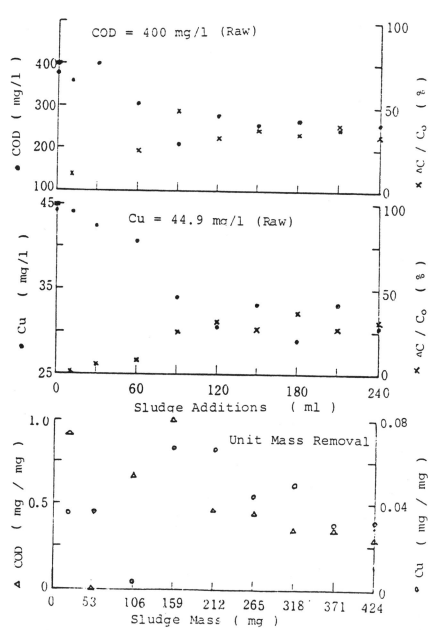

Figure 5 Removal Efficiencies and Unit Mass Removal of Combined Wastewater with Cu Ions Dominant

Table 5. Removal Efficiencies of Combined Wastewater with Cr, Fe ions Dominant.

Items	Sludge addition (mg) MLSS = 2170 mg/l Raw waste	0	10	30	60	90	120	150	180	210	240	AVE
Sludge mass (mg)	-	0	21.7	65.1	130	195	260	326	391	456	521	-
COD (mg/l)	36.3	36.3	18.2	18.2	36.3	18.2	18.2	22.7	9.08	13.6	27.2	-
Unit Mass** removal	-	-	0.834	0.278	0	0.093	0.070	0.042	0.070	0.050	0.017	-
relative* percentage removal	-	-	49.9	49.9	0	49.9	49.9	37.5	75.0	62.5	25.1	40.0
Cr (mg/l)	58.2	57.3	54.5	50.8	39.9	39.5	37.3	37.3	26.2	30.6	41.6	-
Unit Mass** removal	-	-	0.129	0.100	0.134	0.091	0.077	0.061	0.079	0.059	0.030	-
relative* percentage removal	-	-	4.89	11.3	30.4	31.1	34.9	34.9	54.1	46.6	27.4	30.6
Fe (mg/l)	82.9	82.1	79.3	71.3	69.0	64.9	62.7	57.7	59.5	53.5	54.9	-
Unit Mass** removal	-	-	0.133	0.166	0.100	0.088	0.075	0.075	0.058	0.063	0.052	-
relative* percentage removal	-	-	3.41	13.2	16.0	21.0	23.6	29.7	27.5	34.8	33.1	22.5
Cu (mg/l)	0.319	0.665	0.734	0.390	0.414	0.516	0.443	0.389	0.408	0.332	0.517	-
Mn (mg/l)	0.509	0.609	0.565	0.562	0.572	0.546	0.476	0.560	0.604	0.502	0.607	-
Zn (mg/l)	1.633	1.623	5.466	1.400	3.008	1.147	1.56	1.236	6.004	1.304	1.407	-

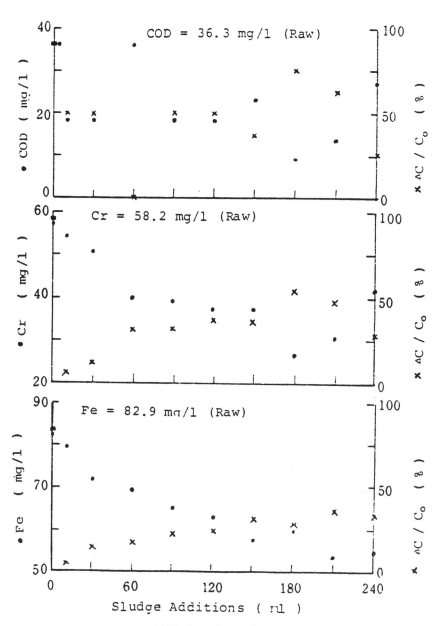

Figure 6 Removal Efficiencies of Combined Wastewater With Cr Fe Ions Dominant

Table 6 Summarizes all the removal efficiencies are compared to sedimentation without sludge addition. It shows clearly that remarkable heavy metal removal could be achieved in all cases. This indicates that the suggested modification of returning waste sludge to the influent point is technically feasible. Furthermore the additional COD removal leads to the saving of aeration capacity in the aeration unit. This means that about 1/3 of the energy required could be saved.

Table 6 Removal efficiencies as compared to no sludge addition

	COD		Heavy Metal	
	MAX.	AVE.	MAX.	AVE.
Cu (low conc.)	37.9	30.1	50.1	32.0
Cu (high conc.)	56.8	52.1	25.7	44.8
Ni	18.2	10.9	40.9	23.3
Cu dominant (Combined)	44.2	27.9	34.9	20.2
Cr dominant Fe (Combined)	75.0	40.0	54.1 34.8	30.6 22.5
Average		32.2		26.7

CONCLUSIONS

Results of this preliminary study allowed the following points to be drawn:
(1) The suggested modification of the conventional activated sludge unit by recycling the wasted sludge to influent point is shown to be technically feasible.
(2) The adsorptive power of the activated sludge could effect around 32.2% of COD removal and 26.7% of heavy metal ions removal.
(3) The optimum sludge dose could not be found , but generally more added sludge would bring about higher removal efficiency.

(4) Since in the activated sludge system, the amount of waste sludge is determined by the removed BOD, therefore the waste sludge available for adsorption in this modification is limited. Therefore the upper limit of the possible COD and heavy metal ions removal should be tested with continuously operated activated sludge unit.

ACKNOWLEDGEMENT

This investigaftion was supported by a grant from the National Science Council of the Republic of China under the project No. NSC 75-0410-E035-01.

REFERENCES

1. Imhoff, K., Imhoff, K. R., : Taschenbuch der Stadtewtwasserung, 23 Aufl. Vertag R. Oldenbourg, Munchen, 1972.
2. Hunken, K, H, : Untersuchungen uber den Reinigugsver lauf und den sauerstoftverbrauch der Abwasserreinigung durch das Belebtschlammverfahren. Stuttgarter Benchte zur Siedlungswasswewirts, Heft 4, Verlag R. Oldenbourg, Munchen, 1960.
3. Mckinney, R. E. : Microbiology for Sanitary Engineers. McGraw Hill Book Co., Inc., New York, 1962.
4. American Society of Civil Engineers, Manuals and Reports on Engineering practice No. 36, p.268, 1974.

THE EFFECT OF INORGANIC CATIONS ON BIOLOGICAL FIXED-FILM SYSTEMS

Robert W. Peters

Environmental Engineering
School of Civil Engineering
Purdue University
West Lafayette, Indiana

INTRODUCTION

Removal of cations in a biological wastewater treatment system is due primarily to the sorption of both soluble and small colloidal particulates by biological flocs. Of particular interest to the removal of inorganic cations from solution is that involving heavy metals. The presence of heavy metals in wastewater results from their wide usage in the plating and metal finishing industries. Heavy metals appear to present great dangers through their promiscuous release to the environment because they are toxic and relatively accessible. Elements such as Cd and Hg exhibit human toxicity at extremely low concentrations. Elements, such as Ag, Cr, Cu, Pb, and Zn, also exhibit toxic properties to humans, although they are orders of magnitude higher than that required for Cd or Hg toxicity.

The wastewaters from such industries as the plating and metal finishing, the chemicals industry, and the pulp and paper industry, often contain high concentrations of heavy metals. These wastewaters are discharged either to waterways, publicly owned treatment works (POTW's) or industrial waste treatment plants (IWTP's). The most publicized case of industrial heavy metal pollution involved the discharge of the catalyst methylated mercury chloride into Minamata Bay, Japan, from a plastic manufacturing factory. Microorganisms converted the sedimented compound to monomethyl-mercury, which led to an enrichment of this most toxic metal in fish consumed by local people, causing severe chronic mercury poisoning [1]. Removals or reductions of heavy metal concentrations below 10 mg/l are usually desirable prior to any wastewater treatment operation since the presence of many heavy metals adversely affect biological oxidation processes such as trickling filters, activated sludge, and anaerobic digestion [2,3,4]. Only low concentrations of heavy metals are allowed in the effluent discharged to receiving waters due to their toxicity to aquatic and human life. Due to the authors interest in heavy metal separations and because most municipal wastewater treatment plants employ biological treatment processes for stabilization of both organic and inorganic substances in wastewater, a knowledge and appreciation of the effects on heavy metals on biological oxidation processes is necessary. Moulton and Shumate [5] point out that the presence of toxic materials in wastes has two important aspects: the deleterious effects they exert on biological wastewater treatment processes and the potential harmful effects they present to aquatic and terrestrial organisms downstream from the wastewater treatment plant.

High concentrations of heavy metals are toxic to most microorganisms and can often cause upsets in a biological waste treatment system. Toxicity of heavy metals to microbial activity is due primarily to soluble metal ions [6]. The degree of inhibition is reduced when precipitation of heavy metal ions occurs [7]. Given a low concentration of heavy metals along with a proper acclimation process, a biological system may be used to remove a portion of the heavy metals without serious deterioration in the biological treatment efficiency [8,9]. The tolerance of a biological system for heavy metals is greatly enhanced through acclimation. One theory of acclimation, pointed out by Chang et al. [10], has the metallic cation causing damage or inactivation of one or more critical enzymes; additional enzymes can be produced to replace the damaged enzymes. With severe damage, an alternate metabolic pathway can be created. With the narrow spectrum of life forms of anaerobic and nitrifying microorganisms, they are more susceptible to toxicity by heavy metals than are aerobic heterotrophic organisms [11]. With aerobic treatment processes, heavy metals can adversely affect several species of microorganisms without causing a loss in the overall efficiency of the biological treatment [8,10].

The degree of metal inhibition varies with different biological species. Toxicity generally takes the form of enzyme inhibition [12]. A pure culture involving a simple substrate may show a different behavior than that demonstrated with a heterogeneous system of microorganisms. Metals exhibit inhibition when the metals are present in the soluble form; controlling pH and promoting the formation of metal precipitates minimizes this inhibition. Most insoluble metals, present as metal oxides and metal hydroxides, are removed by sedimentation prior to biological treatment.

Heavy metals may adversely affect microbial metabolism by inhibiting enzyme catalysis. Metals can bind at the enzyme-active sites or cause conformational changes in the enzyme [13]. Although numerous studies have been performed on the inhibitory effect of heavy metals on various microorganisms, the results are difficult to compare for a variety of reasons. Oftentimes, researchers have investigated metal inhibition on unacclimated cultures while other investigators have used acclimated cultures. Generally, studies on pure cultures do not adequately represent the mixed population of microorganisms present in activated sludge systems. Poon and Bhayani [14] in a study employing separate bacteria and fungi cultures, showed that the toxic behavior of heavy metals is different for various biological species. Heavy metals were added to unacclimated cultures in slug loads. Silver and nickel were less toxic to Geotrichum candidum cultures than to the Zooglea ramigera cultures while zinc, copper, and chromium were more toxic to the Geotrichum candidum culture. With mixed microorganism cultures, a pattern of mixed inhibition may result with the tolerant species predominating [13]. In aerobically produced sludge, Cameron and Koch [15] point out that the metals associated with the sludge mass are usually organically bound via formation of metal-organic complexes. High molecular weight exocellular polymers of the biofloc (including such things as proteins, ribonucleic acid, polysaccarides, and deoxyribonucleic acid) provide many functional groups that can act as binding sites for the metals [16].

With anaerobic sludges on the other hand, metals are generally considered to be distributed between the precipitated insoluble forms and organically bound fractions [15]. As much as 30 to 60% of the metal accumulation in digested sludges are organically bound, contained in the intracellular portions of the biomass [17]. Hayes and Theis concluded that microbial uptake activity competes with precipitation for removal of heavy metals from digester supernatant.

It should be pointed out that biological treatment processes do not alter or destroy inorganics. In fact, soluble inorganic concentrations should be maintained at a low level so that enzymatic activity is not inhibited [12]. Trace inorganic concentrations may be lowered through adsorption onto the microbial cell coating. Johnson [12] noted that microorganisms generally have a net negative charge enabling cation exchange to be performed with the metal ions existing in solution. Dissolved metals and fine metal particulates are concentrated in the biomass, primarily through adsorption onto the activated sludge surface [18,19]. The biomass surface has been suggested [20] as being coated with a polysaccharide slime consisting of polymers of glucuronic acid and neutral sugars. The metal ions form salts with the carboxyl groups present in the slime coating and are electrostatically attached to the hydroxyl groups [20].

The removal of heavy metals in biological treatment systems is attributed to the sorption of both soluble and colloidal metal particulates by biological flocs [18]. Any material that is adsorbed or absorbed by bacterial flocs is removed from the water in the course of wastewater treatment operations. Coagulation/flocculation and settling are the primary mechanisms by which metal removal is achieved in biological treatment operations such as activated sludge [9]. Brown and Lester note that factors such as loading rate, feed composition, sludge volume index, and mixing strength affect the capacity to remove metals. Chang et al. [21] noted that the microbial removal of heavy metals involves an initial rapid uptake followed by a slow consistent long-term uptake phase. The solution pH greatly affects this rate of uptake. Sludge age (as well as the extent of acclimation) can also affect the degree of metal removal in activated sludge systems.

Although numerous research studies have been conducted to

investigate the effects of heavy metals on biological wastewater treatment processes, most of the emphasis has involved suspended growth systems such as activated sludge [5,6,9,16,18,19,20,22-31]. This paper does not have as its intent to review all those metal inhibition/metal toxicity studies involving the activated sludge systems. Such a review will be pursued in a future paper on the subject. The interested reader is referred to the relevant references cited above.

OBJECTIVES

The goals and objectives for this study are summarized below:

1. Review and investigate the effects of heavy metals on the performance of fixed-film biological treatment processes.

2. Determine the effects of heavy metals on the attached growth biofilm ecology where possible.

3. Investigate the removal efficiency of the heavy metal(s) in the fixed-film processes.

Although numerous studies have been conducted to examine the toxic effects of heavy metals on suspended growth biological treatment processes, relatively few studies have been performed using fixed-film biological processes, as noted by Chang et al. [21] and Russell [32]. The fewer number of studies involving trickling filters and rotating biological contactors (RBC's) has been attributed to the difficulty in studying a trickling filter's biofilm and to the relatively recent development of the RBC process [32]. This paper has as its goal to perform a critical review of effect of heavy metals on these attached growth systems.

RESULTS AND DISCUSSION

The effects of heavy metals on the performance of attached growth systems have been addressed in terms of the various unit operations involved: trickling filters, rotating biological contactors (RBC's), and anaerobic filters.

Trickling Filters

A trickling filter is a bed of packing such as broken rock or other course aggregate onto which sewage or industrial waste is sprayed intermittently and allowed to trickle through, leaving organic matter on the surface of the media, where it is oxidized and removed by biological growths. Hammer [33] points out that the term "trickling filter" is actually a misnomer; a better term is a "biological bed" because the process is one of biological extraction rather than filtration. With the development of synthetic (plastic) media, the term "biological tower" has been introduced.

Atkinson and Swilley [34] investigated the effect on ferric chloride on the removal efficiency of a laboratory scale trickling filter (biological film reactor). Ferric chloride had an immediate deleterious effect on the system performance for the concentrations studied (100 and 250 mg/l). Figure 1 shows the results of their study. During exposure to the ferric chloride, the organisms attained a brown color; the coloration diminished as the chloride dosing was completed. After the end of the experiments, the color ccould be removed through repeated washings. This loss of coloration on washing in conjunction with the rapid recovery of removal efficiency following the ferric chloride dosing led the authors to conclude the iron was physically adsorbed at the surface interface. The poor performance was thought to result from a partial blockage of the liquid-microorganism interface (i.e., the microorganism's surface) hindering the mass transfer process. After the entire dosage of ferric chloride was fed to the unit, the chemical oxygen demand (COD) removal efficiency rapidly returned to its previously higher level.

When the trickling filter at the Clayton County Water Authority in Morrow, Georgia became clogged with aluminum hydroxide slime, no BOD reduction occurred; the filter developed ponding [35]. The aluminum hydroxide/oxide gel was speculated as filling the voids within the trickling filter displacing the bacterial slime and air from the filter.

Without chemical precipitation, a highly loaded trickling

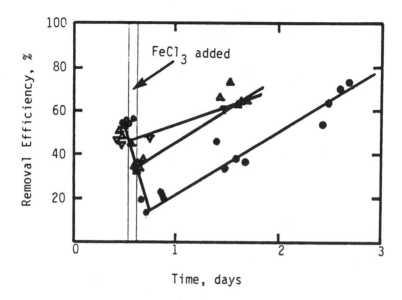

Figure 1. Effect of FeCl$_3$ solution on removal efficiency [34].

Inlet concentration = 200 ppm
Flow rate = 300 gm/min
Support surface angle = 5°
Symbols for FeCl$_3$ concentration:

▽ -   0 ppm
▲ - 100 ppm
● - 250 ppm

filter plant [36] had a removal efficiency of ~60 - 80% BOD_5, 55-75% COD, and a reduction of ~30% total phosphorus. Using pre-precipitation with ferric salts, the removal efficiency increased to 95%, 82%, and 63%, for BOD_5, COD, and total phosphorus, respectively, with similar results obtained using lime at pH ~9 to perform the pre-precipitation. Formation of calcium carbonate crusts necessitated a reduction in the lime dosage lowering the pH to ~8.8, resulting in slightly lower treatment efficiencies.

Two disturbances were reported in the operation of trickling filters at the Sjölunda Sewage Plant [37] resulting in a black colored sewage. Särner [37] reported the mean and range of concentration of various heavy metals (including Fe, Mn, Cu, Cr, Ni, Zn, Pb, Al, Ag, Cd, and Hg) obtained from 24-hour composite samples obtained during the upsets. The removal of heavy metals was small; the concentration of heavy metals in the sludge was thus low.

Reid et al. [38] investigated the effect of chromium on the performance of a trickling filter pilot plant located at Tinker Air Force Base. The filter was 1.2 m (4.0 ft.) in diameter, 1.8 m (6.0 ft.) deep filled with 5.1 cm (2.0 in.) rock. The filter was designed to act as a conventional high rate filter with a normal flow of 21800 l/day (4.0 gallons/min.) and a designed organic loading of 1190 gm/m^3-day (2.0 lb. BOD/yd^3/day). The BOD removal efficiency was established using sewage containing no heavy metals. After introducing sewage containing Cr^{+6}, the BOD removal efficiency was again determined. In both runs, one with 1 mg Cr^{+6}/l and one with 2 mg Cr^{+6}/l, an initial reduction in efficiency was noted the day after the introduction of chromium. Their results are shown in Figure 2. The inhibitory effect reached a maximum on the second day and was subsequently followed by a rapid recuperation of the system; full recovery was obtained by the end of eight days. Reid et al. [38] concluded, based upon the results of the pilot plant studies, that a trickling filter exposed to domestic sewage containing no more than 2 mg Cr^{+6}/l was not significantly affected in terms of the BOD removal efficiency. The Cr^{+6} was capable of being removed by the trickling filter. The more effective the trickling filter's removal of domestic BOD, the more effective its removal of the heavy metal. High concentrations of nickel were not removed as readily as those involving chromium.

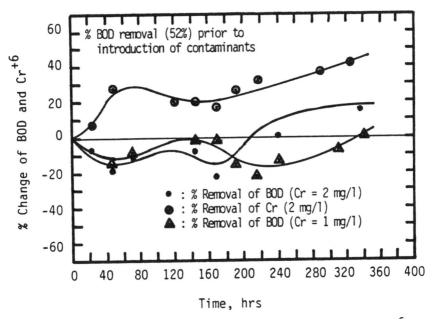

Figure 2. Effect of continuous dosage of Cr^{+6} (1 and 2 mg/l) on BOD removal of Tinker Air Force Base pilot plant trickling filter. On the graph is a second plot, that of % removal of Cr^{+6} versus time [38].

Anaerobic Filter

The anaerobic filter is an oxygen-free, media filled bed reactor in which anaerobes grow not only in the void spaces between the media but also on the entire surface of the media. The wastewater can be distributed over the top of the bed (stationary type reactor) or it can be fed across the bottom of the filter (suspended type reactor). This latter type is termed an "upflow reactor" and is more popular than the downflow reactor. Both reactor types have completely submerged filter media arranged either in a packed bed or in a fluidized bed mode.

Wu et al. [39] noted that anaerobic treatment processes are very effective for removal of heavy metals from waste streams by adsorption and precipitation. DeWalle et al. [40] investigated the anaerobic filter for removal of various heavy metals (such as Fe, Zn, Cu, Cr, Ni, Pb, and Cd). The anaerobic filter was subjected to four different loadings (2,4,6, and 16 l/day) resulting in detention times of 34, 17, 11.3, and 4.2 days, respectively. The recirculation rate was 30.8 l/day resulting in upflow rates of 1.17, 1.24, 1.31, and 1.67 m/day, respectively at the above detention times. The concentrations of heavy metals in the influent and effluent of the anaerobic filter are listed in Table 1. Metal removals as high as 95.5% were achieved; when the removals were computed on a soluble concentration basis, the percentage removal increased to 97.1% indicating significant quantities of heavy metals were associated with the suspended solids leaving the filter. DeWalle et al. [40] concluded the completely mixed anaerobic filter was effective for removal of heavy metals. As the metal concentration increased, the effectiveness of the filter increased. The metals were precipitated as sulfides, carbonates, and hydroxides; these precipitates were removed from the filter as a slurry that accumulated in the bottom solids collection device. Decreasing the hydraulic detention time caused the metal removal to decrease while increasing the metal content in the bottom slurry.

Friedman et al. [41] conducted a study to evaluate the feasibility of using an anaerobic process for roughing treatment of beamhouse wastewaters and the sequential use of conventional aerobic processes for final polishing treatment. Bench scale anaerobic filters were used to treat "standard" beamhouse wastewaters, resulting from the beaming of cowhides

Table 1. Concentration of heavy metals in influent and effluent of the anaerobic filter [40].

Metal	Influent Concentration, mg/l	Maximum Soluble Concentration in Filter, mg/l	Average Soluble Concentration in Filter, mg/l	Average Effluent Concentration, mg/l	Average Soluble Effluent Concentration, mg/l	Average % Metal in Effluent Present as Suspended Solids	Average Metal Removal, %	Average Soluble Metal Removal, %
Fe	430	45	15.1	19.4	12.6	35	95.5	97.1
Zn	16	15	1.29	1.3	0.95	27	92.0	94.1
Cu	5.6	11	1.25	1.47	0.67	54	73.8	88.1
Cr	1.7	0.43	0.18	0.24	0.15	27	85.9	91.2
Ni	1.2	1.0	0.22	0.33	0.20	39	72.5	83.7
Pb	0.33	0.75	0.08	0.08	0.06	25	78.9	84.2
Cd	0.027	0.020	0.012	0.019	0.013	32	29.6	51.9
Average Element	---	---	---	---	---	---	75.4	84.3

and pigskins. Laboratory scale rotating biological contactor (RBC) units were used to compare aerobic treatment of both the pretreated beamhouse wastewaters and the anaerobic filter effluent. At the completion of the study, the filters were drained; the void volumes were found to be about 43% of the original void volume. All the surfaces were coated with a dark gelatinous material. All of the Bio-Rings examined were at least partially plugged. The dark material consisted of gritty sand-like particles that was resistant to dissolution by strong acid or strong base. Approximately one third of the nonvolatile matter was calcium. This was not unexpected, as pointed out by Friedman et al. [41] considering the raw beamhouse wastewaters were saturated with lime. The massive solids deposition suggested that anaerobic filters treating wastes likely to form precipitates should be operated in the downflow mode to assist in washing solids from the system and minimize the potential for plugging.

In a similar study, Wu et al. [42] investigated the feasibility of using a two stage upflow anaerobic fixed-film reactor to treat acidic leachate wastewater. Chemical analyses of the raw leachate determined the concentrations of nitrogen, phosphorus, alkalinity, iron, zinc, acetic acid, and suspended solids to be 100 mg/l as NH_3 - N, 20 mg/l as PO_4 - P, 1000 mg/l as $CaCO_3$, 20 mg/l as Fe^{+++}, 20 mg/l as Zn^{++}, 2480 mg/l as acetic acid, and 2100 mg/l as TSS. The pH of this wastewater was approximately 5.0. Due to the deficiency of nitrogen and phosphorus in the raw leachate, ammonium chloride and monobasic potassium phosphate were added to maintain a ratio of COD:N:P between 100:1:0.1 and 100:2:0.2 in the feed solution. The soluble BOD and COD of this influent wastewater was approximately 13,900 and 20,800 mg/l, respectively. Reductions as high as 96% for COD and 98.4% for BOD were achieved, provided the organic loading is controlled so as not to exceed 1204 gm COD/m^3-day (75.2 lb COD/1000 ft^3-day) or 903 gm BOD/m^3-day (56.4 lb BOD/1000 ft^3-day) with a reactor detention time of 16.7 days. All the phases studied provided excellent removal of both iron and zinc; removals exceeded 92%. The results from their study are shown in Table 2. The highest influent metal concentrations were 115.3 mg/l Fe^{+++} and 10.4 mg/l Zn; for these concentrations, the anaerobic filter was able to remove 98.7% Fe^{+++} and 98.5% Zn^{++} without influencing the filter's ability to degrade BOD and COD. In addition to the efficient treatment of soluble BOD, COD, and heavy metals, the anaerobic filter requires no effluent recy-

Table 2. Metal removal* from the anaerobic treatment of landfill leachate using an upflow two-stage biological filter [42].

Parameter	Phase I		Phase II		Phase III	
	Filter No. 1	Filter No. 2	Filter No. 1	Filter No. 2	Filter No. 1	Filter No. 2
Organic Loading						
lb COD/1000 ft^3-day	141.8	93.9	72.2	4.90	60.7	5.80
lb BOD/1000 ft^3-day	106.7	60.0	53.4	3.90	47.2	1.40
Detention Time, days	8.0		17.5		8.7	
Soluble Influent Concentration, mg/l						
Fe^{+++}	32.4	2.37	30.1	3.71	115.3	1.95
Zn^{++}	0.85	0.09	0.71	0.04	10.4	0.19
Soluble Effluent Concentration, mg/l						
Fe^{+++}	2.37	2.16	3.71	2.37	1.95	1.45
Zn^{++}	0.09	0.06	0.04	0.02	0.19	0.15
Metal Removal						
Fe^{+++}, %	92.7	28.0	87.7	36.1	98.3	26.3
Zn^{++}, %	89.2	32.9	94.3	50.0	98.1	26.9
Overall Metal Removal, %						
Fe^{+++}	93.3		92.1		98.7	
Mn^{++}	92.8		97.1		98.5	

*Average of last 30 days data.

cle or sludge return. The filter also reduced the suspended and volatile solids present in the leachate by over 90% due to the long detention time. The anaerobic filter further produces usable methane gas which can be used to heat the reactors. The volume of methane produced is 5.01 - 5.84 ft^3/lb COD removed, depending upon the organic loading condition and the reactor detention time.

Rotating Biological Contactors (RBC's)

A rotating biological contactor (RBC) consists of a series of closely spaced circular discs, usually made of polystyrene or polyvinyl chloride. The discs are mounted on a horizontal shaft. The discs are partially submerged in wastewater and rotate slowly through it.

In an early study, Reid et al. [38] used a rotating drum apparatus to culture attached sewage slimes. The effects of copper, cadmium, zinc, and chromium on the aerobic treatment efficiency were studied. The concentrations of the metals in the influent sewage found to reduce the aerobic treatment efficiency under continuous dosing conditions were: $Cr^{+6} \geq 10$ mg/l, $Cu \geq 1$ mg/l, Ni: 1-2.5 mg/l, and Zn: 5-10 mg/l.

Olem and Unz [43] used an RBC to treat acid mine drainage. Iron-oxidizing bacteria, presumably Thiobacillus ferro-oxidans colonized the disk surfaces with an average population of 70,000 cells/cm^2 and mediated the transformation from the Fe^{++} state to the less soluble Fe^{+++} state. The Fe^{++} oxidation kinetics with the RBC followed a concentration dependent first order relationship at all hydraulic loadings and disc rotation rates examined. For a peripheral disc velocity of 19 m/min, more 90% of the Fe^{++} was transformed to the Fe^{+++} state for mine waters containing Fe^{++} in the range of 16-313 mg/l. Olem and Unz [43] concluded that the RBC was potentially useful as a pretreatment step in the total treatment of acid mine drainage.

Singhal [44] described the full-scale plant operating experiences at Cadillac, Michigan, in using RBC's for ammonia nitrogen removal and addition of ferric chloride in the aeration system for phosphorus removal. Samples of the biomass analyzed for their metal concentrations showed significant concentrations of iron, calcium, phosphorus, silicon, and manganese. Singhal speculated that phosphates of manganese, calcium, and iron existed as crystalline species in the underlying mineralizing layer. The high concentration of manganese in stage 4 suggested the accumulation (removal) mechanism for manganese is different from that of calcium, iron, and silica. Singhal [44] suggested this might be the result of bacterial activity of the nitrifying organisms. Another explanation involves the fact that manganese exhibits a much slower (less kinetically favored) precipitation than that of the other cations involved.

In a recent study Russell [32] studied the effect of iron and manganese on the performance of a 4 stage laboratory scale RBC unit used to treat a synthetic feed solution. The concentrations of iron and manganese used in the various phases of the study were 0, 5, 7.5, 15, and 30 mg/l; various combinations of these cation concentrations were employed. The metal accumulation within the biofilm was localized differently for the two metals, iron accumulating to the largest extent in the first stage discs while manganese accumulated primarily in the second and third stage biofilm. This result was attributed to differences in the solubilities between the two metals [32], although the precipitation kinetics and driving forces involved in the precipitation may be another explanation for this behavior. The soluble metal removals over the course of the study averaged 98.5% and 86.0% for iron and manganese, respectively. Increased metal concentrations accelerated the settling behavior and caused a change in the floc morphology from a filamentous to a dispersed form. The COD treatment capability may have been influenced by the direct precipitation and blockage of the surface of the biofilm. The iron and manganese accumulated within the biomass to maximum levels of 34% and 78% on a weight to weight basis, respectively. Russell [32] observed an increase in the true biofilm weight as the concentration of metals added to the system increased. Russell postulated that other inert substances entering the feed, such as phosphates and carbonates, precipitated in conjunction with the metals resulting in added biofilm weight. Calcium and magnesium concentrations were not measured in the course of this study; it is conceivable that $CaCO_3$ and $Mg(OH)_2$ precipitates could have formed which adhered to the biofilm surfaces (due to their large surface areas) forming a tougher biofilm. Whatever the cause Russell extrapolated her results to that of a typical full scale system, and determined the additional contribution of weight from these two metals could mean an extra load on the media and shaft of 6840 kg (7.5 tons). Russell speculated that this increase in load might push the system beyond its fatigue limit resulting in the system's failure.

Kang and Borchardt [45] studied the inhibition of nitrification by chromium using a biodisc system. Their study showed no accumulation of nitrite indicating the Nitrobacter organisms were not inhibited. Only Nitrosomonas was inhibited by the chromium. The concentration of chromium in the liquid (initial concentration was 1, 3, or 10 mg/l in their experiments) decreased rapidly in the first two stages (1.1 to 0.6 mg/l) and slowly in the subsequent stages (to 0.4 mg/l). Kang and Borchardt [45] postulated that the major mechanism for chromium removal was adsorption to the biomass. Most of the COD was removed in the first two stages, while nitrification of ammonia began at the third stage and continued in the subsequent stages. Hexavalent chromium can be adsorbed to the biomass up to its adsorptive capacity estimated to be 2% for heterotrophic organisms or less for a mixture of heterotrophic and autotrophic organisms. The rates of adsorption and the resulting inhibition to microorganisms were fast.

Rest et al. [46] investigated the use of an RBC for polishing anaerobic process effluents for tannery beamhouse wastewaters. Their studies indicated that an RBC system was capable of achieving removals of 96% BOD_5 and 75% COD. Higher levels of COD removal were not possible due to the extended time required to biodegrade the complex proteins contained in the wastewater. The materials and growth attached to the discs in the first four RBC stages became very heavy at the high loadings. The attached substances were similar in consistency and color to that of hardened mortar; the material adhered very firmly to the discs. Rest et al. [46] assumed the primary component of the inorganic portion of the attached materials was due to calcium in the form of calcium carbonate and calcium sulfate. Due to the buildup of mineral salts on the RBC discs, the RBC was not considered applicable to beamhouse wastewater treatment [46].

Chang et al. [8,10,21] studied the toxic effects of Cd^{++} and Cu^{++} on a rotating biological contactor treating a sugar waste. Different levels of cadmium (5 and 20 mg/l) and copper (1, 5, 10, 25, and 50 mg/l) were dosed to a three stage RBC unit treating an influent sugar solution (300 mg/l). The treatment efficiency was not adversely affected by the presence of copper at concentrations of 10 mg/l or less. Increasing the copper concentration to 25 and 50 mg/l caused the removal of dissolved organic carbon to decrease by 7 and 10%, respectively. The biological treatment efficiency decreased by 8% for both cadmium concentrations (5 and 20 mg/l). The efficiency of metal removal in the treatment system varied from 85-95% for cadmium and 30-90% for copper, depending on the initial concentration in the feed solution. Both cadmium and copper were effectively retained by the biofilm.

SUMMARY

The effects and inhibitions associated with heavy metals on the performance of fixed-film biological processes such as trickling filters, anaerobic filters, and rotating biological contactors, have been addressed. High concentrations of heavy metals are toxic to most microorganisms. With proper acclimation, the biological system can be used to remove a certain amount of the metal(s) without being adversely affected. The removal of cations, particularly heavy metals, is primarily due to sorption of both soluble and fine particulates by the biofilm. The anaerobic filter is quite efficient for removal of heavy metals and retention of the metals in the biofilm. Due to the relatively recent development of fixed-film processes, much less attention has been paid on these processes (as compared to suspended growth systems) particularly in terms of heavy metal inhibition.

NOMENCLATURE

BOD   Biological oxygen demand
COD   Chemical oxygen demand
pH $-\log [H^+]$
RBC   Rotating biological contactor
TSS   Total suspended solids

ACKNOWLEDGEMENTS

The author wishes to acknowledge the support of the School of Civil Engineering at Purdue University is performing this critical review. The author also wishes to acknowledge the support of the Consortium for Biological Waste Treatment Research and Technology.

REFERENCES

1. Anon., "Trace Metals: Unknown, Unseen Pollution Threat," Chem. Eng. News, 49(29): 29, 30, 33, July 19, (1971).

2. Bowers, A. F., Chin, G., and Huang, C. P., "Predicting the Performance of a Lime Neutralization/Precipitation Process for the Treatment of Some Heavy Metal Laden Industrial Wastewaters," Proc. 13th Mid-Atlantic Indus. Waste Conf., 13: 51-62, (1981).

3. Callander, I. J., and Barford, J. P., "Precipitation, Chelation, and the Availability of Metals as Nutrients in Anaerobic Digestion, II. Applications," Biotech. Bioeng., 25: 1959-1972, (1983).

4. Patterson, J. W., and Minear, R. A., "Physical Chemical Methods of Heavy Metals Removal," pp. 261-276 in Heavy Metals in the Aquatic Environment, Krenkel, P. A., ed., Pergamon Press, Oxford, England, (1975).

5. Moulton, E. Q., and Shumate, K. S., "The Physical and Biological Effects of Copper on Aerobic Biological Waste Treatment Processes," Proc. 18th Purdue Indus. Waste Conf., 18: 602-615, (1963).

6. Suyarittanonta, S., and Sherrard, J. H., "Activated Sludge Nickel Toxicity Studies," J. Water Pollut. Control Fed., 53 (8): 1314-1322, (1981).

7. Jenkins, S. H., Keight, D. G., and Ewins, A., "The Solubility of Heavy Metal Hydroxides in Water, Sewage, and Sewage Sludge - II. The Precipitation of Nickel in Sewage," Internat. J. Air, Water, Pollut. (G.B.), 8 (11/12): 679-693, (1964).

8. Chang, S.-Y., Huang, J.-C., and Liu, Y.-C., "Effects of Cd and Cu on a Biofilm Treatment System," *Proc. 39th Purdue Indus. Waste Conf.*, 39: 305-312, (1984).

9. Brown, M. J., and Lester, J. N., "Metal Removal in Activated Sludge: The Role of Bacterial Extracellular Polymers," *Water Res.*, 13: 817-837, (1979).

10. Chang, S.-Y., Huang, J.-C., Liu, Y.-C., and Huang, J.-S., "Tolerance for Heavy Metals in a Fixed-Film System," pp. 1077-1094 in *Proc. 2nd Internat. Conf. Fixed-Film Biological Processes*, Vol. 2, Bandy, J. T., Wu, Y. C., Smith, E. D., Basilico, J. V., and Opatken, E. J., ed., Arlington, VA, July 10-12, (1984).

11. Kugleman, I. J., and McCarty, P. L., "Cation Toxicity and Stimulation in Anaerobic Waste Treatment, II., Daily Feed Studies," *Proc. 19th Purdue Indus. Waste Conf.*, 19: 667-686, (1964).

12. Johnson, S. L., "Biological Treatment: Overview," pp. 168-191 in *Unit Operations for Treatment of Hazardous Industrial Wastes*, De Renzo, D. J., ed., Noyes Data Corp., Park Ridge, NJ, (1978).

13. Johnson, S. L., "Biological Treatment: Activated Sludge," pp. 192-208 in *Unit Operations for Treatment of Hazardous Industrial Wastes*, De Renzo, D. J., ed., Noyes Data Corp., Park Ridge, NJ, (1978).

14. Poon, C. P. C., and Bhayani, K. H., "Metal Toxicity to Sewage Organisms," *J. Sanit. Eng. Div. (ASCE)*, 97 (SA2): 161-169, (1971).

15. Cameron, R. D., and Koch, F. A., "Trace Metals and Anaerobic Digestion of Leachate," *J. Water Pollut. Control Fed.*, 52(2): 282-292, (1980).

16. Cheng, M. H., Patterson, J. W., and Minear, R. A., "Heavy Metals Uptake by Activated Sludge, *J. Water Pollut. Control Fed.*, 47(2): 362-376, (1975).

17. Hayes, T. D., and Theis, T. L., "The Distribution of Heavy Metals in Anaerobic Digestion," *J. Water Pollut. Control Fed.*, 50(1): 61-72, (1978).

18. Oliver, B. G., and Cosgrove, E. G., "The Efficiency of Heavy Metal Removal by a Conventional Activated Sludge Treatment Plant," Water Res., 8(11): 869-874, (1974).

19. Esmond, S. E., and Petrasek, A. C., "Trace Metal Removal," Indus. Water Eng., 11: 14-17, (1974).

20. Steiner, A. E., McLaren, D. A., and Forster, C. F., "The Nature of Activated Sludge Flocs," Water Res., 10(1): 25-30, (1976).

21. Chang, S. -Y., Huang, J. -C., and Liu, Y. -C., "Effects of Cd(II) and Cu (II) on a Biofilm System," J. Environ. Eng. Div. (ASCE), 112(1): 94-104, (1986).

22. Carter, J. L., and McKinney, R. E., "Effects of Iron on Activated Sludge Treatment," J. Environ. Eng. Div. (ASCE), 99(EE2): 135-152, (1973).

23. Heukelekian, H., and Gellman, I., "Studies of Chemical Oxidation by Direct Methods. IV. Effects of Toxic Metals Ions on Oxidation," Sewage Indus. Wastes, 27(1): 70-84, (1955).

24. McDermott, G. N., Moore, W. A., Post, M. A., and Ettinger, M. B., "Effects of Copper on Aerobic Biological Sewage Treatment," J. Water Pollut. Control Fed., 35(2): 227-241, (1963).

25. Chen, K. Y., Young, C. S., Jan, T. K., and Rohatgi, N., "Trace Metals in Wastewater Effluents," J. Water Pollut. Control Fed., 46(12): 2663-2675, (1974).

26. Rossin, A. C., Sterritt, R. M., and Lester, J. N., "The Influence of Process Parameters on the Removal of Heavy Metals in Activated Sludge," Water, Air, and Soil Pollut., 17(2): 185-198, (1981).

27. Sawyer, C. N., Frame, J. D., and Wold, J. P., "Revised Concepts of Biological Treatment," Sewage Indus. Wastes, 27(8): 929-938, (1955).

28. Ghosh, M. M., and Zugger, P. D., "Toxic Effects of Mercury on the Activated Sludge Process," J. Water Pollut. Control Fed., 45(3): 424-433, (1973).

29. Nelson, P. O., Chung, A. L., and Hudson, M. C., "Factors Affecting the Fate of Heavy Metals in the Activated Sludge," J. Water Pollut. Control Fed., 53(8): 1323-1333, (1981).

30. Kodukula, P. S., and Patterson, J. W., "Distribution of Cadmium and Nickel in Activated Sludge Systems," Proc. 38th Purdue Indus. Waste Conf., 38: 439-448, (1983).

31. Fristoe, B. R., and Nelson, P. O., "Equilibrium Chemical Modeling of Heavy Metals in Activated Sludges," Water Res., 17(7): 771-778, (1983).

32. Russell, E. K., "A Study of Iron and Manganese Removal in a Rotating Biological Contactor," M.S. Thesis, Purdue University, West Lafayette, IN, (1985).

33. Hammer, M. J., Water and Waste-Water Technology/SI Version, John Wiley & Sons, New York, NY, (1977).

34. Atkinson, B., and Swilley, E. L., "The Effect of Ferric Chloride on the the Removal Efficiency of a Biological Film Reactor," Water Res., 1: 687-693, (1967).

35. Ingols, R. S., and Huie, E. L., "Restoring a Dead Trickling Filter," Water and Sewage Works, 120(2): 64-65, (1973).

36. Wolf, P., and Nordmann, W., "Vergleichende Untersuchungen mit Kalk- und Eisen-Vorfällung an einer Tropfkörpenanlage," gwf-Wasser/Abwasser, 123: 255-259, (1982).

37. Särner, E., Plastic-Packed Trickling Filters: A Study of High-Rate Plastic-Packed Trickling Filters in a Post-Precipitation System, Ann Arbor Science Publishers, Ann Arbor, MI, (1980).

38. Reid, G. W., Nelson, R. Y., Hall, C., Bonilla, U., and Reid, B., "Effects of Metallic Ions on Biological Waste Treatment Processes," Water and Sewage Works, 115(7): 320-325, (1968).

39. Wu, Y. C., Kennedy, J. C., Gaudy, A. F., Jr., and Smith, E. D., "Treatment of High-Strength Organic Wastes by Submerged Media Anaerobic Reactors State-of-the-Art Review,"

Proc. 1st Internat. Conf. Fixed-Film Biological Processes, Vol. II: 1212-1238, Kings Island, OH, April 20-23, (1982).

40. DeWalle, F. B., Chian, E. S. K., and Brush, J., "Heavy Metal Removal with Completely Mixed Anaerobic Filter," J. Water Pollut. Control Fed., 51(1): 22-36, (1979).

41. Friedman, A. A., Kowalski, D. P., and Bailey, D. G., "Tannery Effluent: A Challenge Met by Anaerobic Fixed-Film Treatment," Proc. 1st Internat. Conf. Fixed-Film Biological Processes, Vol. III: 1437-1455, Kings Island, OH, April 20-23, (1982).

42. Wu, Y. C., Kennedy, J. C., and Smith, E. D., "Anaerobic Treatment of Landfill Leachate by an Upflow Two-Stage Biological Filter," Proc. 1st Internat. Conf. Fixed-Film Biological Processes, Vol. III: 1495-1520, Kings Island, OH, April 20-23, (1982).

43. Olem, H., and Unz, R. F., "Acid Mine Drainage Treatment with Rotating Biological Contactors," Biotech. Bioeng., 19: 1475-1491, (1977).

44. Singhal, A. K., "Phosphorus and Nitrogen Removal at Cadillac, Michigan," J. Water Pollut. Control Fed., 52(11): 2761-2770, (1980).

45. Kang, S. J., and Borchardt, J. A., "Inhibition of Nitrification by Chromium in a Biodisc System," Proc. 1st Internat. Conf. Fixed-Film Biological Processes, Vol. II: 990-1006, Kings Island, OH, April 20-23, (1982).

46. Rest, G. B., Friedman, A. A., and Bailey, D. G., "RBC and Activated Carbon Treatment of Tannery Beamhouse Wastewater," Proc. 14th Mid-Atlantic Indus. Waste Conf., 14: 190-201, (1982).

PHYSICAL-CHEMICAL AND ANAEROBIC FIXED FILM TREATMENT OF LANDFILL LEACHATE

Dhandapani Thirumurthi

Department of Civil Engineering
Technical University of Nova Scotia
Halifax, Canada

Shahid M. Rana

Department of Civil Engineering
University of Regina
Saskatchewan

Thomas P. Austin

Porter-Dillon Ltd.
Halifax

INTRODUCTION

A Regional (solid waste) Landfill site which receives approximately 600 metric tonnes of municipal refuse per working day from the cities of Halifax, Dartmouth, and Bedford and the County of Halifax, Nova Scotia, Canada, covers about 145 hectares of land area, of which approximately 30 hectares had been cleared by 1982. The landfill is underlain by glacial till having a permeability of about 10^{-6} cm/sec or less. The bedrock consists of quartzite and slates of the Goldenville formation. (1) It is estimated that by the time this research study ended (June 1984), about 1×10^6 metric tonnes of wastes had been deposited to a maximum depth of about 17 m. (2)

Leachate is collected by means of a system of 150 mm diameter perforated poly vinyl chloride (PVC) collectors which underlie the filled areas. These collectors ultimately drain into two anaerobic lagoons (which can be operated either in series or in parallel) followed by four holding ponds (which are operated in series).

The qualities of the raw leachate are summarized in Table I.

Table I. Characteristics of Raw Leachate

Parameter	Mean Concentration*		Concentration Range		
Alkalinity (as CaCO₃)	3,850.	(12)	3,450.	–	4,580.
Aluminum	1.95	(10)	1.5	–	2.7
Ammonia (as N)	370.	(12)	285.	–	406.
Antimony	0.18	(10)	<0.05	–	<0.5
Arsenic	0.03	(10)	0.009	–	<0.05
Barium	0.2	(10)	0.16	–	0.23
Beryllium	<0.005	(10)	<0.005	–	<0.050
BOD5	16,118.	(12)	12,400.	–	17,800.
Boron	6.5	(10)	5.1	–	8.9
Cadmium	0.004	(10)	0.002	–	0.007
Calcium	1,741.	(9)	1,584.	–	1,901.
Chloride	1,108.	(9)	930.	–	1,320.
Chromium	0.2	(10)	0.1	–	0.28
COD	22,845.	(12)	19,900.	–	24,100.
Cobalt	0.40	(10)	<0.010	–	1.78
Copper	0.03	(10)	<0.010	–	0.060
Hardness (as CaCO₃)	5,423.	(12)	4,873.	–	5,916.
Humic Acid	416.	(10)	340.	–	500.
Iron	937.	(10)	767.	–	1,088.
Kjeldahl Nitrogen	490.	(10)	410.	–	762.
Lead	0.10	(10)	<0.002	–	0.33
Magnesium	257.	(8)	223.	–	284.

Table I. Continued

Parameter	Mean Concentration*	Concentration Range
Manganese	58.9 (10)	47.5 – 68.1
Nickle	0.48 (10)	0.25 – 0.75
Nitrate + Nitrite (as N)	<0.5 (12)	<0.5 – <0.5
Orthophosphate	0.5 (12)	0.05 – 2.0
pH	– (12)	5.4 – 5.7
Potassium	420.7 (9)	370. – 470.
Selenium	0.29 (10)	<0.1 – 1.0
Sodium	916.7 (9)	750. – 1,010.
Suspended Solids	467.6 (12)	110. – 1,150.
Sulphate	831. (12)	750. – 930.
Tin	<0.03 (10)	<0.03 – <0.30
Total Dissolved Solids	15,297. (12)	13,342. – 16,840.
Total Organic Carbon	8,136. (12)	7,000. – 9,900.
Total Phosphate	9.1 (10)	6.6 – 17.0
Total Solids	15,734. (12)	13,879. – 17,286.
Total Volatile Acids	10,100. (2)	9,700. – 10,500.
Total Volatile Solids	6,187. (2)	6,030. – 6,344.
Vanadium	0.25 (10)	0.16 – 0.54
Zinc	68.3 (10)	57. – 80.

*All concentrations, except pH, are expressed in mg/L
The numbers in the bracket represent the number of grab samples

CURRENT TREATMENT

The two anaerobic lagoons are presently operated in the series mode. The smaller, northern lagoon is 80 m x 51 m in size and the second lagoon is 92 m x 51 m. The liquid depth is about 4 m in both the units, which are lined with a 30 mm thick chlorinated high density polyethylene liner (3). The effluent from the second lagoon is pumped into the holding ponds, which overflows into a swamp.

Leachate is now produced at an average rate of less than 1 L/s, but the generation rate is expected to increase to 1.6 L/s once the site is fully utilized. These flow estimates do not include the short-duration increases subsequent to storm-induced infiltration (2).

No evidence of pollution of ground water on the nearby Sackville River has been established so far by the limited monitoring of the water samples. However, the owners of the landfill site, concerned about the long-term pollution potential of the leachate (the quantity of which could increase in course of time), initiated a laboratory-model preliminary treatability study, in concert with H. J. Porter and Associates (a consulting firm currently incorporated as Porter-Dillon Ltd.) and Technical University of Nova Scotia (TUNS), to develop preliminary design parameters for a new treatment plant.

LITERATURE REVIEW

A review of literature indicated that, in view of the high concentration of organics (Table I) in the raw leachate, anaerobic biological treatment is much more advantageous than aerobic treatment; however, to enable successful anaerobic stabilization, the raw leachate required the following three pretreatment steps: (i) pH increase to at least 7, (ii) supplementation with phosphate and (iii) precipitation of potentially toxic Zn (4, 5, 6 and 7).

The limited available information on the minimum amount of PO_4 required to sustain anaerobiosis in a fixed film reactor varies between a low requirement of 1 mg/L of total P to 20,000 mg/L of COD and a high requirement of 1 mg/L of P to 4360 mg/L of COD (5, 8). A study is currently (Sept. 1985 to June 1986) in progress at the Technical University of Nova Scotia to estimate the minimum concentration and form of PO_4 (ortho, condensed and organically-bound) to sustain anaerobiosis in fixed film reactors (8).

Boyle and Ham (9) chemically treated leachate with lime and sodium sulfide as precipitants, potassium permanganate and chlorine gas as oxidants and ferric chloride and alum as coagulants. Sodium sulfide was ineffective in improving quality. Extremely high doses of the two oxidants were required to bring about even modest COD reductions of 20%. However, good removals of multivalent cations, color and suspended solids can be achieved by lime, alum and ferric chloride although large amounts of sludge were produced due to such treatment.

Steiner, et. al. (10) evaluated the abilities of three types of lime (Table II) in raising the pH of leachate from 5.5 to 10.0.

Table II. Lime Doses to Increase Leachate pH
(Steiner, et. al., 1977)

Type of Lime	Required Dosage	
	lb/1000 gal of Leachate	Kg/cu. m of Leachate
High Magnesium Lime	125	15.0
High Calcium Quick Lime	52	6.2
High Calcium Hydrated Lime	50	6.0

The authors concluded that it was not economically practical to use the high magnesium lime because it requires a larger dose. Although the remaining two limes were required at nearly equal doses, the high calcium hydrated lime was chosen because the slaking characteristics of the quick lime were expected to cause problems with pumping of the slurry. Lime, at a pH of 9.25 removed 49% of COD, 78% of Cu, 98% of Fe, 66% of Pb, 51% of Ni, 72% of SS, 35% of Cd, 47% of Cr and 94% of Zn.

Keenan and co-workers (11) used lime slurry to precipitate inorganic chemicals from leachate by coagulation followed by sedimentation in an up flow solids contact reactor-clarifier. They concluded that removal of Zn and Fe increased as pH increased (and maximum precipitation occurred at pH 10.4 to 12.2). The optimum pH ranges for removals of Cr, Cu and Hg were, respectively, 10.2 to 11.2,

10 to 11 and 9.8 to 10.8. The plots of effluent concentrations (X axis) and pH for these three elements were characterized by "U-Shaped" curves. When Ni removal tests were conducted at leachate temperatures ranging from about 3°C to 27°C, it was observed that the removal was lowered when temperature was below 16°C.

Coprecipitation of priority trace elements (As, Se, Cd, Cr, Cu, Pb, Ag and Zn) with amorphous iron oxyhydroxide was recommended by Merrill, et. al., (12) as a treatment procedure for liquid wastes from coal-fired electric power plants. To demonstrate the technology, a pilot-scale plant (consisting of rapid mixing, flocculation and sedimentation) was monitored by Roxboro Power Station (Carolina Power & Light Company). Iron Oxyhydroxide is formed when a ferric salt (ferric chloride) is added to the waste and the trace elements (As and Se were the elements of concern in this study) are adsorbed onto and trapped within the precipitate. Iron dose and pH were the two of the important governing factors in precipitating As and Se. While the Se removal was optimized at pH 6.2 and below, As precipitation was concluded to be less dependent on pH; good As removal was obtained at pH values up to 6.5.

A previous study at TUNS consisted of laboratory model anaerobic (upflow) fixed film reactors (4) and a subsequent study consisted of a pilot-scale investigation (13) of treatment of leachate from the same landfill site. The laboratory study used randomly packed "tellerete" loops, the shapes of which are described by the manufacturer (Pennwalt Corporation) as 'filamentous toroidal helix'. The pilot-scale investigation employed an anaerobic downflow stationary fixed film reactor designed by National Research Council (NRC) of Canada. The medium was a 100% polyester geotextile fabric supported on stainless steel stretchers suspended within the reactor. The results of these studies are discussed elsewhere in this paper.

OBJECTIVES

The objectives of the model study were: (i) to estimate the abilities of lime, alum, ferric chloride, potassium hydroxide, and sodium carbonate in increasing the pH and alkalinity of the low-pH leachate and precipitating the potentially toxic zinc, and (ii) to assess the biological treatability of leachate by an anaerobic fixed film reactor (AFFR), subsequent to pretreatment (physical-chemical treatment).

Although Zn was the only chemical that was of concern, data were collected on removals of 30 parameters by physical-chemical treatment. Out of these 30 parameters, primary focus was on nine elements (Al, Co, Cr, Fe, Mn, Ni, Sb, V and Zn). However, the primary focus of the biological treatment unit (AFFR) was on organics (COD, TOC and BOD), volatile acids, solids and nitrogen.

EXPERIMENTAL PROCEDURE

Physical-Chemical Treatment: The standard, six-position jar test apparatus with a multiple stirrer was used to coagulate, flocculate, and settle the leachate. Desired chemical doses were added to six beakers (of one L capacity) containing 800 mL of raw leachate. The contents were rapidly mixed for one minute at 90 to 100 rpm, gently mixed (flocculation) for 15 minutes at 30 to 40 rpm, and allowed to settle for 30 minutes. Temperature and pH were measured during flocculation. At the completion of the settling step, the volume of settled sludge was recorded and the supernatant in each beaker was analyzed for Zn, Fe, pH, and other parameters.

Biological Treatment: A laboratory-model, upflow, anaerobic fixed film reactor (AFFR), as shown in Figure 1, was used. It had a total height of 3 ft. (0.91 m), a liquid column height of 2.638 ft. (0.80 m), the effective medium height of 2.0 ft. (0.61 m), and a square cross-sectional area of 1.0 ft.^{2} (0.093 m^2). Additional dimensions are given in Table III. The medium consisted of two modular blocks constructed from cross-stacked corrugated plastic sheets (Biodek Models 19060 and 12060, The Munters Corp., Fort Myers, Florida). The lower block (Model 19060), a cube of 1-ft. (30.48-cm) side, had the openings with nominal size of about 0.7 in (17.8 mm) and the upper block (Model 12060) had the smaller openings, with nominal size of about 0.4 in (10.2 cm). The reactor had nine ports on the sides. The unit was kept inside a walk-in incubator maintained at a temperature of 32 $\pm$ 2°C

Procedure: The feed was manually added, twice a day, which entered the unit at port 6 and displaced the effluent at port 9. The contents of the AFFR were kept mixed with a peristaltic pump which created a suction at port 2 and delivered the liquid back into the reactor at port 5. The pump was kept on for about 15 hours/day and turned off at nights due to a concern about overheating problem which could

166 Biotechnology for Degradation of Toxic Chemicals

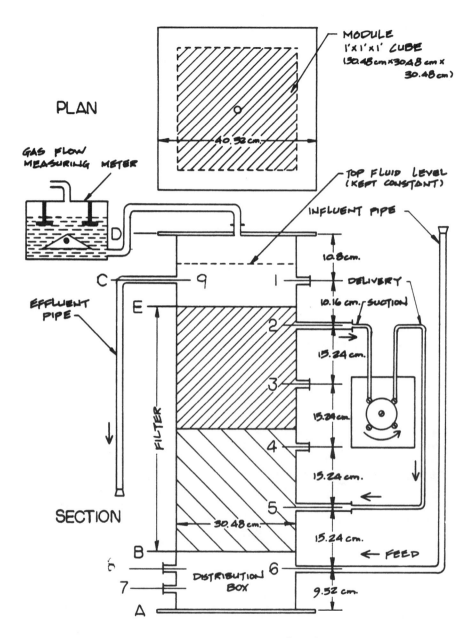

Figure 1. Anaerobic Fixed Film Reactor

Table III. The Dimensions and the Other Physical Features of the Model Reactor (AFFR) Used in this Study

Description	Designation in Figure 1	Height (cm)	Height (inches)	Volume (L)
Distribution box	AB	16.2	6.4	15.05
Reactor medium	BE	61.0	24.0	56.67
Liquid column above the medium	EC	3.2	1.3	2.97
Total liquid column	AC	80.4	31.7	74.69
Space for gas storage	CD	10.8	4.3	10.03
The total unit	AD	91.2	36.0	84.72
Solid volume of the medium	--	--	--	0.32
Void Volume within BE	--	--	--	56.35

Note: Porosity = $\dfrac{\text{Void Volume in BE}}{\text{Total Volume of BE}} = \dfrac{56.35}{56.67} = 0.994$

have lead to a potential fire hazard. Ports 2, 3, 4, 7 and 8 were used to sample the reactor contents for estimating the biomass. A gas flow meter was used to estimate the total volume of biogas produced during the study period. An acclimatized microbial culture from the previous study (4) was used to initiate the reaction. An incubatory phase, consisting of about nine weeks, enabled the microbial system to attain a state of maturation.

During the study period, the reactor was subject to a volumetric organic load of 1.6 $kg \cdot d^{-1} m^{-3}$ by manually adding feed leachate twice a day, once between 9 and 10 a.m. and once between 4 and 5 p.m. The overflow at port 9 (effluent) was collected, and its volume, temperature and pH were recorded immediately. Seven-day composites of the effluent were stored in a refrigerator (at $4 \pm 1°C$) and subject to weekly tests for 13 parameters: COD, BOD, TOC, TS, SS, VSS, TVA, TKN, NH_4-N, $NO_3 + NO_2$, TP, OP and Zn. Less frequent tests were conducted for twenty other parameters, shown in Table I. The total experimental period lasted for about 17 weeks, consisting of nine weeks of incubatory phase and eight weeks of study period.

RESULTS

Tables IV to IX summarize the results of the jar tests and Tables X and XI describe the quality of lime used in the present study. The performance of the AFFR is summarized in Tables XII to XV and Figures 2 to 7. (Effluent qualities were not monitored during the first five weeks of the incubatory phase.)

DISCUSSION OF RESULTS

<u>Physical-Chemical Treatment</u>: The results of jar tests are discussed in the following paragraphs.

Effects of Alum: Lime slurry was added to all beakers during the first three jar tests (Figures 8 to 10 and Tables IV and V) to increase the leachate pH from about 5.5 to a range between 6.5 and 9.5. Alum (10 to 80 mg/L) was added to about half of the beakers in attempting to enhance coagulation and precipitation. The rate of settling (of interface between the supernatant and sludge) was slower in beakers with alum than in the beakers without it. Alum addition, as shown in Table IV, resulted in increased sludge volume and in higher concentrations of total solids (TS) and total dissolved solids (TDS) in the settled sludge. Addition of alum, therefore, could lead to higher cost of sludge disposal. Alum had very little effect on precipitating zinc. In some beakers, zinc removal was slightly increased due to alum, in some it decreased insignificantly and in others there was no change. pH was the single most important factor that governed the precipitation of Al, Cr, Fe, Mn, V and Zn. Typical relationships for selected few metals are shown in Figures 8 to 10. An alum dose of 40 mg/L (Table V) decreased the flocculation pH as well as the percent removals of some of these metals. Neither the organics nor the nitrogen were removed by lime and alum. Total phosphate, however, was precipitated as pH increased from 5.5 to 7.75 (Table V). Use of alum cannot be justified because it increased the sludge volume, could increase the cost of disposal of sludge and was not effective in substantially increasing the precipitation of metals (Figures 8 to 10), unless the dose was increased to 80 mg/L (Figure 10).

Lime Slurry Vs KOH: At the time of fourth and fifth jar tests, KOH was added to the first three beakers and lime slurry to the others. Within the pH range of 7 to 9, lime produced more sludge than KOH (Figure 11). In both cases, sludge volume increased as flocculation pH increased. At a

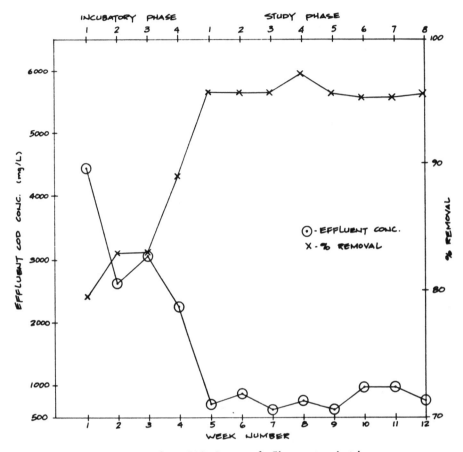

Figure 2. COD Removal Characteristics

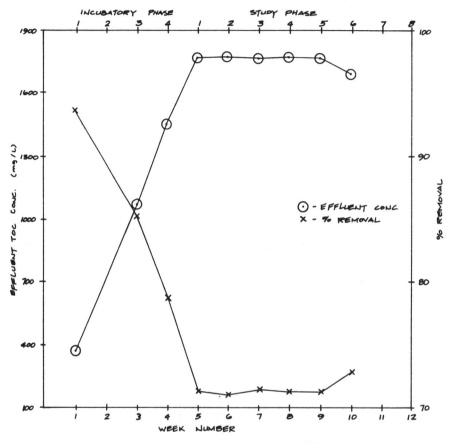

Figure 3. TOC Removal by AFFR

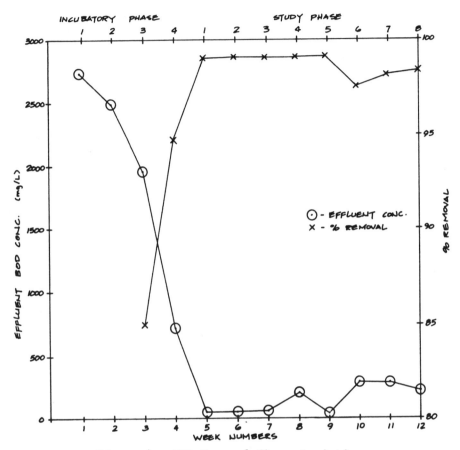

Figure 4. BOD Removal Characteristics

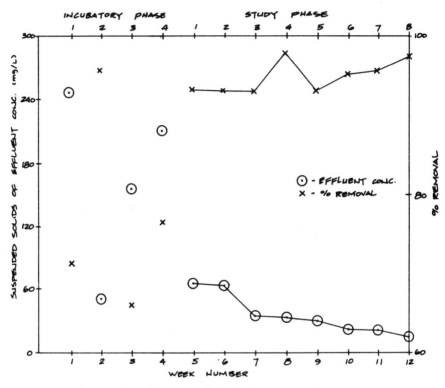

Figure 5. Suspended Solids Removal by AFFR

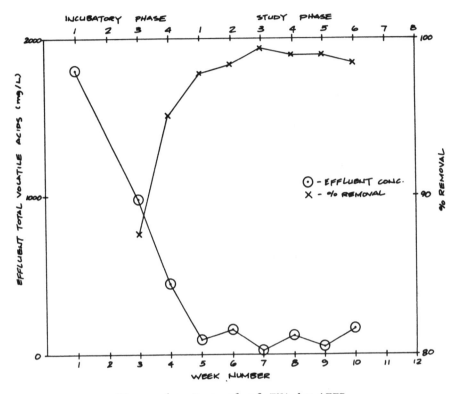

Figure 6. Removal of TVA by AFFR

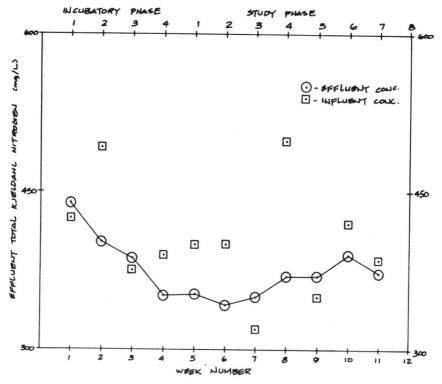

Figure 7. Influent and Effluent TKN

Table IV. Results of Jar Test No. 1

Parameter	Jar 1	Jar 2	Jar 3	Jar 4	Jar 5	Jar 6
Chemicals added:						
Lime slurry (ml)	12	18	18	25	25	25
Alum (mg/L)	0	0	40	0	60	80
Temperature (°C) during flocculation	23	23	23	23	23	23
Flocculation pH	7.2	7.8	--	9.25	8.75	8.65
Removal (%)						
Al	92	97	97	>98.1	98.1	95.8
Cr	33	42	47	75	58	50
Fe	24	58	72	99	95	94
Mn	4	22	40	93	74	69
V	36	71	69	96	>96	93
Zn	65	96.7	99.4	99.8	99.6	99.3
Volume of settled sludge (ml)*		157	195	205	220	225
Total Solids in sludge (%)		--	--	2.09	2.39	2.53
TDS in sludge (%)		--	--	0.96	1.27	1.44

*The initial leachate volume was 800 ml in each case. Therefore dividing these values by 8 will give sludge volume as % of initial leachate volume.

Table V. Results of Jar Test No. 2

Parameters	Raw Leachate	Jar 1	Jar 2	Jar 3	Jar 4	Jar 5	Jar 6
Chemical Doses							
Lime slurry (ml)	--	15	20	25	15	20	25
Alum dose (mg/L)	--	0	0	0	40	40	40
Flocculation pH	--	7.25	7.6	7.85	7.2	7.45	7.75
A. General Parameters		concentrations in supernatants (mg/L)					
Alkalinity (CaCO$_3$)	3950	--	--	--	--	--	--
pH	5.6	7.1	7.4	7.7	7	7.35	7.65
SS	765	660	650	620	680	720	290
TDS	18700	19750	20280	20500	19780	19950	20940
B. Metals							
Al	5.8	<0.5	<0.5	<0.5	.65	<0.5	<0.5
Sb	.13	<0.5	<1.0	<0.5	<0.5	<0.5	<0.5
Ba	.23	.34	.38	.42	.34	.34	.42
Be	<0.005	<0.05	<0.05	<0.05	<0.05	<0.05	<0.05
Cd	<0.002	<0.002	<0.002	<0.002	<0.002	<0.002	<0.002
Cr	.19	.2	.2	.13	.22	.2	.15
Co	1.66	<0.1	<0.1	<0.1	<0.1	<0.1	<0.1
Cu	.02	<0.1	<0.1	<0.1	<0.1	<0.1	<0.1
Fe	1038	823	506	240	827	459	227
Pb	.01	<.002	<.002	<.002	<.002	<.002	<.002
Mn	73	65	60	54	65	60	55
Ni	.4	.51	.43	.35	.51	.45	.41
Se	<0.1	<1.0	<1.0	<1.0	<1.0	<1.0	<1.0
Sn	<0.03	<0.3	<0.3	<0.3	<0.3	<0.3	<0.3
V	.31	.23	.12	<0.1	.23	.15	<0.1
Zn	95	35	7.4	1.3	33	6.7	1.3
C. Nonmetals							
As	<0.05	<.025	<.025	<.025	<.025	<.025	<.025
B	6.7	7	6.5	6	6.8	6.3	5.9
SO$_4$	974	940	906	885	955	910	824
D. Nutrients							
NH$_4$-N	411	406	406	403	407	400	370
(NO$_3$+NO$_2$)-N	<0.5	<0.5	<0.5	<0.5	<0.5	<0.5	<0.5
PO$_4$(ortho)	.6	<0.2	<0.2	<0.2	<0.2	<0.2	<0.2
PO$_4$(total)	20	5.3	3.7	3.4	5.3	7.2	3.7
E. Organics							
COD	24500	25800	26700	27000	24300	24600	26100
TKN	508	490	480	485	480	485	496
TOC	8800	8600	8600	8300	8600	8600	8600
TVA	11400	10800	11400	11250	11400	12200	10600
TVS	7928	--	--	--	--	--	--
VSS	350	--	--	--	--	--	--

Table VI. Results of Jar Test No. 6

Parameters	Raw Leachate	Jar 1	Jar 2	Jar 3	Jar 4	Jar 5	Jar 6
Chemical Doses							
Ferric Chloride (mg/L)	--	0	0	20	20	0	40
Lime slurry (ml)	--	15	12	15	12	20	20
Flocculation pH	--	8.05	7.7	8.1	7.4	8.4	8.1
A. General Parameters		supernatant concentration (mg/L)					
Alkalinity ($CaCO_3$)	3950	5370	5280	5340	5100	--	--
pH	5.6	8	7.5	7.95	7.4	8.3	8.0
SS	765	608	537	396	508	--	--
TDS	18700	20554	19968	21384	19414	--	--
B. Metals							
Al	5.8	0.5	0.5	0.5	.54	0.5	0.5
Sb	.13	0.5	0.5	0.3	0.5	--	--
Ba	.23	.38	.33	.38	.33	--	--
Be	<0.005	<0.05	<0.05	<0.05	<0.05	--	--
Cd	<0.002	<0.02	<0.02	<0.02	<0.02	--	--
Cr	1.9	.15	.15	.11	.16	0.11	0.10
Co	1.66	<0.1	<0.1	<0.1	.17	--	--
Cu	.02	<0.1	<0.1	<0.1	<0.1	--	--
Fe	1038	201	513	198	529	--	--
Pb	.01	<.02	<.02	<.02	<.02	--	--
Mn	73	49	63	49	62	35	34
Ni	.4	.45	.53	.46	.6	--	--
Se	<0.1	<0.1	<0.1	<0.1	<0.1	--	--
Sn	<0.03	<0.3	<0.3	<0.3	<0.3	--	--
V	.31	<0.1	<0.1	<0.1	<0.1	--	--
Zn	95	2.6	7.6	2.7	8.4	1.9	1.7
C. Nonmetals							
As	<0.05	<.05	<.05	<.05	<.05	--	--
B	6.7	6.5	8	6.7	7.1	--	--
SO_4	974	830	832	818	828	--	--
D. Nutrients							
NH_4-N	411	--	--	--	--	--	--
(NO_3+NO_2)-N	<0.5	--	--	--	--	--	--
PO_4(ortho)	.6	<0.2	.3	<0.2	1.1	--	--
PO_4(total)	20	--	--	--	--	--	--
E. Organics							
COD	24500	25500	24400	24400	20100	--	--
TKN	508	--	--	--	--	--	--
TOC	8800	9600	9200	9200	9400	--	--
TVA	11400	--	--	--	--	--	--
TVS	7928	8930	8506	9420	8446	--	--
VSS	350	--	--	--	--	--	--

Table VII. Results of Jar Test No. 7

Parameter	Raw Leachate	Jar 1	Jar 2	Jar 3	Jar 4	Jar 5	Jar 6
Chemical Doses							
Ferric Chloride (mg/L)	--	0	0	0	200	200	200
Lime Slurry (ml)	--	15	20	25	20	25	25
Flocculation pH	5.5	7.65	8.25	8.80	7.75	8.35	7.85
Temperature °C	--	11	11	11	11.5	12	12
Metals	Raw Leachate Concentration	concentrations in supernatants (mg/L)					
Al	4.1	0.15	0.11	0.07	0.12	0.09	0.10
Sb	0.06	<0.05	<0.05	<0.05	<0.05	<0.05	<0.05
Cr	0.14	0.08	0.08	0.07	0.09	0.07	0.08
Cu	<0.01	<0.01	<0.01	<0.01	<0.01	<0.01	<0.01
Fe	990.0	450.0	168.0	44.0	233.0	44.0	147.0
Mn	58.1	48.9	40.0	23.2	41.8	28.2	36.2
V	0.30	0.11	0.04	0.01	0.06	<0.01	0.04
Zn	47.0	4.80	1.1	0.59	1.7	0.52	1.2
Settled sludge volume (ml)	--	116	158	163	163	205	193
TS (%) in sludge	--	--	2.40	2.78	2.42	2.45	2.45
TDS (%) in sludge	--	--	1.18	1.28	1.31	1.37	1.28

Table VIII. Results of Jar Test No. 8

Parameter	Raw Leachate	Jar 1	Jar 2	Jar 3	Jar 4	Jar 5	Jar 6
Chemical Doses							
Ferric Chloride (mg/L)	--	400	400	150	400	400	400
Lime slurry (ml)	--	8	12	18	8	14	18
Flocculation pH	5.5	7.55	8.9	10.75	7.55	9.3	10.5
Temperature °C	10	11	11	11	11	11	11
Metals	Raw Leachate Concentration	concentrations in supernatants (mg/L)					
Al	3.1	0.16	0.08	0.10	0.15	0.07	0.06
Sb	<0.05	<0.05	<0.05	<0.05	<0.05	<0.05	<0.05
Cr	0.13	0.08	0.08	0.03	0.09	0.07	0.03
Co	<0.01	<0.01	<0.01	<0.01	<0.01	<0.01	<0.01
Fe	734.0	216.0	3.5	2.0	217.0	2.1	0.70
Mn	58.1	38.0	5.0	0.20	38.0	2.1	0.12
V	0.25	0.07	<0.01	<0.01	0.07	<0.01	<0.01
Zn	49.0	2.2	0.24	0.14	2.6	0.20	0.11
Settled sludge volume (ml)	--	179	270	305	163	226	272

Table IX. Results of Jar Test No. 9

	Jar 1	Jar 2	Jar 3	Jar 4	Jar 5	Jar 6
Chemical(s) added						
Lime slurry (ml)	0	0	0	20	0	40
Na_2CO_3 (mg/L)	2000	3000	4000	3000	3000	3000
$FeCL_3$ (mg/L)	0	0	0	0	200	0
Flocculation pH	6.2	6.6	7.1	11.1	6.5	12.0
Temperature (°C)	24	24	24	24	24	24
Removal (%)						
Al	70	91	94	99	92	97
Cr	8	17	25	92	25	99
Fe	24	--	--	>99	57	>99
Mn	7	>33	--	>99	>36	>99
V	21	36	50	93	43	93
Zn	24	66	74	>99	--	99
sludge volume (ml per L of leachate)	46	61	74	225	80	325
TS in sludge (mg/L)	3.55	3.64	3.36	4.32	3.55	4.83
TDS in sludge (mg/L)	3.29	2.33	2.13	2.87	2.69	3.52

Table X. Composition of Lime Slurry Used in this Study

Parameter	Range
slurry pH	12.1 to 12.7
pH of Supernatant (after 30-minute settling)	12.0 to 12.5
Total Solids	40,000 to 70,000 mg/L
Suspended Solids	20,000 to 60,000 mg/L

Table XI. Chemical Composition of Dry Slurry (by weight)

Parameter	Concentration
Calcium	50.6%
Carbonate	4.1%
Hydroxyl	35.4%
Aluminum	6470 - 7020 ppm
Silica	6000 - 19,766 ppm
Iron	802 - 1,210 ppm
Magnesium	363 - 613 ppm
Manganese	13 - 97 ppm
Chromium	12 - 46 ppm
Lead	10 - 49 ppm
Copper	13 - 25 ppm
Zinc	5 - 10 ppm
Cadmium	0.05 - 5 ppm

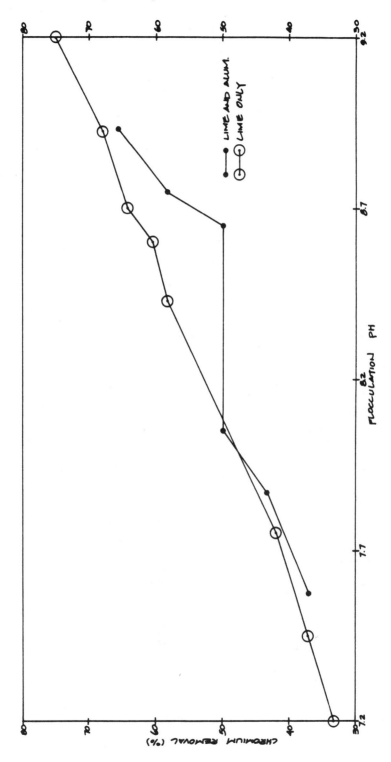

Figure 8. Effects of pH and Alum on Chromium Removal (Jar Tests 1 to 3)

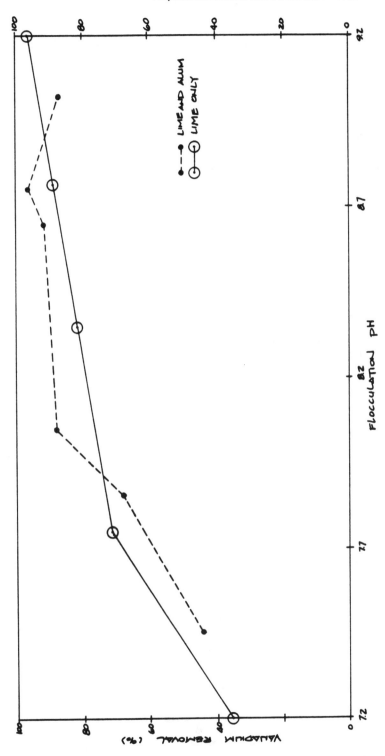

Figure 9. Effects of pH and Alum on Vanadium Removal (Jar Tests 1 to 3)

182 Biotechnology for Degradation of Toxic Chemicals

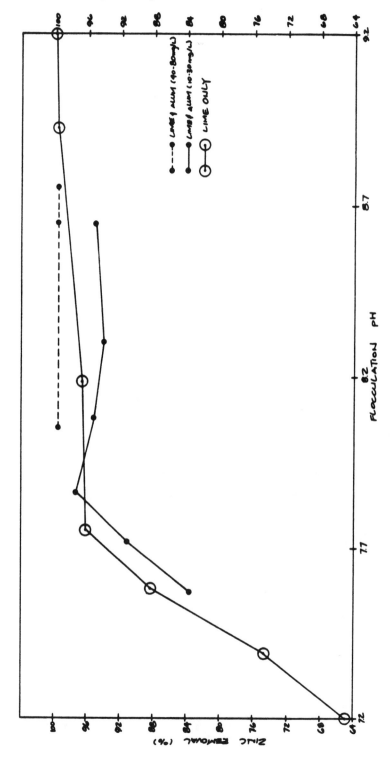

Figure 10. Effects of pH and Alum on Zinc Removal (Jar Tests 1 to 3)

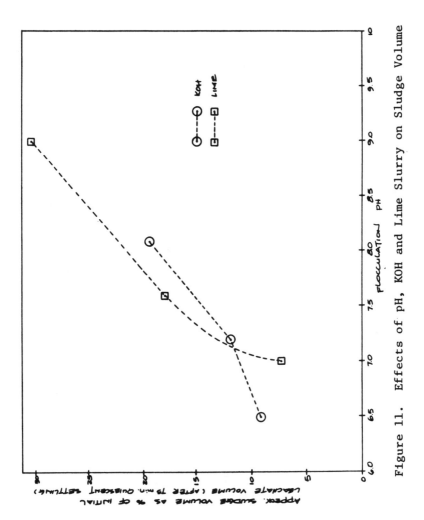

Figure 11. Effects of pH, KOH and Lime Slurry on Sludge Volume

given pH, lime slurry was slightly more efficient than KOH in precipitating the metals (Figures 12 and 13). (The supernatants were filtered using a membrane filter during these two jar tests before analyzing for Fe and Zn. However, for the remaining jar tests the supernatants were tested directly, without filtration.)

Effects of $FeCl_3$: During the sixth, seventh and eighth jar tests, lime slurry was added to leachate to bring the flocculation pH to a range between 7 and 9.6 in different beakers. Ferric chloride doses of 20 to 400 mg/L were added to half of the beakers, but not to the others (Tables VI and VII). A low dose range of 20 to 40 mg/L of $FeCl_3$ had no significant effect (during test 6) on removals of the metals (Table VI). Therefore the dose was increased to 200 mg/L for jar test 7. At this dose, ferric chloride (Table VII) enhanced the removals of Al, Fe, Mn, and V, but the removals of Cr and Zn were not improved. $FeCl_3$ increased the sludge volume as well as the concentrations of TS and TDS in the settled sludge.

For the eighth jar test (Table VIII) a $FeCl_3$ dose of 400 mg/L was added to all beakers and different lime slurry volumes were added to maintain a pH range of 7.5 to 10.7. The results showed that as pH increased, the percent removals of all metals increased; however, the increase was not substantial above pH 9.3 for Al, 7.5 for Fe and 8.9 for Mn, V and Zn. Removal of Cr increased gradually as the flocculation pH increased from 7.5 to 10.7.

Effects of Na_2CO_3: An uneconomically high dose of 4000 mg/L of sodium carbonate was required to increase the leachate pH to about 7.1 (Table IX). During jar test 9, Na_2CO_3 alone was added to four jars and Na_2CO_3 and lime were added to two jars (Table IX). The results of this test indicate that Na_2CO_3 alone is not effective in removing the metals. Moreover, the results confirm those of the previous tests, namely, pH at the time of flocculation (and settling) has an overwhelming influence on the rate of removal of the metals as well as the quantity and quality of sludge.

Biological Treatment: The data presented in Figures 2 to 7 indicate that a near-steady-state condition was reached during the last four or five weeks. At the time of such a steady state, the efficiencies of treatment (as shown in Table XII) were about 96% of COD removal, 97% TOC removal, 98% BOD removal, 90% VSS removal, 95% SS removal and 99% TVA removal. However, TKN reduction was very low at 9%. NH_4-N concentration increased by 5%.

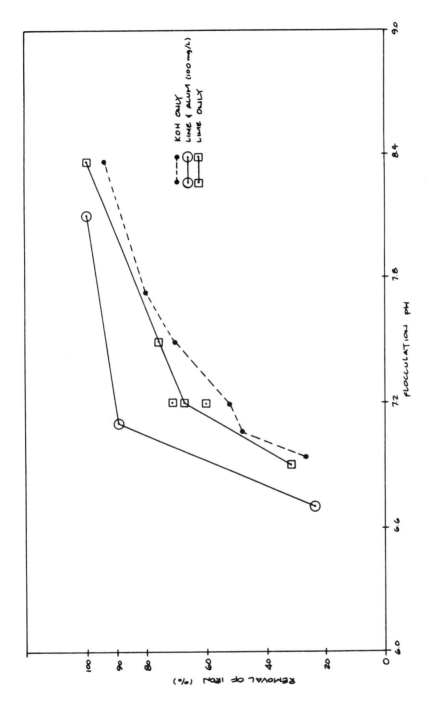

Figure 12. Precipitation of Iron by Lime Slurry and KOH

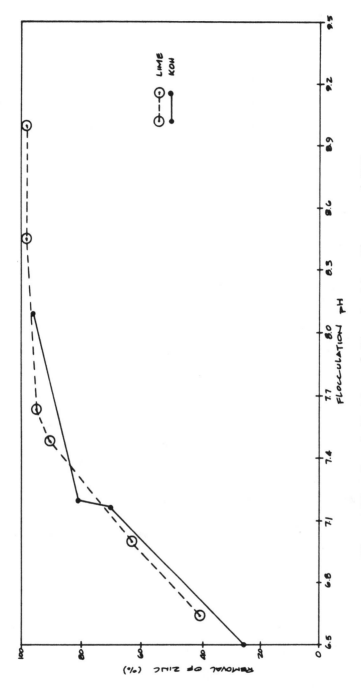

Figure 13. Precipitation of Zinc by Lime Slurry and KOH

Table XII. Performance Summary of AFFR

Parameter	Influent (feed) (mg/L)	Effluent (mg/L)	% Removed
COD	21,800	780	96
TOC	7,680	200	97
BOD	14,444	186	98
SS	756	36	95
VSS	237	24	90
TVA	9,780	106	99
TKN	349	319	9
Amm. N	339	357	(5)

Average Values shown above.

Biomass Estimation: On the last day of the experiment, samples were collected at ports 1, 3, 4, 6, and 7 when the recirculating pump was mixing the reactor contents. The concentrations of VSS, as shown in Table XIII, indicate the effectiveness of

Table XIII. Concentration of VSS (biomass) in the liquid phase of the reactor

Sampling port number (see Figure)	Concentration of VSS (mg/L)	
	During the pumping cycle	After one hour of settling (no pumping)
1	643	--
2	--	108
3	630	85
4	656	48
5	--	328
6	753	1900
7	824	3680

mixing. The concentrations of the samples taken from ports 1, 3 and 4 were used to estimate the biomass in the liquid phase of reactor BC, held in the interspaces of, and around, the medium (Table XIV). However, the concentrations of the

Table XIV. Summary of data of biomass held in the liquid phase.

Details	At the time of pumping	After one hour of settling (no pumping)
Biomass (g of VSS) in the liquid phase of reactor, BC	37.04	7.08
Biomass (g of VSS) in the distribution box, AB (this space had no solid medium)	11.83	41.85
Total biomass (g of VSS) in the liquid phase of the unit, AC	48.87	48.93

the samples of ports 6 and 7 were used to estimate the biomass in the distribution box AB. Later on, the recirculating pump was turned off and the samples were taken from ports 2 to 7, after one hour of settling. The data (Table XIII) indicated that a pronounced quality gradient existed in the absence of mixing. Subsequently, the remaining liquid was drained out and distilled water was added to the reactor, and the pump was turned on to slough off and flush out the biomass sticking to the solid medium. Subsequent to 10 hours of mixing, the wash water (the first wash, Table XV) was sampled to estimate the biomass which was originally attached to the medium. The process was repeated two more times (second and third wash) to slough off the remaining biofilm.

While the biomass in the liquid phase (of the unit from A to C) was estimated to be 48.87 g of VSS, the biomass sticking to the solid surface of the medium (biofilm) was calculated to be 14.58 g of VSS. In other words, 23% of the total biomass was in the biofilm, 58% in the liquid phase held in the interspaces of, and around, the medium, and 18% in the liquid phase within the distribution box (when the pump was operating, as shown in Tables XIII and XIV.

Table XV. Data of biomass sticking to the solid medium
(held as a part of the biofilm)

Biomass sloughed off of the solid medium (at the end of the test)	g of VSS
After the first wash	8.51
" " second "	5.92
" " third "	0.15
Total biomass of the biofilm	14.58

Substrate (removal) utilization rate: Estimation of biomass within a reactor is an essential prerequisite to calculate the substrate utilization rate (U), expressed as g of COD utilized per g of biomass per day. While the literature (16) points out the importance of U as a very useful parameter, very little information is available on a precise way to estimate the value of biomass in a given reactor. One uncertainty posed is whether the VSS in the distribution box should be included or not. Moreover, should the volume of the box be included in calculating the reactor volume? In this study, the box volume, as well as the biomass within it, has been included. Another uncertainty lies in the fact that, in the middle of a study, biomass growing within the biofilm cannot be estimated. Therefore, in this study, the biomass was determined after terminating the experiment. The biomass sticking to the solid medium (as a biofilm) had to be washed out three times to slough off all the solids. A need for a standardized procedure to estimate the biomass is recognized.

Comparison With Previous Studies: Table XVI compares the performance of the reactor used in this study with those of the previous studies.

The summaries of the three types of media are given for information purpose only. A precise comparison is not possible because the AFFR units were of different depths and shapes, they were subject to six different organic loads and were operated at slightly varying temperature ranges (because the laboratory walk-in incubator had a minor problem in maintaining precise temperature control). Moreover, the initial seed used in the pilot scale study was very much different from that used in the laboratory-model studies.

Table XVI. A Comparison of the Performances of the Three Types of Reactor Media

Description	The Present Study (Lab-Model)	The Previous Study* (Lab-Model)		Pilot-Scale Study** by others
Column (1)	Column (2)	Column (3)		Column (4)
Study period	May/August, 1984	Oct. 83 to Apr. 84		Jan/July, 1984
Direction of leachate flow	upflow	upflow		downflow
The medium	modular blocks with cross-stacked corrugated plastic sheets	randomly packed TORO1 DAL loops (rings)		100% polyester geotextile fabric
Reactor (liquid) temperature (°C) range	31-35	31-33		--
Average temperature (°C)	33	32		35
Number of reactors	1	2 in parallel		1
		AFFR1	AFFR2	
Reactor volume (L)	56.67	10.5	17.0	640.0
Solid volume of the medium (L)	0.32	0.96	1.53	70 to 80
Void Volume (L)	56.35	9.54	15.47	560 to 570
Medium height (cm)	61.0	68.6	108.0	--

Note - * Reference (4)
     ** Reference (13)

Table XVI (cont'd)

Column (1)	Column (2)	Column (3)		Column (4)		
Initial seed	digested municipal sludge and culture acclimatized to leachate	digested municipal sludge and culture acclimatized to leachate		microbial culture from a fish-processing waste treatment study		
		AFFR1	AFFR2	Phase I	Phase II	Phase III
Organic load (g of COD. $d^{-1} \cdot L^{-1}$)	1.60	2.0	1.6	1.08	1.80	3.61
Duration of the study (days) excluding the incubation and transition phases	56	56	56	30	30	--
Incubation period (d)	72	--	--	--	--	--
% COD (total) removed	96	97	97	95	95	93
% COD (soluble) removed	--	--	--	95	96	94
% TVA removed	99	99	--	98	98	97
% TOC removed	97	98	98	95	96	90
% change in ammonia	+5	+0.5	-2.0	+152	+29	+10
% change in TKN	-9	-16.0	-17.4	+116	+13	-7
% change in Humic acids	--	-9.1	--	+19	+11	+9
% Zn removed	--	87	86.5	86	77	93
% Fe removed	--	92	94	96	91	97

CONCLUSIONS

The following conclusions can be made, based on this laboratory-model study:
1. The single most important factor that overwhelmingly governs the precipitation of metals such as Al, Cr, Fe, Mn, Ni, V and Zn from the leachate is pH.
2. Low doses of alum and $FeCl_3$ (up to 40 mg/L in each case) have very little effect on metals removal.
3. Though alum at 80 mg/L and $FeCl_3$ at 200 mg/L enhanced the precipitation of the metals, the expected higher costs associated with the chemicals and the disposal of excess sludge produced prohibit the use of the chemicals.
4. The volume of settled sludge increased as the flocculation pH increased.
5. At a given pH, lime slurry was slightly more efficient than KOH in precipitating the metals.
6. Sodium carbonate, even at a high dose of 4000 mg/L, was not effective in removing the metals.
7. Subsequent to physical-chemical treatment and phosphate supplementation, the leachate was successfully treated biologically.
8. The upflow anaerobic fixed film reactor (AFFR) with modular blocks, at an organic load of 1.6 kg/day/m^3 and at $32 \pm 2°C$, reduced the COD of the pre-treated leachate from 21,800 to 780 mg/L, TOC from 7680 to 200 mg/L, VSS from 237 to 24 mg/L and SS from 756 to 36 mg/L.
9. When the biomass within the reactor was estimated on the last day of the experiment, it was concluded that 77% of the biomass (VSS) was in the liquid phase of the reactor held in the interspaces of the medium while the balance was on the biofilm of the solid medium.
10. The substrate removal rate was estimated to be 1.77 g of COD per day per g of VSS in the liquid phase of the reactor. When the VSS sticking to the solid medium were included in the biomass estimate, however, the rate decreased to 1.37 g of COD per day per g of "total" VSS. A need for a standardized procedure to estimate the biomass is recognized because depending on the procedure used to estimate the biomass, the value of the substrate removal rate would change. For instance, it is very difficult to evaluate the biomass sticking to the surface of an active reactor. It can be evaluated only at the termination of an experiment.
11. A precise value of the substrate removal rate can be calculated only when a precise definition of the biomass

can be standardized. For instance, it is not well established in the literature if the VSS within the distribution box should be considered a part of the biomass.

ACKNOWLEDGEMENTS

This study was supported, in part, by an Operating Grant from the Natural Sciences and Engineering Research Council (NSERC) of Canada and, in part, by a contract with H. J. Porter and Associates Ltd. (currently Porter-Dillon Ltd.). The encouragement and support of Metropolitan Authority of Halifax, Dartmouth and the County of Halifax who operate the landfill site and granted the contract to H. J. Porter and Associates Ltd. is gratefully acknowledged. The authors appreciate the assistance of Theresa Innis who typed the manuscript and Alvin Yee who drafted the figures.

REFERENCES

1. Porter, H. J. and Associates Ltd., "Highway 101 Regional Landfill Environmental Analysis - Phase I," A Report Submitted to Metropolitan Authority of Halifax, Dartmouth and the County of Halifax, Feb., 1982, p. 2-1.
2. Porter-Dillon Ltd., "Highway 101 Landfill Site Leachate Treatability Study," A Report Submitted to Metropolitan Authority of Halifax, Dartmouth and the County of Halifax, July, 1984, pp. 1.1-5.3.
3. Porter, H. J. and Associates Ltd., "Highway 101 Regional Landfill Environmental Analysis - Phase II," A Report Submitted to Metropolitan Authority and the County of Halifax, Sept., 1982, pp. 1.1-6.2.
4. Thirumurthi, D., Austin, T. P., Ramalingaiah, R., and Khakhria, S., "Anaerobic/Aerobic Treatment of Municipal Landfill Leachate," *Water Pollution Research Journal of Canada*, Vol. 21, No. 1, Jan., 1986, pp. 8-20.
5. Chian, E. S. K., and F. B. DeWalle, "Treatment of High Strength Acidic Wastewater With A Completely Mixed Anaerobic Filter," *Water Research*, Vol. 11, 1977, pp. 295-304.
6. DeWalle, F. B., E. S. K. Chian and J. Brush, " Heavy Metal Removal With Completely Mixed Anaerobic Filter," *Journal WPCF*, Vol. 51, 1979, pp. 22-36.
7. Kirsch, E. J. and Sykes, R. M., "Anaerobic Digestion in Biological Waste Treatment," *Progress in Industrial Microbiology*, Vol. 9, 1971, pp. 155-237.
8. Groskopf, G., "The Minimum Concentration and Form of Phosphate for Anaerobic Film Treatment of Leachate," Masters Thesis in Progress, Technical University of Nova Scotia, Halifax, Canada, 1986.
9. Boyle, W. C. and Ham, R. K., "Treatability of Leachate from Sanitary Landfills," Proceedings of 27th *Industrial Waste Conference*, Purdue University, Vol. 27, 1972, pp. 687-704.
10. Steiner, R. L., Keenan, J. D. and Fungaroli, A. A., "Demonstration of a Leachate Treatment Plant - Interim Report," USEPA Report No. 9/76-10/77, 1977, pp. 1-58.
11. Keenan, J. D., Steiner, R. L., and Fungaroli, A. A., "Chemical-Physical Leachate Treatment," *Journal of Environmental Engineering Div.*, ASCE, Vol. 109, No. 6, Dec., 1983, pp. 1371-1384.

12. Merrill, D. T., et. al., "Field Evaluation of Arsenic and Selenium Removal by Iron Coprecipitation," *Journal Water Pollution Control Federation*, Vol. 58, No. 1, Jan., 1986, pp. 18-26.
13. Wright, P. J., Austin, T. P., Kennedy, K. and Robson, D. R., "Utilization of an Anaerobic Reactor Pilot Plant to Assess Methane Gas Production and Treatability of a Landfill Leachate," Presented in June, 1985, at an International Conference on 'New Directions and Research in Wastewater Treatment and Residuals Management', held at the University of British Columbia, Vancouver, Canada.
14. Vanderborght, J. P. and Wollast, R., "Elimination of Micropollutants By $NaAlO_2$ Flocculation During Primary Treatment of Mixed Wastewater," *Water Science Technology*, Vol. 18, Antwerp, 1986, pp. 67-74.
15. Licskó, I. and Takács, I., "Heavy Metal Removal in the Presence of Colloid-Stabalizing Organic Material and Complexing Agents," *Water Science Technology*, Vol. 18, Antwerp, 1986, pp. 19-29.
16. Henze, M. and Harremoës, P., "Anaerobic Treatment of Wastewater in Fixed Film Reactors - A Literature Review," *Water Science Technology*, Vol. 15, Antwerp, 1983, pp. 1-87.

TREATMENT OF LEACHATE FROM A HAZARDOUS WASTE LANDFILL SITE USING A TWO-STAGE ANAEROBIC FILTER

Y.C. Wu

Department of Civil and Environmental
Engineering
New Jersey Institute of Technology
Newark, NJ

O.J. Hao

Department of Civil Engineering
University of Maryland
College Park, MD

K.C. Ou

Department of Environmental Resources
Pittsburgh, PA

INTRODUCTION

Solid waste land disposal sites can be sources of soil, surface water, and groundwater contamination from leachate generated by water percolating through the refuse (1-3). The types, quantity and strength of leachate depend on the solid waste characteristics, the age of the landfill, local weather, geology and hydrology conditions, site management, and the soundness of the landfill design (4-8). Leachate normally contains high levels of chemical oxygen demand (COD), biochemical oxygen demand (BOD), ammonia nitrogen (NH_3-N), dissolved solids and metals such as iron, manganese, calcium and zinc (9,10).

Proper landfill design and site management can significantly reduce the quantity and strength of leachate, but never completely eliminate it. Leachate is often pumped to lagoons for storage prior to further treatment. The treated leachate can then be discharged to a nearby municipal sewage system, or a receiving water body, or onto grass land in reclaimed areas of the landfill site (11).

Leachate treatment techniques can be roughly divided into four

lysimeter. In this paper, the use of two stage upflow anaerobic filters without recycle for the treatment of leachate from a hazardous waste landfill site is described. The effects of leachate concentrations, mass organic loadings, and the addition of metals on the performance of the anaerobic filters were evaluated.

The landfill site, which was 16 years old at the beginning of this study, contains approximately 60 acres. The average waste loading is approximately 200 tons/day. The records of the Pennsylvania Department of Environmental Resources (DER) show that municipal solid wastes and hazardous wastes have been disposed on this site. The hazardous wastes consist of tar/slug mixtures, coal tar decanter sludge, waste lime, ammonia still lime sludge, and oil separator sludge. Concentrations of some of the priority toxic pollutants in the raw leachate are listed in Table 1. Concentrations of these pollutants in a downstream creek are slightly higher than those from upstream samples, indicating possible leachate contamination. Consequently, the site has been classified as a hazardous waste site by the Pennsylvania DER.

MATERIALS AND METHODS

Leachate

Raw leachate was collected periodically from the solid waste landfill site in Forward Township, Allegheny County, Pennsylvania, and stored in a refrigerator. The typical composition of this raw leachate is shown in Table 2.

System Description

Two laboratory scale anaerobic submerged fixed-film reactors were constructed in series (Fig. 1). Each reactor was made of Plexiglass, and had dimensions of 30.5 cm X 30.5 cm X 61 cm. The reactors were individually packed with corrugated plastic media (Munters Biodek 19060 and B.F. Goodrich Koro-z). The plastic media had similar void ratios (about 96.5%) and specific surface areas per media volume (141 M^2/M^3). However, flow was redistributed horizontally in the Biodek and vertically in the Koro-z packed reactors.

The raw leachate solution was pumped continuously from a 40-L feed reservoir through the bottom of the first reactor by a Cole Parmer's UltraMasterflex tubing pump. The effluent from the top of the reactor was then circulated into the bottom of the second

categories: physio-chemical treatment, biological treatment, leachate recycle and combined treatment processes. During chemical treatment, lime, activated carbon, oxidants and coagulants are used for chemical precipitation (9,12-16), coagulation (12,16), adsorption (6,12-14,17), and oxidation (12-14). Chemical treatment is costly and generally produces large quantities of chemical sludge. Chemical precipitation and coagulation processes are inefficient for the removal of organic matter, despite the almost total removal of metals (18). Consequently, they are used in conjunction with biological processes (16), or as a polishing unit for treatment of effluent from biological processes (12).

Biological processes for treating leachate consist of anaerobic (11,19), aerobic (11,19-22), or combined anaerobic-aerobic processes (19). Boyle and Ham (19) reported that anaerobic treatment of raw leachate achieved COD removals of 90 to 96% at a loading of 0.43 - 2.2 kg COD/day.M^3. Hydraulic retention times were 5 - 20 days and temperature was in the range of $23°$ to $30°C$.

Through simulated landfill cell studies, Pohland and other investigators (6,23,24) concluded that leachate recycling to landfills could greatly accelerate the biological stabilization of organic matter in the refuse, and reduce the pollutant strength in the leachate to such levels that it might be discharged directly to a fresh water body or only required minimal treatment. The use of pH adjustment for the recycled leachate might be necessary to provide better environmental conditions to chemical reactions and and biological activities within the landfill itself.

Often, a combination of physical, chemical and biological processes is required for complete leachate treatment. For example, a system consisting of equalization, chemical precipitation and coagulation, sedimentation, air stripping, neutralization, nutrient supplementation, activated sludge units, and chlorination is reported in a recent study (5). The COD, BOD and NH_3-N were reduced to 945 (95% removal), 153 (99% removal), and 75 mg/L (90% removal), respectively. Concentrations of chromium, lead, mercury, nickel and zinc in the effluent were insignificant.

Anaerobic filters have received considerable attention as a viable alternative in treating high strength waste (25), following initial work by Young and McCarty (26). The literature on the use of anaerobic filters for the specific treatment of toxic leachates is sparse. Chian and DeWalle (9) reported its use with effluent recycling for treating leachate from a laboratory landfill

Table 1: Concentrations of Priority Toxic Pollutants in Raw Leachate and River[a]

Parameters[b]	Leachate	Collection Box	Fallen Timber Run, upstream	Fallen Timber Run, downstream
Benzene	156	57	-	-
Ethyl benzene	350	107	-	-
n-butyl benzene	95[c]	13	0.6	0.6
n-propyl benzene	-	14	0.5	0.8
Cumene	98	17	0.2	0.4
Indene	114	22	-	-
Mesitylene	305[c]	97	1.4	2.2
Naphthalene	420	43	0.3	0.5
1-methyl Naphthalene	11	3.2	-	-
2-methyl Naphthalene	18	3.5	0.2	0.3
Phenanthrene/Anthracene	6	2.4	0.1[c]	0.2
Bi-n-butyl-phthalate	-	1.5	0.1[c]	0.2[c]
Toluene	230	37	0.1	1.8
Xylene				
meta	446	129	0.04	0.05[c]
ortho	670	92	0.54	0.81
para	269	97	-	0.03[c]

a. From Pennsylvania DER
b. All values in ppb
c. Estimated values

reactor. The treated liquid was discharged from the top of the second reactor into a Plexiglass holding basin. An effluent-gas overflow type port on the top portion of each unit provided access to effluent and gas samplings. Gas was discharged into displacement vessels for volume production measurements. Gas samples for analysis of methane content were taken from the entrapped gas.

System Start-up

Anaerobic digester sludge collected from the Pleasant Hills

Table 2: Leachate Characteristics[a]

Chemical Characteristics	Summary of 20 Samples[b]		Summary of 12 Samples[c]		This Study	
	Median	Range	Median	Range	Average	Range
COD	8100	40-89520	12000	30-71700	19470	11760-24600
BOD	5700	80-3336	8100	4-57600	13070	6590-17860
Ammonia as N	220	0-1110	300	2-1030	290	60-600
SS	220	10-26500	200	55-920	2740	320-13900
Alkalinity as $CaCO_3$	3050	0-20850	460	40-3520	2710	50-7770
Volatile Acid as CH_3COOH	-	-	-	-	4070	250-12590
Phosphate as PO_4	10	0-130	40	3-300	120	-
Iron, Total	90	0-2820	560	2-2200	30	0-140
Zinc,	3.5	0-370	-	-	0.8	0-6
pH, unit	5.8	3.7-8.5	6.0	5.1-7.3	-	4.4-8.8
VSS/SS, %					65	61-69

a. All units are in mg/L except pH and VSS/SS
b. From municipal solid wastes (10)
c. From Chian and DeWalle (9)

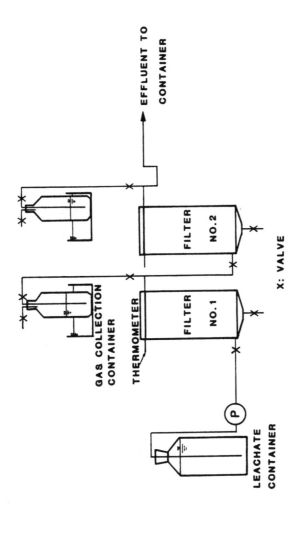

Figure 1: Schematic Diagram of Upflow Anaerobic Fixed Film System.

sewage treatment plant at Pleasant Hills, Pennsylvania was used as a seed. Each media equipped reactor was dosed with 10 L of digested sludge. A mixture of raw leachate and a synthetic medium (sodium acetate, glycine, acetic acid, pyrogallol and ferrous sulfate) was fed continuously to the system for aproximately five weeks at room temperature. During this period, very little gas production from the reactors was observed. Consequently, an additional 20 L digested sludge was added to each reactor. The reactors were heated to $37 \pm 2°C$. Raw leachate was fed daily into the reactors at a rate of 3 L/day for approximately 6 more weeks. During the final three weeks of this acclimation period, steadily increasing gas production was observed.

System Operation

The study consisted of six operational phases (Fig. 2). As the scattering data in Fig. 2 indicate, flow could not be maintained at a pre-determined rate due to solids clogging. Following system start-up, Phase I was initiated under the same feeding conditions. Both flow rate (total HDT = 35.8 days) and COD concentration (21.4 g/L) were relatively high during Phase I. The hydraulic flow rate was reduced by half in Phase II. To maintain a relatively constant feed concentration, the raw leachate was diluted to a COD of approximately 10 g/L after this phase. In Phase III, the flow rate was increased in order to compare the effect of the same COD loading with the results obtained during Phase II. For the next three phases, the flow rate was increased gradually to observe its effect on system performance. Also, the iron and zinc concentrations were increased gradually by adding metal salts ($FeCl_2$ and $ZnCl_2$) to the feed reservoir after Phase III to evaluate metal toxicity.

Analytical Methods

Samples were taken three times a week from the feed solution (influent) and the effluents from the first and second reactors. Soluble samples were obtained by filtering samples through a 1.2 um membrane filter (Gelman Membrane Filter GA-3) and measured for COD, BOD, volatile acids (VA), iron and zinc concentrations. Additionally, liquid samples were tested for residues, pH, and alkalinity. These parameters were basically measured in accordance with Standard Methods (27, Table 3). The iron and zinc ions were measured using a Norelco SP. 90 Atomic Absorption Spectrophotometer. The gas production was determined by a water displacement technique using two calibrated clear acrylic cylinders. The percentage of methane in the gas was measured by a combustible gas indicator (Mine Safety and Applicance Co.).

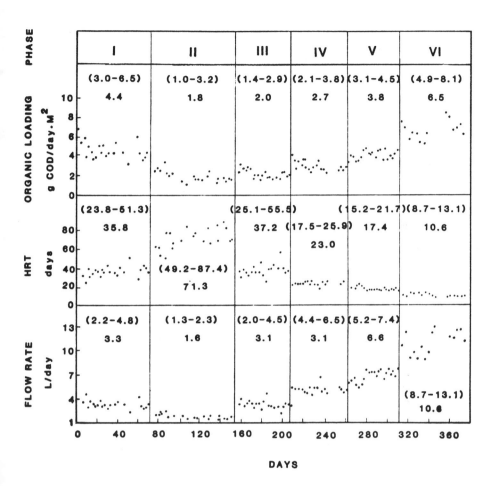

Figure 2: Flow Rate, Hydraulic Retention Time, and Organic Loading in Phase 1 Through VI.

Table 3: Analytical Methods of Parameters

Parameter	Section No.[a]
COD (s)[b]	508A
BOD (s)	507
MLSS	209B
MLVSS	209E
Amminia as N	417E
Volatile Acid (s) as CH_3COOH	504C
Alkalinity as $CaCO_3$	403
pH	423
Fe(s) and Zn(s)	303A

a. Section No. referring to Standard Methods[27]
b. soluble

RESULTS AND DISCUSSION

Start up and Acclimation

The system start-up process was lengthy and arduous. The total start-up time was approximately 80 days (Table 4). Typically, a large quantity of biomass seed was needed to accelerate microorganism development on the media surface. Unstable and low pH values (below 6.5) may have delayed the starting time. The slow growth of methane formers even at room temperatures greatly increased the biomass build-up time. After a significant COD reduction and gas generation, Phase I was initiated with the same high influent COD concentration of 21 g/L. At about day 30 during Phase I, the COD, BOD and VA concentrations in effluents from both reactors peaked. Effluent quality began to improve after day 40, and became relatively constant after about day 90. Apparently, the reactors were in a state of acclimation before day 90 which extended over Phase I and part of Phase II. Thus, the total time required to reach a pseudo-steady state was approximately 170 days. The organic loading after the start-up period probably should be reduced, and then gradually increased to avoid any possible overloading.

Table 4: Seeds Used for Anaerobic Filter Start-Up

Investigation	Seed	Start-up, days
El-Shafie and Bloodgood[28]	cotton-filtered supernatant from anaerobic digester	31
Young and McCarty[26]	two additions of 30 g volatile solids from anaerobic digester	40
Jennett abd Dennis[30]	30 g of digested sludge to the bottom of filter	20
Plummer and Malina[31]	digested sludge screened through 25" mesh screen; units were purged with N_2; feed contained 80% sludge and 20% waste and continuously recycled	8
Khan and Siddigi[31]	cow-dung slurry	75
This Study	20 liters of digested sludge (twice)	77

Influent Leachate Characteristics

Raw leachate quality varied considerably during Phase I and II (Table 2) because of waste attenuation and local weather conditions. Chain and DeWalle (9) report that the time lapse between sample collection and analyses is an important factor in determining parameter concentrations. In general, the organic matter in this particular leachate was relatively high. The COD or BOD concentration was more than twice that of the median value for 32 other leachate samples. Also, the leachate organic matter was highly biodegradable (BOD/COD = 70%, VSS/SS = 65%). The comparatively different leachate data in Table 2 also indicate that the raw leachate used in this study contains a relatively high concentration of phosphate, a low concentration of zinc and iron, and approximately the same amount of NH_3-N.

Figure 3 exhibits the influent organic substrate concentration throughout the entire study. Because of the variable waste quality, raw leachate was diluted to a COD of approximately 10 g/L after the Phase II study. Due to imperfect dilution and biological

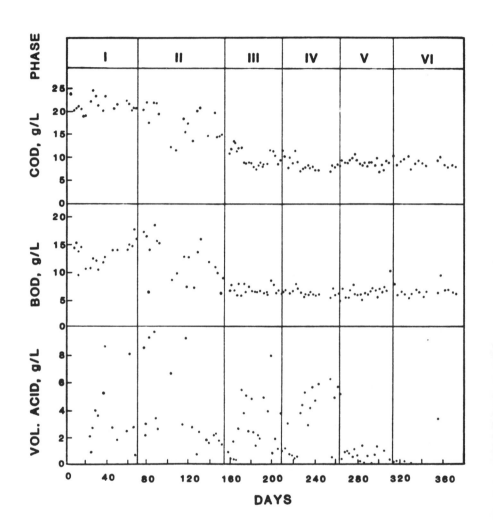

Figure 3: COD, BOD, and VA Profiles of Influent Leachate Wastewater.

activity in the feed reservoir, the average COD during Phases III-VI differed slightly from 10 g/L (Fig. 3). Nevertheless, the variations in BOD concentrations generally resembled those for COD after Phase I. The VA concentration fluctuated throughout the entire study. Extremely low volatile organics during Phases V and VI resulted from the lack of VA in the raw leachate. The influent VA/COD ratio averaged approximately 0.16, 0.26, 0.29, 0.38, 0.10 and 0.04, respectively, for the 6 phases.

Despite the relatively consistent COD or BOD, the influent leachate wastewater showed a remarkable variation in alkalinity, pH, NH_3-N, and solids (Table 5). Apparently, seasonal variations not only influence the leachate characteristics but also change its chemical composition. The pH was usually observed within 5 and 7, while the alkalinity concentrations ranged from 1,000 to 5,000 mg/L as $CaCO_3$. Similar to with VA, the concentrations of alkalinity and pH were lowest during Phases V and VI.

Table 5: Overall Influent Leachate Characteristics[a]

	Phase					
	I	II	III	IV	V	VI
COD	21470	17470	10140	8790	9280	9460
BOD	13590	12540	7110	6650	6910	7120
Alk.[b]	2600	2830	1810	3180	1100	550
pH	6.19	5.82	5.92	6.88	5.86	5.70
NH_3-N	300	290	645	150	250	230
SS	3110	2380	1890	1570	1590	2360
VSS	1890	1630	1020	620	610	740
Volatile acid[c]	3530	4610	2960	3370	950	400
Iron	25.5	33.0	67.7	101	185	219
Zinc	0.5	1.2	6.5	19.2	39.5	54.8

a. all in mg/L, except pH.
b. as $CaCO_3$.
c. as acetic acid.

With the exception of Phase III, the ammonia content in all influent samples ranged from 45 mg/L to 600 mg/L. The average concentrations of NH_3-N were approximately 300, 290, 650, 150, 250, and 230 mg/L for Phases I through VI, respectively. A relatively high average concentration of NH_3-N occurred in Phase III. Excluding the data from Phase III, COD/NH_3-N ratio ranged from 0.37 to 0.73.

The average influent SS and VSS concentrations were calculated to be in the ranges 1,570 and 3,110 mg/L, and 610 to 1,890 mg/L, respectively (Table 5). The fluctuations in solids concentration among phases was considered to be primarily due to the varying amounts of solids in raw leachate and to the microbial growth in the feed tank. The ratio of VSS/SS decreased from 0.6 - 0.7 for Phases I and II to 0.3 - 0.4 during Phases V and VI. Similar variations of VSS/SS in leachate were also noted by others (9,32).

Influent metals measurement during Phases I and II revealed low concentrations. Iron and zinc salts were added to the feed solution during Phases III through Phase VI to observe their removal and toxicity in the system. During the preparation of the feed solution, the additional metals appeared to act as coagulant resulting in solids precipitation. The influent metal concentrations were lower than the initially added amount. This phenomenon may partially explain the lower ratio of VSS/SS for Phases IV-IV.

Waste Stabilization

Organic matter removal efficiency in the fixed-film process is a function of surface organic loading. The results (Fig. 4 and Table 6) indicate that higher COD removal efficiencies can be obtained at lower loading rates. Phases III, IV and V, with loadings of 2.0, 2.7, and 3.8 kg COD/day.M^2, respectively, achieved 91, 89 and 83% of COD reduction rates. Similar results with anaerobic treatment of leachate were reported by Chian and DeWalle (33). These investigators used a completely mixed anaerobic filter (raw leachate is diluted with recirculated effluent) to reduce raw COD from 20-30 g/L to below 1.5 g/L at a loading of 3.8 g COD/day.M^2.

COD removal, including the transient acclimation period prior to steady state, was approximately 75% during Phase II. Removal amounts to 82% when the Phase II transient data are ignored. Owing to a rapid increase in the feed rate and a concomitant increase in organic loading, the reactors only achieved an average 40% COD

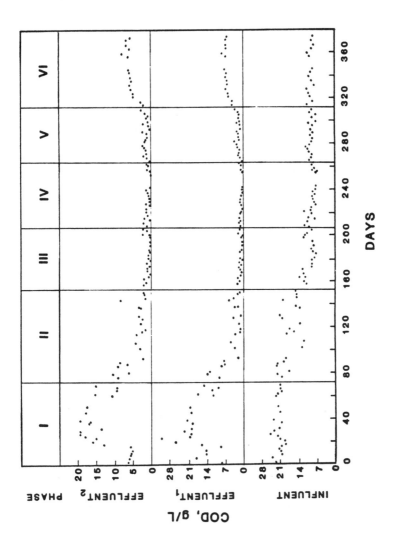

Figure 4: Soluble COD Concentration Profile in Influent and Effluents.

Table 6: Biofilter Effluent Characteristics

Leachate Waste[a]	Phase Number (organic loading in g COD/day·M^2)											
	I (4.4)		II (1.8)		III (2.0)		IV (2.7)		V (3.8)		VI (6.5)	
	Stage 1	Stage 2	Stage 1	Stage 2	Stage 1	Stage 2	Stage 1	Stage 2	Stage 1	Stage 2	Stage 1	Stage 2
COD	17450	12740	4850	4590	1250	930	930	960	2110	1580	6410	5660
BOD	10200	7090	2300	1920	170	130	150	130	680	400	2200	2080
VA[b]	8080	7410	1400	1080	340	280	310	320	410	250	3300	2720
NH$_3$-N	280	310	370	290	470	420	300	380	230	270	180	190
pH	6.7	8.0	8.1	8.7	7.8	8.6	7.8	8.5	7.9	8.5	-	6.4
Alk.[c]	5490	5670	8340	7890	5780	6020	4660	4400	4470	4560	1990	2910
Fe	12	12	1.9	1.5	1.3	1.2	2.2	1.2	15	1.8	4.5	1.8
Zn	0.6	0.5	0.1	0.1	0.3	0.2	0.7	0.3	1.6	0.5	1.7	1.0
Gas, L/day	18.4	6.2	25.6	4.2	23.1	2.6	32.7	1.9	33.0	1.7	26.8	1.5

a. Concentrations in mg/L, except noted
b. as acetic acid
c. as CaCO$_3$

reduction during Phase VI of the study. The low efficiency was unlikely due to the toxic effects of the priority pollutants in the raw leachate, since the same waste was successfully treated in Phase V. The concentrations of toxic pollutants shown in Table 1 are low enough not to inhibit anaerobic culture (34). Again, Phase I occurred during the transient state resulting in the average of 41% COD removal, while the loading rate was lower than that during Phase VI. In this study, the first filter usually accounted for nearly of all the COD reduction, which agrees with the observation of other investigators (26,35,36), who found that the first few feet of filter depth are responsible for most of the biological activities. The second filter evidently contains little methane forming bacteria. Consequently, in order to separate the two-stage reaction, the volume of the first reactor should be smaller than that of the second stage reactor.

The strength and reduction of the effluent BODs were proportional to those of the CODs. BOD reduction was high during Phases II through V and low during Phases I and VI. The lowest effluent BOD concentration (150 - 170 mg/L) occurred during Phases III and IV (Table 6), with a reduction of approximately 98%. Phase II (adjusted) and Phase V achieved 93% and 95% of BOD reduction, respectively. As would be expected from the biological process, the organic matter removal percentage in terms of BOD was higher than that calculated from COD concentrations. Also, the second filter provided insignificant BOD removal, except for Phase V. The effluent COD/BOD ratio ranged from 2.5 to 7.2 as opposed to an influent COD/BOD ratio of 1.4, indicating that only several hundreds mg/L of nonbiodegradable organic matter were present in the effluent. This refractory organic matter mainly consisted of fulvic acid-like materials with a molecular weight between 500-10,000 (37,38). The accumulation of large quantities of VA under unfavorable conditions (Phases I and VI, Table 6) is attributed to the lower COD removal. As the first reactor was overloaded (Phase VI), the second filter was not capable to utilize the rapidly accumulated VA. Despite an increase in alkalinity during Phases I and VI, the ratio of alkalinity/VA was still approximately 0.8. Conversely, alkalinity/VA ratios for the satisfactory COD removal stages were 7.5, 22, 14 and 19, respectively, for Phases II to V. For anaerobic sludge digestion, the alkalinity/VA ratio is normally maintained above 4.0 (39).

The net increase in ammonia reflects the degradation of nitrogenous organic compounds. Bull et al. (11) also reported that free NH_3-N concentration increased (from 300 to 770 mg/L) with anaerobic treatment of leachate. The NH_3-N concentration during

212 Biotechnology for Degradation of Toxic Chemicals

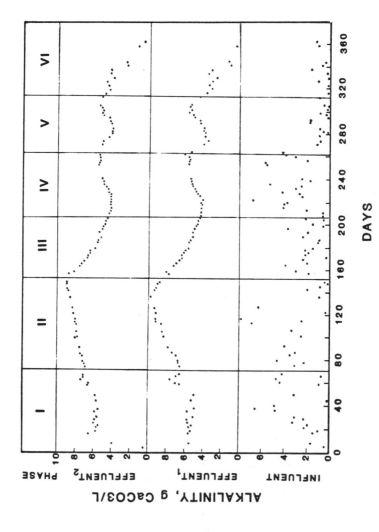

Figure 5: Alkalinity Concentration Profile in Influent and Effluents.

Phase III (465 mg/L from stage 1 effluent) was well below the toxicity level of 1,500 to 2,500 mg/L for anaerobic sludge digestion (39); hence, it exhibited no adverse effects on COD/BOD removal efficiency. Profile of alkalinity concentrations (Fig. 5) indicates that the variations in the influent are generally alleviated in filter effluents as soon as pseudo-steady state conditions are reached after phase changes. The relatively constant effluent pH and alkalinity remained suitable for biological growth until day 310 (i.e., the beginning of Phase VI) which corresponded to increases in influent flow rate, organic loading, and effluent VA concentration. As indicated in Tables 5 and 6, the effluent alkalinity for each phase increased an average of two to three times of its initial value. Alkalinity concentrations in both first and second stage effluent were above 4500 mg/L, except during Phase VI. The low alkalinity (2000 mg/L) and high VA concentrations (3300 mg/L) in this phase were resulted from organic overloading.

The results of this study show that biofilters are capable of treating acidic waste with a pH of 6.0 or less (45% of influent samples). For experiments during Phases II to V, the pH of the first reactor effluent increased by about 1-2 pH units as compared to influent pH. The destruction of VA and sulfate and the formation of ammonia, coupled with the increase in alkalinity, caused the increase in pH. Similarly, the effluent pH values from the second biofilter were higher than those from the first one.

The addition of zinc and iron salts did not affect the system performance, since the effluent soluble metal concentrations were generally below the toxic level for Zn for anaerobic treatment (9). This finding was partially attributed to the fact that metal additions were increased gradually in order to allow the biofilters to acclimate to the new environment. The soluble iron and zinc concentrations in effluent remained relatively stable, although they did fluctuate with influent metal concentrations. Iron removal ranged from 53 to 99%, and zinc removal was 29 to 98%. The highest removal for both metals occurred during Phase V (Table 6). The resulting average effluent concentrations during this Phase were 1.8 mg/l iron and 0.5 mg/l zinc, while the influent metal concentrations were 185 and 40 mg/l for iron and zinc, respectively. The chemical precipitation, coagulation and adsorption onto biofilm probably accounted for the metal removal. Significant increases in influent pH occurred after the first stage, indicating that most of the metal reduction was accomplished in the first biofilter. During Phase VI, pH decreased due to the accumulation of large quantities of VA resulting in more soluble metal concentrations.

Gas was produced mainly from the first filter, in which there were higher COD, BOD and VA removals. Also, gas production from the first bioreactor essentially followed the same trend of organic loading rate, with the exception of the transient condition in Phase I and the overloading condition in Phase VI. An increase of 40% in organic loading in Phase IV resulted in an almost equivalent percentage increase in gas production. The gas production in Phase V, however, was the same as that for Phase IV. Apparently, methane formers were being washed out as indicated by an increase in BOD.

The methane percentage ranged from 56 to 78% of gas produced from the first filter and 11 to 51% from the second filter. The total methane production from the first filter was 15.0 to 27.7 liter/day; production for the second filter was 0.2 to 5.3 liter/day. The relatively high methane purity in the gas phase from the first biofilter infers that a significant portion of methane was derived from CO_2.

Kinetics

The effect of substrate inhibition was investigated using the following model (40):

$$U = \frac{U_{max}}{1 + \frac{K_s}{S} + (\frac{S}{K_i})^n} \tag{1}$$

where

U = utilization rate, g COD/day·M^2
U_{max} = maximum utilization rate, g COD/day·M^2
K_s = Monod half saturation coeffficient, g/L
S = substrate concentration, g/L
K_i = inhibition constant, g/L
n = inhibition response coefficient

Without any inhibition, Eq. 1 reduces to a typical Monod model. At $n = 1$, Eq. 1 is the Haldene relationship between substrate utilization rate and substrate concentration which was used by

Rozich et al. (41) for the growth rate model of phenol degradation.

The relationship between overall utilization rate and second stage COD concentration was shown in Fig. 6. The kinetic coefficients U_{max}, K_s, K_i, and n of the data shown in Fig. 6 were calculated to be 8.0 g COD/day.M^2, 2.6 g/L, 5 g/L, and 3, respectively, using non-linear regression analysis. Plots of Eq. 1 with several different n values were also shown in Fig. 6 for comparison. Clearly, experimental data could not be best represented by Haldene relationship (n = 1).

Without inhibition, an increase in organic loading (Phase V) resulted in an increase in utilization rate. As loading further increased (Phase VI), the VA product accumulated in reactors and caused process inhibition. As a result, there exists the optimum utilization rate which corresponds to an organic loading above which the utilization rate is reduced. Mathematically, it can be proven that all plots with different n values intercept at point of the same utilization rate where $S = K_i$. Below this K_i value, high n results in a greater utilization rate. Above this K_i value, the degree of inhibition is proportional to the magnitude of n, i.e., high n results in severe inhibition. The substrate concentration corresponding to the optimum utilization can be solved as follows:

$$S = \left(\frac{K_s K_i^n}{n}\right)^{\frac{1}{n+1}} \quad (2)$$

Sunstituting n = 3, K_s = 2.6 g/L, and K_i = 5 g/L, the S was determined to be 3.2 g/L which corresponds to 3.9 g COD/day.M^2.

SUMMARY AND CONCLUSIONS

Leachate from a hazardous waste landfill site contained a high concentration of organic matter and low metal concentrations. COD, BOD and volatile acid were approximately 21, 14, and 1.9 g/L, respectively. Both the strength and chemical composition of raw leachate varied with sampling time. The high ratio of BOD to COD of raw leachate is amenable to biological treatment.

The start-up of the anaerobic filter system was highly time-consuming. Larger quantities of seed material and higher temperatures should reduce the start-up time.

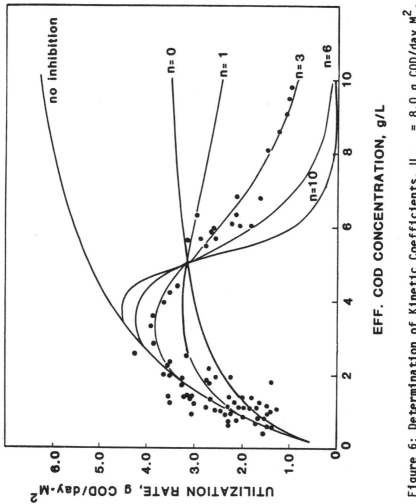

Figure 6: Determination of Kinetic Coefficients, U_{max} = 8.0 g COD/day·M², K_s = 2.6 g/L, K_i = 5.0 g/L, and n = 3.

With an organic loading up to 4 g $COD/day \cdot M^2$, COD/BOD removal was satisfactory, even though the leachate contained several toxic pollutants. BOD as low as 130 mg/L or a 98% reduction could be achieved. Thus, the anaerobic filters can be used to treat concentrated leachate without the effluent recycle and without any pretreatment.

The biofilters were very effective for metal removal. However, the anaerobic process did not remove ammonia and the effluent NH_3-N concentration often increased due to degradation of nitrogenous organic matter. Either chemical process or biological nitrification may be employed to remove residual BOD and ammonia to an acceptable level.

Due to the large volume of the first reactor, almost all of the organic matter reduction and gas production occurred in the first stage reactor. Once the first stage was upset, the second reactor was unable to degrade VA present in the first stage effluent. The overloading was characterized by the accumulation of large quantities of VA, a low ratio of VA/alkalinity, and reduction of gas.

ACKNOWLEDGMENTS

This study was funded by the Army Construction Engineering Research Laboratory, Champaign, Illinois (Grant No. DACA 888 EC 0013). At the time of this research, Y.C. Wu and K.C. Ou were Professor and PhD graduate student, respectively, in the Department of Civil Engineeering, University of Pittsburgh. Dr. Wu presently is Professor and Executive Director, Consortium for Biological Waste Treatment Research and Technology. Dr. Ou is staff engineer, Department of Environmental Resources, Pittsburgh, Pennsylvania. O.J. Hao is Assistant Professor at the University of Maryland, Department of Civil Engineering. Correspondence should be addressed to Dr. Wu at Department of Civil and Environmental Engineering, New Jersey Institute of Technology, Newark, NJ 07102.

REFERENCES

1. Robertson, J.M., "Organic Leachate Threatens Groundwater Quality." Water and Sewage Works, 123, 58 (1976).
2. Frost, R.R., and Griffin, R.A., "Effect of pH on Adsorption of Arsenic and Selenium from landfill Leachate by Clay Minerals." J. Soil Sci., 41, 53 (1977).

3. Frost, R.R. and Griffin, R.A., "Effect of pH on Adsorption of Copper, Zinc, and Cadmium from Landfill Leachate by Clay Minerals." J. Environ. Sci. and Health, A12, 139, (1977).
4. Vesilind, P.A., and Rimer, A.E., Unit Operation in Resource Recovery Engineering, Prentice-Hall, Inc., Englewood Cliffs, New Jersey (1981).
5. Keenan, J.D., Steiner R.L., and Fungaroli, A.A., "Landfill Leachate Treatment." J. Water Pollut. Control Fed., 56, 27 (1984).
6. Pohland, F.G., and Kang, S.J., "Sanitary Landfill Stabilization with Leachate Recycle and Residual Treatment." American Institute of Chemical Engineers Symposium Series, 71, 308 (1975).
7. Chian, E.S.K., and DeWalle, F.B., "Characterization and Treatment of Leachates Generated from Landfills." American Institute of Chemical Engineers Symposium Series, 71, 319 (1975).
8. Bookter, T.J., and Ham, R.K., "Stabilization of Solid Waste in Landfill." J. Environ. Eng. Div., ASCE, 108, 1089 (1982).
9. Chian, E.S.K. and DeWalle, F.B., "Sanitary Landfill Leachates and Their Treatment." J. Environ. Eng. Div., ASCE, 102, 411 (1976).
10. Solid and Hazardous Waste Research Laboratory. National Environmental Research Center, "Summary Report: Gas and Leachate from Land Disposal of Municipal Solid Waste," Tech. report, U.S. Environmental Protection Agency, 1974, 62 pp.
11. Bull, P.S., Evans, J.V., Wechsler, R.M., and Cleland, K.J., "Biological Technology of the Treatment of Leachate from Sanitary Landfills." Water Res., 17, 1473 (1983).
12. Ho. S., Ham, R.K., and Boyle, W.C., "Chemical Treatment of Leachates from Sanitary Landfills," J. Water Pollut. Control Fed., 46, 1776 (1974).
13. Karr, P.R., "Treatment of Leachate from Sanitary Landfills." Tech. report, School of Civil Engineering, Georgia Institute of Technology, (Oct. 1972).
14. Cook, E.N., and Foree, E.G., "Aerobic Biostabilization of Sanitary Landfill Leachate." J. Water Pollut. Control Fed., 46, 380 (1974).
15. Rogers, W.P., Treatment of Leachate from a Sanitary Landfill by Lime Precipitation Followed by an Anaerobic Filter, PhD dissertation, Clarkson College of Technology, Potsdam, N.Y. (1973).
16. Thornton, R.J., and Blanc, F.C., "Leachate Treatment by Coagulation and Precipitation." J. Environ. Eng. Div., ASCE, 99, 535 (1973).

17. Van Fleet, S.R., Judkins, J.F., and Molz, F.J., "Discussion, Aerobic Biostabilization of Sanitary Landfill Leachate." J. Water Pollut. Control Fed., 46, 2611 (1974).
18. Slater, C.S., Uchrin, C.G., and Ahlert, R.C., "Physiochemical Pretreatment of Landfill Leachates Using Coagulation." J. Env. Sci. & Health 18, 125 (1983).
19. Boyle, W.C., and Ham, R.K., "Biological Treatment of Landfill Leachate." J. Water Pollut. Control Fed., 46, 860 (11974).
20. Robinson, H.D., and Maris, P.J., "The Treatment of Leachates from Domestic Wastes in Landfills -I. Aerobic Biological Treatment of a Medium-Strength Leachate." Water Res., 17, 11 (1983).
21. Irvine, R.L., Sojka, S.A., and Colaruotolo, J.F., "Enhanced Biological Treatment of Leachates from Industrial Landfills." Proc. 37th Ind. Waste Conf., pp. 861, Ann arbor Science, (1982).
22. Marinic, D.S., "Kinetics of Carbon Utilization Treatment of Leachate." Water Res., 18, 1279 (1984).
23. Pohland, F.G., "Leachate Recycle as Landfill Management Option." J, Environ. Eng. Div., ASCE, 106, 1057 (1980).
24. Tittlebaum, M., "Organic Carbon Content Stabilization Through Landfill Leachate Recirculation." J. Water Pollut. Control Fed., 54, 428 (1982).
25. Henze, M., and Harremoes, P., "Anaerobic Treatment of Wastewater in Fixed Film Reactors - A Literature Review." Water Sci. Tech., 15, 1 (1983).
26. Young, J.C., and McCarty, P.L., "The Anaerobic Filter for Waste Treatment." Proc. 22nd Ind Waste Conf., pp. 559, Ann Arbor Science, (1967).
27. APHA, Standard Method for the Examination of Water and Wastewater, 14 Ed., American Public Health Association, Washington, D.C., (1976).
28. El-Shafie, A.T., and Bloodgood, D.E., "Anaerobic Treatment in a Multiple Upflow Filter System." J. Water Pollut. Control Fed., 45, 2345 (1973).
29. Jennett, C.J., and Dennis, N.D. Jr., "Anaerobic Filter Treatment of Pharmaceutical Waste." J. Water Pollut. Control Fed., 47, 104 (1975).
30. Plummer, A.H. Jr., Malina, J.F. Jr., and Eckenfelder, W.W., "Stabilization of a Low Solids Carbohydrate Waste by an Anaerobic Submerged Filter." Proc. 23rd Ind. Waste Conf., pp. 462, Ann Arbor Science (1968).
31. Khan, A.N., and Siddigi, R.H., "Wastewater Treatment by Anaerobic Contact Filter." J. Environ. Eng. Div., ASCE, 102, 102 (1976).

32. Qaism, S.R., and Burchinal, J.C., "Leaching from Simulated Landfills." J. Water Pollut. Control Fed., 42, 371 (1970).
33. Chian, E.S.K., and DeWalle, F.B., "Treatment of High Strength Acidic Wastewater with a Completely Mixed Anaerobic Filter," Water Res., 11, 295 (1977).
34. Johnson, L.D., and Young, J.C., "Inhibition of Anaerobic Digestion by Organic Priority Pollutants." J. Water Pollut. Control Fed., 55, 1441 (1983).
35. Mosey, F.E., "Anaerobic Filtration: A Biological Treatment Process for Industrial Effluents." Waste Pollut. Control, 77, 370, (1978).
36. Witt, E.R., Humphrey, W.J., and Roberts, T.E., "Full-Scale Anaerobic Filter Treats High Strength Wastes." Proc. 34th Ind.Waste Conf., Ann Arbor Science, 333, (1979).
37. Chian, E.S.K., and DeWalle, F.B., "Characterization of Soluble Organic Matter in Leachate." Environ. Sci. & Technol., 11, 158 (1977).
38. Chian, E.S.K., and DeWalle, F.B., "Evaluation of Leachate Treatment; Vol. 1: Characterization of Leachate." Tech. report EPA-600/2-77-186a, US Environ. Prot. Agency (Sept. 1977).
39. Operations Manual, Anaerobic Sludge Digestion, EPA 430/976-00, Office of Water Program Operation, Washington, D.C. (1976).
40. Yang, R.D. and Humphrey, A.E., "Dynamic and Steady State Studies of Phenol Biodegradation in Pure and Mixed Cultures." Biotechnol. and Bioeng., 17, 1211 (1975).
41. Rozich, A.F., and Gaudy, A.F., "Response of Phenol-Acclimated Activated Sludge Process to Quantitative Shock Loading." J. Water Pollut. Control. Fed., 57, 795 (1985).

EFFECTS OF EXTENDED IDLE PERIODS ON HAZARDOUS WASTE BIOTREATMENT

A. Scott Weber, Mark R. Matsumoto, John G. Goeddertz and Alan J. Rabideau

Department of Civil Engineering
State University of New York at Buffalo
Buffalo, New York

INTRODUCTION

During a site investigation and evaluation of a former hazardous waste facility in New York, groundwater underlying the site and leachate from the site were found to be highly contaminated by a wide variety of hazardous organic substances. Characteristics of the combined groundwater/leachate are presented in Table 1. The recommended remedial plan for site cleanup includes the construction of a leachate system, installation of a groundwater recovery system, and the construction of an on-site treatment facility [1].

Due to the projected low groundwater pumping rates and leachate collection volumes expected during remediation, it

Table 1
POLLUTANTS IDENTIFIED FROM GC/MS SCAN
AND CHEMICAL ANALYSES OF GROUNDWATER/LEACHATE

Pollutant identified above detectable limit	Concentration, ug/L
Positively Identified	
Methylene chloride	11,000
Acetone	42,000
trans-1,2-dichloroethane	3,500
2-Butanone	14,000
Benzene	1,500
4-Methyl-2-Pentanone	12,000
Toluene	4,300
Total Xylenes	1,600
Phenol	5,700
Aniline	5,600
2-Methylphenol	2,300
4-Methylphenol	15,000
2,4-Dimethylphenol	3,400
Iron, mg/L	96,000
Nickel, mg/L	2,600
COD, mg/L	1,950
TOC, mg/L	1,020
BOD_5, mg/L	1,090
Tentatively Identified	
Ethyl benzene	620
Oxirane,2,3-Diethyl	28,000
2-Pentanol,4-Methyl	6,400
Benzene, Methyl	8,500
Formamide,N,N-Dimethyl	5,400
Butanoic acid	14,000
Ethanol,1-Methyoxy-,Acetate	2,900
Benzene, ethyl-	940
Benzene,1,4-Dimethyl-	2,400
Pentanoic acid	6,300
Butanoic acid, 2-Ethyl-	2,300
2-Pyrrolidinone,1-Methyl	7,900
Benzenamine,N,N-Dimethyl	4,700
Hexanoic acid,2-Ethyl	4,200
Phenol,2,3-Dimethyl	20,000
Benzene acetic acid	1,700
Benzoic acid,3-Methyl-	5,200

is expected that on-site treatment would be most cost effective if carried out on an intermittent basis (i.e. once per week or longer). To determine the treatability of the contaminated groundwater and leachate collected from the site, a preliminary study was conducted by the Department of Civil Engineering at the State University of New York at Buffalo (UB). Based on the findings of the preliminary analyses, a coupled biotreatment/activated carbon system was identified as the most promising on-site treatment alternative [2].

As originally planned, the proposed onsite system was to consist of a batch biological reactor followed by granular activated carbon columns for removal of non-biodegradable organics. The potential of using batch biological treatment processes, commonly referred to as sequencing batch reactors (SBR), for waste treatment has been demonstrated previously by Irvine and coworkers [3-5]. However, in the studies conducted by Irvine, the time between operation cycles, or the idle time, was relatively short; typically less than one day. Because the operating constraints and degree of treatment achieved by biological processes subjected to very infrequent operation are unknown, a pilot study was initiated to assess the technical feasibility of the proposed process. Results of this study are presented in this paper.

EXPERIMENTAL APPROACH

As previously noted, treatment operations at the former hazardous waste site are likely to be required only on an intermittent basis because of expected low site groundwater and leachate collection rates. Periods between operation may be greater than one week. The effects of periodic operation on biological treatment process performance are unknown. To determine the effects of the extended idle periods on organic carbon removal rates and treatment efficiency, experimental studies were conducted. The bioreactors used in these experiments were operated on the following schedules: 1) one reaction cycle per week, or an idle period of 6 days; 2) one reaction cycle per two weeks, or an idle period of 13 days, and 3) one reaction cycle per three weeks, or an idle period of 20 days.

The site is located in Upstate New York, and as such, onsite treatment facilities may be required to operate under cooler temperatures. To investigate the effects of temperature on intermittent biological treatment operations, study bioreactors were operated under temperatures of 5, 15, and 25°C using a once per week operating protocol.

As will be noted in the results section, significant degradation of performance was observed when the bioreactors were operated under extended idle or cold temperatures. To decrease operation time and improve overall performance under these conditions, the addition of powdered activated carbon (PAC) to the bioreactors was evaluated. Three PAC dosages were studied, 1000, 2000, and 5000 mg/L, based on the influent feed volume. The highest concentration was chosen to achieve a treatment level comparable to that obtainable by a combination of biotreatment followed by a granular activated carbon column.

The glass reactors used in the study had an operating volume of 1 liter and were operated in a batch mode with a 50 percent fill volume, in which 500 mL of waste were fed to each reactor during an operating cycle. All of the reactors were seeded with activated sludge obtained from the Town of Amherst, New York POTW at an initial MLVSS concentration of 2000 mg per liter of total reactor volume. Based on an influent COD concentration of 2000 mg/L and a 50 percent feed volume, the F/M ratio was 0.5.

The operating protocol for the bioreactors was as follows: 1) waste fed to the reactors, 2) aeration started, 3) mixed liquor samples taken for total and volatile suspended solids determinations (MLSS and MLVSS) in the bioreactor, 4) bioreactors aerated for 23 hours, 5) bioreactors settled for 1 hour, 6) supernatant samples drawn for SS and chemical oxygen demand (COD) determinations, 7) supernatant decanted to the original idle reactor volume of 500 mL, 8) reactor left idle (without aeration) for 6, 13, or 20 days as needed.

ANALYTICAL TESTS

Analytical measurements made during this study include five-day biochemical oxygen demand (BOD_5), chemical oxygen demand (COD), total organic carbon (TOC), total suspended solids (SS), volatile suspended solids (VSS), and turbidity. Tests were conducted according to procedures outlined in Standard Methods [6]. Unless noted otherwise, all COD determinations were performed on samples that had been filtered through a glass fiber filter paper with an average pore size of 1.2 micron.

RESULTS AND DISCUSSION

The results of experimental studies to assess the feasibility of using intermittent bioreactors for the treatment of contaminated groundwater and leachate are presented in this section.

Effect of Extended Idle Time Between Bioreactor Operation

Because treatment operations at the proposed on-site facility would be expected to be intermittent, bioreactor efficiency when operated with extended idle times was evaluated. The idle time is the period between times of bioreactor operation. Three conditions were studied in the extended idle study: one react cycle per week, one react cycle per 2 weeks, and one react cycle per 3 weeks. Five bioreactors were used for the study: one for the 6 day idle condition while duplicate reactors were run for the 13 and 20 day idle conditions. Duplicate bioreactors were used for the 13 and 20 day idle conditions to evaluate reproducibility of the data obtained from the study bioreactors.

Presented in Figure 1 are the COD values obtained at the end of the react cycle as a function of time. Conclusions drawn from inspection of the data presented in Figure 1 are: 1) there appears to be only minor differences in COD removal efficiencies obtained after 23 hours of aeration when extended idle times are imposed on reactor operation, 2) there is excellent reproducibility between the duplicate bioreactors used for the 13 and 20 day idle conditions and 3) the 6 day idle bioreactor consistently achieved the lowest effluent COD values after the first week of operation.

To determine if extended idle periods affect the time required to achieve a desired percent COD removal, COD removal as a function of react time was measured for each idle condition. As shown in Figure 2, the imposition of extended idle periods significantly impacted the rate of COD removal. Although not measured during the study, the probable cause for the degradation in reactor performance, is a reduction in the viable biomass present in the extended idle bioreactors. The major importance of the information presented in Figure 2 is the implication that, as the idle period is extended in the intermittent biological treatment process, the aeration time required to remove biodegradable COD will also increase.

226 Biotechnology for Degradation of Toxic Chemicals

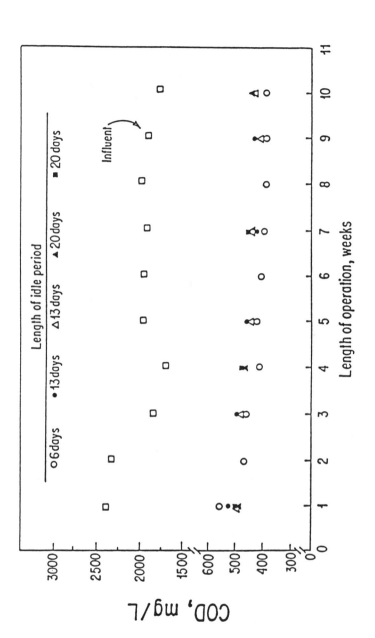

Figure 1. EFFLUENT COD CONCENTRATIONS AS A FUNCTION OF IDLE PERIOD DURATION AND LENGTH OF OPERATION

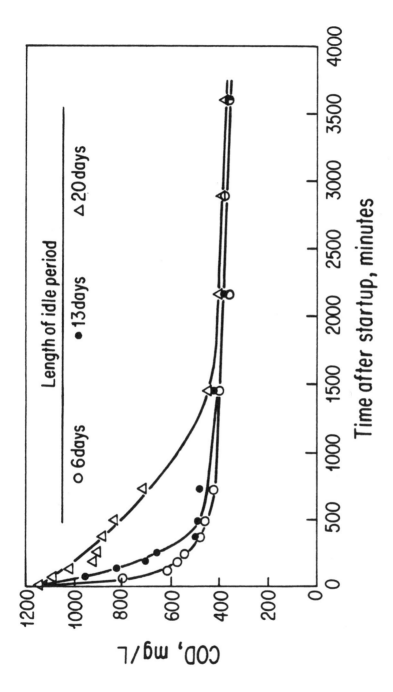

Figure 2. COD REMOVAL AS A FUNCTION OF TIME AND IDLE PERIOD DURATION

To determine whether COD removal was a result of assimilation and subsequent oxidation, or uptake and storage, oxygen uptake rates were measured for the bioreactors operating during the extended idle study and compared to the COD removal trends presented in Figure 2. As presented in Table 2, the oxygen uptake rate is greatest in the initial phases of the react cycle and drops off rapidly during the aeration period for the bioreactor operating with a 6 day idle period. This rapid drop is common for batch operated processes and follows the trend in COD removal for that reactor. As the length of the idle period is increased, the rate of O_2 uptake during the initial phase of the react cycle decreases. This decrease in initial O_2 uptake is indicative of a loss of acclimation which was verified by the slower rates of COD removal in the 13 and 20 day idle reactors. For the 20 day idle period reactor the rate of O_2 uptake is essentially constant for the 1 and 6 hrs. reading because of the slow rate of COD removal observed in that bioreactor.

Oxygen uptake data can also be used to gage oxygen supply equipment needs. If the aeration equipment is sized to meet the initial demand, some ability to reduce the supply later in the cycle may be desirable from the standpoint of energy savings. If the oxygen supply equipment is sized based on a lower rate than needed initially, then longer aeration times may be required to remove biodegradable COD.

Effect of Temperature on Bioreactor Performance

A study of temperature effects on the performance of the intermittent bioreactor were performed to evaluate the impact of climatic differences expected at the superfund site. For this purpose, bioreactors were operated at 5, 15, and 25°C, to determine the effect of temperature on overall effluent quality and COD removal rates. Five bioreactors were utilized for the temperature effects study. One bioreactor was operated at 25°C (room temperature), while duplicates were operated at the 5 and 15°C conditions. The temperature effects study was run for 6 consecutive weeks.

Influent and effluent COD values for each weekly react cycle are presented in Figure 3 as a function of bioreactor temperature. Observations that can be made from the data presented in Figure 3 are: 1) during the initial phase of the study effluent COD values were found to increase as the bioreactor temperature decreased; 2) as the study continued the effluent COD in all bioreactors improved; 3) performance

Table 2
OXYGEN UPTAKE RATES (mg O_2/g MLVSS-hr.) AS A FUNCTION OF REACT CYCLE TIME FOR EXTENDED LAG STUDY BIOREACTORS

Time in react cycle	Time between bioreactor operation		
	6 days	13 days	20 days
1 hour	185	62	36
6 hours	14	12	34
23 hours	7.6	5.8	4.4

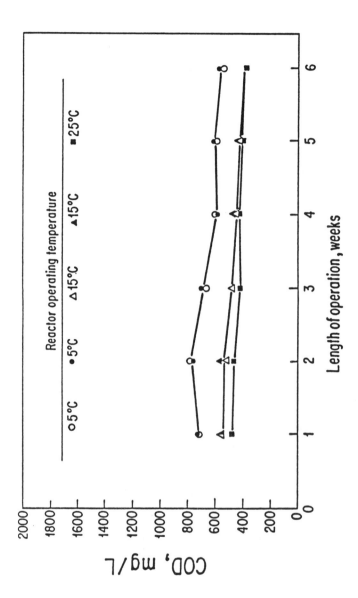

Figure 3. EFFLUENT COD CONCENTRATIONS AS A FUNCTION OF REACTOR OPERATING TEMPERATURE AND LENGTH OF OPERATION

of the two 5°C bioreactors improved at a faster rate than the 15 and 25°C bioreactors; 4) as the study continued, effluent COD in the 15°C bioreactor approached the same level of performance achieved in the 25°C bioreactor; and 5) there was excellent reproducibility between the duplicate reactors.

To determine the rate of COD removal, samples from each reactor were taken for COD analysis throughout the react cycle of the sixth operating week. As shown in Figure 4, the rate of removal was similar in the 15 and 25°C reactors while a somewhat slower rate was observed in the 5°C reactor.

To further gage effluent quality, effluent suspended solids were monitored as part of the temperature study. Average suspended solids for the bioreactor based on eight observations were measured to be 262, 99, and 82 mg/L for the 5, 15, and 25°C systems, respectively.

Based on the data collected as part of the temperature study the following conclusions can be made: 1) COD removal is affected by cooler temperatures; 2) the exact temperature that retards the removal or COD is unknown but lies somewhere between 5 and 15°C; 3) the biomass does acclimate to cooler temperatures as evidenced by the data in Figure 3; 4) the rate of acclimation is insufficient to negate longer aeration period requirements for cool weather operation; and 5) significant deterioration in effluent suspended solids quality occurs as bioreactor temperature is decreased.

Effects of Powdered Activated Carbon (PAC) Addition to the Bioreactors

Because of the long aeration times required to remove biodegradable COD under conditions of extended idle periods and reduced temperatures, further experimental studies were conducted to determine what rate advantages might be realized by supplementing the bioreactors with powdered activated carbon. In addition, depending on dosages used, elimination of the planned subsequent activated carbon columns might be made possible with the use of powdered activated carbon.

To determine the effect of PAC dosage on filterable COD removal, three concentrations were studied and compared to a control reactor operating without PAC addition. The bioreactors were operated with 50 percent feed volumes, with one react cycle per week. The PAC dosages studied were 1000, 2000, and 5000 mg/L, based on the influent feed volume. The PAC was added simultaneously with the influent waste. The PAC

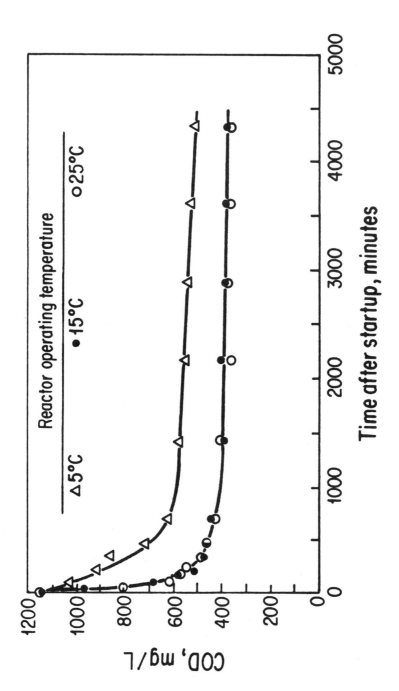

Figure 4. COD REMOVAL AS A FUNCTION OF TIME AND REACTOR OPERATING TEMPERATURE

study was conducted over 7 weeks during which time there was no wasting of solids other than those lost in the effluent.

Weekly effluent COD values from the PAC study are presented in Figure 5. Observations drawn from inspection of Figure 5 include: 1) enhanced removal of COD was observed with the addition of PAC; 2) a higher COD removal percentage was observed with increasing PAC dosages; 3) COD removal increased with study duration until an approximate equilibrium had been reached; 4) equilibrium COD values of approximately 170, 130, and 90 mg/L COD were achieved with PAC dosages of 1000, 2000, and 5000 mg/L, respectively. Equilibrium COD values plotted as a function of PAC dosage are presented in Figure 6.

Presented in Figure 7 are the data collected during the PAC study delineating the effect of PAC addition on the rate of COD removal. Removal of COD is quite rapid with baseline levels occurring in the first 2 to 3 hours of the react cycle for the 5000 mg/L dosage. The time to reach the COD baseline increases with decreasing PAC dosage. Based on these results, it is hypothesized that the addition of PAC may significantly reduce the react time requirements for COD removal when the bioreactors are operating under periods of extended idle or extreme cold. This hypothesis is based on the assumption that a significant portion of initial COD removal observed in the PAC reactors is brought about by adsorption of organics rather than by bacterial degradation. Mechanistically, both adsorption and bacterial degradation are occurring simultaneously.

To test the above hypothesis, oxygen uptake rates were measured in a control reactor with no PAC addition, and the PAC reactors, at different times throughout the react cycle. If carbon adsorption is the major mechanism of COD removal in the initial phases of the react cycle, the observed oxygen uptake rates should decrease as the PAC dosage is increased. In addition, O_2 uptake rates in the PAC reactors should be higher than endogenous rates after baseline COD values in the bioreactor are approached as adsorbed organics on the carbon continue to be biodegraded. Eventually, after the influent substrate is biodegraded, all bioreactors should approach the same endogenous O_2 uptake rate.

The depletion of oxygen, measured as mg O_2/L-min for the PAC study bioreactors is presented as a function of time in Table 3. The reading at one hour was selected to record O_2 uptake during periods of rapid uptake, a measurement at 6 hours was selected to determine O_2 uptake rates in a time period when COD had reached baseline levels in the PAC

234 Biotechnology for Degradation of Toxic Chemicals

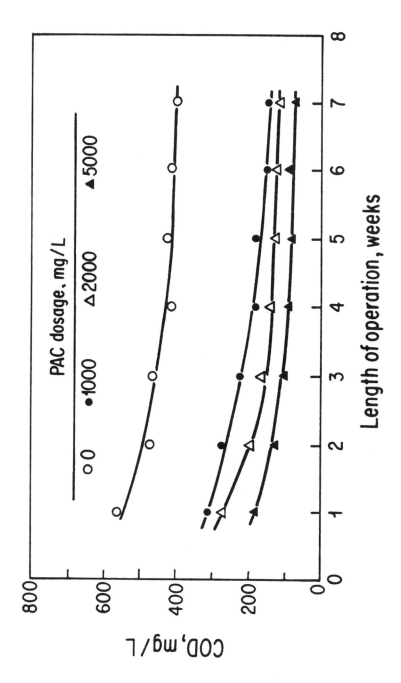

Figure 5. EFFLUENT COD CONCENTRATIONS AS A FUNCTION OF PAC DOSAGE AND LENGTH OF OPERATION

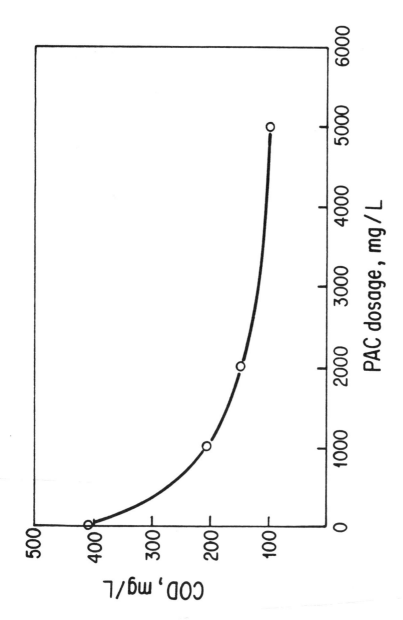

Figure 6. EFFLUENT EQUILIBRIUM COD CONCENTRATIONS AS A FUNCTION OF PAC DOSAGE

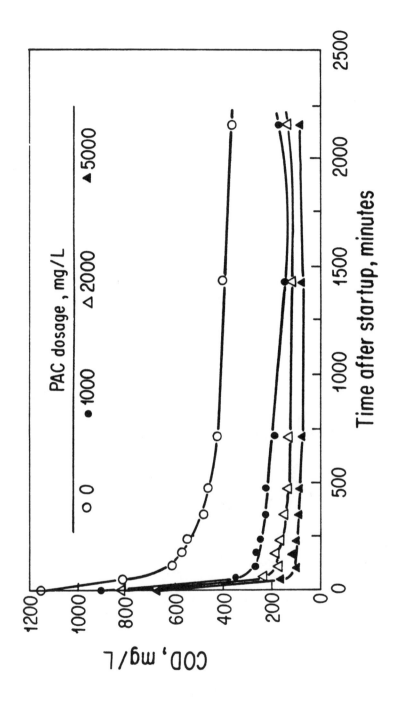

Figure 7. COD REMOVAL AS A FUNCTION OF TIME AND PAC DOSAGE

Table 3
OXYGEN UPTAKE RATES (mg O_2/L-min.) AS A FUNCTION
OF REACT CYCLE TIME FOR PAC STUDY BIOREACTORS

Time in react cycle	PAC Dosage, mg/L			
	0	1000	2000	5000
1 hour	11.1	7.8	3.9	2.8
6 hours	0.48	0.51	0.43	0.40
23 hours	0.15	0.15	0.15	0.15

reactors, but not in the control and to compare midcycle O_2 uptake rate with endogenous rates, and the last measurement was taken to establish endogenous O_2 uptake rates.

As hypothesized, the initial oxygen uptake rates for the PAC study reactors were found to be inversely proportional to the carbon dosage. This phenomenon occurs as a result of a greater fraction of the substrate partitioning on the carbon at the higher PAC dosages, reducing the amount of available substrate for bacterial degradation in the bulk fluid. At 6 hours, O_2 uptake rates for both the control and PAC bioreactors are nearly identical and are higher than the endogenous rate measured at 23 hours. Based on these results, one may postulate that the rate of bacterial degradation is similar in the control and PAC reactors even though the COD removal rates are different. This assumption seems reasonable since it is unlikely that the addition of PAC enhances the biodegradation capacity of the biomass.

The significance of these results from the point of system design is twofold: 1) because enhanced COD removal rates are likely due to adsorption of organics on the activated carbon, addition of PAC to bioreactors operating under conditions of extended lag and low temperature should improve treatment performance, and 2) because increased removal rates result from adsorption, a period continued aeration of the settled PAC bioreactor after decanting would be desirable to bioregenerate the carbon for the next react cycle.

Oxygen uptake rates for each PAC dosage at 23 hours were identical. Assuming that all biodegradable organics had been removed, it is probable that the biomass levels in the bioreactors were nearly identical. Ideally, oxygen uptake rates, should be normalized to account for the amount of biomass present in the reactor. Unfortunately, measurement of bacterial solids levels, recorded as MLVSS, is difficult in the PAC reactors because of the interference of adsorbed organics which are volatilized during the MLVSS procedure.

To further evaluate the effects of PAC addition, solids levels were monitored in the effluent from the PAC reactors. The average effluent suspended solids measured were 82, 69, 65, and 57 mg/L for 0, 1000, 2000, and 5000 mg/L PAC dosages. Qualitatively the effluent from the PAC reactors was clear, void of color, and had low turbidity. measured values for turbidity were 12, 3, 3, and 1.5 NTUs for 0, 1000, 2000, 5000 mg/L PAC dosages.

Comparison of Effluent Quality Measures

Chemical oxygen demand (COD) was used as the primary measure of process performance and effluent quality. To determine the quality of the intermittent bioreactor effluent with other indicators of organic carbon, total organic carbon (TOC) and five day biochemical oxygen demand (BOD_5) were measured also.

Presented in Table 4 is a summary of COD, TOC, and BOD_5 values for selected samples. It appears from the data presented in Table 3 that a bioreactor effluent of less than 30 mg/L BOD_5 can be achieved regardless of reactor protocol. Bioreactors with PAC addition produce the highest quality effluent in terms of BOD_5 with values typically less than 5 mg/L in the effluent. It is hypothesized that the majority of BOD_5 measured in the bioreactor effluents is brought about by respiring bacteria present in the bioreactor supernatant.

Comparison of COD and TOC values for influent and bioreactor effluent with no PAC addition produce COD/TOC ratios which range from approximately 2.5 to 3.0. This range is normal for domestic wastewater. Values for COD/TOC increase with the PAC reactors. To determine if COD/TOC ratios change throughout the react cycle for non-PAC bioreactors, C/C_o values for COD and TOC were plotted as a function of time in the react cycle. As shown in Figure 8, the C/C_o ratios are nearly identical for COD and TOC during the cycle. Therefore, it would appear that for the influent and bioreactors without carbon addition a nearly constant ratio exists for COD and TOC. Chemical oxygen demand is recommended as the process evaluation parameter for two reasons: 1) values for BOD_5 are not indicative of the true organic content; and 2) analyses for TOC require skilled personnel and expensive analytical equipment.

SUMMARY

An experimental study was conducted to assess the technical feasibility of biological processes treating hazardous waste when extended periods of inoperation are imposed on operation protocol. The impetus for this study was to develop a process that could provide cost effective treatment of leachate and contaminated groundwater from a former hazardous waste handling facility in Upstate New York. Waste collection rates at the site are expected to be low and

240 Biotechnology for Degradation of Toxic Chemicals

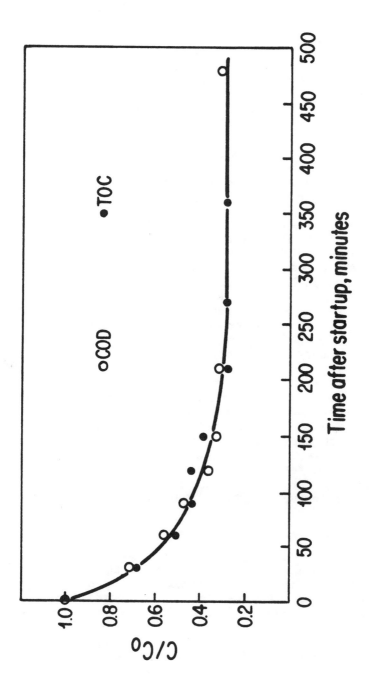

Figure 8. COMPARISON OF COD AND TOC REMOVAL AS A FUNCTION OF TIME

Table 4
COMPARISON OF EFFLUENT QUALITY IN TERMS
OF COD, TOC, AND BOD_5

Bioreactor sample	Concentration, mg/L		
	COD[a]	TOC[a]	BOD_5[b]
Waste influent	2704[c]	944[c]	
	2865[d]	1018[d]	
	1980[d]		1052[d]
	1889[d]		1130[d]
Bioreactor effluent			
– 1 React cycle per week	355	140	
	395		21.9
	412		20.7
	392		12.0
– 1 React cycle per 2 weeks	385	128	17
– 1 React cycle per 3 weeks	375	128	
– 1000 mg/L PAC, Weekly operation	197	57.5	3.5
	173		<1.0
	151		5.4
– 2000 mg/L PAC, Weekly operation	141	37.4	3.4
	110		2.3
	125		3.5
– 5000 mg/L PAC, Weekly operation	79		1.7
	90	24.1	1.5
	91		2.2

[a]Samples filtered with Whatman GF/C filter.

[b]Non-filtered samples.

[c]Pretreated influent sample.

[d]Non-pretreated influent sample.

as such, intermittent process operation will be desirable or necessary.

Based on the reported findings of the experimental study, intermittent biological treatment can achieve significant removal of waste pollutants. In this study the rate of COD removal was adversely affected by decreases in operation frequency and decreased operating temperatures. However, if reaction periods are extended, all biodegradable COD removal can be removed in bioreactors using long lag periods. Addition of powdered activated carbon was tested under one operating scenario and hypothesized to be advantageous in reducing aeration times for conditions of low temperature and extended lag periods.

While the feasibility of intermittent biological treatment was studied using a groundwater contaminated with hazardous waste substances, other potential applications exist. They include: 1) leachate treatment of municipal and hazardous waste landfills, industrial waste flows released on an intermittent basis, and other waste sites requiring remediation. It is hoped that the results presented in this paper, will alert others to the potential that biological systems may offer even when operating under conditions that are less than ideal.

ACKNOWLEDGMENTS

This study was conducted in cooperation with URS Company, Inc., Buffalo, New York and the New York State Department of Environmental Conservation (NYSDEC).

A. Scott Weber and Mark R. Matsumoto are Assistant Professors, and John G. Goeddertz and Alan J. Rabideau are graduate students in the Department of Civil Engineering at the State University of New York at Buffalo. Correspondence should be addressed to A. Scott Weber at the Department of Civil Engineering, State University of New York at Buffalo, Buffalo, NY 14260.

REFERENCES

1. "Site Investigations and Remedial Alternative Evaluations at the _____, New York, Draft Final Report," prepared for New York State Department of Environmental Conservation by URS Company, Inc., January 1984.

2. "Experimental Investigation of Treatment Process Options for Remediation of _____ Groundwater," prepared for URS Company, Inc. by Department of Civil Engineering, State University of New York at Buffalo, January 1985.
3. Irvine, R.L., and W.B. Davis, "Use of Sequencing Batch Reactors for Waste Treatment - CPC International, Corpus Christi, Texas," Proceedings, 26th Industrial Waste Conference, Purdue University, Indiana, May 1973.
4. Irvine, R.L., S.A. Sojka, and J.F. Colaruotolo, "Enhanced Biological Treatment of Leachates from Industrial Landfills," Proceedings, 37th Industrial Waste Conference, Purdue University, Indiana, May 1982.
5. Herzbrun, P., M.J. Hanchak, and R.L. Irvine, "Treatment of Hazardous Wastes in a Sequencing Batch Reactor," Proceedings, 39th Industrial Waste Conference, Purdue University, Indiana, May 1984.
6. Standard Methods for the Examination of Water and Wastewater, 16th Edition, American Public Health Association, 1985.

A TECHNIQUE TO DETERMINE INHIBITION IN THE ACTIVATED SLUDGE PROCESS USING A FED-BATCH REACTOR

Andrew T. Watkin and W. Wesley Eckenfelder, Jr.

Department of Civil and Environmental Engineering
Vanderbilt University
Nashville, Tennessee

INTRODUCTION

A need exists for a better understanding of the response of activated sludge plants to variations in loadings of toxic organic compounds. While many toxic organic compounds have been successfully treated in laboratory activated sludge units receiving steady-state inputs, difficulty arises in plants when unsteady-state or slug inputs occur(1). Modeling of activated sludge systems' response to unsteady state inputs of a known inhibitory compound will be the primary focus of this paper. Glucose and 2,4-dichlorophenol (DCP) acclimated sludges will be used in assessing unsteady-state inhibition kinetics of DCP on glucose removal. The unsteady-state inhibitory and non-inhibitory responses of DCP and glucose will be modeled in a specific semi-batch reactor configuration known as a fed-batch reactor.

FED-BATCH REACTOR METHODOLOGY

Fed-batch reactors have previously been used to determine nitrification kinetics (2) and specific pollutant kinetics (3,4) in activated sludge. The essential characteristics of a fed-batch reactor are that 1) substrate is continuously introduced at a sufficiently high concentration and low flow rate so that the reactor volume is not significantly changed during the test, (2) the feed rate exceeds the maximum utilization rate, 3) the test duration is short enough or the protocol allows for simple modeling of biological solids growth and 4) acclimated activated sludge is used.

Fed-Batch Reactor Testing Protocol

A schematic diagram of the fed-batch reactor (FBR) configuration used in all phases of this work is provided in Figure 1. All parts of the FBR which came in contact with the test sludge were either pyrex glass or teflon. Feed solution for each FBR test was made immediately prior to the beginning of the test. Nitrogen was added to the feed solution in a 1:10 ratio of N:C, phosphate buffer was added to yield a 0.029 M concentration, and trace inorganic nutrients were added. The phosphate buffer was sufficient to maintain the FBR pH's between 7.0 and 7.2 and consisted of 58 mole percent of K_2HPO_4 and 42 mole percent of KH_2PO_4.

Two liters of test sludge were taken from a steady-state chemostat and immediately placed in the FBR. After an initial sample was taken for oxygen utilization rate, mixed liquor volatile and total suspended solids, and specific compound analyses, the feed flow was started. The feed flow rate was preset at approximately 100 ml per hour and introduced just below the liquid surface of the FBR. A peristaltic pump with silicon pump tubing and teflon feed tubing was used to regulate the feed flow. Samples for DCP and/or glucose analysis were withdrawn from the FBR every 20 minutes for the duration of the three hour test. Samples for oxygen utilization rate were withdrawn 10 and 25 minutes into the test, and every 20 minutes thereafter. Suspended solids determinations were made every hour during the test.

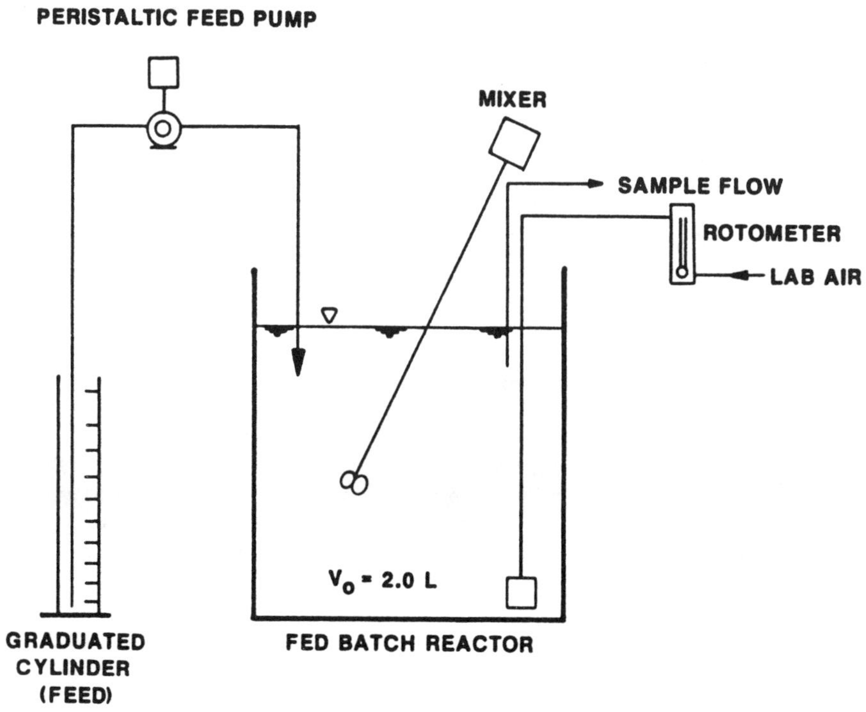

FIGURE 1 FED-BATCH REACTOR CONFIGURATION

pH and temperature were periodically monitored and recorded during the test. Initial and final total dissolved solids concentrations were also determined. The non-volatile total dissolved solids concentration was never found to increase more than 100 mg/l from the acclimated concentration of approximately 4,300 mg/l, over the three hour test period. Air flow rates were held constant at a typical rate of 100 cc/min using rotameters. The FBR's were completely mixed using laboratory mixers and diffused air. The feed flowrate was determined by taking initial and final readings, over the three hour period, from the graduated cylinder containing the feed solution. The sampling flow rate was calculated knowing the change in reactor volume over the test period and the feed flow rate. Evaporation was assumed to be negligible.

Theoretical Substrate and Oxygen Utilization Response in a Fed-Batch Reactor

This section is included to help the reader visualize the expected theoretical response in a fed-batch reactor to both inhibitory and non-inhibitory substrates. In the specific case that substrate is added at a sufficiently high mass and low volumetric flow rate, the maximum substrate utilization rate will be exceeded and reactor volume change will be insignificant. If the basic premise of the fed-batch reactor holds, i.e., volume change is significant and the biological solids concetration is constant, and if the then at mass feed rate exceeds the maximum substrate utilization, then a rate substrate concentration build-up will result in the reactor over time. Both inhibitory and non-inhibitory substrate responses are depicted in Figure 2.

Non-inhibitory substrate response results in a linear residual substrate buildup in the reactor over time. The maximum specific substrate utilization is simply calculated as the difference in slopes between the substrate feed rate and the residual substrate build-up rate divided the biomass concentration. As seen in Figure 2 a maximum oxygen utilization rate should be quickly attained corresponding to maximum substrate utilization.

In the case of inhibition substrate utilization would rapidly decrease resulting in an upward deflection of the substrate response curve as shown in Figure 2. As

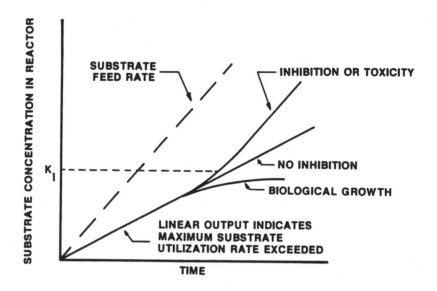
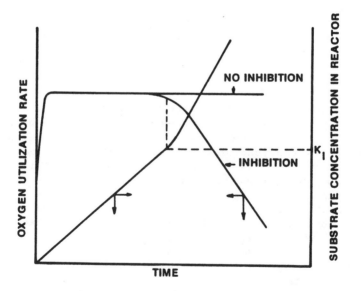

FIGURE 2 THEORETICAL FED-BATCH REACTOR RESPONSE WITH INFLUENT SUBSTRATE MASS FLOW RATE GREATER THAN $q_{MAX} \cdot X_v$ AND INHIBITION EFFECTS.

inhibition progresses and total toxicity occurs, then the residual substrate response should become parallel to the substrate feed rate line. The inhibition constant, K_I, may be approximated by identifying the inhibitor concentration at the midpoint of the curvilinear section of the substrate response. Additionally, a rapid decrease in the oxygen utilization rate should correspond to the decrease in substrate utilization as seen in Figure 2.

ANALYSIS OF EXPERIMENTAL RESULTS

The following equations were used to model either DCP or glucose removal in the fed-batch reactor, respectively.

$$\frac{dDCP}{dt} = \frac{Q_1 DCP_0}{V} - \frac{Q_2 \cdot DCP}{V} - r_{DCP} - K_v \cdot DCP \quad (1)$$

$$\frac{dGLU}{dt} = \frac{G_1 \cdot GLU_0}{V} - \frac{Q_2 \cdot GLU}{V} - r_{GLU} \quad (2)$$

where:

DCP = reactor DCP concentration ML^{-3}
DCP_0 = reactor DCP concentration, ML^{-3}
GLV = reactor glucose concentration, ML^{-3}
GLV_0 =
Q_1 = feed flow rate, $L^3 T^{-1}$
Q_2 = sampling flow rate, $L^3 T^{-1}$
V = reactor volume, L^3
r_{DCP} = DCP biological removal rate, T^{-1}
r_{GLV} = glucose biological removal rate, T^{-1}
K_v = first order volatilization constant for DCP, T^{-1}

The biological removal of DCP was found to be non-inhibitory in all FBR tests conducted in this work. DCP concentrations up to 150 mg/l were tested and no signs of substrate inhibition were distinguished (3). A simplified Monod kinetic expression was used to model DCP removal in the FBR. Starting with the Monod equation for DCP removal:

$$r_{DCP} = \frac{q_{max,DCP} \cdot X_v \cdot DCP}{K_s + DCP}$$

where:

$q_{max,DCP}$ maximum specific substrate removal rate for DCP, T^{-1}
K_s = saturation coefficient for DCP, ML^{-3}
X_v = mixed liquor volatile suspended solids, ML^{-3}

Since the substrate build-up rate is rapid in all FBR tests reported in this work, K_s quickly becomes negligible and Equation 3 may be rewritten as follows:

$$r_{DCP} = -q_{max,DCP} \cdot X_v \quad (4)$$

Equation 4 was used to model all DCP data collected in this work.

Glucose removal was found to be inhibited by DCP in all FBR tests. Classical rate expressions for both noncompetitive and competitive inhibition as reported by Segel (5) were found to be inadequate in modeling the inhibitory glucose response. This inadequacy was reflected in poor fits of the respective models to experimental data. A modified competitive inhibition model was developed and reported by Johnson, et. al (6) in the following general form.

$$r_s = \frac{-q_{max,s} \cdot X_v \cdot S}{(K_s+S)(1+K_I I^n)} \quad (5)$$

where:

S = substrate concentration, ML^{-3}
I = inhibitor concentration, ML^{-3}
k_I = inhibition constant, ML^{-3}
n = number of inhibitor molecules which combine with the active enzyme.

Again using the simplification that K_s will quickly become negligible, due to the rapid build-up of substrate in the fed-batch reactor, we can rewrite Equation 5 as follows:

$$r_s = \frac{-q_{max,s} \cdot X_v}{1+K_I I^n} \quad (6)$$

Attempts were made to model glucose inhibition by DCP using Equation 6, however, two problems were encountered. First, nonlinear curving-fitting was extremely inefficient in terms of required computer time to determine the best fit values of K_I and n. Secondly, in both Equations 5 and 6 the classical definition of K_I is violated. The inhibition constant, K_I, was first defined by Haldane (7) as the inhibitor concentration at which the substrate utilization rate is reduced to one half of its maximum rate. For these reasons, the following rate expression was derived from Equation 6 by redefining K_I.

$$r_s = \frac{-q_{max,s} \cdot X_v}{1+(I/K_I)^n}$$

Equation 7 was used in all modeling of glucose removal inhibition by DCP because 1) it retains the classical definition of K_I and 2) computer modeling was much more time efficient.

The following simultaneous equations were used in modeling inhibitory substrate removal in the fed-batch reactor.

$$\frac{dGLU}{dt} = \frac{Q_1 \cdot GLU_0}{V} - \frac{Q_2 \cdot GLU}{V} - \frac{q_{max,GLU} \cdot X_v}{1 + [DCP/K_I]^n} \tag{8}$$

$$\frac{dDCP}{dt} = \frac{Q_1 \cdot DCP_0}{V} - \frac{Q_2 \cdot DCP}{V} - q_{max,DCP} \cdot X_v - K_v \cdot DCP \tag{9}$$

$$\frac{d_v}{dt} = Q_1 - Q_2 \tag{10}$$

$$\frac{dX_v}{dt} = m \tag{11}$$

$$\frac{dGLU}{dt} = \frac{Q_1 \cdot GLU_0}{V} - \frac{Q_2 \cdot GLU}{V}$$

where:

m = the best fit slope of a linear regression analysis of the biological solid response in the FBR. $ML^{-3}T^{-1}$

In all cases, the $q_{max,GLU}$ value was determined independently and then fixed in the given rate expression. Once $q_{max,DCP}$ had been determined and fixed the remaining kinetic parameters, K_I and n, were determined by the nonlinear curve fitting technique. The value of $q_{max,GLU}$ was fixed and determined independently because all of the above rate expressions resulted in unrealistically high $q_{max,GLU}$ valves when computer fitted. Fixing the q_{max} valve prior to computer fitting the K_I valve was also found to be necessary by Grady, et. al (4). The value used for $q_{max,GLU}$ was determined by feeding glucose alone to the sludge at the same feed rate used in the inhibition tests and using the non-inhibitory rate expression similar to Equation 4.

A typical example of the noninhibitory substrate removal model is shown in Figure 3 for DCP. As seen in Figure 3, the upper line in this figure represents the DCP build-up in the reactor assuming no biological removal, but including removal due to stripping. The bottom line represents the best fit model response including biological and non-biological removal, and the triangles represent the experimental data. Hence, the difference between the rate of increase between the upper and lower lines shown in Figure 3 represents the biological removal alone.

A typical example of the inhibitory model described by Equations 8 through 12 is shown in Figure 4. In both Figures 3 and 4 the best-fit parameters are given followed by the smallest increment search (DELTA) used in the curve-fitting algorithm. Additionally, the smallest sum of the squared deviations (SSD), found by the algorithm, between the model's prediction and the experimental data is provided.

Differences Between Inhibition Constants of Various Sludges

Triplicate FBR tests were conducted separately on three different activated sludges and inhibition was modeled using the rate expression given in Equation 7. The three chemostat sludges tested were all acclimated to glucose and 1) acclimated to DCP, 2) previously, but not presently acclimated to DCP, and 3) never acclimated to DCP. All data was analyzed using the simultaneous differential equations given in Equations 8 through 12 which incorporates the modified inhibition model.

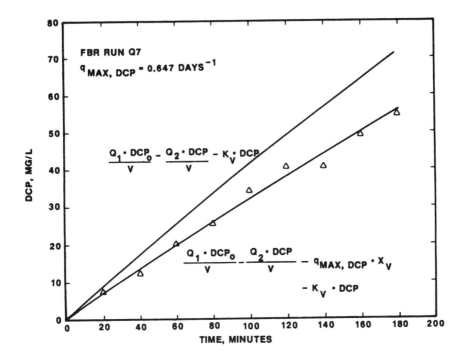

FIGURE 3 TYPICAL NON-INHIBITORY DCP RESPONSE IN THE FED-BATCH REACTOR.

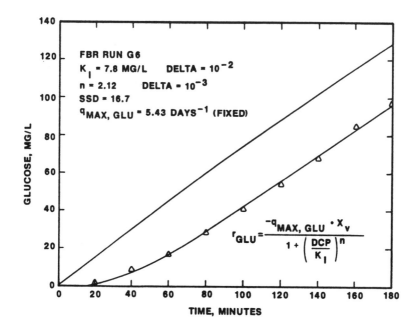

FIGURE 4 GLUCOSE RESPONSE AS PREDICTED BY THE MODIFIED INHIBITION MODEL.

The FBR runs reported in Table I received nominal glucose input rates of 45 mg/l-hr. The acclimated to DCP sludge received a nominal DCP input rate of 30 mg/l-hr; the previously, but not currently, acclimated to DCP sludge received a nominal DCP input rate of 10 mg/l-hr; and the never acclimated to DCP sludge received a nominal DCP input rate of 45 mg/l-hr. The DCP input rates were set differently for each sludge due to their respective "sensitivities" to DCP; or in other words, to produce an inhibitory glucose response which was sufficiently curvilinear to facilitate precise modeling.

Table I: Statistical Differences in Numerically Determined Inhibition Constants for Different Sludges

Sludge	Mean Values[1] + Standard Deviations	
	K_I	n_2
Acclimated to DCP	17.3 ± 1.5	5.3 ± 0.3
Previously Acclimated to DCP	6.5 ± 1.3	2.5 ± 0.6
Never Acclimated to DCP	40.4 ± 6.5	2.8 ± 0.3

[1]Sample Population = 3

Tukey intervals were determined for both sets of K_I and n values. Tukey intervals allow for pairwise comparison of the mean values of the three sample populations. In each case, if the Tukey confidence interval includes zero, then there is no significant difference between the means compared. If the Tukey confidence interval does not include zero, then the null hypothesis is rejected that he two means being compared are equal. The Tukey method is a more stringent method for comparing means than simply using analysis of variance between each pairwise combination because the overall or "global" level of significance is maintained (8). For the analysis in this section we will maintain the global Type I error at a 5 percent significance level. A Type I error is associated with rejecting the null hypothesis when it is true.

Tukey intervals are compared for the K_I values for the three different sludges (Table I) in Table II. Again, the Tukey intervals represent a 5 percent significance level. It is clearly seen in Table II that all of the Tukey intervals do not include zero and therefore in every case the null hypothesis that the means are equal is rejected. The direct implication is that the K_I values are significantly different for each of the sludges tested.

Table II: Comparison of Mean K_I Values for Different Sludges Using Tukey's Method

$i - j^1$	$x_i - x_j$	95% Tukey Confidence Interval	Conclusion
G - N	-33.9	(-43.8, -24.0)	Reject H_0'
G - Q	-10.81	(-20.7, -0.9)	Reject H_0'
N - Q	-23.1	(-13.2, -33.0)	Reject H_0'

[1] Q is the population mean for the sludge acclimated to DCP; G is the population mean for the sludge previously, but not currently, acclimated to DCP; N is the population mean for the sludge never acclimated to DCP.

It should be noted that the sludge with the highest K_I value was never acclimated to DCP and consisted of a predominantly filamentous population. Additionally, the DCP acclimated sludge had a significantly higher K_I, value than the previously but not currently acclimated sludge which makes intuitive sense due to its acclimation.

Tukey intervals are compared for the n values for the three different sludges in Table III. The Tukey intervals indicate that the mean n values determined for the previously but not currently DCP acclimated sludge and the non-acclimated sludge are equal. While the mean n value for the DCP acclimated sludge is not equal to the mean n values of either of the two other sludges. As the value of n increases, the rate of change in the upward deflection of the inhibitory glucose response versus time curve increases.

Table III: Comparison of Mean n Values for Different Sludges Using Tukey's Method

$i - j$[1]	$x_i - x_j$	95% Tukey Confidence Interval	Conclusion
G - N	-0.31	(-1.31, 0.69)	Accept H_0'
G - Q	-2.81	(-3.81, -1.81)	Reject H_0'
N - Q	-2.50	(-3.50, -1.50)	Reject H_0'

[1] Q is the population mean for the sludge acclimated to DCP; G is the population mean for the sludge previously, but not currently, acclimated to DCP; N is the population mean for the sludge never acclimated to DCP.

DISCUSSION

A method for rapidly determining both inhibitory and non-inhibitory biological degradation parameters was discussed in detail. The fed-batch reactor technique is believed to provide an excellent method to rapidly determine unsteady-state inhibition effects of toxic organics in activated sludge. The test is quite precise and requires minimum amount of time.

Distinctly, different inhibition constants were for three sludges with different histories. These differences were shown to be statistically significant by Tukey's Method at a global significance level of 5 percent. A sludge acclimated to DCP and glucose demonstrated a mean K_I value of 17.3 mg/l. This value was significantly higher than for sludge which was previous but not currently acclimated to DCP, demonstrating an inhibition constant of 6.5 mg/l. A third sludge which like the others was acclimated to glucose, but which had never been acclimated to DCP demonstrated the highest mean K_I value of 40.4 mg/l. From these results it is apparent that there is no universal inhibition constant for DCP on glucose removal. The inhibition constant for DCP is likely to be highly dependent on the specific enzyme system involved, which in turn is dependent on the history and population dynamics of the sludge. Several aerobic pathways for glucose metabolism have been demonstrated (9,10). The inhibition constant, K_I may be highly dependent on which metabolic pathway(s) is (are) present in any given mixed microbial population such as those studied in this work.

REFERENCES

1. Tischler, L. F. and Kocurek, D., "The CMA/EPA Five Plan Study-Biological Treatment of Toxic Organic Pollutants," WPCF Conference, St. Louis, MO., 1982.

2. Williamson, K. J. and McCarty, P. L., "Rapid Measurement of Monod Half-Velocity Coefficients for Bacterial Kinetics", Biotechnol. and Bioeng., 17, 1975, p. 915-924.

3. Philbrook, D. M. and Grady, C. P. L., "Evaluation of Biodegradation Kinetics for Priority Pollutants", Presented at the 40th Industrial Waste Conference, Purdue University, West Lafayette, IN, 1985.

4. Grady, C. P. L., Loven, L. E., and Philbrook, D. M., "Assessment of Biodegradation Potential for Toxic Organic Pollutants", Presented at the 58th WPCF Conference, Kansas City, MO, 1985.

5. Segel, I. H., Enzyme Kinetics-Behavior and Analysis of Rapid Equilibrium and Steady State Enzyme Systems, John Wiley and Sons, New York, 1975.

6. Johnson, F. H., Eyring, H., and Stover, B. J., The Theory of Rate Processes in Biology and Medicine, John Wiley and Sons, New York, 1974.

7. Haldane, J. B. S., Enzymes, M. I. T. Press, Cambridge, Massachusetts, 1965.

8. Larsen, R. J. and Marx, J. L., An Introduction to Mathematical Statistics and Its Applications, Prentice-Hall, Inc., Englewood Cliffs, NJ, 1981.

9. Martin, D. W., Mayes, P. A., Rodwell, V. W., and Granner, D. K., Harper's Review of Biochemistry, 12th Edition, Lange Medical Publications, Los Altos, 1985.

10. McGilvery, R. W. and Goldstein, G. W., Biochemistry - A Functional Approach, W. B. Saunders Co., Philadelphia, 1983.

A COMPARISON OF THE MICROBIAL RESPONSE OF MIXED LIQUORS FROM DIFFERENT TREATMENT PLANTS TO INDUSTRIAL ORGANIC CHEMICALS

G. Lewandowski, D. Adamowitz, P. Boyle, L. Gneiding, K. Kim, N. McMullen, D. Pak and S. Salerno

Dept. of Chemical Engineering, Chemistry, and Environmental Science
New Jersey Institute of Technology
Newark, NJ

As the first stage of an effort to determine the ability of publicly owned treatment works (POTW's) to detoxify industrial organic chemicals, several compounds were added individually to the mixed liquors from two very different treatment plants - the Livingston (NJ) municipal wastewater treatment plant, and the Passaic Valley Sewerage Commissioners (PVSC) plant in Newark, N.J. The former handles 2.5 million gallons per day of domestic sewage, while the latter handles 250 MGD of wastewater with a 55% industrial component (on a BOD basis). An aerated batch reactor was used, and the chemicals (and feed concentrations) were: phenol (100 ppm), 2-chlorophenol (20 ppm), 2,6-dichlorophenol (10 ppm), and 2,4-dichlorophenoxyacetic acid (10 ppm). Substrate disappearance was determined by gas chromatography, and the microbial systems characterized by microscopy and plating techniques.

Although the two mixed liquors came from very different systems, their response to the industrial chemicals added, as well as the initial and final microbial populations, were very similar. This suggests that the phenomena observed might be generalized to many other POTW's.

BACKGROUND

Recent studies sponsored by the U.S. Environmental Protection Agency [1,2,3] have demonstrated the ability of secondary treatment plants to remove industrial organic pollutants from wastewater. These studies have generally involved pollutant concentrations of 50 ppb to 10 ppm, with the majority of the data at the lower concentration. Such removal can follow three mechanisms: adsorption, air stripping, and biodegradation.

Biodegradation is clearly the most preferable mechanism of organic removal. Here two important questions arise: (1) what are the products of the reaction, and (2) how general is the applicability of biodegradation as a treatment technology? An attempt to answer the second of these questions, at relatively high pollutant concentrations (10 to 100 ppm), forms the basis of this study.

Mixed liquors from two very different treatment plants were used in this study. The Livingston (N.J.) municipal treatment plant handles 2.5 million gallons per day of wastewater that is almost exclusively domestic (>99%). By contrast, the Passaic Valley Sewerage Commissioners (PVSC) plant handles 250 million gallons per day of wastewater that is 55% industrial on a BOD basis (18% by volume), including approximately 500 lb/day of phenolics (at an average influent concentration of about 0.25 ppm). Thus the PVSC organisms have long been exposed to industrial chemicals, while those at Livingston have not.

PROCEDURES

All experiments were conducted in batch reactors. While it's recognized that continuous flow studies should be used when making design extrapolations, the purpose of the present work was to determine if general conclusions could be reached concerning the applicability of biological treatment to POTWs operated at very different conditions. For that purpose, batch reactors were easier to use.

The reactors were loosely covered 5-liter clear plastic cylinders containing 2 liters of mixed liquor. There was no mechanical agitation. Aeration was provided at the rate of 0.5 to 1.0 liter per minute through a diffuser. This also created sufficient agitation to keep the reactors well-mixed.

All data are for room temperature (approx. $26°$).

Two types of experiments were performed - one in which the mixed liquor was pre-acclimated to phenol, and one in which there was no pre-acclimation. For the pre-acclimated experiments, the mixed liquor was spiked to 100 ppm phenol immediately upon receipt in the laboratory. After the phenol concentration fell below the detection limit (approx. 1 ppm by GC analysis), it was respiked to 100 ppm, and the procedure repeated twice more for a total of 3 successive spikes (usually accomplished over a 2 to 3 day period). Once the acclimation procedure was completed, the rate experiments proper began.

Prior to the rate experiments, the mixed liquors contained a nitrogen equivalent of about 20 to 120 ppm ammonia, as measured by an ammonia electrode. During the course of the kinetic runs, no additional nitrogen or phosphorus were added, and the nitrogen concentration declined about 5 to 15 ppm ammonia (never falling below about 15 ppm). Sufficient buffering was present to maintain the pH in the range 6.5 to 7.5.

In the rate experiments, the reactor was shock-loaded with the compound of interest, and 15 ml samples of mixed liquor periodically collected from the reactor, until a preliminary analysis indicated that the substrate had disappeared. The initial concentrations were nominally: 100 ppm phenol, 20 ppm 2-chlorophenol, 10 ppm 2,6-dichlorophenol, and 10 ppm 2,4-dichlorophenoxyacetic acid.

For the phenolic compounds, the method of analysis employed direct aqueous injection into a gas chromatograph. To 10 ml of the centrifuged mixed liquor were added 0.5 ml of a 1000 ppm thymol solution (as internal standard), and 0.5 ml of 20,000 ppm copper sulfate solution as a biocide. The samples were then refrigerated until they could be analyzed (with an accuracy of approximately 1 ppm).

For 2,4-D, liquid chromatography was employed. Samples, without internal standard or copper sulfate, were filtered through 0.2 µm filters to sterilize them, and then were injected onto an octadecyl column. Acetonitrile and water were the solvents. The accuracy in the analysis was within 0.4 ppm.

MICROBIAL CHARACTERIZATION

Several different growth media were used for plating bacteria and fungi. For bacterial cultures, plate count agar was the general growth medium, while selective media were: floc-forming, nitrifying, and cellulose-utilizing. After growing on a particular medium, major gram negative bacterial colonies were subjected to the oxidase test, and identified using Oxiferm tubes and Enterotubes (manufactured by Hoffmann-LaRoche). Confirmatory tests included color of bacterial colonies, location of flagella, sensitivity to antibacterial agents, etc. Gram positive colonies were subjected to catalase and coagulase tests, and inoculated onto blood agar, dextrose tryptone agar, phosphatase media, and Enterotubes. Fungi were grown on the following media: yeast malt, yeast nitrogen base with sucrose, Czapek with rose bengal, oxgall with antibiotic, and Sabouraud. Yeast colonies were then identified using GBE and SAM tubes, and Uni-Yeast-Tek plates (manufactured by Flow Laboratories), while molds were identified by morphology and color. Protozoa were identified by morphology and type of locomotion.

RESULTS AND DISCUSSION

The rate of air stripping from distilled water was negligible for the compounds tested. Furthermore, based on their solubilities, octanol-water partition coefficients, and rates of disappearance, adsorption was considered to be negligible for the compounds tested.

The primary removal mechanism was biodegradation.

Although mixed liquor suspended solids (MLSS) concentration was measured, this indication of the active biomass concentration is generally poor. Dead organisms and debris are measured along with viable organisms. Furthermore, the initial biomass concentrations were high. As a result, most of the MLSS data simply indicated an approximately constant value during the course of an experiment (generally between 1000 and 2400 mg/l).

Biodegradation data are plotted in Figures 1 to 7, and the average degradation rates for each compound are compared in Table 1.

These data clearly indicate the close similarity between the results with Livingston and PVSC mixed liquors. This in spite of the considerable differences in the prior history of the microbial mixtures.

Table 1
Batch Degradation Rates (ppm/hr)*

Compound	Livingston Sludge	PVSC Sludge
phenol	4.8 (80)	3.9 (94)
2-chlorophenol	0.3 (11)	0.3 (9)
2,6-dichlorophenol	0.07 (.09)	0.09 (.10)
2,4-D **	--- (0.6)	0.6 (---)

* Shown in parenthesis are the degradation rates after prior acclimation to phenol.

** On initial exposure, there was a lag time of about 120 hours before biodegradation could be observed. By the second exposure, the lag time disappeared.

FIGURE 1 - PHENOL DEGRADATION WITH PVSC MIXED LIQUOR

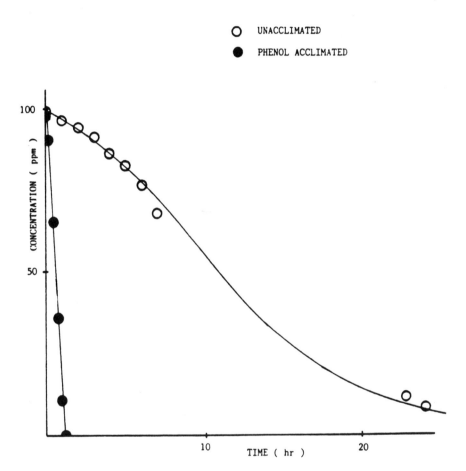

266 Biotechnology for Degradation of Toxic Chemicals

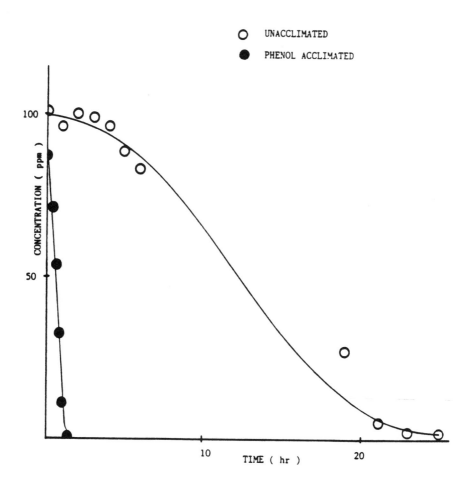

FIGURE 2 - PHENOL DEGRADATION WITH LIVINGSTON MIXED LIQUOR

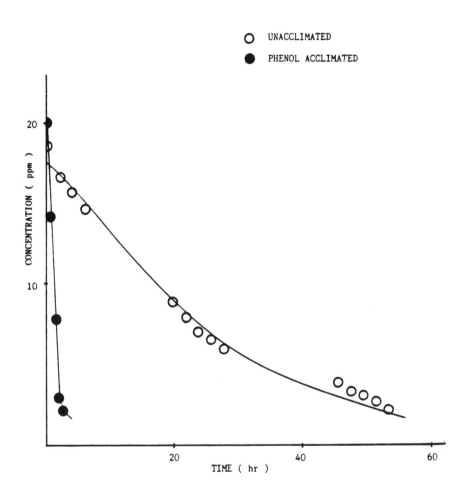

FIGURE 3 - 2-CHLOROPHENOL DEGRADATION WITH PVSC MIXED LIQUOR

268 Biotechnology for Degradation of Toxic Chemicals

FIGURE 4 - 2-CHLOROPHENOL DEGRADATION WITH LIVINGSTON MIXED LIQUOR

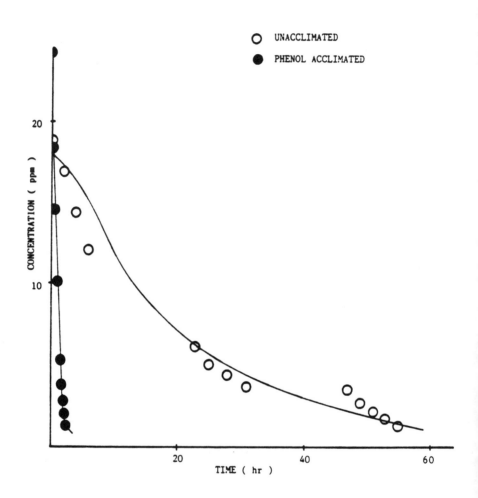

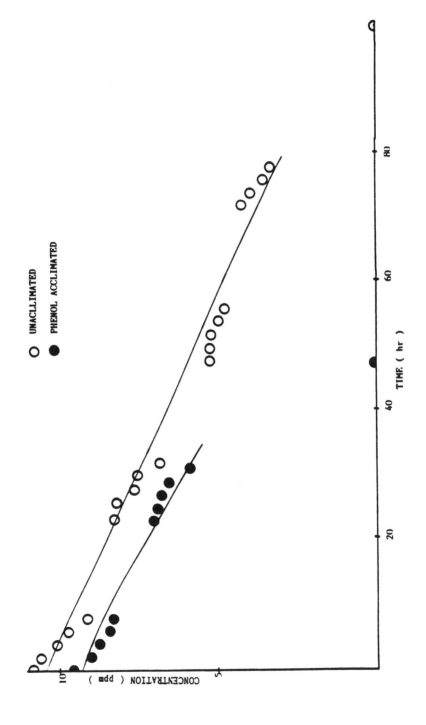

FIGURE 5 — 2,6-DICHLOROPHENOL DEGRADATION WITH PVSC MIXED LIQUOR

○ UNACCLIMATED
● PHENOL ACCLIMATED

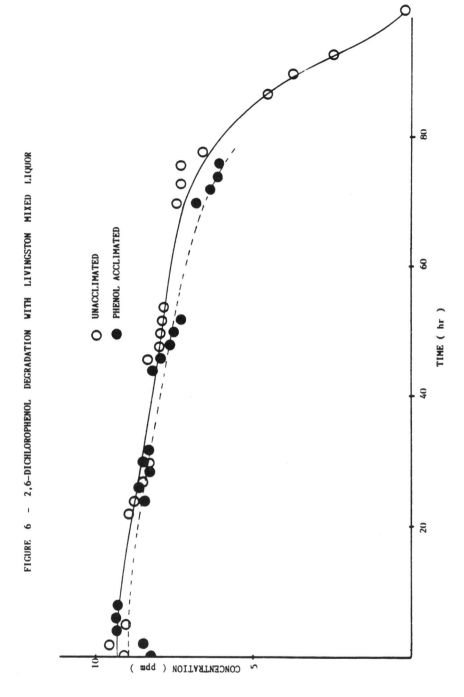

FIGURE 6 — 2,6-DICHLOROPHENOL DEGRADATION WITH LIVINGSTON MIXED LIQUOR

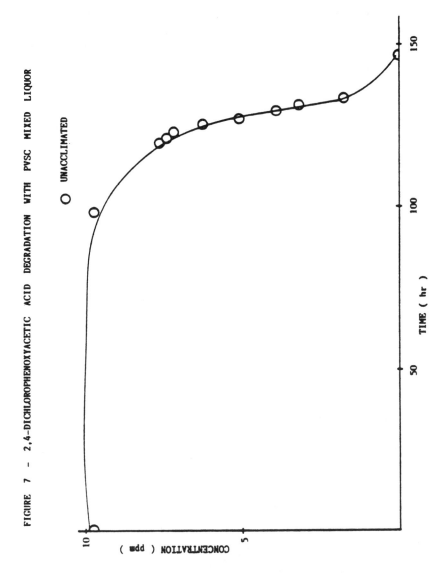

FIGURE 7 - 2,4-DICHLOROPHENOXYACETIC ACID DEGRADATION WITH PVSC MIXED LIQUOR

Results of the microbial characterization tests are shown in Tables 2 and 3. These show surprising similarity between the Livingston and PVSC mixed liquors before as well as after phenolic exposure. Therefore, not only the overall rates of disappearance, but also the microbial responses are quite similar between the two treatment plants. The surprising capacity of the Livingston mixed liquor to adjust to drastic changes in its environment is undoubtedly due to the wide variety of microorganisms present.

These results strongly suggest that the phenomena observed might be generalized, and that POTW's have a considerable capacity to biodegrade toxic chemicals.

ACKNOWLEDGEMENT

This research was sponsored by the NSF Industry/University Cooperative Center for Research in Hazardous and Toxic Substances (Project # BICM-6).

REFERENCES

1. Kincannon,D.F., Stover, E.L., Nichols,V., and Medley,D.,"Removal Mechanisms for Toxic Priority Pollutants", J. Water Pollut. Control Fed., **55**, 157-163 (1983).

2. Gaudy,A.F., Kincannon,D.F., and Manickam,T.S., "Treatment Compatibility of Municipal Waste and Biologically Hazardous Industrial Compounds", vol. I and II, EPA-600/2-82-075 (June,1982).

3. Petrasek,A.C., et.al., "Fate of Toxic Organic Compounds in Wastewater Treatment Plants", J. Water Pollut. Control Fed., **55**, 1286-1296 (1983).

TABLE 2

PREDOMINANT MICROBIAL GENERA IN LIVINGSTON MIXED LIQUOR

Fresh	Phenol-acclimated[*]	Phenol-acclimated 2-Chlorophenol[**]
10^{10} bacteria/cm^3	10^9 bacteria/cm^3	10^9 bacteria/cm^3
<u>gram positive</u>/gram negative 1.3	0.5	0.3

Gram positive rods (Bacillus) ————————————————)
Gram positive cocci (Micrococcus, Staphylococcus) ————)
Pseudomonas ——————————————————————)
Acinetobacter ——————————————————————)
Enterobacter ————————————)
Alcaligenes ——————————)
Serratia ————————————)
Escherichia coli.

| 10^6 yeast cells/cm^3 | 10^6 yeast cells/cm^3 | 10^6 yeast cells/cm^3 |

Candida ———————————————————————————)
Cryptococcus ————————————————————————)
Trichosporon ————————————)
Debaromyces ——————————————)

Penicillium ——————————————————————————)
Gliocladium ————————————)
Streptomyces ——————————————)
Trichophyton.

| 10^5 protozoa/cm^3 | 10^5 protozoa/cm^3 | 10^5 protozoa/cm^3 |

Epistylis ——————————————————————————)
Vorticella ——————————————————————————)
Paramecium ————————————)
Peranema.
Polychaos.
Carchesium.
(also, rotifers and nematodes——))

* 100 ppm for 10 days

** 100 ppm phenol for 10 days, followed by 20 ppm 2-chlorophenol for 10 more days

TABLE 3

PREDOMINANT MICROBIAL GENERA IN PVSC MIXED LIQUOR

	Fresh	Phenol-acclimated[*]	Phenol-acclimated 2-Chlorophenol[**]
	10^9 bacteria/cm^3	10^8 bacteria/cm^3	10^8 bacteria/cm^3
gram positive / gram negative	1.0	0.3	0.3

Gram positive rods (Bacillus) ----------------------------->
Gram positive cocci (Micrococcus, Staphylococcus) ------->
Pseudomonas --->
Acinetobacter --->
Enterobacter ------------------>
Providencia ------------------>
Pasturella.
Alcaligenes.

| | 10^6 yeast cells/cm^3 | 10^6 yeast cells/cm^3 | 10^6 yeast cells/cm^3 |

Candida --->
Cryptococcus -->
Trichosporon -->
Debaromyces --->
Saccharomyces.

Penicillium --->
Aspergillus --->
Streptomyces -->
Trichophyton.
Geotrichum.
Rhodotorula.

| | 10^5 protozoa/cm^3 | 10^5 protozoa/cm^3 | 10^5 protozoa/cm^3 |

Epistylis --->
Opercularia --->
Peranema -->
Colpidium --->
Stylonichia.
Carchesium.
Paramecium.
Podophyra.

* 100 ppm for 10 days

** 100 ppm phenol for 10 days, followed by 20 ppm 2-chlorophenol for 10 more days

SBR TREATMENT OF HAZARDOUS WASTEWATER—
FULL-SCALE RESULTS

Kenneth L. Norcross III
Process Research & Development
Jet Tech, Inc.

Robert L. Irvine
Department of Civil Engineering
University of Notre Dame

Philip A. Herzbrun
Cecos International

ABSTRACT

A preliminary pilot study was completed to evaluate the possibility of utilizing the sequencing batch biological reactor process to pretreat landfill leachate and hazardous waste streams prior to polishing with the existing activated carbon system. The motivation for the study was a desire to reduce carbon regeneration energy and make-up costs.

The results of the pilot study were successful and a full-scale, 550,000 gallon SBR reactor was designed and installed in the winter of 1983/84 in Niagara Falls, New York. Qualitative parameters associated with oxygen tranfer (alpha, beta) were evaluated during the pilot study. The data indicated that alpha increased with increasing aeration turbulence and shear rate. Process performance was shown to be highly temperature sensitive, decreasing rapidly at low temperatures. A submerged jet aeration system was selected for high efficiency, high heat conservation, independent aeration and mixing, and alpha resistance.

After one year of operation, the plant has proven to be stable and effective. The system maintained a mixed liquor temperature 10°C above the influent temperature. Oxygen transfer alpha values were measured to range between 0.75 and 2.0, while beta values ranged from 0.6 to 0.75 (TDS was 30,000 mg/l). The system operated at roughly half the design power requirements. The system maintained stability throughout the entire year except for two heavy metal toxicity strikes which decreased treatment efficiency for several days. The effluent decanting system and rapid settleability of the mixed liquor (SVI averaged 47 ml/g) resulted in minimum effluent suspended solids values.

As a result of the organic and phenol removal efficiencies achieved by the SBR, carbon changes were reduced by about 50%. Annual carbon cost savings are estimated to be $200,000 per year.

*Presented at the 58th Annual Conference of the Water Pollution Control Federation October, 1985

INTRODUCTION

Removal of priority pollutants and hazardous waste compounds is an expensive and vital task of increasing importance. Current techniques are primarily physical-chemical treatment processes which are often exceedingly costly and relatively difficult to operate. Conventional biological activated sludge treatment of these wastes has not proven efficacious. However, two projects completed in 1984 demonstrated the ability of the Sequencing Batch Reactor (SBR) process to achieve removal and destruction of priority pollutants in a cost effective manner (1 & 2). Results of these investigations were based upon bench scale testing of the actual wastewaters involved. No attempt was made to develop accurate aeration system and decant system (effluent withdrawal) design values from these bench-scale studies because scale-up factors are largely unknown. However, qualitative parameters were measured and utilized along with standard indicators, such as oxygen uptake rates, to develop the design parameters upon which the full-scale installations were based.

This paper presents the results of initial operation of a full-scale SBR treating a combined flow consisting of leachate from a hazardous waste landfill and various industrial wastewaters discharged directly to the plant. Aeration system performance parameters will be presented along with flow and organic loading data, effluent quality characteristics, and energy considerations.

BACKGROUND

The Sequencing Batch Reactor (SBR) process is a modification of the activated sludge process wherein each of the unit processes required is completed within a single tank in a sequence of timed events. The process basically consists of five unit processes: FILL, REACT, SETTLE, DECANT, IDLE. Reactions initiated during FILL (when influent enters the tank) are completed during REACT, with no flow entering or leaving the tank. Solids separation is accomplished during a similar, ideal quiescent SETTLE period. Clarified supernatant is discharged during DECANT. While waiting for the start of the next FILL cycle, the system is in IDLE. The SBR process differs from conventional systems in that time is used to separate unit process steps as opposed to multiple dedicated process tanks. This provides powerful flexibility with obvious inherent design, process, and operation advantages. Any unit process operation or sequence can be altered after startup by simply changing time allotments to affect an increase, decrease, or restructuring of any part of the process. Due to the complex and unpredictable nature of this hazardous waste, flexibility is of vital importance to the long term reliability of the treatment system.

Treatment of landfill leachates, hazardous wastes, and priority pollutants is of ever increasing significance as more sites, sources, and compounds are identified, and mandated for cleanup. The SBR modification of the activated sludge process represents a new, unique, and powerful approach to the treatment of these wastes. Alleman et. al. (3) first discussed the application of SBRs to industrial wastes. The SBR is particularly

SBR Treatment of Hazardous Wastewater 277

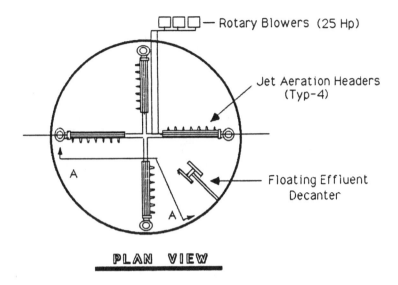

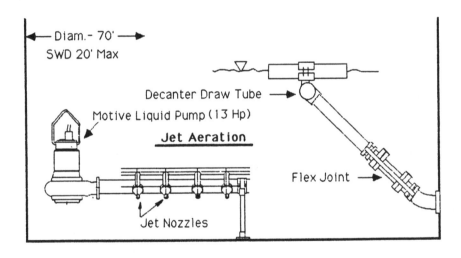

Figure 1 — Single Tank SBR- Cecos International Niagara Falls, NY.

appropriate for this purpose because the process inherently offers influent flow equalization, blending, permanent time-based flexibility, quiescent settling, and the elimination of short-circuiting. Irvine and Richter (4) demonstrated through kinetic analysis and computer simulation that the SBR process offered significant reductions in process volume as compared to a conventional continuous flow system due to increased substrate-removal kinetics achievable in the SBR.

Herzbrun et. al. (5 & 6) demonstrated bench-scale SBR treatment of the hazardous waste at CECOS with hydraulic retention times varying from 2.0 to 10 days. The system achieved TOC degradation from 55% to 81%, and phenol degradation ranging from 97% to 99%. Subsequent bench studies and full-scale tests of aeration mass transfer characteristics are reported in this paper.

Current standards call for treatment of hazardous waste leachates by conventional activated carbon technology. At the CECOS plant, excessive costs and operational difficulties of the carbon system dictate the need to investigate additional treatment techniques. The capital cost for carbon replacement alone was estimated at $21 million over the next ten years. The most significant problem is that adsorption of primary priority pollutants was grossly inefficient because of competition for adsorptive sites by less hazardous organic compounds. Various strategies were investigated for removing the majority of organic carbon prior to filtration. The SBR was selected as the most cost-effective.

DESCRIPTION OF FACILITY

CECOS International operates a 14.9 square kilometer site in Niagara Falls, New York, which is one of only two fully permitted treatment, storage, and disposal facilities in New York State. The treatment plant typically receives 9,500 - 11,300 m^3 (2.5 - 3.0 million gallons) of wastewater each month. Approximately 50% of this is leachate pumped from both active and inactive landfills on the CECOS site, approximatley 30% is pumped as part of a remedial groundwater program from an unrelated facility bordering the CECOS site, and the remaining 20% is received in bulk or drums from various industries in a 640 kilometer radius. The treatment plant is a two-phase operation, with Phase I dedicated to oxidation-reduction reactions, acid neutralization, heavy metals precipitation, and dewatering. Phase II is dedicated to pH adjustment, biological degradation, carbon adsorption, and batch discharge.

The primary purpose of the Phase II operation is to remove organic compounds. This was previously accomplished by adsorption with activated carbon. Because of the high costs associated with the regeneration of activated carbon, CECOS decided to study alternative methods for removing organic carbon. Biological removal in the SBR was selected as the most cost effective alternative. Because of the large storage tanks available for influent holding at the CECOS site, a single tank SBR system was designed. The full-scale plant design and operational parameters were determined from the bench-scale studies by Herzbrun, et. al. (5 & 6)

The SBR installed at the CECOS facility is a 1900 m³ (500,000 gallon) reactor, consisting of a single 21 m (70 ft.) diameter, 6.4 m (21 ft.) tall, insulated steel tank (Figure 1). The aeration and mixing system consists of four directional submerged jet aeration headers, mounted at 90° radii; four 10 Hp non-clog jet aeration pumps, and three 40 Hp rotary positive displacement blowers. Jet aeration was selected for high efficiency, high heat conservation, inherent ability to mix independent of aeration, and efficient operation over varying liquid depths.

Each of the four jet aerators (Figure 1) includes a jet motive liquid pump. The pump recirculates mixed liquor through the aerators and discharges a high velocity stream through the inner nozzle of each compound aeration jet in the direction shown. Compressed air is delivered by any of three rotary blowers to the outer nozzle of each jet. The air is dispersed by the liquid jet stream into the mixed liquor. Since the liquid jet provides mixing and the blowers provide aeration, control of dissolved oxygen is independent of mixing. This feature allows the operator to minimize energy consumption, control substrate to microorganism tension, and achieve anoxic mixed conditions. Further, the conduct of in-process aeration tests is greatly simplified as will be discussed. In addition, a large volume is available for sludge storage. This capacity might be utilized if: sludge disposal was interrupted; increased sludge age was desired; process upset caused poor sludge settling; increased flow rates necessitated more rapid decanting. The decanter draw tube orifices are covered by an FRP hood and flexible diaphram to keep out solids during mixing or aeration. Effluent is discharged to a filter feed tank.

Control of the various process steps, pumps and blowers is provided by programmable controller. The controller continuously monitors and records the mixed liquor dissolved oxygen concentration, various motor statuses, etc.

OXYGEN TRANSFER THEORY

A thorough discussion of oxygen transfer theory is provided in a publication by the Joint Committee of the EPA and the American Society of Civil Engineers (7). A few simplified concepts follow.

For clean water oxygen transfer analysis, the mass transfer of oxygen from air to water is expressed as:

$$dC/dt = K_L a(C_s - C_t)$$ 1.

where:

$K_L a$ = Oxygen mas transfer coefficient, $hr.^{-1}$.

C_s = Dissolved oxygen saturation concentration, mg/l.

C_t = Dissolved oxygen concentration at time t, mg/l.

In wastewater, the effects of contaminents on the mass transfer process must be considered. Surface active agents affect the surface tension at the bubble air/liquid interface. This can in turn affect the aeration rate, typically suppressing, but sometimes augmenting oxygen transfer. For a given installation, the ratio of transfer in wastewater to that in tap water is referred to as the alpha factor:

$$\text{Alpha } (\alpha) = \frac{K_L a_T(\text{waste})}{K_L a_T(\text{water})} \quad \quad 2.$$

where: T = Liquid temperature, °C.

The presence of dissolved solids reduces the saturation concentration of dissolved oxygen in the wastewater. The "beta factor" is defined as:

$$\text{Beta}(\beta) = \frac{C_s(\text{waste})}{C_s(\text{water})} \quad \quad 3.$$

Each dissolved compound or element affects saturation differently. Accurate determination of the beta value requires either direct measurement or a detailed analysis of the dissolved solids. Direct measurement is done either by the Winkler Titration procedure or with meter and probe (8). Both methods, especially the titration, are susceptible to chemical interference and subsequent loss of accuracy. The analysis would be prohibitively costly, and at CECOS, meaningless since the influent is so variable. Further, the results would be of dubious value since the beta effects of most isolated compounds are unkown. The effects on beta of various compounds combined are totally obscure.

Bass and Shell (9) suggest a rough approximation of beta by assuming that the TDS be equated to chloride concentration. The effect of chloride on oxygen saturation is well known. The correlation between chloride and most compounds is obscure or unknown. However, as Bass and Shell correctly state, direct measurement of beta in industrial applications is often erroneous because of chemical interference with the analytical techniques. Where substantial concentration of dissolved solids exist, as at CECOS (up to 30,000 mg/l), estimation of beta is a crude guess at best. The practical method of Bass and Shell was utilized here.

Adjusted for those parameters then, the mass transfer equation for process conditions becomes:

$$dC/dt = \alpha K_L a (\beta C_s - C_t) - r \quad \quad 4.$$

where: r = mixed liquor oxygen uptake rate, mg/l-hr.

For steady state, batch influent systems, the rate of change in dissolved oxygen concentration is zero and the process oxygen transfer expression becomes:

$$\alpha K_L a = \frac{r}{\beta C_s - C_t} \qquad 5.$$

While the SBR is clearly not a steady state system, the behavior over a small time period can be modeled as steady state.

C_s is typically an assumed value for subsurface aeration due to the varied effects of pressure, transfer, and stripping which occur during a bubbles rise to the surface. However, in process conditions, bubble oxygen depletion reduces the value of C_s. In order to double-check the results from equation 5, the value of $\alpha K_L a$ can in turn be measured and the value of βC_s determined from in-process batch unsteady state reaeration data. The technique requires that the process aeration be stopped while complete mixing continues. Fortunately, the jet aerator provides this capability of independent mixing and aeration. With the data obtained, equation 5 can be rearranged:

$$\beta C_s = C_R + \frac{r}{\alpha K_L a} \qquad 6.$$

where: C_R = dissolved oxygen residual concentration, mg/l

OXYGEN TRANSFER EVALUATION PROCEDURE

Each test was initiated by stopping the supply of air (blowers off) and maintaining operation of the jet motive liquid pumps (Fig. 1). The decrease in dissolved oxygen with time was monitored with an in-situ, self-cleaning D.O. probe and recorded by the process controller. After the D.O. approached zero, the blower was restarted with the D.O. monitored.

Each test yields a curve similar to that in Figure 2. The slope of the linear decrease in D.O. is the oxygen uptake rate, r. The D.O. value C_t in equation 5, is C_r, in Figure 2. The reaeration curve in Figure 2 can be analyzed by the non-linear regression technique to determine the value of $\alpha K_L a$, then with the D.O. residuals C_{r1} and C_{r2}, equations 5 and 6 can be solved separately for $\alpha K_L a$ and βC_s, and the results compared.

OXYGEN TRANSFER DISCUSSION

During the initial stages of design, a series of bench-scale oxygen transfer tests was completed on both clean water and process water to establish a design alpha value. It was found that alpha varied dramatically, increasing with increasing aeration intensity, as shown in Figure 3. The high alpha values are typical of jets in liquid with high dissolved solids concentration, as this wastewater contains (15,000 -

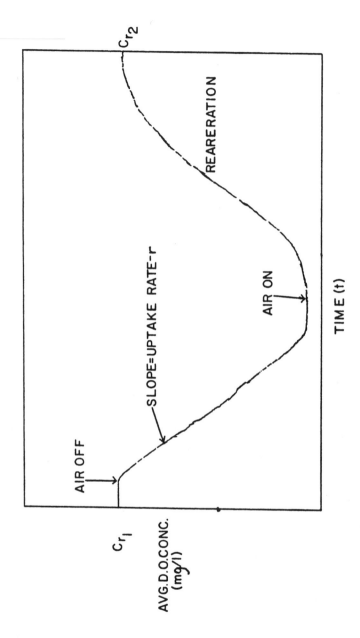

FIGURE 2- OXYGEN TRANSFER TEST DATA, D.O. VS TIME CURVE

35,000 mg/l NFR). While high alpha values were anticipated, a value of 0.9 was used in the design of the aeration system. There is as yet no reliable method for scale-up of alpha values.

The data in Figure 4 summarize the results of about a dozen of the full-scale oxygen transfer tests completed. The results of these tests vary widely, as would be expected with daily influent truck dumping and corresponding variations in feed characteristics. However, the trend indicated was consistent in defining high alpha values.

The data in Table 1 summarize the data from several of the tests completed, and illustrate the variability of the waste stream character, and its effect on aeration performance. The two methods of calculating C_s yield somewhat different results. However, the method based on equation 6 results in the most consistent oxygen transfer calculations and appears to be more accurate. The average alpha value observed is about 1.6.

Interestingly, the apparent alpha value often seemed to decrease as air flow increased, contrary to the bench test. However, this is due more to the effect of decreased bubble rise rate at higher air flows than to alpha. The jet system is oriented to rotate the basin liquid to increase mixing and transfer efficiency, (Figure 1 - plan view). At high air flow rates, the bubbles are not sheared as fine at the jet. The turbulent rise of these bubbles can reduce spin velocity. At lower air flow rates, the action of the jets with the high TDS liquid generates extremely fine bubbles at high shear. Spin velocity, and higher efficiencies, are maintained. The enhanced aeration performance is evident from the data. As a result of this increased efficiency, the system was operated with only one instead of the anticipated two 40 Hp blowers most of the time. D.O. values still approached saturation by the end of most REACT cycles.

PROCESS RESULTS

As mentioned previously, the influent flow to the CECOS SBR is highly variable in both flow and character. Figure 5 shows the variation in flow rate and T.O.C. concentration over a typical two week priod. The nature of leachate is extremely variable. The truck dump is even more so. The influent to the SBR includes such compounds as phenol and many chlorinated, nitrated, or other phenol-based comounds; heavy metals such as copper, nickle, barium, zinc; high concentrations of ammonia; and miscellaneous compounds many of which are not identifiable from GC/MS analyses.

In spite of the difficult nature of the waste, the SBR has maintained high levels of treatment efficiency. The data in Table 2 summarize the first three months of operations data. The average TOC removal was 76% at an average flow rate of 69,000 gpd. Phenol removal exceeded 97%. Settling behavior of the MLSS was excellent as indicated by the low SVI value observed. The plant operator and supervisor reports the process to be stable and resistant to upset.

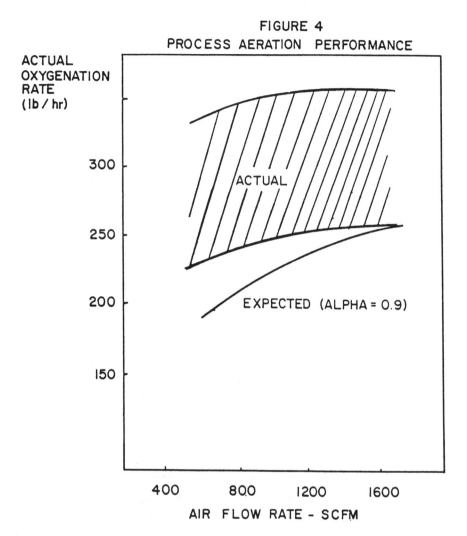

FIGURE 4
PROCESS AERATION PERFORMANCE

TABLE 1 - AERATION PERFORMANCE PARAMETERS FROM SEVERAL TESTS

TEST	DATE	TEMP °C	TDS mg/l	AIR SCFM	r mg/l-hr	$K_L a_t$ hr^{-1}	D.O. mg/l	$C_{S_{cal}}$* mg/l	$C_{S_{est}}$* mg/l	AOR_{cal}* lb O_2/hr	AOR_{est}* lb O_2/hr	ALPHA
1**	10/15	11.4	5,000	800	----	7.3	---	13.0	12.3	355	230	1.72
2**	10/15	11.5	5,000	1,600	----	6.8	---	12.4	12.1	314	310	1.13
3	6/15	28.0	12,000	800	3.9	9.4	7.4	7.9	8.4	267	203	1.46
4	7/3	32.0	26,000	800	4.4	8.6	6.2	6.7	6.3	256	160	1.78
5	7/3	32.0	26,000	800	17.7	7.3	4.8	7.2	6.3	231	160	1.60
6	7/3	32.0	26,000	1,600	17.6	13.1	4.5	5.8	6.2	343	218	1.75

* $C_{S_{cal}}$ and AOR_{cal} are the values calculated from Equations #5 and #6, $C_{S_{est}}$ and AOR_{est} are estimated using standard clean water assumptions, a beta corresponding to the TDS as chloride value, and an alpha of 0.9. AOR values are calculated for maximum driving force or zero D.O. conditions.

** Results from unsteady state reaeration test on influent in full-scale tank.

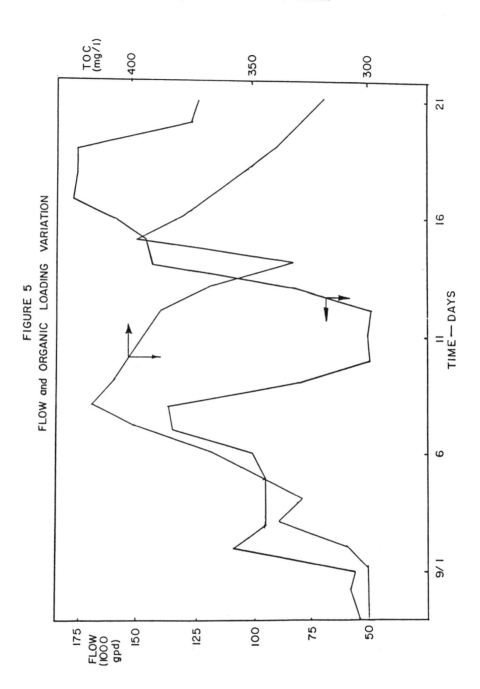

FIGURE 5
FLOW and ORGANIC LOADING VARIATION

TABLE 2

PROCESS PARAMETERS

Average Value for Initial 3 Month Operation

FLOW	TOC		PHENOL		MLSS	MLVSS	SVI
	Influent	Effluent	Influent	Effluent			ml/gm
69,000	1330	320	34.2	0.9	2540	1650	47

All concentrations are in mg/l

A classic upset response is shown in Figure 6. An accidental release of copper to the SBR influent resulted in the mixed liquor copper concentration exceeding 12 mg/l. Biological activity, as evidenced by the spiked oxygen uptake rate, virtually halted. However, over the next several weeks, the activity steadily improved. Within four weeks full activity was restored. During the entire period, the system treatment efficiency was maintained at reduced feed rates. The effluent never missed permit levels.

One of the primary concerns arising from the preliminary pilot studies was the drastic effect of reduced temperature on treatment efficiency. The system was designed to take full advantage of the high thermal conservation efficiency reported for submerged jet aerators (10), but concerns persisted. The data of Figure 7 show that maximum heat savings were indeed realized. The SBR mixed liquor temperature was maintained 8°C to 11°C warmer than the influent temperature with no additional heating.

As a result of the success of the SBR, the frequency of carbon filter recharge was greatly reduced. The data in Table 3 show the required number of carbon changes and flows treated at the facility over a 5 year period. Note that prior to SBR startup in 1984, carbon changes were required after every 630,000 gallons of throughput, roughly. After the SBR startup, this was reduced to less than one recharge for each 2.5 million gallons, or less than 25% as often. CECOS estimates savings in excess of $200,000 annually on carbon alone.

SUMMARY

The jet aerated SBR installed at CECOS Ineternational, Niagara Falls Hazardous Waste Facility has operated with stability in spite of drastic variations in influent flow rates and content. Phenol removals exceeded 97%, T.O.C. removals averaged 76%, and carbon regeneration frequency was reduced 75%, for the initial period of operation. Carbons and power savings realized by the SBR is estimated to save the owner over $200,000 annually. The system recovered from a heavy metal toxicity strike quickly and completely.

The jet aeration system has operated with alpha values of about 1.6; heat loss so low that mixed liquor temperatures were maintained 8°C to 11°C above the influent; and high aeration efficiency which resulted in approximately half the design power draw required.

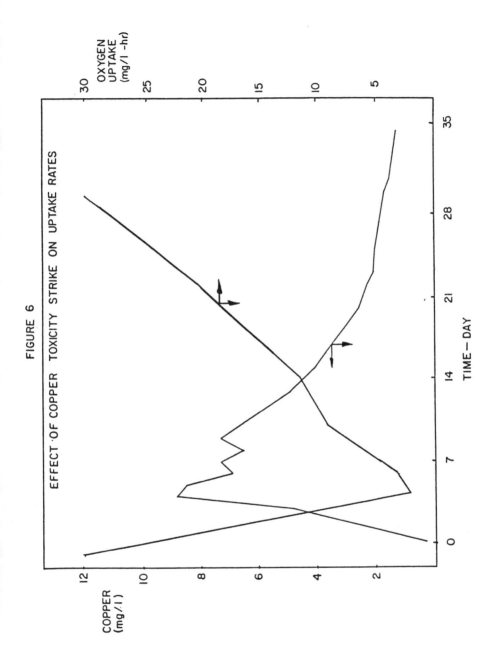

FIGURE 6
EFFECT OF COPPER TOXICITY STRIKE ON UPTAKE RATES

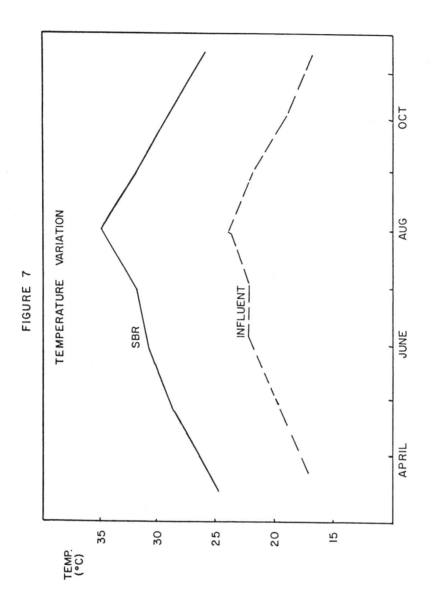

FIGURE 7. Temperature variation.

TABLE 3 - CARBON FILTER REGENERATION HISTORY

Year	Million Gallons Processed	Carbon Changes	Gal x 10^6/Change
1980	2.92	5	0.58
1981	2.02	3	0.67
1982	4.95	8	0.62
1983	9.29	14	0.66
1984	9.95	4	2.50

REFERENCES

1. Irvine, R.L., Sojka, S.A., and Colaruotolo, J.F., "Enhanced Biological Treatment of Leachate from Industrial Landfills", Hazardous Waste, 1, 1, 1984.

2. Bell, B.A., and Hardcastle, G.J., "Treatment of a High-Strength Industrial Waste in a Continuously Fed, Intermittently Operated, Activated Sludge System", Journal Water Pollution Control Federation, 56, 1160, 1984.

3. Alleman, J.E., Irvine, R.L., and Dennis, R.W., Sequencing Batch Reactor Application to Industrial Waste Treatment", Water Research, Date Unknown.

4. Irvine, R.L., and Richter, R.O., "Comparative Evaluation of Sequencing Batch Reactors", Journal of the Enviromental Engineering Division, ASCE, 104, #EE3, 1978.

5. Herzbrun, P.E., Irvine, R.L., and Hanchak, M.J., "Treatment of Hazardous Wastes in a Sequencing Batch Reactor", Presented at the 39th Purdue Industrial Waste Conference, 1984

6. Herzbrun, P.E., Irvine, R.L., and Malinowski, K.C., "Biological Treatment of Hazardous Waste in the SBR", Journal of the Water Pollution Control Federation, 57, 1163, 1985.

7. ASCE/EPA, "Development of Standard Procedures for Evaluating Oxygen Transfer Devices", EPA-600/2-83-102, October 1983.

8. "Standard Methods for the Examination of Water and Wastewater", 16th Edition, American Public Health Association, 1985.

9. Bass, S.J. and Shell, G., "Evaluation of Oxygen Transfer Coefficients in Complex Wastewaters", Proceedings, 32nd Purdue Industrial Waste Conference, 1977, pp 953-967.

10. Norcross, K.L., "A Full-Scale Evaluation of Thermal Efficiency for a Mechanical Surface and a Submerged Jet Aeration System", Presented at the 58th Annual Conference of the Water Pollution Control Federation, 1985.

THE USE OF PURE CULTURES AS A MEANS OF UNDERSTANDING THE PERFORMANCE OF MIXED CULTURES IN BIODEGRADATION OF PHENOLICS

G. Lewandowski, B. Baltzis and C. Peter Varuntanya

Department of Chemical Engineering, Chemistry and Environmental Science
New Jersey Institute of Technology
Newark, New Jersey 07102

In an effort to gain a more fundamental understanding of the performance of mixed microbial cultures in the biodegradation of toxic organic chemicals, studies have been initiated using pure species from the same mixed population. Three phenol degrading species have been isolated from the secondary treatment tanks at the Passaic Valley Sewerage Commissioners (PVSC) plant in Newark. The rate of phenol degradation was investigated for: (1) each of the three individual species, (2) different mixtures of the pure species, and (3) the mixed culture from PVSC. Experimental results, as well as kinetic modelling of the data, are presented.

BACKGROUND

Any kinetic model which adequately fits the experimental data of a waste treatment process must be used with caution. No safe predictions can be made before a detailed study is performed of the phenomena occuring in the system. As has been argued [1] "any formal model even if it 'works' presents just a more or less reasonable approximation to real processes and phenomena". Actually, when a mixed culture is used, one expects that there are a number of interactions taking place between microbial species (such as competition and/or synergistic and antagonistic effects), and consequently an overall kinetic model is only an apparent one. The result has been a plethora of kinetic expressions, which are largely empirical in nature. For example, a number of expressions have been proposed for phenol biodegradation, including the Haldane [2,3,4] , Monod [5,6,7], simple first order [8], and second order [9] models. These studies exhibit an entire spectrum of different conditions, ranging from the composition of the culture (types of microorganisms present) to the initial phenol concentration and the prevailing pH, which makes any effort at comparison almost impossible.

It was therefore the purpose of the present work to attempt a more systematic construction of a mixed liquor model using data from pure species studies (as described by Monod kinetics).

PROCEDURES

Two liters of PVSC mixed liquor were aerated at 0.5 to 1.0 liter/min. in a clear plastic batch reactor (see Figure 1). Phenol was added to the reactor to bring the concentration to 100 ppm. When the phenol concentration fell below 1 ppm (the accuracy of the GC analysis using direct aqueous injection), phenol was again added to bring it back up to 100 ppm. After the third time, 10 ml of inoculum from the reactor were taken and diluted to 100 ml with distilled water. The diluted solution was then inoculated on nutrient agar plates, and incubated for 24 hours at 37 $^\circ$C.

After 24 hours, representative bacterial colonies showing morphological differences were selected and streaked onto fresh nutrient agar plates. They were allowed to grow at 37 $^\circ$C for another 24 hours.

Successive species isolations were needed in order to assure that the isolated cultures were purified.

Once purified, the individual species were maintained in nutrient broth, and identified using a modification of the standard taxonomic techniques

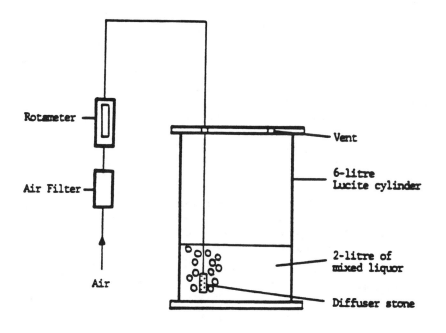

Figure 1

(see Figure 2). (Note: the Enterotubes and Oxiferm tubes are manufactured by Hoffmann-LaRoche, and allow up to 15 biochemical tests to be performed conveniently on a bacterial colony.)

Figure 3 shows the predominant genera in PVSC mixed liquor. Three categories of microbial organisms are indicated: bacteria, yeast and fungi, and protozoa. They are shown in two groups: those which were isolated from the fresh mixed liquor, and those which were isolated after phenol acclimation. Of these, only the phenol-acclimated bacteria were of interest in this research. The population density of the bacteria decreased on phenol exposure (as did the population diversity), and gram negative bacteria predominated.

Figure 4 shows the results of three investigators, using various batches of the PVSC mixed liquor. Some differences in dominant species are noted, but in general the results are very similar. This paper will concentrate on the third group of results.

PHENOL-DEGRADING SPECIES

Each isolated bacterial colony was tested for its ability to degrade phenol over a two-week period. Only three of the species listed for Experimentor III (Figure 4) consumed phenol: Klebsiella pneumoniae, Serratia liquefaciens, and Pseudomonas putida. The presence of other species in the phenol-acclimated liquor is apparently a result of their utilization of metabolic products from the primary phenol degraders.

The three phenol degrading species are also available commercially (see Figure 5). However, none of the commercially obtained species were able to degrade phenol in the laboratory. This indicates that there are strains of the same species which have very different responses to a given carbon source.

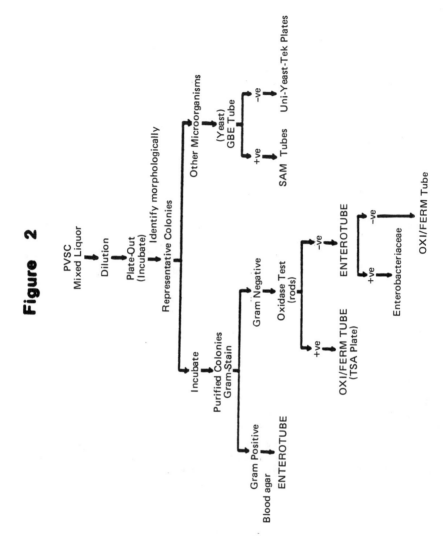

Figure 2

Figure 3

PREDOMINANT MICROBIAL GENERA IN PVSC MIXED LIQUOR

Fresh	Phenol-acclimated[*]	Phenol-acclimated 2-Chlorophenol[**]
10^9 bacteria/cm^3	10^8 bacteria/cm^3	10^8 bacteria/cm^3
gram positive / gram negative 1.0	0.3	0.3

Gram positive rods (Bacillus) ─────────────────────────⟩
Gram positive cocci (Micrococcus, Staphylococcus) ─────⟩
Pseudomonas ───⟩
Acinetobacter ───⟩
Enterobacter ──────────────────⟩
Providencia ──────────────⟩
Pasturella.
Alcaligenes.

10^6 yeast cells/cm^3	10^6 yeast cells/cm^3	10^6 yeast cells/cm^3

Candida ───⟩
Cryptococcus ──⟩
Trichosporon ──⟩
Debaromyces ───⟩
Saccharomyces.

Penicillium ───⟩
Aspergillus ───⟩
Streptomyces ──⟩
Trichophyton.
Geotrichum.
Rhodotorula.

10^5 protozoa/cm^3	10^5 protozoa/cm^3	10^5 protozoa/cm^3

Epistylis ───⟩
Opercularia ───⟩
Peranema ──⟩
Colpidium ───⟩
Stylonichia.
Carchesium.
Paramecium.
Podophyra.

* 100 ppm for 10 days

** 100 ppm phenol for 10 days, followed by 20 ppm 2-chlorophenol for 10 more days

Figure 4

MICROBIAL SPECIES IN PHENOL-ACCLIMATED PVSC MIXED LIQUOR

EXPERIMENT I

Achromobacter specied biotype 2
Acinetobacter lwoffii
Bacillus
Enterobacter agglomerans
Enterobacter gergoviae
Escherichia coli
Micrococcus
Moraxella species
Group M-4 Moraxella-like
Pseudomonas cepacia
Pseudomonas fluorescens
Pseudomonas maltophilia
Pseudomonas species
Group 2K-1 Pseudomonas-like
Serratia marcescens
Staphylococcus

EXPERIMENT II

Acinetobacter anitratus
Acinetobacter lwoffii
Alcaligenes faecalis
Bacillus
Enterobacter agglomerans
Micrococcus
Providencia stuarti
Pseudomonas aeruginosa
Pseudomonas capacia
Pseudomonas fluorescens
2K-1 Pseudomonas-like
5E-1 Pseudomonas-like

EXPERIMENT III

Acinetobacter lwoffii
Aeromonas hydrophilia
Bacillus cereus
Enterobacter cloacae
Escherichia coli
• Klebsiella pneumoniae
Pseudomonas cepacia
Pseudomonas fluorescens
• Pseudomonas putida
Pseudomonas species
• Serratia liquefaciens

Figure 5

TESTED COMMERCIALLY PREPARED PURE CULTURES

1. Bacillus cereus (ATCC 8100)
2. Klebsiella pneumoniae (ATCC 11778)
3. Serratia marcescens (ATCC 13883)
4. Serratia marcescens (Pigmented)
5. Enterobacter cloacae (Carolina 15-5032)
6. Escherichia coli (Carolina 15-5065)
7. Pseudomonas fluorescens (Carolina 15-5255)
8. Pseudomonas putida (Carolina 15-5265)

PURE-CULTURE EXPERIMENTS

Stock cultures of the three phenol degrading species were maintained by periodic subculture on Difco-Bacto nutrient broth and incubated at 37 °C for 10-14 hours (see Figure 6). The primary culture was prepared by transferring a loop of stock culture to 10 ml of nutrient broth and again incubated for 10-14 hr. The secondary culture was prepared by transferring 2.5 ml of primary culture to 50 ml of a diluted defined medium solution [10] (Figure 7) containing different phenol concentrations. Phenol is the rate limiting substrate. Tertiary cultures were also inoculated to insure that the cells had fully developed and adapted to growth on the phenolic medium solution. The tertiary cultures were grown in either the defined medium solution, or in a mixture of filtered supernatant solution from the original mixed liquor plus defined medium solution.

The cells were grown in 250 ml nephelometric shaker flasks, with the initial phenol concentrations ranging from 10 to 200 mg/liter. The cultivation time was 10-14 hours. There were no baffles, and no aeration other than that transferred by shaking. The course of growth in each flask was assessed optically as percent transmittance at 540 nm, and converted to biomass concentration (see Figure 8). The initial slope of the optical density plot on semi-log paper is the specific growth rate (μ) and is representative of the exponential growth phase. Figure 9 shows the results from seven batch reactors using different initial phenol concentrations, with Klebsiella pneumoniae.

These results were then used with the Monod model to determine the maximum specific growth rate and saturation constant for each of the three phenol-degrading species (Figure 10).

Figures 11 indicates the rate of phenol degradation for one of the batch experiments, and Figures 12 shows the relationship between biomass growth and phenol utilization (with the slope [Y] equal to the yield coefficient).

The Use of Pure Cultures 301

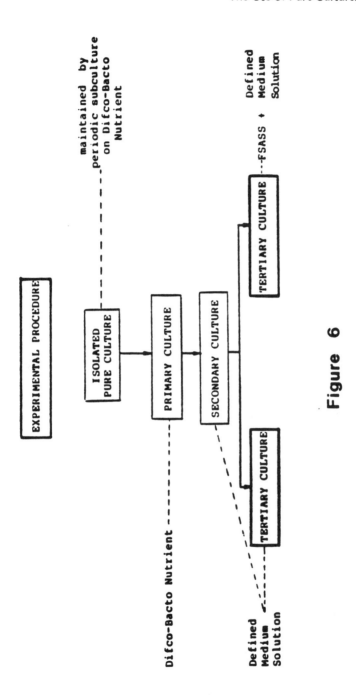

Figure 6

Figure 7

DEFINED MEDIUM SOLUTION

Phenol	1000	mg
Ammonium Sulfate	500	mg
Magnesium Sulfate	100	mg
Ferric Chloride	0.5	mg
Magnesium Sulfate	10	mg
1.0 M. Potassium Phosphate Buffer Solution (pH 7.2)	30	ml
Tap Water	100	ml
Distilled Water	to volume of 1 litre	

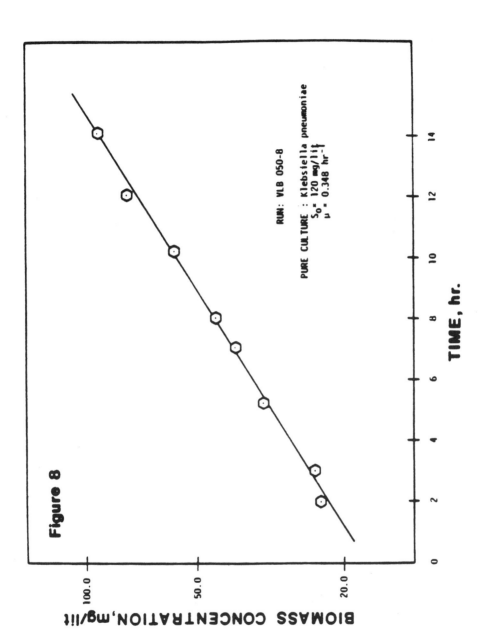

Figure 8

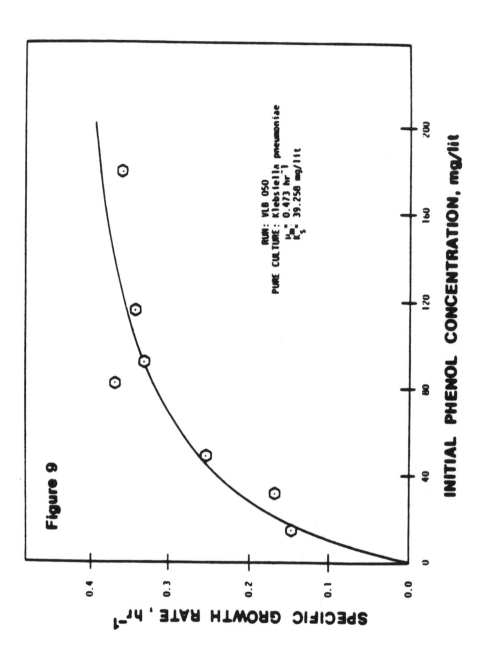

Figure 9

Figure 10

MONOD PARAMETERS
FOR THREE PURE PHENOL-DEGRADING SPECIES

	μ_{max} (hr^{-1})	K_s (mg/liter)
Klebsiella pneumoniae	0.473	39.3
Serratia liquefaciens	0.401	42.4
Pseudomonas putida	0.117	11.0

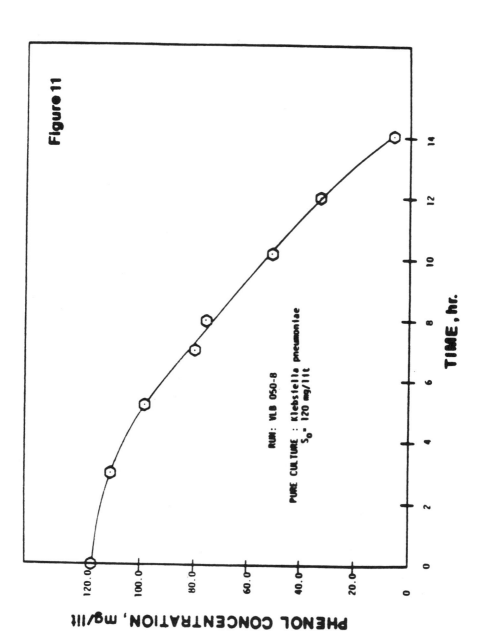

Figure 11

The Use of Pure Cultures 307

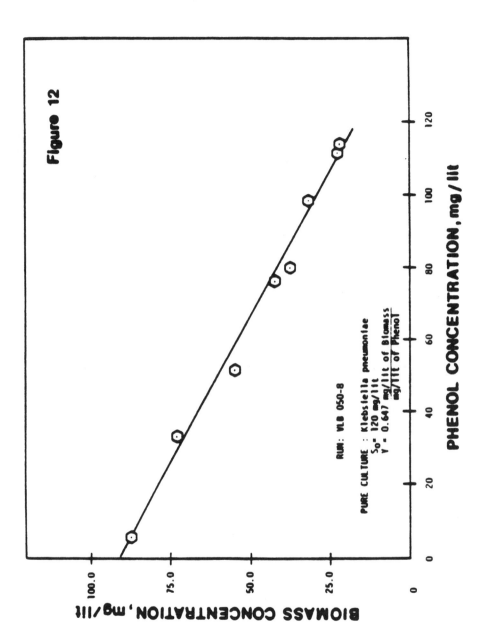

Figure 12

RUN: VLB 050-8
PURE CULTURE: Klebsiella pneumoniae
S_0 = 120 mg/lit
Y = 0.647 $\frac{mg/lit \text{ of Biomass}}{mg/lit \text{ of Phenol}}$

MIXED-CULTURE EXPERIMENTS AND KINETICS

Figure 13 shows a simple competitive model for two species degrading the same substrate. This model was then used to calculate the theoretical rate of phenol degradation for the mixed cultures.

Figures 14 to 17 are examples of mixed-culture data, with the theoretical curve superimposed. These indicate that the simplified model is adequate for some conditions, but not for others. There may be other interactions between the species that the simplified model does not take into account (such as secondary metabolite utilization, or inhibition). These additional interactions will continue to be explored in an effort to improve and expand upon the kinetic model.

CONCLUSIONS

(1) Eleven dominant bacterial species were isolated from a phenol-acclimated mixed liquor obtained originally from the Passaic Valley Sewerage Commissioners wastewater treatment plant in Newark, N.J. However, of these eleven species, only three (Klebsiella pneumoniae, Serratia liquefaciens, and Pseudomonas putida) were able to degrade phenol. Therefore, the remaining eight species must have survived by utilizing the metabolic products of the three primary phenol degraders.

(2) Regarding the three primary phenol degraders, when the same species were purchased from commercial suppliers, they could not degrade phenol, which underlines the importance of the strain as well as the species.

(3) Using the kinetic parameters from the single species experiments, a simple competitive model was tested for phenol utilization by any two of the three primary phenol degraders. This model was able to predict the rate of total biomass growth fairly well, but was much less accurate in predicting the rate of substrate utilization. This indicates that simple competition for the same substrate is not an adequate physical model for the mixed culture

$$\frac{dC_1}{dt} = \frac{\mu_{m1} S}{K_{s1} + S} C_1$$

$$\frac{dC_2}{dt} = \frac{\mu_{m2} S}{K_{s2} + S} C_2$$

$$\frac{dS}{dt} = -\frac{1}{Y_1} \frac{\mu_{m1} S}{K_{s1} + S} C_1 - \frac{1}{Y_2} \frac{\mu_{m2} S}{K_{s2} + S} C_2$$

Figure 13

310   Biotechnology for Degradation of Toxic Chemicals

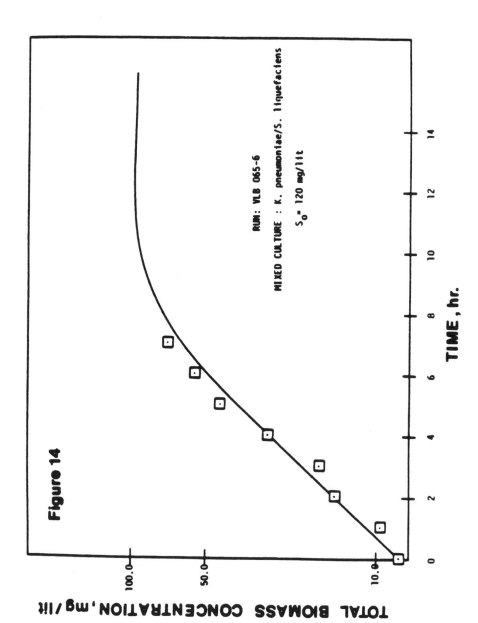

Figure 14
RUN: VLB 065-6
MIXED CULTURE : K. pneumoniae/S. liquefaciens
$S_o = 120$ mg/lit

The Use of Pure Cultures 311

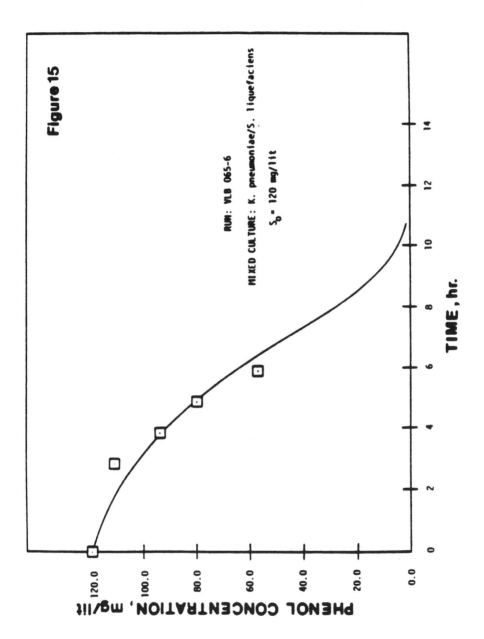

Figure 15

RUN: VLB 065-6

MIXED CULTURE: K. pneumoniae/S. liquefaciens

S_0 = 120 mg/lit

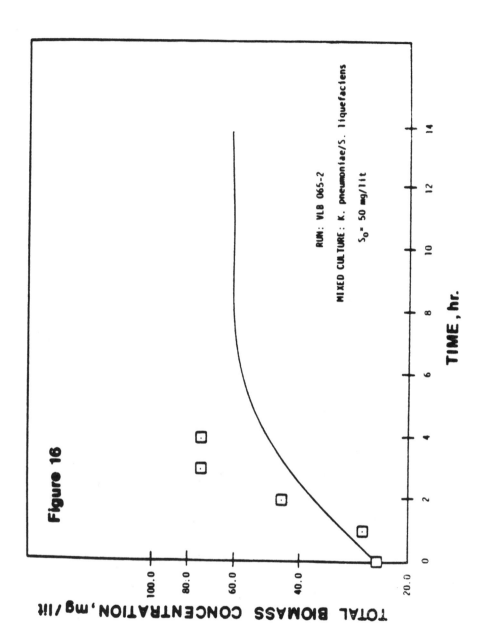

Figure 16

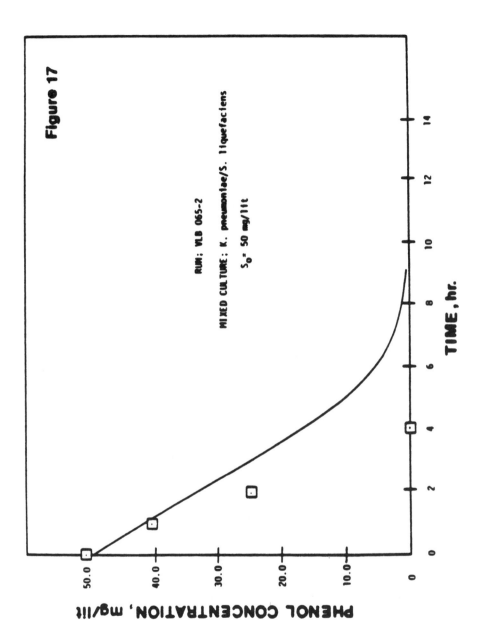

Figure 17

RUN: VLB 065-2

MIXED CULTURE: K. pneumoniae/S. liquefaciens

S_o = 50 mg/lit

system. It may be, for example, that one (or both) of the organisms is producing an inhibitory agent for the growth of the competing organism.

REFERENCES

(1) Vavilin, V.A., "Dependence of Biological Treatment Rate on Species Composition in Activated Sludge or Biofilm, II: From Models To Theory", Biotech. and Bioeng., **25**, 1539(1983).

(2) Hill, G.A., and C.W. Robinson, "Substrate Inhibition Kinetics: Phenol Degradation by Pseudomonas putida", Biotech. and Bioeng., **17**, 1599 (1975).

(3) Pawlowsky, U., et.al., "Mixed Culture Biooxidation of Phenol, I: Determination of Kinetic Parameters", Biotech. and Bioeng., **15**, 889 (1973).

(4) Yang, R.D., et.al., "Dynamic and Steady State Studies of Phenol Biodegradation in Pure and Mixed Cultures", Biotech. and Bioeng., **17**, 1211 (1975).

(5) Beltrame, P., et.al., "Kinetics of Phenol Degradation by Activated Sludge in a Continuous Stirred Reactor ", JWPCF, **52**, 126 (1980).

(6) Drummond, C.J., et.al., "Biological Oxidation of Coal Conversion Wastewaters", AIChE Symp. Series, **76**, 209 (1979).

(7) Kim, J.W., and N.E. Armstrong, "A Comprehensive Study on the Biological Treatabilities of Phenol and Methanol", Water Research, **15**, 1221 and 1233 (1981).

(8) Luthy, R.G., et.al., "Treatment of Coal Coking and Coal Gasification Wastewaters", JWPCF, **53**, 325 (1981).

(9) Paris, D.F., N.L. Wolfe, and W.C. Steen, "Development of Quantitive Relationships between Microbial Degradation Rates and Chemical Structures ", paper presented before Div. Environ. Chem., ACS Meeting New York, N.Y. (August 1981).

(10) Peil, K.M., and Gaudy, A.F., "Kinetic Constants for Aerobic Growth of Microbial Populations Selected with Various Single Compounds and with Municipal Wastes as Substrates ", Appl. Microbiol., **21**, 253 (1971).

A COMPARTMENTALIZED ONE SLUDGE BIOREACTOR FOR SIMULTANEOUS REMOVAL OF PHENOL, THIOCYANATE, AND AMMONIA

Jeffrey H. Greenfield

Department of Civil and Environmental
Engineering
Florida International University
Miami, Florida

Ronald D. Neufeld

Department of Civil Engineering
University of Pittsburgh
Pittsburgh, Pennsylvania

INTRODUCTION

There are many "advanced wastewater" treatment processes based on physical-chemical and biological techniques for the removal of ammonia, however, these processes are not often employed in industry due to economics, marginal overall removal efficiencies and questionable technological application to specific industrial operations. Of the processes available for the removal of ammonia from steel industry wastewaters, engineered biological systems are the most pragmatic. Data from the steel industry indicate that a properly designed and operated biological nitrification facility is capable of reducing ammonia levels to less than 10 mg/L over extended periods of time. Biological nitrification processes, however, are known to exhibit unaccounted for upsets, and thus are considered in some industrial sectors as unreliable (1).

Effluents from coking, coal gasification, and other coal carbonization processes often contain high levels of ammonia, as well as a variety of additional constituents such as polynuclear aromatic hydrocarbons, phenolics, thiocyanates, cyanides, sulfides, sulfates, and carbonates. Concentrations of individual constituents from such effluents vary from process to process due to differing conditions of coal carbonization, gas scrubbing temperatures, and type of feedstock used (2).

BAT guidelines for coke plants indicate that while a two-stage (two-sludge) biological process may be used for carbonaceous and nitrogenous removal respectively, it is possible that economical and stable stage combined systems may be designed for the same application. While this is common for municipal systems, it is uncommon in practice for industrial systems. Understanding of fundamentals for the proper design and operation of single stage systems for nitrification is important for companies that already own and operate such systems as originally constructed to meet BPT phenol effluent limitations, and are now being required to upgrade these facilities to accommodate recent ammonia and phenol effluent limitations.

LITERATURE REVIEW

Numerous research studies have been conducted to investigate industrial one stage nitrification process experiments. This literature review will present information on these process experiments.

Olthof (3) performed treatability studies on coke oven wastewater using a two stage system and a one stage system. Results of this study showed that the removal rate for ammonia is a function of the BOD/N ratio. The two stage system resulted in a better effluent quality than a one stage system when the total detention time in the biological treatment system was the same.

Wong-Chong and Caruso (4) examined a single stage reactor for the control of nitrogen compounds. The kinetics of the biological reactions involved in the treatment of coke plant wastewaters were examined in an effort to understand the different reactions, interactions, and conditions under which the process functions.

Wong-Chong and Caruso (5) also investigated a single stage, activated sludge process for the treatment of coke plant wastewater for the control of ammonia, free cyanide, phenol, and thiocyanate. A laboratory scale study demonstrated that if the treatment system is designed and operated for ammonia oxidation, treatment of free cyanide, phenol, and thiocyanate are achieved; treated effluent concentrations for those parameters were observed in this study to be less than 0.1, 0.1, and 1.0 mg/L respectively.

One of the most comprehensive laboratory scale research studies of single stage, phenol oxidation-nitrification, activated sludge treatment of coke plant wastewater was conducted by Wong-Chong and Hall (6). The results of this study show that the single stage, phenol oxidation-nitrification process, operated at 85°F can produce high degrees of treatment for ammonia, free cyanide, phenol, thiocyanate, and sulfide but was ineffective in treating complex cyanide. The data from this study indicated that phenol was easily degraded but thiocyanate and ammonia effluent concentrations were still greater than 10 mg/L when the system was operated at HRT's of 3-5 days and at extremely long sludge ages attained by no deliberate sludge wasting. This process was also effective in controlling priority organic pollutants found in coke plant wastewater. Sudden changes in the reactor loadings of conventional pollutant constituents resulted in neither toxic nor prolonged inhibitory effects. However, the process was sensitive in responding to abrupt changes in feed and reactor composition. Wong-Chong and Hall (6) also stated that operation of a one stage bioreactor at sludge ages of 30-50 days would cause the nitrifiers and thiocyanate organisms to be washed from the reactor.

Jones et al. (7) studied one stage biological treatment of high strength, coke plant wastewater using a bench-scale, complete-mix activated sludge process. These investigators concluded that high influent loadings of coke plant wastewater may lead to the development of filamentous growth. Once these filamentous growths occur, treatments such as alkaline pH, exposure to hydrogen peroxide, and wasting at 4% or greater do not reduce the filament population. They also reported that feeding a reactor increased organic loadings (above what is normally found in ammonia still effluent) while wasting at a rate of 10% results in loss of nitrification, while wasting at 2% while feeding intermediate levels of TOC (350 mg/L), results in successful treatment.

Adams (8) investigated the biological treatment for a weak ammonia liquor from a coke plant containing only ammonia and phenolic compounds. He evaluated both single and multi-stage activated sludge systems by performing tests on one and two sludge systems. Based on the results of his research, he concluded that nitrification could be effectively maintained in a combined system. However, the limiting organic loading, above which nitrification ceases, appeared to be 0.3 lb BOD/day/lb MLVSS. Nitrification removal efficiencies, considering the available ammonia-nitrogen, were as high as 73.5 in a single stage system.

Argamon (9) reported on the single-sludge nitrification and denitrification process including the economic advantages of the single-sludge system over the multi-sludge systems, a description of the processes, a discussion of the kinetics, and two examples of the treatment process.

Medwith and Lefelhocz (10) examined a single stage, high rate, hybrid, suspended-growth, fixed-film biological treatment process for simultaneous carbonaceous and ammonia-nitrogen oxidation of coke plant wastewaters by combining the advantages of the rotating biological surface and the suspended-growth activated sludge technologies. Finely divided coke breeze or coal dust was used as the inert substrate material which acted as a site for biological growth and a "weighting" agent for increasing the site and density of the biomass. The increase in aeration basin bacterial population and residence time produced effective phenol, thiocyanate, and ammonia oxidation under conditions of a 12 to 24 hour hydraulic retention time, temperatures of 20-30°C, pH range of 6.5 to 8.5, and dissolved oxygen concentration of 1-2.5 mg/L.

Nutt et al. (11) conducted pilot-scale treatability studies and demonstrated that the two-stage fluidized bed is an effective system for the treatment of coke plant wastewater alone and in combination with blast furnature blowdown water. Greater than 90 percent removal of total nitrogen from undiluted coke plant wastewater was achieved at a system hydraulic detention time of 16 hours. Removal of conventional contaminants including filtered organic carbon, phenolic compounds and thiocyanate approached or exceeded 90 percent consistently using coke plant wastewater with or without blast furnace blowdown water.

METHODS AND MATERIALS

Experimental System

Four rectangular Plexiglas aeration tanks with outside dimensions of 17.5 inches x 12.5 inches x 12 inches deep with 6 inches of water were constructed for this research. The aeration basins had a working volume of 20 liters. These tanks were subdivided into 1,2,3, and 4 compartmented bioreactors by Plexiglas baffles. These baffles had a 1 inch hole located 2.5 inches from the bottom which allowed for flow from compartment to compartment. The baffles were secured by silicone rubber adhesive sealant so the baffles could be removed and repositioned. Mixing was achieved by the use of Marineland bubble wand diffusers in each compartment so that each compartment acted as a completely mixed reactor. Each reactor had an external clarifier which was a 1 liter styrene-acrylonitrile Imhoff settling cone. Recirculation was achieved by the use of Masterflex pumps with Model 7015 drives attached to a Veeder-Root Industrial timer, Model CM-8, with a Model A-12 gear rack. Return sludge was pumped to the first compartment every hour at a rate of approximately 140 ml/min for three minutes. This provided a recirculation ratio of 0.5. Each aeration tank was also equipped with a Horizon Ecology pH controller, Model 5997-20, which maintained the pH of the aeration tanks at 7.0 ± 0.2 by the use of Masterflex pumps using dilute solutions of sodium bicarbonate. The feed pumps, Gorman-Rupp Model M14250, were used to deliver the synthetic feed to the first compartment at a rate of 14 ml/min. This flow rate provided a hydraulic detention time of 1 day.

Operation of Bioreactors

The phenol and thiocyanate bacteria were obtained from a previous research project. These bacteria were acclimated to a synthetic feed solution containing reagent phenol and potassium thiocyanate for approximately six months prior to this study. The nitrifier organisms had been grown and cultivated in pure culture by this investigator for four years prior to this research. At the beginning of this project, these two different cultures were mixed creating the one sludge bioreactors.

Synthetic coke plant wastewater was used for this research. The main constituents were ammonium chloride, potassium thiocyanate, and phenol. The feed solution also contained sodium bicarbonate in a ratio of alkalinity to ammonia by weight of approximately 7:1 as well as potassium phosphate (monobasic) for cell maintenance. This feed solution was prepared daily to a volume of 19 liters using tap water. By preparing the feed solution daily, the effect of bacterial degradation in the influent was minimized. All the bioreactors for this study were operated at room temperature ($22 \pm 2°C$).

Experimental Methods

The influent and compartment effluents were measured for ammonia, phenol, and thiocyanate. Ammonia determinations were performed daily on the reactor effluents using an Orion research ammonia electrode, Model 95-12, attached to an Orion research Model 601A digital Ionalyzer. Nitrate and nitrite determinations were also performed using an Orion Model 93-07 nitrate electrode and an Orion Model 95-46 Nitrogen Oxide probe with a standard reference probe. Effluent samples from each compartment as well as influent samples were taken twice a week, where possible, and filtered through a Millipore pressure filter, Model YY22 142 00, equipped with 0.45 um filter paper, for ammonia, phenol, and thiocyanate determinations. Phenol (4-AAP) (Method 510.6) and thiocyanate determinations (413 K) were made in accordance with Standard Methods (12).

Mixed liquor suspended solids and volatile suspended solids determinations were made twice a week according to the procedures outlined in Standard Methods (Methods 208 D and E) (12). Effluent suspended solids determinations were also made to determine the concentration of solids lost in the effluent. Sludge volume index tests were performed following the procedure in Standard Methods (Method 213 E) (12). Dissolved oxygen measurements were monitored by the use of a YSI Model 54 A oxygen meter.

In addition, the influent and recycled sludge flow rates were measured and adjusted when necessary. Excess sludge was wasted from the last compartment and was replaced with an equal volume of tap water.

RESULTS AND DISCUSSION

All four bioreactors were operated under similar conditions. First, the mixed cultures were fed ammonium chloride to establish nitrification. When this was established, reagent phenol was added to the feed solution in low levels. After the phenol effluent was analyzed, thiocyanate was added to the feed. This acclimation pattern was followed for each bioreactor.

In order to meet the objectives, the bioreactors were designed to be 1,2,3, and 4 compartment bioreactors. The only difference between the reactors was the number of compartments. For this study, the individual compartment volumes within each individual reactor were equal.

One Compartment Bioreactor

The data from the one compartment bioreactor demonstrated that the biota in this configuration were capable of nitrification when the phenol, thiocyanate, and ammonia influent was 100 mg/L, the HRT was 1 day, and when no sludge was wasted except for the sludge lost in the effluent. However, when the sludge was wasted from the aeration tank and the influent thiocyanate and phenol were raised to 200 mg/L/day, both ammonia and thiocyanate breakthrough were experienced coupled with bulking conditions.

Two Compartment Bioreactor

The removal efficiencies for all three substrates were excellent for the two compartment bioreactor as demonstrated on Figure 1. The two compartment reactor was extremely stable and demonstrated greater than 99% removal efficiencies for all three substrates while 2% sludge wasting was practiced. The actual sludge age based on the wasting policy as well as the sludge lost in the effluent was 42 days. The sludge in this configuration demonstrated excellent settleability characteristics with a SVI from 30-60 cc/g. Table I shows the experimental data for this reactor. This data shows that the removal efficiencies within the first compartment were over 98% with the second compartment acting as a polishing basin for thiocyanate and ammonia.

Table I. Two Compartment Experimental Data

"Steady-State" Period:Day 78-112

Reactor Influent:

Thiocyanate = 348 $\pm$ 13 mg/L
Phenol = 298 $\pm$ 30 mg/L
Ammonia = 196 $\pm$ 21 mg/L

Hydraulic Detention Time = 1 Day
Actual Sludge Age = 42 Days

Effluent Compartment #1 (mg/L)
Thiocyanate = 4.31 $\pm$ 2.69
Phenol = 0.07 $\pm$ 0.04
Ammonia = 4.20 $\pm$ 1.12

Effluent Compartment #2 (mg/L)
Thiocyanate = 0.40 $\pm$ 0.33
Phenol = 0.05
Ammonia = 0.61 $\pm$ 0.35

Average Reactor Volatile
Suspended Solids = 3600 $\pm$ 385 mg/L

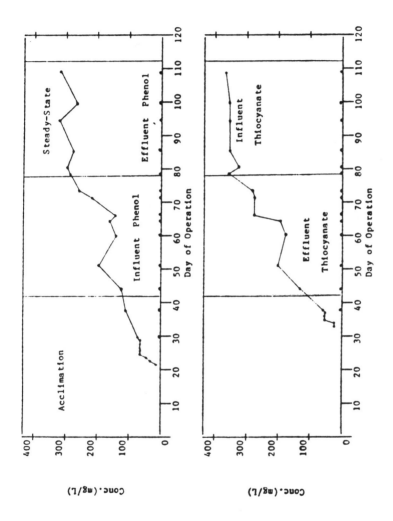

Figure 1. Two Compartment Bioreactor Performance Characteristics.

324 Biotechnology for Degradation of Toxic Chemicals

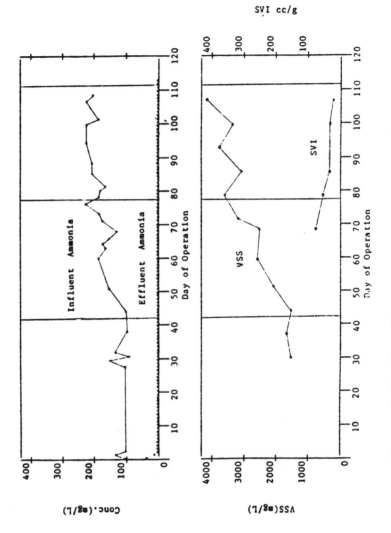

Figure 1 (Continued).

Three Compartment Bioreactor

The three compartment reactor demonstrated excellent removal efficiencies when the system was operated while wasting at a rate of only 1%. The actual sludge age during this steady-state period considering the wasting rate and the sludge lost in the effluent was 63 days. After the wasting policy was modified to 2%, the reactor began to fail. Bulking conditions were easily diagnosed during daily operation by observing the settling in the clarifier for this configuration. Bulking became so severe, that one could easily see the bacterial solids being washed out over the clarifier.

There are several possible explanations for the sludge bulking that was observed and documented in this system. Since the reactor was subdivided into three equal compartments and the feed was directed to the first compartment, possibly the organic loading was too high and the compartments too small. This research has shown that the effluent from the first compartment demonstrated greater than 95% removal for all three substrates. It is possible that the bacteria in the second and third compartment had insufficient quantities of substrate thereby causing the bulking conditions.

The experimental data for this system showed that the three compartment bioreactor experienced severe sludge settleability problems when the influent feed solution contained over 350 mg/L thiocyanate and 300 mg/L phenol and when 2% sludge wasting was practiced.

It was predicted that the greater the number of compartments, the lower the reactor effluent substrate concentration. Data from this three compartment reactor demonstrated that no significant benefit was gained over the two compartment bioreactor by the addition of the third compartment. Even when the two compartment system was operated under similar conditions as the three compartment configuration, the final effluent from the two compartment reactor was nearly equal.

Four Compartment Bioreactor

The four compartment reactor data was very similar to the three compartment reactor. The effluent substrate concentrations for phenol, thiocyanate, and ammonia were excellent until the sludge wasting policy was increased to 2% and the influent phenol and thiocyanate were increased. This resulted in sludge bulking coupled with thiocyanate and ammonia breakthrough.

The experimental results for this reactor showed that there was no benefit gained over the three compartment bioreactor by the addition of the fourth compartment. The three compartment reactor was even more stable than the four compartment configuration when both were operated under similar conditions. The four compartment reactor demonstrated instability for thiocyanate biodegradation.

CONCLUSIONS

Based upon laboratory data obtained, the following conclusions may be drawn:

1. The one compartment bioreactor displayed 99% removal efficiencies for phenol, thiocyanate, and ammonia when the influent concentrations for all three substrates was approximately 150 mg/L and when the reactor was operated at an actual sludge age of 43 days. When the influent substrate concentrations were raised to values greater than 200 mg/L/day, both ammonia and thiocyanate breakthrough were experienced coupled with bulking conditions.

2. The two compartment bioreactor displayed greater than 99% removal efficiencies for thiocyanate, phenol, and ammonia when the reactor consisted of $SCN^- $ = 350 mg/L, phenol = 300 mg/L, and NH_3 = 200 mg/L at an actual sludge age of 42 days. The biota in this reactor demonstrated excellent settleability characteristics having a SVI equal to 30-60 cc/g.

3. Experimental results from the three and four compartment reactors at sludge ages of 63 and 57 days respectively, exhibited similar effluent phenol and ammonia concentrations. Bulking was experienced with the three and four compartment configurations possibly due to the lack of sufficient substrates in the latter compartments.

4. The steady-state results from the 1-4 compartment reactors displayed greater than 90% removal of all three substrates in the first compartment with the remaining compartments functioning as final polishing tanks.

5. By utilizing compartmentalized reactors, the working volume of the aeration basin can be reduced by as much as one half without a loss in efficiency.

ACKNOWLEDGEMENTS

This research was carried out at the Department of

Civil Engineering, University of Pittsburgh, Pittsburgh, Pennsylvania, with R.D. Neufeld as Principal Investigator. At the time of this study, J.H. Greenfield was a graduate student within the Environmental Engineering Program at the University of Pittsburgh.

REFERENCES

1. Costle, D.M. et al., Development Document for Effluent Limitations Guidelines and Standards for the Iron and Steel Manufacturing Point Source Category, Volume II, EPA/440/1-80/024-b, Dec., 1980.
2. Forney, A.J., Haynes, W.F., Gasior, S.J., Johnson, G.E., and Strakey, J.P., Analysis of Tars, Chars, Gases, and Water Found in Effluents from the Synthane Process, U.S. Bureau of Mines, TPR, Jan., 1974, p. 4.
3. Olthlof, M., "Nitrification of Coke Oven Wastewater With High Ammonia Concentration," Pro. of the 34th Annual Purdue Industrial Waste Conference, May, 1979, pp. 22-35.
4. Wong-Chong, G.M., and Caruso, S.C., "Biological Oxidation of Coke Plant Wastewaters for the Control of Nitrogen Compounds in a Single Stage Reactor," Proc. of the Biological Nitrification/Denitrification of Industrial Wastes Workshop, Burlington, Oct., 1977, 38p.
5. Wong-Chong, G.M., and Caruso, S.C., "Biological Treatment of By-Product Coke Plant Wastewater for the Control of BAT Parameters," U.S. EPA and the American Iron and Steel Institute, EPA-600/9-82-021, 1982, pp. 446-459.
6. Wong-Chong, G. and Hall, J.D., "Single Stage Nitrification of Coke Plant Wastewater," U.S. EPA and the American Iron and Steel Institute, EPA-600/9-81/017, 1981, pp. 395-456.
7. Jones, D.D., Speake, J.L., White, J., and Gauthier, J.J., "Biological Treatment of High Strength Coke-Plant Wastewater," Proceedings of the 38th Annual Purdue Industrial Waste Conference, 1983.
8. Adams, C.E., Jr., "Treatment of a High Strength Phenolic and Ammonia Wastestream by Single and Multi-Stage Activated Sludge Processes," Proceedings of the 29th Annual Purdue Industrial Waste Conference, 1974, pp. 617-629.
9. Argamon, Y., "Single Sludge Nitrogen Removal from Industrial Wastewater," Wat. Sci. Tech., Vol. 14, 1982, pp. 7-20.

10. Medwith, B.W. and Lefelhocz, J.F., "Single-Stage Biological Treatment of Coke Plant Wastewaters With a Hybrid-Suspended Growth-Fixed-Film Reactor," <u>Proc. of the 36th Annual Purdue Industrial Waste Conference</u>, 1981, pp. 68-76.
11. Nutt, S.G., Melcer, H., Marvan, I.J., and Sutton, P.M., "Treatment of Coak Plant Wastewater With or Without Blast Furnace Blowdown Water in a Two-Stage Biological Fluidized Bed System," U.S. EPA and the American Iron and Steel Institute, EPA-600/9-83-016, 1983, pp. 300-315.
12. <u>Standard Methods for the Examination of Water and Wastewater</u> (14th Edition), American Public Health Association, American Water Works Association, and Water Pollution Control Federation, 1975.

HIGH-RATE BIOLOGICAL PROCESS FOR TREATMENT OF PHENOLIC WASTES

Alan F. Rozich, Richard J. Colvin, and Anthony F. Gaudy, Jr.

Department of Civil Engineering
University of Delaware
Newark, DE

INTRODUCTION

The application of biological processes to the treatment of toxic and hazardous wastes has the potential for economical improvement of technology and disposition and destruction of these poisonous materials. Currently three general approaches are utilized for the treatment of toxic organic wastes: physical-chemical processes, incineration, and biological processes. Physical-chemical processes such as adsorption are probably most economically effective when contaminated waste streams or waters contain very dilute ($<$ 1 mg/L) concentrations of undesirable organic constituents. In contrast, incineration is likely to be practicable for those waste materials that are either in semi-solid form or in high ($>$ 1% by weight) concentrations because lower concentrations of waste would not burn without inordinate energy expenditures. Because a significant portion of toxic waste streams fall in an intermediate concentration range, where either physical-chemical or incineration systems are not economically feasible, biological processes have been suggested as a potential treatment alternative. The purpose of this report is to describe the results of bench-scale pilot plant work which indicate the potential for significant enhancement of the rate of biological treatment of phenol wastes.

When the topic of enhanced biodegradation rates of toxic wastes is discussed, scientists and engineers inevitably consider genetically modified microorganisms. Although such an approach may eventually yield results, there are a number of crucial microbiological, biochemical, and physiological questions that must be addressed before the application of these organisms to field treatment situations can be contemplated. An alternative to the genetic engineering approach consists of utilizing the engineering parameters which are available for controlling a biological reactor to select for populations which have enhanced biodegradative ability for toxic organics. Given the diversity which characterizes natural populations, it should be possible to manipulate a reactor environment for the purpose of maximizing the rate of degradation of toxic organics. The practical questions which are associated with achieving this goal concern the identification of the environmental characteristics which select for organisms with rapid degradative ability and the concomitant control strategy that will permit the maintenance of these cultures under continuous flow conditions.

Previous results (1) of studies on the biodegradation of phenol, using a relatively novel enrichment culture technique, indicated that the maintenance of relatively high (~ 100 to 200 mg/L) steady-state phenol concentrations selected for microbial populations which exhibited rapid growth on this target organic. This work suggested the feasibility of developing a continuous flow biological process which would be characterized by significantly higher removal rates for phenol wastes. The current work presents the results of subsequent bench-scale studies which indicate the potential for using such a process for treating phenolic waste streams with higher efficiency than would otherwise be achieved with conventional biological technology. Also, suggested process configurations for field applications are discussed, as well as a potential approach for developing a design and operational model for the new process. Finally, considerations and suggestions are outlined regarding developmental needs that should be addressed before this technology is applied in field situations.

BACKGROUND

The mechanistic basis for suggesting the high-rate process modification for the treatment of phenol wastes is the fact that it is possible to select organisms which are capable of rapid phenol degradation. Recent results from our laboratories (1) indicated the efficacy of continuously culturing high-rate biodegraders on phenol. The key to selecting these cultures evidently consisted of growing heterogeneous populations at relatively high steady-state phenol concentrations. The growth of microbial cultures at high steady-state substrate concentrations is normally achieved by increasing the dilution rate (reducing the detention time) in a chemostat; however, this is not practicable for compounds such as phenol because they elicit substrate inhibition (as opposed to Monod) biodegradation kinetics. Figure 1 compares the difference between Monod kinetics and inhibitory kinetics; the latter are often represented by the Haldane function. Chemostats that are treating substances such as phenol that are characterized by inhibition kinetics become unstable and wash out once the dilution rate causes the growth rate to reach the peak growth rate, μ^*, (see Figure 1) which is designated by the inhibition function. More details regarding these aspects of the treatment of inhibitory wastes are given elsewhere (2,3,4). The instability problems associated with the continuous cultivation of organisms at high phenol concentrations can be eliminated by growing cells in the second stage of a two-stage system such as that depicted in Figure 2. This continuous culture technique has been successfully employed for growing a pure bacterial culture at high phenol concentrations (5) and was employed to demonstrate the inhibition characteristics of ammonia-limited nitrification (6).

Figure 3 shows typical steady state data for heterogeneous populations growing on phenol in a two-stage system such as the one depicted in Figure 2. The biomass and substrate concentrations in both reactors were relatively steady. The results of five steady-state runs such as that shown in Figure 3 (1) indicated that continuous cultivation of organisms on phenol in the inhibitory range of the growth curve was possible. The continuous flow studies were augmented by batch growth studies using seed from both reactors as inocula; this was done in order to compare the batch growth rates on phenol of cells cultured under low substrate concentrations in reactor 1 with those of cells grown at high concentrations of phenol in reactor 2. The results from seventeen batch growth studies, seven for the R1 cells and ten for the R2 cells, were averaged and are plotted in Figure 4. The data in this figure indicate that the cells cultivated in R2 at the high phenol concentrations are capable of much higher growth and substrate

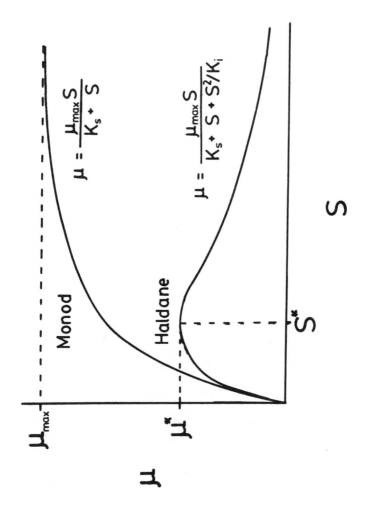

Figure 1. Comparison of Monod (non-inhibitory) and Haldane (inhibitory) growth kinetics.

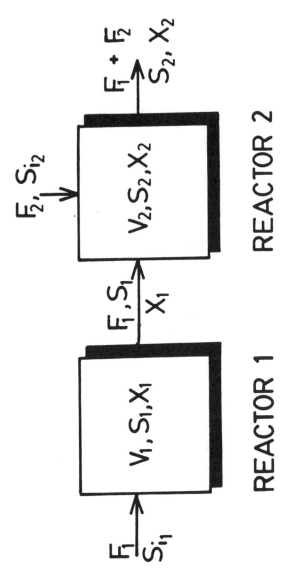

Figure 2. Schematic of two-stage continuous culture system.

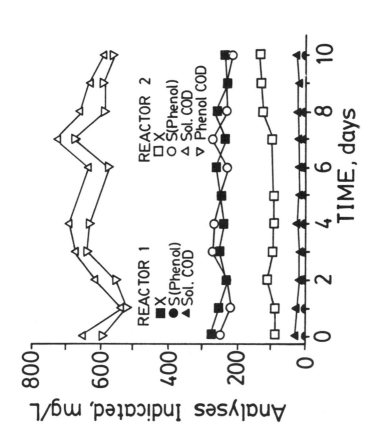

Figure 3. Steady-state operational characteristics for a two-stage continuous culture system. S_1 and S_2 were both 500 mg/L phenol (1190 mg/L COD). Additional details are given in reference (1).

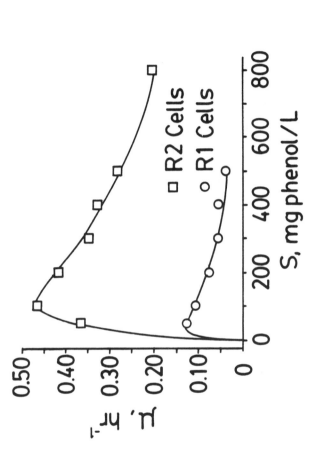

Figure 4. Plot of average growth rates detected in batch growth studies for cells cultured at low (R1) and high (R2) steady state phenol concentrations in a two-stage continuous culture system. Values for μ_{max}, K_s, and K_i were 0.19 h^{-1}, 7.9 mg/L, and 139 mg/L, respectively for the R1 cells and 1.07 h^{-1}, 79 mg/L and 172 mg/L, respectively for the R2 cells. Additional details are given in reference (1).

utilization rates than are those organisms cultured at low concentrations. Analysis of comparative continuous flow data yields similar conclusions (1). Thus, it was concluded that the continuous cultivation of heterogeneous populations at high steady-state phenol concentrations selected for organisms which exhibited enhanced ability to degrade toxic organics.

It is interesting to compare the performance of a reactor which is treating phenol and contains the high-rate biodegraders with that of one that contains a relatively low-rate or "normal" population. In previous work (3), we described the results of research which dealt with predicting the operating point at which completely-mixed activated sludge reactors would fail when treating a phenol waste. As part of this effort, a peak growth rate for this system was predicted using batch tests and verified in the continuous flow system. The observed substrate utilization rate of the activated sludge system immediately prior to failure was 3.63 mg COD/mg X-d (7); it should be stressed that operation at this rate is not advocated since it resulted in failure of the system, but it does represent the absolute upper limit at which a single-stage system could perform. The substrate utilization rates which are calculated for the high-rate (second stage) reactor from the data of Colvin and Rozich (1) were in excess of 30 mg COD/mg X-d which is almost a tenfold increase. A comparison of the μ^* values which were measured in batch tests (1,3) also indicates large differences in degradative ability. Thus, ample evidence had been gathered which suggests that it is feasible to enhance phenol degradation by selectively culturing the appropriate populations.

There are many applications in biological waste treatment that require that a portion of a reactor or system be utilized as a selector to cultivate a particular microbial population in order to meet specific treatment goals. For example, operation of an activated sludge reactor at relatively long θ_c's (> 5 days) is recommended to select for nitrifying populations in order to remove ammonia from wastewaters via nitrification (8). Also, the Bardenpho process advocates the utilization of an anoxic zone in order to achieve denitrification (9) while the A/O process uses an anaerobic zone to encourage the proliferation of cells that can assimilate and store phosphorus (10). Chiesa, et al. (11) suggest the use of a variety of environmental conditions that are subject to control in a sequencing batch reactor to encourage feast/famine conditions in order to select for populations with good settling characteristics. For realizing high-rate treatment of phenol wastes, one reactor can be employed as a selector to grow high-rate organisms. It should be noted that the observation that cells can be selected for rapid phenol biodegradation is mechanistically analogous to the reports of other workers with regard to microbial growth on noninhibitory substrates. For instance, several researchers (12,13,14) have reported that acclimated cells which were grown in chemostats at high dilution rates and high substrate concentrations exhibited higher μ_{max} values than cells which were cultured at low dilution rates and correspondingly low substrate concentrations. Thus, the basis for the high-rate phenol process rests on relatively sound principles and observations that have been made in microbial ecology.

MATERIALS AND METHODS

Bench-Scale Apparatus

A flow diagram of the bench-scale pilot plant apparatus which was used in the experimental work is given in Figure 5. The system consisted of two reactors, Reactor 1 and Reactor 2, which had volumes of 1.25 L and 4.0 L, respectively. For this work, Reactor 1 operated at high steady-state phenol concentrations and Reactor 2 was utilized to reduce phenol concentrations to low ($<$ 100 µg/L) levels.

The composition of the synthetic phenolic waste which was utilized in this work is given elsewhere (1); phenol (C_6H_5OH) was the sole carbon source in this waste. Ammonia nitrogen, phosphorous, and other nutrient salts were added in excess to insure that phenol was the growth-limiting substrate. Reactor pH was maintained at 7.2 by using automatic pH control and temperature was kept at 22 ± 1°C.

The system was started by inoculating each reactor with phenol-acclimated biomass which had been growing in continuous flow reactors. Continuous-flow operation started with an influent flow rate, F_1, to the system of 0.5 L/h and an influent phenol concentration of 500 mg/L (1190 mg/L COD), while the recycle flow rate, F_2, was set at 0.1 L/h. It is interesting to note that the development of a high-rate population in Reactor 1 required only approximately 2 days and was achieved with relatively little difficulty.

When the system reached steady state, cells were harvested from each reactor and utilized as inocula in separate batch growth studies for the purpose of quantifying the growth kinetics of the cells cultured in the two reactors; this would provide a comparison of the batch growth response of the cells cultured at the different steady-state phenol concentrations. The methodology for performing the batch growth studies is

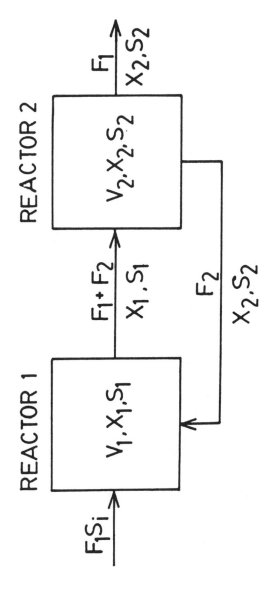

Figure 5. Schematic of bench scale system for high rate treatment of phenolic wastes.

given elsewhere (2,15).

Analyses

A number of chemical and physical analyses were performed during the steady-state studies. Phenol concentrations were assessed using the 4-amino-antipyrine technique (16). Low phenol concentrations (0 to 1.0 mg/L) were analyzed by a procedure similar to that employed by Yang and Humphrey (17); in the range of 0 to 1.0 mg/L a 5.0 ml sample size was used. Chemical oxygen demand (COD) was measured using the Hach method (18). Biomass concentration in the reactors was evaluated as suspended solids concentration by filtration through a 0.45 µm filter, drying at $103^{\circ}C$, and weighing; for batch growth studies, biomass concentration was measured as optical density against a distilled water blank on a Bausch and Lomb Spectronic 20 spectrophotometer at a wavelength of 540 nm.

PROCESS DESCRIPTION

Bench Scale Results

The basic idea behind the high-rate process is to reverse the flow of the system depicted in Figure 2. That is, in order to use the high-rate cells in a treatment process, it is necessary to grow them in the first stage of the system and use the second stage to achieve effluent discharge requirements. With this goal in mind, it seemed appropriate to assess the efficacy of running such a system using a bench-scale apparatus. The bench-scale system followed the flow scheme which is depicted in Figure 5. This scheme "reversed" the flow of the two-stage system that is depicted in Figure 2 in order to determine whether it is feasible to utilize high-rate biodegraders in a treatment process.

A bench-scale pilot system which utilized the flow scheme that is given in Figure 5 was employed to treat a synthetic phenol waste with a concentration of 500 mg/L phenol (1190 mg/L COD). The results from this run are depicted in Figure 6 while a summary of averaged data is provided in Table 1; these data suggest a number of interesting points regarding the feasibility and utility of the high rate technology. The soluble substrate data clearly show that, over a relatively long term, (20 days) it was possible to maintain Reactor 1 at relatively high (~ 300 mg/L phenol) steady-state phenol concentrations while achieving low concentrations (~ 30 µg/L phenol) in Reactor 2; in effect, this demonstrated that the two-stage

340 Biotechnology for Degradation of Toxic Chemicals

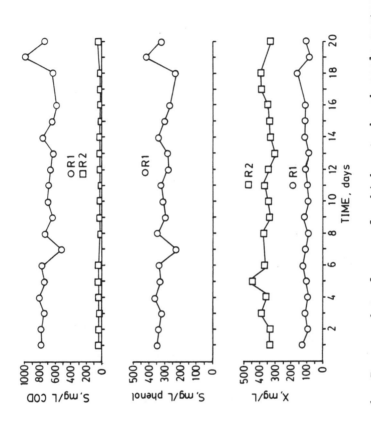

Figure 6. Process data for run for high rate bench scale system depicted in Figure 5. Note: all S_2 phenol values were < 100 µg/L.

scheme (1) which was previously utilized to select for high-rate phenol degraders could be reversed and employed as a treatment process.

The averaged data given in Table 1 can be utilized to calculate the growth and substrate utilization rates which were

Table 1. Summary of Continuous Flow Data for High Rate Bench Scale System (Figure 5)

S_i mg/L	F_1 L/hr	F_2 L/hr	μ_1^+ hr^{-1}	μ_2^{++} hr^{-1}	Analysis Indicated mg/L	No. of Data Points	Mean Value	Standard Deviation	Coefficient of Variation %
					$\bar{X}_1$	19	111	17.6	15.9
					$\bar{X}_2$	18	352	42.8	12.2
500	0.5	0.1	0.226	0.103	$\bar{S}_1$ (Phenol)	19	316	44.1	13.9
					$\bar{S}_2$ (Phenol)	18	0.027	0.032	118.5
					$\bar{S}_1$ (Soluble COD)	19	717	101.2	14.1
					$\bar{S}_2$ (Soluble COD)	18	37	8.7	23.4

$$+ \quad \mu_1 = \frac{F_1 + F_2(1 - X_2/X_1)}{V_1}, \quad V_1 = 1.25 \text{ L}$$

$$++ \quad \mu_2 = \frac{(F_1 + F_2)(1 - X_1/X_2)}{V_2}, \quad V_2 = 4.0 \text{ L}$$

realized in the two reactors; expressions for calculating these rates can be derived by using a mass balance analysis and they are given in the Appendix. Calculation of the growth rates indicates that in R1, which was run at higher phenol concentrations, the growth rate, μ, was more than double that observed in R2 (0.22 h^{-1} vs. 0.10 h^{-1}), while substrate utilization rates computed on the basis of both phenol and COD were almost four times higher (10.4 mg ∅/mg X-d vs. 3.2 mg ∅/mg X-d and 29.1 mg COD/mg X-d vs. 6.9 mg COD/mg X-d, respectively). It should be noted that data from previous work (1) indicated that the fastest growth rates could be obtained at phenol concentrations of approximately 200 mg/L; thus, this suggests that the high-rate process can be optimized to realize even higher rates since, for the run reported herein, the first stage was maintained at an average concentration of 316 mg/L phenol.

It is interesting to compare the overall process performance in terms of effluent phenol concentration, of the two-stage system with that of a single-stage chemostat treating phenol. In other work (1, 19), we reported that a chemostat fed a synthetic waste containing 250 mg/L of phenol reduced phenol concentrations to 53 µg/L at a detention time of 18 h which, for this series of runs, was the lowest observed effluent phenol concentration. By contrast, the run for the two-stage system in the current work was performed at a total (sum of Reactors 1 and 2) detention time of 10.5 h (5.25 L/0.5 L/h) with an influent phenol concentration of 500 mg/L; the effluent from Reactor 2 of this system had an average phenol concentration of 27 µg/L. This comparison indicates that, based on these data, the two-stage system was capable of reducing the soluble phenol concentration to lower values than did a single-stage system which had a longer detention time and a lower influent phenol concentration. This result would be expected because the cells, selected in the first reactor for maximum resistance to the toxic effects of phenol and maximum ability to degrade phenol, are afforded an opportunity to recover from inhibition in the second reactor. In this reactor, growth rate is much slower and the cells are able to remove phenol to a very low concentration. Recycle of cells from the second reactor prevents the wash-out of the first reactor, which would occur under the same conditions in a single-stage reactor without recycle. Thus, the two-stage system provides more rapid degradation of the toxic organic than can be achieved in a single stage process and also guards against wash-out of the system by providing an opportunity for the cells to grow more slowly and recover from the damage caused by the toxic substrate. The results of these studies are important because of the technological need to develop cost-effective systems which can reduce the concentrations of toxic pollutants to the lowest possible levels to attenuate the dispersal of these hazardous constituents into the environment. Consequently, the development of the two-stage high-rate process may not only be attractive because it has the potential to provide more cost-effective treatment systems but also because it may be able to deliver better effluents, insofar as target organic concentration is concerned, than other forms of biological treatment.

In addition to realizing higher biodegradation rates, data collected to date indicate that the two-stage system has the potential to realize significantly lower sludge production rates. Initial work (1) on the high rate biodegraders had indicated that growth at the higher phenol concentrations selected for populations which, in addition to exhibiting faster growth rates, had lower cell yields. An examination of the data given in Table 1 shows that the yield for the cells in Reactor 1 (high phenol concentrations) was about one-half that of the organisms grown in Reactor 2 at low phenol concentration (0.19 versus 0.35 mg cells/mg COD for R1 and R2, respectively). Thus, the results of the bench-scale run also show that application of these systems to field treatment situations has the potential for significant reduction of waste sludge production rates for biological treatment processes that treat phenol wastes.

Potential Field Applications

A schematic of a potential configuration for a high-rate process for treatment of phenolic wastes is given in Figure 7. The operational strategy for such a system would be to force relatively high growth rates in Stage 1 and use Stage 2 to bring effluent concentrations to acceptable levels. It seems prudent to recommend the use of some device or technique (e.g., constant X_R dosing tank (20)) to insure that both stages receive a steady input of recycle cells since this control strategy has shown to impart significant stability to activated sludge reactors treating both toxic and nontoxic wastes (3,20,21). The use of this control technique is especially crucial for the high-rate process because the goal is to utilize Stage 1 as a selector for growing the high rate cells while Stage 2 will be operated at more "conventional" rates in order to meet effluent requirements.

In order to optimize the design and operation of the proposed process, it is essential to utilize a predictive model. Such a model can be formulated by performing mass balances for biomass and substrate concentrations on each stage of the system and using the Haldane equation to relate growth rate to exogenous substrate concentrations. Mass balances for Stage 1 would be:

$$V_1 \frac{dX_1}{dt} = F_{R_1} X_R + \mu_1 X_1 V_1 - k_{d_1} X_1 V_1 - (F + F_{R_1}) X_1 \qquad (1)$$

$$V_1 \frac{dS_1}{dt} = FS_i + F_{R_1} S_R - (F + F_{R_1}) S_1 - \frac{\mu_1}{Y_{t_1}} X_1 V_1 \qquad (2)$$

By inserting the Haldane equation for growth rate in Stage 1, letting $D_1 = F/V_1$, and assuming steady-state conditions,

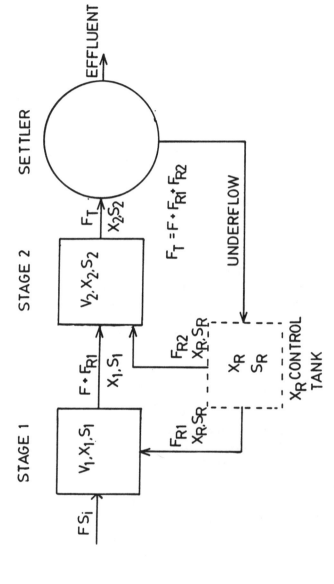

Figure 7. Proposed process flow scheme for two-stage biological system for high rate treatment of phenolic wastes.

predictive equations for Stage 1 can be obtained; details of the derivation of these equations are given in the Appendix.

$$a_1 S_1^3 + b_1 S_1^2 + c_1 S_1 + d_1 = 0 \tag{3}$$

$$a_1 = (F + F_{R_1})(D_1 + \frac{F_{R_1}}{V_1} + k_{d_1})/K_{i_1}$$

$$b_1 = (D_1 + \frac{F_{R_1}}{V_1} + k_{d_1})\left[F + F_{R_1} - \frac{(FS_i + F_{R_1} S_R)}{K_{i_1}}\right] - (F + F_{R_1})\mu_{max_1}$$

$$c_1 = \mu_{max_1}(FS_i + F_{R_1} S_R + \frac{F_{R_1} X_R}{Y_{t_1}}) + (D_1 + \frac{F_{R_1}}{V_1} + k_{d_1}) \\ ((F + F_{R_1})K_{s_1} - (FS_i + F_{R_1} S_R))$$

$$d_1 = -(D_1 + \frac{F_{R_1}}{V_1} + k_{d_1})(FS_i + F_{R_1} S_R)K_{s_1}$$

The predictive equation for X_1 is:

$$X_1 = \frac{Y_{t_1}(FS_i + F_{R_1} S_R - (F + F_{R_1})S_1)}{\mu_1 V_1} \tag{4}$$

It should be pointed out that the biokinetic constants, μ_{max}, K_i, K_s, Y_t, and k_d are given the subscript "1" in Equations (3) and (4) because they may have different values for the populations which are grown in each stage.

Mass balances can also be made around Stage 2 for biomass and substrate concentrations.

$$V_2 \frac{dX_2}{dt} = (F + F_{R_1})X_1 + F_{R_2} X_R + \mu_2 X_2 V_2 - k_{d_2} X_2 V_2 \tag{5}$$

$$- (F + F_{R_1} + F_{R_2})X_2$$

$$V_2 \frac{dS_2}{dt} = (F + F_{R_1})S_1 + F_{R_2}S_R - (F + F_{R_1} + F_{R_2})S_2 - \frac{\mu_2}{Y_{t_2}}X_2 V_2 \quad (6)$$

By assuming steady-state conditions and substituting the Haldane expressions for μ_2, steady state predictive equations for Stage 2 are obtained; details of the derivation of the Stage 2 predictive equations are also given in the Appendix.

$$a_3 S_2^3 + b_3 S_2^2 + c_3 S_2 + d_3 = 0 \quad (7)$$

where $a_3 = \dfrac{c_2 d_2}{K_{i_2}}$

$b_3 = d_2(c_2 - \mu_{max_2}) - \dfrac{b_2 c_2}{K_{i_2}}$

$c_3 = (a_2 + b_2)\mu_{max_2} + c_2(d_2 K_{s_2} - b_2)$

$d_3 = -b_2 c_2 K_{s_2}$

and $a_2 = ((F + F_{R_1})X_1 + F_{R_2}X_R)/Y_{t_2}$

$b_2 = (F + F_{R_1})S_1 + F_{R_2}S_R$

$c_2 = (F + F_{R_1} + F_{R_2})/V_2 + k_{d_2}$

$d_2 = F + F_{R_1} + F_{R_2}$

$$X_2 = \frac{Y_{t_2}[(F + F_{R_1})S_1 + F_{R_2}S_R - (F + F_{R_1} + F_{R_2})S_2]}{\mu_2 V_2} \quad (8)$$

The purpose of presenting process equations for the

high-rate system is to illustrate the process controls which are available in operating such a system and how they may be utilized in concert with a predictive model. That is, once the biokinetic constants for each stage are known, via appropriate kinetic methodologies, the model can be employed to set values for the engineering control parameters in order to meet specific treatment goals. It should be emphasized that, before advocating the use of the system in full-scale treatment situations, developmental work is needed to assess the utility of, or make modifications to, the predictive model in an attempt to realize optimal application of the system.

Regarding general operational strategies, it seems reasonable to suggest that the first stage be employed to oxidize most of the influent COD and the second be utilized for reducing target organic concentrations to the lowest levels possible. With this goal in mind, Stage 1 will, of course, be operated at relatively high rates to achieve rapid degradation of phenol; it should be pointed out that it is likely that, if most of the COD is metabolized in the first stage, sludge production for the system as a whole will be reduced because of the lower yields which have been reported in this work and elsewhere (1) for the high-rate cells. In order to reduce phenol concentrations to low levels (< 100 µg/L), it would be appropriate to operate the second stage at extremely low growth rates, i.e., Stage 2 should be run essentially in an extended aeration mode. This should be readily achievable since target operating conditions for low growth rate operation in Stage 2 can be obtained by use of the predictive model. Operation of the second stage in an extended aeration mode for the purpose of achieving low target organic concentrations seems to be technologically appropriate, since it has been reported that these systems exhibit the highest percentage removals for priority pollutants (22).

Finally, it should be stressed that the application of the high-rate process is not contingent on the design and/or construction of new facilities. Rather, it seems reasonable to suggest that this technology can be applied by providing relatively minor modifications to current biological process facilities. In other words, it should be possible to renovate treatment works with a relatively small capital outlay for the purpose of providing the appropriate controls and flow scheme in order to convert a system to the high-rate process.

Summary and Conclusions

A proposed process flow scheme for a high-rate fluidized

biological system for treating phenol wastes was presented. The key feature of this system is the use of two reactors each of which has different operating conditions. The first stage of the system is utilized for selective culture of organisms which are characterized by high-rate phenol degradation kinetics; the critical parameter for selecting these populations appears to be the maintenance of relatively high ($\geq$ 100 mg/L) steady-state phenol concentrations in this reactor. The second stage of the system is operated in an extended aeration mode (extremely low growth rates) in order to achieve low effluent organic concentrations. The advantages of this process over conventional technology are significantly increased phenol degradation rates and lower sludge production rates; the results of the bench-scale pilot study indicate that the process may also be capable of delivering lower phenol concentrations than can a conventional single-stage technology.

Equations were also derived which may be utilized, subsequent to additional developmental work on the process, for providing guidelines for design and operation. Provided the appropriate process controls are available, it should be feasible to use the process model to achieve treatment goals and optimize operation.

Finally, it was pointed out that application of the high-rate process technology is not contingent on the design of new facilities. Conversely, it is reasonable to suggest that the application can be made by modifying existing facilities. That is, the process can be applied by insuring that the appropriate level of controls and the two-stage flow scheme are achievable at a particular treatment facility.

ACKNOWLEDGEMENTS

This work was supported by a grant from the University of Delaware Unidel Foundation to Dr. Anthony F. Gaudy, Jr. Richard J. Colvin also received support in the form of a teaching assistantship from the University of Delaware, Department of Civil Engineering. At the time of this study, A. F. Rozich, R. J. Colvin, and A. F. Gaudy, Jr. were Associate Scientist, Research Assistant, and H. Rodney Sharp Professor, respectively in the Department of Civil Engineering. Correspondence should be directed to Alan F. Rozich, University of Delaware, Department of Civil Engineering, Newark, DE 19716.

REFERENCES

1. Colvin, R. J., and Rozich, A. F., "Phenol Growth Kinetics of Heterogeneous Populations in a Two-Stage Continuous Culture System," *Journal of Water Pollution Control Federation*, Vol. 58, No. 4, 1986, pp. 326-332.
2. Rozich, A. F., Gaudy, A. F., Jr., and D'Adamo, P. C., "Selection of Growth Rate Model for Activated Sludges Treating Phenol," *Water Research*, Vol. 19, No. 4, 1985, pp. 481-490.
3. Rozich, A. F., and Gaudy, A. F., Jr., "Response of Phenol-Acclimated Activated Sludge Process to Quantitative Shock-Loading," *Journal of Water Pollution Control Federation*, Vol. 57, No. 7, 1985, pp. 795-804.
4. Rozich, A. F., and Gaudy, A. F., Jr., "Critical Point Analysis for Toxic Waste Treatment," *Journal of the Environmental Engineering Division, ASCE*, Vol. 110, No. SA3, 1984, pp. 562-572.
5. Jones, G. L., Jansen, F., and MacKay, A., "Substrate Inhibition of the Growth of Bacterium NCIB8250 by Phenol," *Journal of General Microbiology*, Vol. 74, Part 1, 1973, pp. 1599-1615.
6. Rozich, A. F., and Castens, D. J., "Inhibition Kinetics of Nitrification in Continuous Flow Reactors," *Journal of Water Pollution Control Federation*, Vol. 58, No. 3, 1986, pp. 220-226.
7. Rozich, A. F., "The Modelling of a Constant Cell Recycle Activated Sludge Reactor Treating the Inhibitory Substrate Phenol," Doctoral Dissertation, University of Delaware, Department of Civil Engineering, December, 1982.
8. U. S. Environmental Protection Agency, "Process Design Manual for Nitrogen Control," U.S.E.P.A. Office of Technology Transfer, Washington, D.C., 1975.
9. Burdick, C. R., Refling, R. R., and Stensel, H. D., "Advanced Biological Treatment to Achieve Nutrient Removal," *Journal of the Water Pollution Control Federation*, Vol. 54, No. 7, 1982, pp. 1078-1086.
10. Tracy, K. D. and Flamino, A., "Kinetics of Biological Phosphorus Removal," presented, 58th Annual Water Pollution Control Federation Conference, Kansas City, MO, October, 1985.
11. Chiesa, S., Irvine, R. L. and Manning, J., "Feast/Famine Growth Environments and Activated Sludge Population Selection," *Biotechnology and Bioengineering*, Vol. 27, 1985, pp. 562-569.

12. Gaudy, A. F., Jr., Ramanathan, M., and Rao, B.S., "Kinetic Behavior of Heterogeneous Populations in Completely Mixed Reactors," *Biotechnology and Bioengineering*, Vol. 9, 1967, pp. 387-411.
13. Ramanathan, M., and Gaudy, A. F., Jr., "Effect of High Substrate Concentration and Cell Feedback on Kinetic Behavior of Heterogeneous Populations in Completely Mixed Systems," *Biotechnology and Bioengineering*, Vol. 11, 1969, pp. 207-237.
14. Chiu, S. Y., et al., "Kinetic Behavior of Mixed Populations of Activated Sludge," *Biotechnology and Bioengineering*, Vol. 14, 1972, pp. 207-231.
15. Peil, K. M., and Gaudy, A. F., Jr., "Kinetic Constants for Aerobic Growth of Microbial Populations Selected with Various Single Compounds and with Municipal Wastes as Substrates," *Applied Microbiology*, Vol. 21, 1971, pp. 253-256.
16. *Standard Methods for the Examination fo Water and Waste-Water*, Am. Public Health Assn., 14th Ed., New York (1975).
17. Yang, R. D., and Humphrey, A. E., "Dynamic and Steady State Studies of Phenol Biodegradation in Pure and Mixed Cultures," *Biotechnology and Bioengineering*, Vol. 17, 1975, pp. 1211-1235.
18. Hach Chemical Co., *Hach Water Analysis Handbook*, Loveland, CO, 2-144-146 (1980).
19. Colvin, R. J., and Rozich, A. F., "Biodegradation Kinetics of a Mixed Phenol/Glucose Waste in Continuous Flow Systems, "Submitted for presentation, 59th Annual Water Pollution Control Federation Conference, Los Angeles, CA, October, 1986.
20. Srinivasaraghavan, R., and Gaudy, A. F., Jr., "Operational Performance of an Activated Sludge Process with Constant Sludge Feedback," *Journal of Water Pollution Control Federation*, Vol. 47, No. 7, 1975, pp. 1946-1960.
21. Saleh, M. A., and Gaudy, A. F., Jr., "Shock Load Response of Activated Sludge at Constant Recycle Sludge Concentrations," *Journal of Water Pollution Control Federation*, Vol. 50, No. 4, 1974, pp. 764-774.
22. Gaudy, A. F., Jr., Kincannon, D. F., and Manickam, T.S., "Treatment Compatibility of Municipal Waste and Biological Hazardous Industrial Compounds," NTIS PB83-105536, Volume I, PB83-105544, Volume II (Appendix), U. S. Department of Commerce (1982).

LIST OF SYMBOLS

D_1 = Dilution rate; ratio of flow, F, to volume of liquor, V_1, in Stage 1 of high rate process (Figure 7), time^{-1}.

F = Rate of flow of incoming wastewater, $\text{vol} \times \text{time}^{-1}$.

F_j = Rate of flow to Reactor j (Figure 5), $\text{vol} \times \text{time}^{-1}$.

F_{R_j} = Recycle sludge flow rate to Stage j (Figure 7), $\text{vol} \times \text{time}^{-1}$.

k_{d_j} = Specific decay rate for biomass in Stage j (Figure 7), time^{-1}.

K_{i_j} = Inhibition constant used in Haldane equation to account for the effect of concentration of inhibitory substrate on specific growth rate of biomass in Stage j (Figure 7), $\text{mass} \times \text{vol}^{-1}$.

K_{s_j} = Saturation constant or shape factor in the rectangular hyperbola form of the Monod relationship for biomass in Stage j; defined as concentration of limiting substrate at which $\mu = 0.5\ \mu_{max}$, $\text{mass} \times \text{vol}^{-1}$.

S^* = Critical substrate concentration for an inhibitory substrate; equal to $(K_s K_i)^{\frac{1}{2}}$, $\text{mass} \times \text{vol}^{-1}$.

S_j = Soluble substrate concentration in reactor or Stage j, $\text{mass} \times \text{vol}^{-1}$.

S_i = Concentration of substrate or waste in inflowing feed to a biological reactor, $\text{mass} \times \text{vol}^{-1}$.

S_R = Soluble substrate in biological solids stream recycled to the reactor or stage, $\text{mass} \times \text{vol}^{-1}$.

U_j = Substrate utilization rate in Reactor j, $\text{mass} \times \text{mass}^{-1} \times \text{time}^{-1}$.

V_j = Volume of reactor or Stage j, vol.

X_j = Biological solids concentration in reactor or Stage j, $\text{mass} \times \text{vol}^{-1}$.

X_R = Biological solids concentration in recycle flow to reactor or stage, $\text{mass} \times \text{vol}^{-1}$.

Y_{o_j} = Observed cell yield in Reactor j, $\text{mass} \times \text{mass}^{-1}$.

Y_{t_j} = True cell yield for biomass in Stage j, $\text{mass} \times \text{mass}^{-1}$.

μ^* = Critical growth rate for reactors treating inhibitory substrates, time^{-1}

μ_j = Specific growth rate in reactor or Stage j, time^{-1}.

μ_{max_j} = Maximum specific growth rate of which a biomass in Stage j is capable under specified conditions in the absence of any restriction on growth rate by limiting nutrient concentration or inhibition, time^{-1}.

APPENDIX

Growth Rate, Substrate Utilization and Observed Yield Expressions for High Rate Bench Scale System (Figure 5)

$$\mu_1 = \frac{F_1 + F_2(1 - X_2/X_1)}{V_1}$$

$$U_1 = \frac{F_2 S_2 + F_1 S_i - (F_1 + F_2)S_1}{V_1 X_1}$$

$$\mu_2 = \frac{(F_1 + F_2)(1 - X_1/X_2)}{V_2}$$

$$U_2 = \frac{(F_1 + F_2)(S_1 - S_2)}{V_2 X_2}$$

$$Y_{o_1} = \frac{(F_1 + F_2)X_1 - F_2 X_2}{F_2 S_2 + F_1 S_i - (F_1 + F_2)S_1}$$

$$Y_{o_2} = \frac{X_2 - X_1}{S_1 - S_2}$$

Derivation of Stage 1 Equations (Figure 7)

Mass balances for biomass, X_1, and substrate, S_1, in Stage 1 are given below.

$$V_1 \frac{dX_1}{dt} = F_{R_1} X_R + \mu_1 X_1 V_1 - k_{d_1} X_1 V_1 - (F + F_{R_1})X_1 \qquad (1)$$

$$V_1 \frac{dS_1}{dt} = FS_i + F_{R_1} S_R - (F + F_{R_1})S_1 - \frac{\mu_1}{Y_{t_1}} X_1 V_1 \qquad (2)$$

Equation 1 can be used to obtain an expression for the steady state growth rate, μ_1, in Stage 1.

$$\mu_1 = \frac{(F + F_{R_1})}{V_1} - \frac{F_{R_1} X_R}{V_1 X_1} + k_{d_1} \qquad (A1)$$

An expression for X_1 can be obtained from the steady state substrate balance by using Equation 2.

$$X_1 = \frac{Y_{t_1}(FS_i + F_{R_1}S_R - (F + F_{R_1})S_1)}{\mu_1 V_1} \quad (4)$$

Substitute 4 for X_1 into A1, let $D_1 = F/V_1$, and solve for μ_1.

$$\mu_1 = \frac{(D_1 + \frac{F_{R_1}}{V_1} + k_{d_1})(FS_i + F_{R_1}S_R - (F + F_{R_1})S_1)}{FS_i + F_{R_1}S_R - (F + F_{R_1})S_1 + F_{R_1}X_R/Y_t} \quad (A2)$$

Substitute the Haldane equation for μ_1, and solve for S_1.

$$\mu_{max_1}S_1(FS_i + F_{R_1}S_R - (F + F_{R_1})S_1 + F_{R_1}X_R/Y_t) =$$

$$(D_1 + F_{R_1}/V_1 + k_{d_1})(FS_i + F_{R_1}S_R - (F + F_{R_1})S_1)(K_{S_1} + S_1 + S_1^2/K_{i_1}) \quad (A3)$$

$$F\mu_{max_1}S_iS_1 + F_{R_1}\mu_{max_1}S_RS_1 - (F + F_{R_1})\mu_{max_1}S_1^2 + \frac{F_{R_1}X_R}{Y_t}\mu_{max_1}S_1 =$$

$$(D_1 + \frac{F_{R_1}}{V_1} + k_{d_1})(FS_i + F_{R_1}S_R)K_{S_1} + (D_1 + \frac{F_{R_1}}{V_1} + k_{d_1})(FS_i + F_{R_1}S_R)S_1$$

$$\frac{(D_1 + \frac{F_{R_1}}{V_1} + k_{d_1})(FS_i + F_{R_1}S_R)}{K_{i_1}}S_1^2 - (D_1 + \frac{F_{R_1}}{V_1} + k_{d_1})(F + F_{R_1})K_{s_1}S_1$$

$$-(D_1 + \frac{F_{R_1}}{V_1} + k_{d_1})(F + F_{R_1})S_1^2 - \frac{(D_1 + \frac{F_{R_1}}{V_1} + k_{d_1})(F + F_{R_1})}{K_{i_1}}S_1^3 \quad (A4)$$

Factor for powers of S_1.

$$\frac{(F + F_{R_1})(D_1 + \frac{F_{R_1}}{V_1} + k_{d_1})}{K_{i_1}} S_1^3 + \left[(D_1 + \frac{F_{R_1}}{V_1} + k_{d_1}) \left[F + F_{R_1} - \frac{(FS_i + F_{R_1}S_R)}{K_i} \right] \right.$$

$$- (F + F_{R_1}) \mu_{max_1} \left] S_1^2 + \left[\mu_{max_1}(FS_i + F_{R_1}S_R + \frac{F_{R_1}X_R}{Y_{t_1}}) \right. \right.$$

$$+ (D_1 + \frac{F_{R_1}}{V_1} + k_{d_1})((F + F_{R_1})K_{s_1} - (FS_i + F_{R_1}S_R)) \right] S_1$$

$$- (D_1 + \frac{F_{R_1}}{V_1} + k_{d_1})(FS_i + F_{R_1}S_R)K_{s_1} = 0 \tag{A5}$$

The predictive equation for S_1 is given in Equation 3.

$$a_1 S_1^3 + b_1 S_1^2 + c_1 S_1 + d_1 = 0 \tag{3}$$

$$a_1 = (F + F_{R_1})(D_1 + \frac{F_{R_1}}{V_1} + k_{d_1})/K_{i_1}$$

$$b_1 = (D_1 + \frac{F_{R_1}}{V_1} + k_{d_1}) \left[F + F_{R_1} - \frac{(FS_i + F_{R_1}S_R)}{K_{i_1}} \right] - (F + F_{R_1})\mu_{max_1}$$

$$c_1 = \mu_{max_1}(FS_i + F_{R_1}S_R + \frac{F_{R_1}X_R}{Y_{t_1}}) + (D_1 + \frac{F_{R_1}}{V_1} + k_{d_1})((F + F_{R_1})K_{s_1} - (FS_i + F_{R_1}S_R))$$

$$d_1 = -(D_1 + \frac{F_{R_1}}{V_1} + k_{d_1})(FS_i + F_{R_1}S_R)K_{s_1}$$

Equation 4 can be used as predictive equation for X_1.

Derivation of Stage 2 Equations (Figure 7)

Mass balances for biomass, X_2, and substrate, S_2, in Stage 2 are given below.

$$V_2 \frac{dX_2}{dt} = (F + F_{R_1})X_1 + F_{R_2}X_R + \mu_2 X_2 V_2 - k_{d_2} X_2 V_2 - (F + F_{R_1} + F_{R_2})X_2 \quad (5)$$

$$V_2 \frac{dS_2}{dt} = (F + F_{R_1})S_1 + F_{R_2}S_R - (F + F_{R_1} + F_{R_2})S_2 - \frac{\mu_2}{Y_{t_2}} X_2 V_2 \quad (6)$$

Equation 5 can be used to obtain an expression for the steady state growth rate, μ_2, in Stage 2.

$$\mu_2 = \frac{(F + F_{R_1} + F_{R_2})}{V_2} - \left[\frac{(F + F_{R_1})X_1 + F_{R_2}X_R}{V_2 X_2}\right] + k_{d_2} \quad (A6)$$

Equation 6 can be used to obtain an expression for the steady state biomass, $V_2 X_2$, in Stage 2.

$$V_2 X_2 = \frac{Y_{t_2}[(F + F_{R_1})S_1 + F_{R_2}S_R - (F + F_{R_1} + F_{R_2})S_2]}{\mu_2} \quad (A7)$$

Substitute A7 for $V_2 X_2$ into A6 and factor out μ_2.

$$\mu_2 \left(1 + \frac{((F + F_{R_1})X_1 + F_{R_2}X_R)/Y_{t_2}}{((F + F_{R_1})S_1 + F_{R_2}S_R - (F + F_{R_1} + F_{R_2})S_2)}\right) = \frac{F + F_{R_1} + F_{R_2}}{V_2} + k_{d_2} \quad (A8)$$

Let $a_2 = ((F + F_{R_1})X_1 + F_{R_2}X_R)/Y_{t_2}$

$b_2 = (F + F_{R_1})S_1 + F_{R_2}S_R$

$c_2 = (F + F_{R_1} + F_{R_2})/V_2 + k_{d_2}$

$d_2 = F + F_{R_1} + F_{R_2}$

Substitute the Haldane expression for μ_2 into A8 and obtain an equation for S_2.

$$\left(\frac{\mu_{max_2} S_2}{K_{s_2} + S_2 + S_2^2/K_{i_2}}\right)\left(1 + \frac{a_2}{b_2 - d_2 S_2}\right) = c_2 \tag{A9}$$

$$\mu_{max_2} S_2 \left(1 + \frac{a_2}{b_2 - d_2 S_2}\right) = c_2 K_{s_2} + c_2 S_2 + \frac{c_2}{K_{i_2}} S_2^2 \tag{A10}$$

$$\mu_{max_2} S_2 (b_2 - d_2 S_2 + a_2) = b_2 c_2 K_{s_2} + b_2 c_2 S_2 + \frac{b_2 c_2}{K_{i_2}} S_2^2$$

$$- c_2 d_2 K_{s_2} S_2 - c_2 d_2 S_2^2 - \frac{c_2 d_2}{K_{i_2}} S_2^3 \tag{A11}$$

$$(a_2 + b_2)\mu_{max_2} S_2 - d_2 \mu_{max_2} S_2^2 = b_2 c_2 K_{s_2} + (b_2 c_2 - c_2 d_2 K_{s_2}) S_2$$

$$+ \left(\frac{b_2 c_2}{K_{i_2}} - c_2 d_2\right) S_2^2 - \frac{c_2 d_2}{K_{i_2}} S_2^3 \tag{A12}$$

$$\frac{c_2 d_2}{K_{i_2}} S_2^3 + \left(d_2(c_2 - \mu_{max_2}) - \frac{b_2 c_2}{K_{i_2}}\right) S_2^2 + \left((a_2 + b_2)\mu_{max_2} + c_2(d_2 K_{s_2} - b_2)\right) S_2$$

$$- b_2 c_2 K_{s_2} = 0 \tag{A13}$$

The predictive equation for S_2 is given in Equation 7.

$$a_3 S_2^3 + b_3 S_2^2 + c_3 S_2 + d_3 = 0 \tag{7}$$

where $a_3 = \dfrac{c_2 d_2}{K_{i_2}}$

$b_3 = d_2(c_2 - \mu_{max_2}) - \dfrac{b_2 c_2}{K_{i_2}}$

$c_3 = (a_2 + b_2)\mu_{max_2} + c_2(d_2 K_{s_2} - b_2)$

$d_3 = -b_2 c_2 K_{s_2}$

The predictive equation for X_2 is obtained from A7.

$$X_2 = \dfrac{Y_{t_2}[(F + F_{R_1})S_1 + F_{R_2}S_R - (F + F_{R_1} + F_{R_2})S_2]}{\mu_2 V_2} \qquad (8)$$

BIOTECHNOLOGY FOR THE TREATMENT OF HAZARDOUS WASTE CONTAMINATED SOILS AND RESIDUES

Mark E. Singley, Andrew J. Higgins, Vijay S. Rajput, Sumith Pilapitiya, Reba Mukherjee, and Ven Mercade

Department of Biological and Agricultural Engineering
Cook College, Rutgers University
New Brunswick, New Jersey

INTRODUCTION

The research reported in this paper deals with the development and use of biotechnology for treating contaminated soils and waste residues at the site of the problem. The major treatment methods include physical/chemical washing of soils and waste residues as a pretreatment followed by biological degradation. The challenge to pursue the use of biological degradation arose from previous work by two of the authors concerning the composting of sewage sludge which frequently contains hazardous organic compounds (1).

Little research has been conducted on the use of land treatment or biodegradation for the detoxification and degradation of hazardous organic wastes. Most of the knowledge pertaining to the effects of hazardous compounds on soils comes from the literature on pesticide degradation.

A comprehensive literature search has revealed that little research work has been conducted on the use of land treatment or biodegradation for the detoxification and degradation of hazardous wastes.

Most of the knowledge pertaining to the effects of hazardous compounds on the soil comes from the literature on pesticide degradation. A number of scientific documents have characterized microbial responses to a wide range of these chemicals (2, 6, 7, 8, 9, 10, 11, 12, 13). Most of these reports deal with the acute effects of these pesticides on the microflora. However, little is known of the chronic effects on the soil microorganisms exposed to low levels of toxic organic compounds.

A study on the use of soil treatment for biodegradation of waste petroleum was conducted by Dibble (5). He studied the allowable loading rates for land treatment of petroleum oils and the techniques used in maximizing the rate of decomposition. The results of his work indicated that loading rates of 1 gr per 20 gr of soil resulted in the greatest removal of both aromatic and aliphatic hydrocarbons.

Research demonstrates that biodegradation of haloaromatic substances has potential. Suflita et al. (21) found reductive dehalogenation of aromatic substrates by anaerobic microflora in sediment and sludge. Alexander and Lustigman (20) found that hydroxyl and carboxyl groups were most favorable and sulfonate and nitro substituents least favorable to microbial degradation of benzene rings by a mixed microflora in soil.

Rose and Mercer (15) found that the insecticides diazinon, parathion, and dieldrin degraded rapidly when biodegraded with cannery wastes. They also reported that DDT was relatively resistant to degradation. Hunter et al. (13) found that PCB's decomposed during sludge composting. Their research also showed that aliphatic hydrocarbons decomposed more readily than aromatic hydrocarbons during composting.

The results of biodegrading petroleum refinery sludges were reported by Deever and White (16) who detected a significant level of degradation of toluene-hexane extractable grease and oil after composting. Epstein and Alpert (17) also conducted studies involving both crude and No. 6 oil, pulp and paper mill wastes, and pharmaceutical wastes. All of these wastes appeared amenable to biodegradation.

Several studies on the treatment of toxic halogenated compounds by biological wastewater treatment have been found in the literature (4, 21). Sharmat and Maier concluded from their study that cultures capable of completely biodegrading chlorinated compounds have been isolated from municipal wastewater treatment plants and that the use of biological waste treatment for removal of some chlorinated organic wastes is feasible. DiGeronimo et al. (21) investigated the microbial transformation of chlorinated benzoates to carbon and energy sources in a sewage microcosm.

A report in Science (14) identified biodegradation as a viable treatment technique for many hazardous wastes. DDT wastes were quoted as having been degraded by 64 percent in as little as 50 days while organo-phosphate pesticide wastes were completely degraded in 2 weeks. Pesticides, phenols, and aromatics were cited as among the other materials suitable for degradation.

In the biodegradation of hazardous wastes, it is quite probable that some compounds may exhibit greater toxicity to microbes than others. In addition, the concentration in the media may also play an important role in determining the level of tolerance. The effect of frequent successive applications at acceptable loading rates compared with a single dose may also be important in sustaining the microbial population.

The use of pretreatment mechanisms such as extraction and chemical conversion may help reduce the levels of these compounds to concentrations amenable to biodegradation. Studies by Rulkens et al. (3) on the extraction of organic bromine compounds by washing agents (NaOH, Na_2CO_3, and soap) indicate a potentially viable pretreatment technique for contaminated soils and residues. Once the levels have been reduced, biodegradation techniques could be used to reduce the contaminate levels to acceptable concentrations.

MICROCALORIMETER

The microcalorimeter is a device that offers a convenient and relatively quick way of determining the interaction of the microbiological community with a contaminant. Each contaminant or combination of contaminants must be checked to learn first if the contaminant is metabolizable and, second, what the maximum level of concentration of the contaminant can be.

A microcalorimeter developed by the Department of Biological and Agricultural Engineering to sustain the aerobic microbiological community has been used to initiate these tests (22). The microbial activity in terms of heat output for a one gram sample is termed a 'thermogram'. A normalization test performed on a sample that is the organic medium to sustain the microbiological community will reflect the endogenous activity of the microorganisms present. The heat output will be proportional to their numerical density. For these tests, dewatered raw sewage sludge was used as the organic medium. To obtain a baseline from which to measure the activity contributed by the microorganisms in the organic medium, a sample of the soil is tested in the microcalorimeter. Only background activity, if there is any in the soil, will be shown by the baseline thermogram. Measuring the area between the baseline and the normalization thermogram for the sewage sludge provides a quantative measure of the heat contributed by the sewage sludge.

Adding the contaminant to the sewage sludge at various rates will determine the input of the toxicant on the microbiological community. If the added compound is a toxicant, the heat output will be less than for the normalized thermogram. If the compound can be metabolized, the heat flux will be greater. The change in heat output may then be considered a measure of the degradation response of the microorganisms.

Figure 1 shows the response of the microorganisms to the addition of 1,2,4 trichlorobenzene to the sewage sludge. At levels below 1.5% the area under the thermogram was increased; beyond 1.5%, the area began to diminish until at 3% the area was less than for the normalized sewage sludge. At 5% the microbial activity had essentially ceased.

Figure 2 shows how the concentration of 1, 2, 4, trichlorobenzene relates to the area produced on the thermogram.

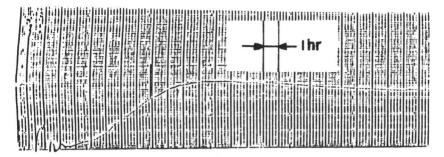

FIG. I.I SEWAGE SLUDGE THERMOGRAM

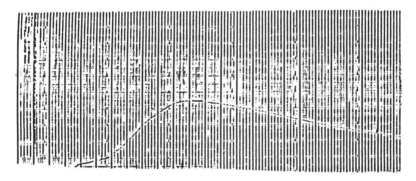

FIG. I.2 0.01% 1,2,4 TRICHLOROBENZENE AND
SEWAGE SLUDGE THERMOGRAM

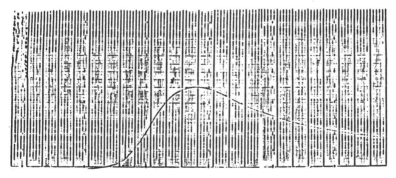

FIG. I.3 0.1% 1,2,4 TRICHLOROBENZENE AND
SEWAGE SLUDGE THERMOGRAM

364 Biotechnology for Degradation of Toxic Chemicals

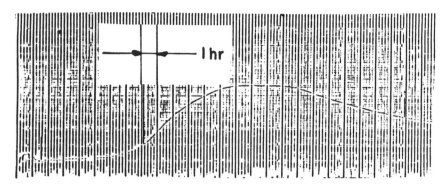

FIG. 1.4 1% 1,2,4 TRICHLOROBENZENE AND SEWAGE SLUDGE THERMOGRAM

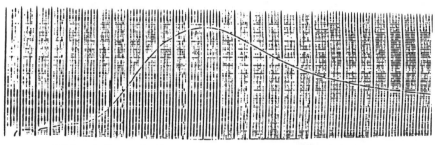

FIG. 1.5 1.5% 1,2,4 TRICHLOROBENZENE AND SEWAGE THERMOGRAM

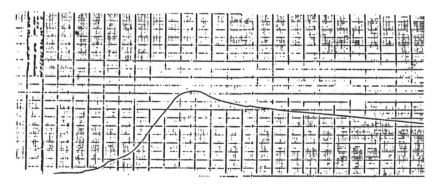

FIG. 1.6 2.0% 1,2,4 TRICHLOROBENZENE AND SEWAGE SLUDGE THERMOGRAM

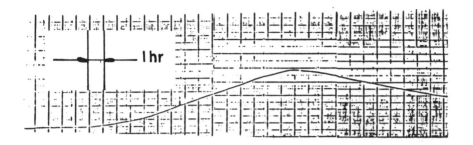

FIG. I.7 3.0% 1,2,4 TRICHLOROBENZENE AND SEWAGE SLUDGE THERMOGRAM

FIG. I.8 5% 1,2,4 TRICHLOROBENZENE AND SEWAGE SLUDGE THERMOGRAM

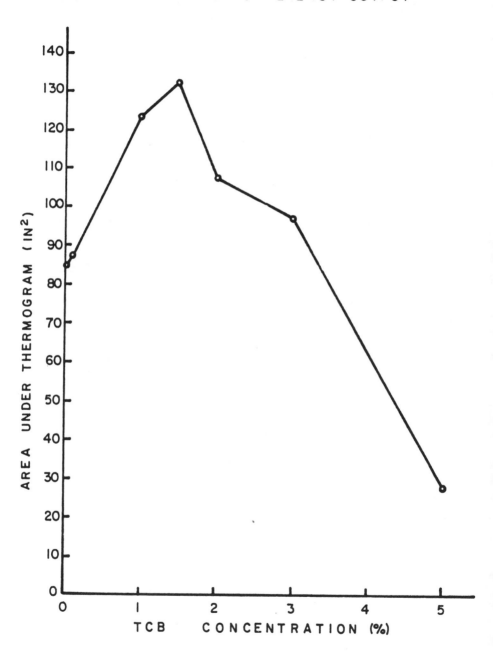

FIG. 2
CONCENTRATION OF 1, 2, 4 TCB VS AREA AS A FUNCTION OF ENERGY OUTPUT

The thermogram is recorded over a one and one-half to two-day period. For analysis it is divided into three phases. The first is the lag phase which occurs in the beginning while the microorganisms are accommodating themselves to the temperature, which is 50°C, and the medium they are in. This is the growth or multiplication phase. The second phase is the sustained phase where the population and heat output are relatively constant. The third phase is a population and heat decline phase.

Indicative of the response to the contaminant added to the sewage sludge is not only the area under the thermogram, but also the shift in the lag phase. For 1,2,4 trichlorobenzene at the 3% contamination level, the lag phase had lengthened considerably indicating a selection or adaptation within the microbiological community. Since the microbiological community is selective for temperature, temperature may also have to be varied to examine the matrix of temperature and microbiological community for a particular contaminant. The results of these tests indicate that microbial degradation of 1,2,4 trichlorobenzene can be used up to at least the 3% contamination level. If that level is exceeded, then an extraction process such as soil washing must be used.

Figure 3 shows the thermograms resulting from the addition of hexadecane to sewage sludge. Hexadecane is a highly energetic contaminant that enhances the thermograms considerably.

Figure 4 shows how the concentration of hexadecane relates to the area produced on the thermogram.

The thermogram obtained from the sample in the microcalorimeter provides the information that shows the feasibility of using the microbiological community to degrade an organic contaminant. To simulate an in-site biodegradation, reactor vessels saturated with air to ensure aerobic microbiological activity and contained in a controlled environment were used. Three vessels were used, each having differentiated substrates. The vessels contained the following: Vessel No. 1 - 50% primary sewage sludge plus 50% sandy loam soil; Vessel No. 2 - 0.25% 1, 2, 4 trichlorobenzene on a dry weight basis in a base of 50% sludge and 50% soil; Vessel No. 3 - 0.75% 1, 2, 4

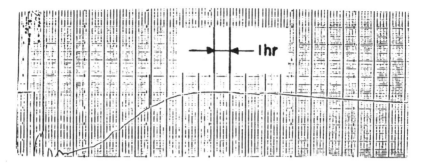

FIG. 3.1 SEWAGE SLUDGE THERMOGRAM

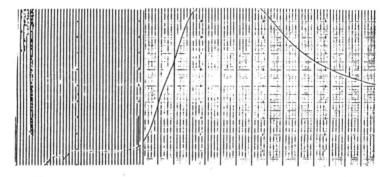

FIG. 3.2 0.1% HEXADECANE AND SEWAGE SLUDGE THERMOGRAM

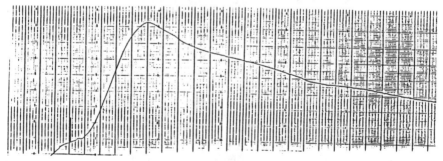

FIG. 3.3 0.2% HEXADECANE AND SEWAGE SLUDGE THERMOGRAM

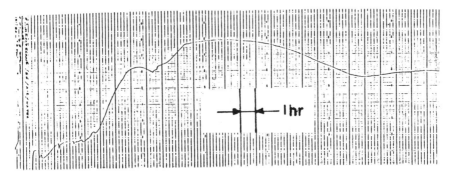

FIG. 3.4 0.5% HEXADECANE AND SEWAGE SLUDGE THERMOGRAM

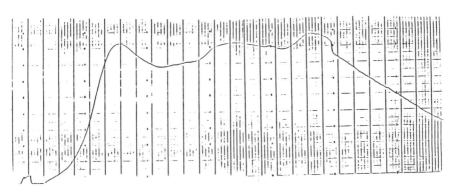

FIG. 3.5 1% HEXADECANE AND SEWAGE SLUDGE THERMOGRAM

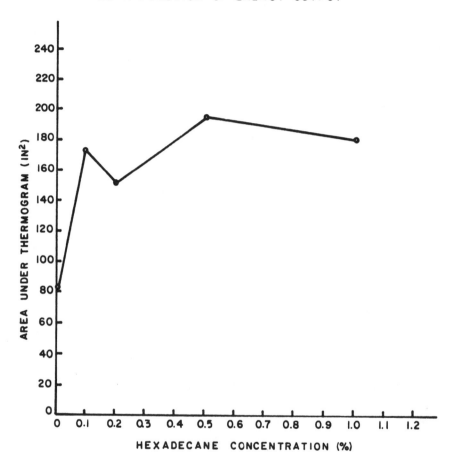

FIG. 4
CONCENTRATION OF HEXADECANE VS AREA AS A FUNCTION OF ENERGY OUTPUT

trichlorobenzene on a dry weight basis in a base of 50% sludge and 50% soil.

The temperature in the environmental chamber was maintained at the thermophilic level (50°C). The air passing through each of the reactor vessels was passed through a charcoal filter to collect any of the contaminant that was volatilized. Samples were collected at intervals of one week for a total of two weeks and analyzed for moisture content, volatile solids content and 1, 2, 4 trichlorobenzene. The analysis of the carbon filter and the soil for 1, 2, 4 trichlorobenzene was made in a gas chromatograph.

The activity that occurred in the reactor vessels is shown in Table 1.

To distribute the 1, 2, 4 trichlorobenzene throughout the soil sludge mixture, methylene chloride was used as a solvent and added at the rate of 50 parts to one. A rapid loss of moisture occurred and it is suspected that the vaporization of the methylene chloride accelerated the water loss. Future experiments will be modified to include a water spray in the reactor vessels to maintain a relatively constant moisture content.

Since the moisture content fell to less than 2% in the first week, the microbiological activity was essentially terminated. During the first week, however, the 1, 2, 4 trichlorobenzene in Vessel No. 2 at a concentration of 0.25% was reduced by approximately 46% and in Vessel No. 3 at a concentration of 0.75% by approximately 82%. The amount that was vaporized and caught in the charcoal filter was approximately 10%.

These small scale experiments show how the feasibility can be determined for using biodegradation to decontaminate soils. However, additional organic toxicants and additional environmental parameters must be explored.

[NJAES Paper No. J-03144-01-86, supported by Toxic Waste Institute, New Jersey Institute of Technology.]

Table 1. Analytical results of reactor vessel experiment, contaminant 1,2,4 trichlorobenzene

	BASE COMPOSITION 50% SEWAGE SLUDGE - 50% SOIL		
	VESSEL NO. ONE CONTROL	VESSEL NO. TWO 0.25% CONT.	VESSEL NO. THREE 0.75% CONT.
INITIAL ANALYSIS			
Moisture (%)	48.39	52.31	54.92
Volatile solids (%)	25.57	24.19	24.45
Concentration of contaminant (ppm)	----	2584	7476
AFTER ONE WEEK			
Moisture (%)	41.64	1.47	1.12
Volatile solids (%)	21.29	24.37	22.39
Concentration of contaminant (ppm)	----	1172	711
Amount of contaminant vaporized (ppm)	----	225	662
percent degraded	----	45.9	81.6

REFERENCES

1. Singley, M.E., A.J. Higgins, M. Frumkin-Rosengaus. 1982. Sludge composting and utilization: A design and operating manual. New Jersey Agricultural Experiment Station Report.

2. Bollag, Jean-Marc. 1974. Microbial transformations in pesticides. Advances in Applied Microbiology. Academic Press.

3. Rulkens, W.H., J.W. Assink, and W.J. Th. Van Gemert. 1982. Development of an installation for on-site treatment of soil contaminated with organic bromine compounds. Proceedings of the National Conference on Management of Uncontrolled Hazardous Waste Sites, HMCRI, Silver Springs, Maryland.

4. Shamat, N.A. and W.J. Maier. 1980. Kinetics of biodegradation of chlorinated organics. journal of Water Pollution Control Federation. 52, 2158.

5. Dibble, J.T. 1978. Stimulated biodegradation of waste petroleum. Ph.D. Thesis, Rutgers University.

6. Bollen, W.B. 1961. Interactions between pesticides and soil microorganisms. Ann. Rev. Phytopathol. 1, 101-126.

7. Kreutzer, W.A. 1963. Selective toxicity of chemicals to soil microorganisms. Ann. Rev. Phytopathol. 1, 101-126.

8. Martin, J.P. 1964. Influence of pesticide residues on soil microbiological and chemical properties. Residue Rev. 4, 96-129.

9. Helling, C.S., P.C. Kearney, and M. Alexander. 1971. Behavior of pesticides in soils. Advan. Agron. 23, 147-240.

10. Goring, C.A.I. 1967. Physical aspects of soil in relation to the action of soil fungicides. Ann. Rev. Phytopathol. 5, 285-318.

11. Guenzi, W.D., and W.E. Beard. 1967. Anaerobic biodegradation of DDT to DDD in soil. Science. 156, 1116-1117.

12. Parr, J.F. and S. Smith. 1973. Degradation of trifluralin under laboratory conditions and soil anaerobiosis. Soil Sci. 115, 55-63.

13. Parr, J.F. and S. Smith. 1976. Degradation of toxaphene in selected anaerobic soil environments. Soil Sci. 121, 52-57.

14. Research News. 1979. Hazardous waste technology is available. Science. 204, 4396, 930.

15. Rose, W.W. and W.A. Mercer. 1968. Fate of Pesticides in Composted Agricultural Wastes. Nation Cancer Association, Washington, DC. 27 pp.

16. Deever, W.R. and R.C. White. 1978. Composting Petroleum Refinery Sludges. Texaco, Inc. Port Arthur, Texas. 24 pp.

17. Epstein, E. and J.E. Alpert. 1980. Enhanced biodegradation of oil and hazardous residues. In Proc. Conf. on Oil and Hazardous Material Spills.

18. Hunter, J.V., M.S. Finstein, D.J. Suler, and R.R. Bobal. 1981. Sludge composting and utilization: Fate of concentrated industrial waste during laboratory scale composting of sewage sludge. New Jersey Agricultural Experiment Station, New Brunswick, NJ. 45 pp.

19. Suflita, J.M., A. Horowitz, D.R. Shelton, and J.M. Tiedje. 1982. Dehalogenation: A novel pathway for the anaerobic biodegradation of haloaromatic compounds. Science. 218, 1115-1117.

20. Alexander, M., and B.K. Lustigman. 1966. Effect of chemical structure on microbial degradation of substituted benzenes. J. Agr. Food Chem. 14, 410-413.

21. DiGeronimo, M.J., M. Nikaido, and M. Alexander. 1979. Utilization of chlorobenzoates by microbial populations in sewage. Applied and Env. Micro. 37, 619–625.

22. Rahman, M.S. 1984. Microcalorimetric measurement of heat production and the thermophysical properties of compost. Ph.D. Thesis, Rutgers University.

BIOLOGICAL DEGRADATION OF POLYCHLORINATED BIPHENYLS

Ronald Unterman, Michael J. Brennan, Ronald E. Brooks, and Carl Johnson

Biological Sciences Branch
General Electric Corporate Research and Development
Schenectady, New York

INTRODUCTION

The polychlorinated biphenyls (PCBs) are a family of compounds (congeners) that were commonly used over the last half century. Their release and accumulation in the environment and possible effects on human health have sparked an intense interest in devising technologies for their destruction or safe disposal. It is for these reasons that we began a research program several years ago to further investigate the bacterial degradation of PCBs (for a recent review of this subject see reference 1. We initially isolated over two dozen new bacterial strains capable of growth on biphenyl (BP) as sole carbon source. These strains exhibited various capabilities for biodegrading PCBs with several demonstrating exceptional and novel PCB-degradative competence (2-5). Our subsequent studies have concentrated on characterizing the biochemistry and genetics of PCB biodegradation in several of these microorganisms.

Recently we expanded our research program in an effort to explore the possibility of using one or more of our new bacterial isolates for the decontamination of PCB-laden soil. <u>Corynebacterium</u> sp. MB1, <u>Alcaligenes eutrophus</u> H850, and <u>Pseudomonas putida</u> LB400 had all been shown to have excellent PCB-degrading capabilities under laboratory assay conditions (2-5), however, little was known about their ability to oxidize PCBs adsorbed to soil. Our initial studies using sand as a model demonstrated that PCBs were biodegradable even when bound to a solid substrate.

We have now begun model biodegradation studies using PCB-laden soil containing either Aroclors or pure PCB congeners. Experimentally, our approach is identical to our previously described resting cell assays (4) except that the PCBs are bound to soil when mixed with the bacteria. For these studies the cultures were grown on PAS buffer (4) with biphenyl as the sole carbon source, then harvested, washed, and resuspended in buffer at a concentration of approximately 10^9 cells/ml. These washed cells were then incubated with one of several PCB/soil formulations: 50 ppm of Aroclor 1242 spiked onto clean (i.e., non-PCB containing) soil; 500 ppm of Aroclor 1242 on clean soil; 50 ppm of Aroclor 1254 on clean soil; an environmental sample (South Glens Falls, NY, dragstrip site 43C) which contains 525 ppm PCB (very similar in composition to Aroclor 1248); and various concentrations of pure PCB congeners spiked onto clean soil. In addition, some assay variations were conducted using unwashed cells, Luria-grown cultures, cultures actively growing on biphenyl, unstirred samples, and limited amounts of water.

BIODEGRADATION OF SOIL-BOUND PCBs

The results to date have been very promising. Our first experiments with soil containing 50 ppm of Aroclor 1242 demonstrated extensive biodegradation of this Aroclor by all three bacterial strains assayed (MB1, H850, LB400). MB1 and LB400 degraded better than 95% of the PCBs in as little as one day. In a parallel assay using BP-grown MB1 the rate of degradation of PCBs bound to soil was the same as that for free PCBs in a non-soil assay. This result was surprising yet encouraging. One might have expected the

PCBs to be irreversibly sequestered in soil particles thereby decreasing the extent of degradation. This was not the case for any of the time points assayed (2h, 1d, 3d). This may be because these soil assays were conducted under water-excess conditions (1.0 ml aqueous culture with 0.1 g soil). Using field conditions (much less water) we may observe a difference between free and bound-PCB biodegradation. Such studies are currently underway.

As we observed for free PCBs, each of our strains also exhibit congener specific degradation of soil-bound PCBs. That is, MB1 is better able to degrade soil-bound double-para substituted chlorobiphenyls (e.g. 4,4'-CB), and H850 and LB400 are better able to degrade soil-bound congeners with blocked 2,3 positions, such as 2,5,2',5'-CB (2,4). This specificity is clearly evident in the gas chromatographic (GC) profiles in Figure 1. LB400 more readily degrades peak 1 and the center peak of the triplet designated by three dots. On the other hand, MB1 more readily degrades peaks 3,4 and the first and third peaks of the triplet. Because the specifities of MB1 and LB400 complement each other, it seemed reasonable to try a biodegradation using a mixed culture consisting of both strains. Figure 1 (third panel) shows the results of such an assay and demonstrates the additive effect of using a mixed culture.

We have extended our Aroclor biodegradation studies to higher concentrations and more difficult Aroclors. LB400 was able to degrade greater than 85% of the PCB in a 500 ppm Aroclor 1242 assay incubated for two days. In a 50 ppm Aroclor 1254 assay, LB400 degraded greater than 65% of the PCB in one day. Studies are currently under way with Aroclors at 500 and 5000 ppm on soil.

In addition to the resting cell assays described above, we are interested in modeling an *in situ* biodegradation using actively growing cultures. For these studies we have introduced a 1.0% LB400 inoculum into an assay flask containing soil-laden PCB, PAS growth medium, and solid biphenyl as carbon and energy source. The assay was monitored for both cell count and PCB biodegradation. In the first of these experiments LB400 grew well (5 cell doublings in one day), and the actively growing cells degraded the PCB. However, the extent of Aroclor 1242 degradation (40% in one day) was significantly less than with resting cells.

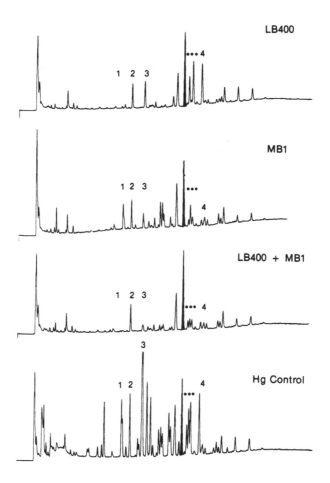

Figure 1. Biodegradation of soil-bound Aroclor 1242 by MB1 and LB400 applied separately or as a mixture. Soil spiked with 50 ppm of Aroclor 1242 was incubated with resting cells for 3 days. Cells (2ml at approximately 10^9 cells/ml) were incubated with 0.4 g of soil. The shaded peak is a nondegradable PCB congener (2,4,6,2',4'-CB) used as an internal standard.

By the next time point (3 days) the viable cell count was down more than 10,000 fold, i.e. the culture had died some time after 24 hours. The degree of PCB degradation for the 3 day and 6 day (also not viable) samples was essentially the same as for the one day. It thus appears that when the culture died, PCB degradation ceased. There are several possible explanations for the lower amount of degradation seen for growing cells as compared to resting cells. One of these is that the biphenyl, which is in vast excess over the PCB, is competing with the PCB for a limited amount of enzyme. To address this question we have tried several resting cell assays in the presence of BP. For example, MB1 showed little or no difference in the degree of PCB degradation (50 ppm Aroclor 1242 on soil) when assayed with and without 0.1% BP present. These results would indicate that the presence of biphenyl is not a problem, however, further comparative rate studies need to be done.

We have recently begun studies with an actual PCB-contaminated soil from New York State. This environmental sample is from the site of a former racing dragstrip where PCB oils were used for dust control. The site contains Aroclor 1242 at concentrations up to 7000 ppm. The sample we obtained was analyzed and contains 525 ppm of a transformed Aroclor 1242. It is depleted in the di- and trichlorobiphenyls and therefore appears similar in composition to Aroclor 1248. We have been unable to determine whether this transformation is evaporative and/or biological, however, we have been able to duplicate an almost exact congener depletion profile by biodegrading a higher concentration of soil-bound Aroclor 1242 which mostly limited the biodegradation to lower chlorinated congeners.

Resting cell biodegradation studies using the dragstrip soil have shown substantial PCB biodegradation. As seen in Figure 2, LB400 degraded 15% of the PCBs in one day and 51% in three days. This compares with 85% degradation of Aroclor 1242 (500 ppm) on laboratory soil. MB1 and H850 also biodegraded the dragstrip PCBs, but to a somewhat lesser extent. Our current studies are focusing on conditions for maximizing the rate and extent of PCB degradation in this environmental sample.

Biological Degradation of Polychlorinated Biphenyls 381

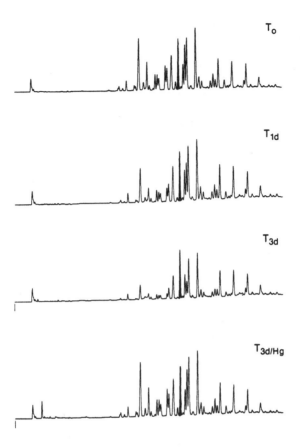

Figure 2. Biodegradation of PCBs in an environmental sample by <u>Pseudomonas putida</u> LB400. Cells (2ml at approximately 10^9 cells/ml) were incubated with 0.2 g of soil from dragstrip site 43C (described in the text). The shaded peak is a nondegradable PCB congener (2,4,6,2',4'-CB) used as an internal standard.

In addition to our Aroclor studies we have also been conducting biodegradation assays using soil-bound pure PCB congeners. Experiments have been initiated to define the rates of degradation of selected PCB congeners and to determine the metabolites produced from these biotransformations. The results have demonstrated that the steps of PCB oxidation that we have observed in our aqueous resting cell assays also occur under soil assay conditions. For example, soil-bound 2,3-CB is completely degraded to 2,3-chlorobenzoic acid. Other soil-bound PCB congeners are degraded to chlorobenzoic acids via transient hydroxylated high molecular weight intermediates.

MODELING A MOCK BIODEGRADATION PROCESS

It has been important throughout our biodegradation process modeling research to demonstrate that our bacterial soil-decontamination results are unequivocally due to biological activity. One pitfall that we, as well as other investigators, must be concerned with is congener depletion in a "biodegradation" process that is due to physical loss of the PCB and not to true biological degradation. With Aroclor studies these alternatives can be easily distinguished because biodegradation processes result in depletion of specific congeners yielding GC profiles which are distinctly different from those of Aroclors. Physical depletion, on the other hand, results in uniform depletion of all congeners (e.g., adsorptive loss) or a depletion of lower chlorinated congeners due to their higher volatility (e.g. evaporative loss). The production of PCB metabolites is of course another unequivocal method for demonstrating the biological basis of PCB depletion.

In order to better evaluate results from open air, aerated, stirring reactors, we set up a model process that appears to be biologically mediated, but is not. A sample of Oakland, CA, soil contaminated with Aroclor 1260 was air dried at room temperature, homogenized with a glass mortar and pestle and then sieved. Argon was bubbled through a water/soil mixture (approximately 125 ml) in a 250 ml round-bottom flask at a flow of approximately 200 ml/min. A florosil sample tube was attached to trap PCBs in the argon as it exited the vessel. Samples (20 ml) were taken periodically while mixing to ensure a homogeneous sample. The volume in the vessel was maintained by adding 20 ml of distilled water after each sampling and the florosil sample tube was replaced each time a sample was taken. The soil was mixed and purged with argon at room temperature for 19 days, after which the vessel was disassembled.

Each soil sample and florosil tube was extracted for GC analysis. The remaining soil and water were removed and pooled for GC analysis. The entire vessel was washed several times with hexane/acetone (1:1) to remove any PCBs bound to the vessel. These extracts were also pooled for GC analysis. Upon disassembly we observed a tar-like substance sticking to the Teflon stirrer. This was removed and added to the soil and water fraction before the hexane/acetone extractions.

Neither oxygen nor bacterial inoculum was introduced in this mock process yet the analytical results (Figure 3) might be mistaken for biodegradation. Although the time course analysis indicated greater than 90% PCB depletion, it is clear from the mass balance calculations (Figure 3) that the aeration and stirring of the soil resulted in the redistribution of PCBs from the soil to unassayed locations in the reactor (glassware, stirrer, coalesced droplets of PCB). The GC profiles also demonstrate that the observed depletion was not due to a biological process, since all GC peaks are depleted proportionally. Therefore, experiments that purport to show biodegradation of PCBs by quantifying total GC peak areas must be carefully evaluated. It is for this reason that we include non-biodegradable PCB internal standards wherever possible. If such standards or dead-cell controls cannot be included, then one must rely on differential congener depletion (or metabolite production) as evidence for the biological basis of PCB "biodegradation" processes.

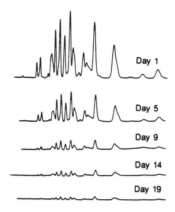

Day	Sample wt.	ppm	mg
1	2.75 g	7800	21
2	2.84 g	5300	15
5	2.75 g	3400	9.4
6	2.06 g	2900	6.0
7	2.16 g	2100	4.5
9	1.73 g	1500	2.5
12	1.57 g	990	1.6
14	1.69 g	830	1.4
16	1.84 g	640	1.2
19	1.53 g	620	0.9

Starting material = 35 g Oakland soil

Total starting PCBs =   273 mg

PCBs Recovered:

Sampling total (for 20.9 g soil) =   64   mg
Glassware =   62   mg
Florosil sample tubes total =   0.4 mg
Final soil + water =   41   mg
Tar balls (in final soil) =   68   mg

Total PCBs recovered =   235 mg (86% recovery)

Figure 3. Depletion of Aroclor 1260 in a mock biodegradation experiment.

SUMMARY

Our research has demonstrated the biodegradation of various soil-bound PCBs under a variety of conditions, using both pure and mixed cultures of MB1, H850, and LB400. We conclude from these results that it should be possible to biologically degrade PCBs on contaminated soil in the environment with appropriate cell concentrations and moisture conditions. Our continuing studies are now focusing on experimental conditions that most closely model an actual site decontamination process. We intend to conduct a laboratory scale-up of a possible *in situ* process and, if feasible, an *in situ* environmental test.

REFERENCES

1. Furukawa, K., 1982. "Microbial Degradation of Polychlorinated Biphenyls," pp. 33-57. A.M. Chakrabarty, (Ed.), Biodegradation and Detoxification of Environmental Pollutants, CRC Press, Inc., Boca Raton.

2. Bedard, D.L., Brennan, M.J., and Unterman, R., 1984. "Bacterial Degradation of PCBs: Evidence of Distinct Pathways in _Corynebacterium_ sp. MB1 and _Alcaligenes eutrophus_ H850", p. 4-101 to 4-118. In G. Addis and R. Komai (Eds.), Proceedings of the 1983 PCB Seminar, Electrical Power Research Institute. Palo Alto, CA.

3. Unterman, R., Bedard, D.L., Bopp, L.H., Brennan, M.J., Johnson, C., and Haberl, M.L., 1985. "Microbial Degradation of Polychlorinated Biphenyls," p. 481-488. In Proceedings: International Conference on New Frontiers for Hazardous Waste Management. U.S. Environmental Protection Agency, Cincinnati, Ohio.

4. Bedard, D.L., Unterman, R., Bopp, L.H., Brennan, M.J., Haberl, M.L. and Johnson, C., 1986. "Rapid Assay for Screening and Characterizing Microorganisms for the Ability to Degrade Polychlorinated Biphenyls." Appl. Environ. Microbiol. $\underline{51}$:761-768.

5. Bopp, L.H., 1986. "Degradation of Highly Chlorinated PCBs by _Pseudomonas_ strain LB400." J. Ind. Microbiol. $\underline{1}$:23-29.

PROCESS DEVELOPMENT AND TREATMENT PLANT STARTUP FOR AN EXPLOSIVES INDUSTRY WASTEWATER

David M. Potter

Environmental Engineering
Hercules Incorporated

INTRODUCTION

This case study covers the process development and treatment plant startup for the Hercules plant in Kenvil, New Jersey. Two treatment processes were modeled on bench-scale, the first being a conventional aerobic activated sludge process. The second study was of a two-stage anaerobic/aerobic activated sludge process. The results of the two bench-scale studies will be covered and compared along with the startup data from the actual wastewater treatment plant which was built in Kenvil, New Jersey.

The Hercules plant in Kenvil produces smokeless gunpowder and military ordnance. The wastewater is a carbohydrate-based wastewater which is deficient in nitrogen and phosphorus. The characteristics of this wastewater are listed in Table 1. It is a strong industrial-strength wastewater with a BOD_5 in excess of 1,000 mg/l. It is also a highly degradable wastewater as noted by the high BOD_5 to COD ratio. The organics are almost entirely solvents. The wastewater also contains significant concentrations of nitroglycerine and cyanide. The nitroglycerine is the result of the production of nitroglycerine on site at the Kenvil plant. The Table I concentrations of nitroglycerine are completely soluble. The cyanide in the wastewater is from an unknown source. Cyanide is not used or produced at this plant. It has been speculated by process engineers that the cyanide is a degradation product of nitrocellulose under alkaline conditions. This has not been confirmed, nor is it necessary to be confirmed. Cyanide is regularly present in the wastewater at this plant, and it has been found to be present in the wastewaters at other similar plants. The raw wastewater is acutely toxic. It has an LC_{50} of approximately 2% by volume. Toxicity may be due to nitroglycerine or cyanide alone. The solvents may also contribute to the toxicity. Obviously, this wastewater requires substantial treatment prior to discharge to surface waters.

Two treatability studies were conducted on this wastewater. The two processes were both modeled on bench-scale. The first process modeled was the conventional aerobic activated sludge process. The second process modeled was a two-stage anaerobic/aerobic process. The first, anaerobic, stage used was the "Bioenergy" version of the anaerobic activated sludge process. The Bioenergy process is a licensed, suspended growth, anaerobic process developed by Biomechanics Limited, of the UK. The second stage of this process was an aerobic polishing stage. Again, conventional aerobic activated sludge was used. These studies were conducted separately several months apart. They were conducted separately because the wastewater strength during the aerobic activated sludge study was much higher than expected. The strength of the wastewater was well above the cut-off of 1,000 mg/l BOD_5 that Hercules has generally used as a practical cut-off for determining whether anaerobic treatment is feasible or not. Since it did appear to be feasible and represented a potential cost savings, an anaerobic study was initiated shortly after the conclusion of the aerobic study.

TABLE 1:

RAW WASTEWATER CHARACTERIZATION

PARAMETER	UNITS	MEAN	RANGE
FLOW	MGD	.135	.083 - .179
BOD_5	MG/L	2,200	68 - 4,400
COD	MG/L	2,600	350 - 6,000
NITROGLYCERINE	MG/L	27	0 - 120
CYANIDE	MG/L	8.6	0.2 - 21

AEROBIC TREATABILITY STUDY

The aerobic activated sludge treatability study was conducted with three bench-scale units. These units had a working volume of just under 30 liters. They were modified 10-gal. aquariums. The aerobic activated sludge study lasted for 10 weeks. The three reactors were run at different loading rates in order to develop the necessary process data and determine the limits of the treatment process. The results of the study are typified by the data presented in Figures 1-4. Figures 1-4 cover data from the No. 2 of the three reactors. Figure 1, which shows COD data, displays a substantial acclimation period associated with this wastewater. By examining the figure, one will note that the soluble effluent COD takes between 25 and 30 days before it finally reaches its steady state effluent value. The same curve is displayed in the soluble COD removal; that is, the COD removal continuously increases until it reaches a plateau somewhere around 28 days. At that time it reaches its high efficiency plateau and stays there. Soluble COD data is presented here because these bench-scale units suffer from high effluent TSS which is not typical of full-scale performance. Consequently, the soluble COD is followed in the effluent in order to filter out the effect of excess effluent solids. Use of soluble COD effluent data also gives a realistic picture of the process kinetics. The COD that is associated with the suspended solids is not available for immediate use in biological growth.

Figure 2 shows BOD_5 data for this reactor. The effluent BOD dropped rapidly during acclimation. In examining Figure 2, it is apparent that the process is very successful in removing BOD.

The results of all three reactors are summarized in Table 2. Based on the results of this study, it is concluded that the aerobic activated sludge process was capable of producing a soluble effluent BOD of 6.4 mg/l with an accompanying removal of 98.8%. COD removal is also high, as is nitroglycerin and cyanide removal. The high NG and cyanide removals are key to effecting the removal of toxicity from this wastewater. Figure

Explosives Industry Wastewater 391

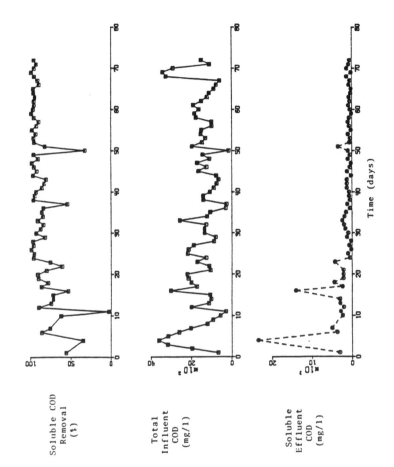

Figure 1: AAS Reactor 2 COD Data

392 Biotechnology for Degradation of Toxic Chemicals

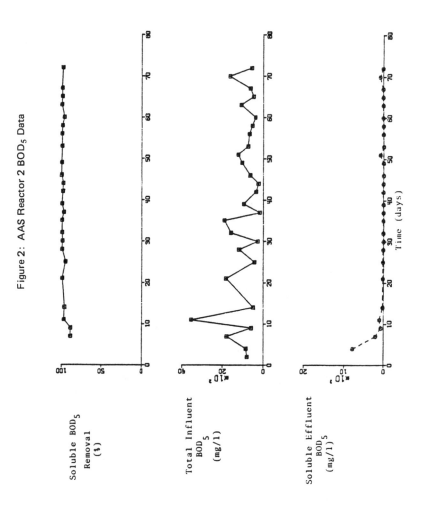

Figure 2: AAS Reactor 2 BOD$_5$ Data

Explosives Industry Wastewater 393

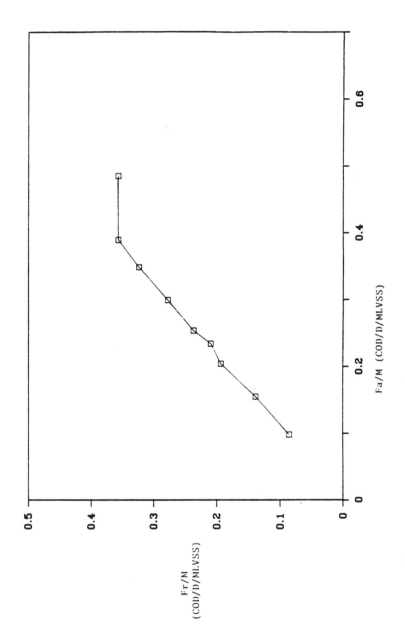

Figure 3: AAS Treatability Limits

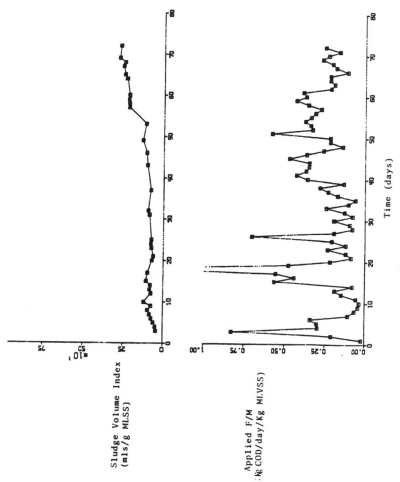

Figure 4: AAS Reactor 2 F/M and SVI Data

TABLE 2:

AEROBIC ACTIVATED SLUDGE BENCH-SCALE RESULTS

PARAMETER	SOLUBLE EFFLUENT	TOTAL EFFLUENT	PERCENT REMOVAL
BOD_5	6.4	-	98.8
COD	111	-	91.1
NG	-	0.93 MG/L	96.5
CN	-	0.19 MG/L	97.9

3 shows the defined limits of the process, in terms of applied F/M, on a COD to MLVSS basis. Above the applied F/M of 0.38, the removal efficiency starts to drop off. It was also observed that the treated effluent became turbid above this loading limit. This obviously defines a limit of the process. It should be noted that the data points presented in Figure 3 are average performance data over several days. This figure does not represent the ability of the plant to handle short term or instantaneous swings and load, but rather the figure defines the ability of the process to handle sustained or continuous loads.

By trial and error, it was determined that the kinetics of this process can best be expressed as $F_r/M = K(S_E/S_o)$, where $K = 3.7$ day^{-1}. This kinetic expression is based on COD. The development of this kinetic expression is purely empirical. Several other kinetic expressions were tried and did not provide a reasonable fit. The process was also marked by a very long SRT. The design SRT from the process developed was 21 days. The hydraulic retention time is 2-3 days. While this SRT is not unusual for an industrial wastewater treatment plant, it is certainly a long SRT and is indicative of the need to develop, culture and retain bacteria that are capable of degrading the recalcitrant components of this wastewater.

The effluent and the waste-activated sludge from the bench-scale units were evaluated for toxicity and other hazards. The treated effluent had an LC_{50} of $\geq$ 68% by volume. This LC_{50} was developed by 96-hour static renewal toxicity testing with fathead minnows. This result was developed under extremely unfavorable operating conditions. Mechanical failures in the lab resulted in a gross overload during the middle of this 4-day test. The logistics of the test were such that nothing could be lost by continuing the test, and these results were interpreted as the minimum toxicity removal. The results were considered satisfactory at the time since we understood that we simply had to meet the normal acute toxicity standard of an LC_{50} of $\geq$ 50%. The waste-activated sludge was thickened and subjected to dewatering tests for use on a rotary vacuum filter and also on a belt filter press. The filter cakes from these tests and the thickened waste-activated sludge were tested as per EP toxicity testing. All were found to be nonhazardous as defined by EP toxicity.

Some unusual results of this bench-scale study should be noted. The most unusual result was that it was the development of a nonfilamentous bulking sludge. Typically, the SVI's ran in the range of 150-200 ml/g. The supernatant from the SVI test was very clear. There were no filaments present. As the process was loaded more highly, the bulking characteristic increased slightly within the loading limits of $F/M \leq 0.38$ COD/D/MLVSS. Above the loading limits bioflocculation characteristics deteriorated rapidly. One other unusual result found on the bench-scale study was a bloom of yeast. This developed overnight and killed off all three reactors. Yeast are in competition with the desirable bacteria. In this case, yeast produced organic acids which lower the pH which favor the growth of yeast over the bacteria. This established a cycle where an uncontrolled pH drop simply overwhelmed the desirable bacteria. The result was a yeast bloom which dropped the reactors' pH to 2.5 overnight. The source of the yeast is unknown. Yeast do not bloom in the raw wastewater collection system because the raw wastewater is nutrient deficient. The episode was unique to our experience in Hercules and demonstrated, quite vividly, the need for very reliable pH control on a full-scale aerobic activated sludge plant.

One other unusual finding, of course, is that there was an extended acclimation period associated with this wastewater. On average, the acclimation period took just about 30 days for all three reactors. This sort of an extended acclimation period is a bit unusual in our experience at Hercules. Typically, with aerobic activated sludge systems, high removal efficiency develops immediately or within a few days. The extended acclimation period is indicative of the toxicity that is present in the wastewater. Apparently, over 30 days, certain bacteria proved incapable of surviving in this environment while those bacteria which were capable of utilizing the organic substrates not only survived, but also grew to sufficient numbers over a 30-day period.

TWO-STAGE ANAEROBIC/AEROBIC TREATABILITY STUDY

The anaerobic/aerobic study was conducted using a four parallel process train, anaerobic treatablity unit. This treatability unit was developed and built by Biomechanics and it modeled on the Bioenergy process. The anaerobic study lasted for 203 days. Following these four anaerobic units were two aerobic reactors. Only two aerobic reactors were used, and they were small because of the volumes involved. A sketch of the anaerobic "Bioenergy" process is shown in Figure 5. The Bioenergy process is described in more detail in a paper presented by Rippon, et al. (1). The Bioenergy process utilizes the thermal shock technique to stun the methanogens and essentially stop the production of methane gas on a temporary basis. When gas production is stopped, the biomass settles very well, just as a conventional aerobic-activated sludge does, and it is easily separated in a conventional clarifier and returned to the reactor. When the sludge is returned to the reactor and warmed back up to normal temperature (35-38° centigrade), the bacteria resume normal activity with no ill effects. Hercules has used the Bioenergy process in several other studies, and Hercules has one highly successful Bioenergy plant installed and operating for almost two years now at a Hercules location in France.

One objective of the anaerobic study was to locate a local source of seed sludge for a full-scale plant. Several local POTW sources of seed sludge were considered. Samples were obtained and tested in the lab. The criteria were the turbidity and the color of the supernatant, settling rate, and bioflocculation characteristics. If these characteristics were acceptable, sludge was then tested in the reactors while being fed a sucrose solution. A gas yield of at least 0.3 m^3/kg COD applied was required before a sludge was accepted. Unfortunately, no ideal local candidates were found, but two marginally acceptable seed sludges were started in the study.

Explosives Industry Wastewater 399

FIGURE 5

BIOENERGY PROCESS FLOW DIAGRAM

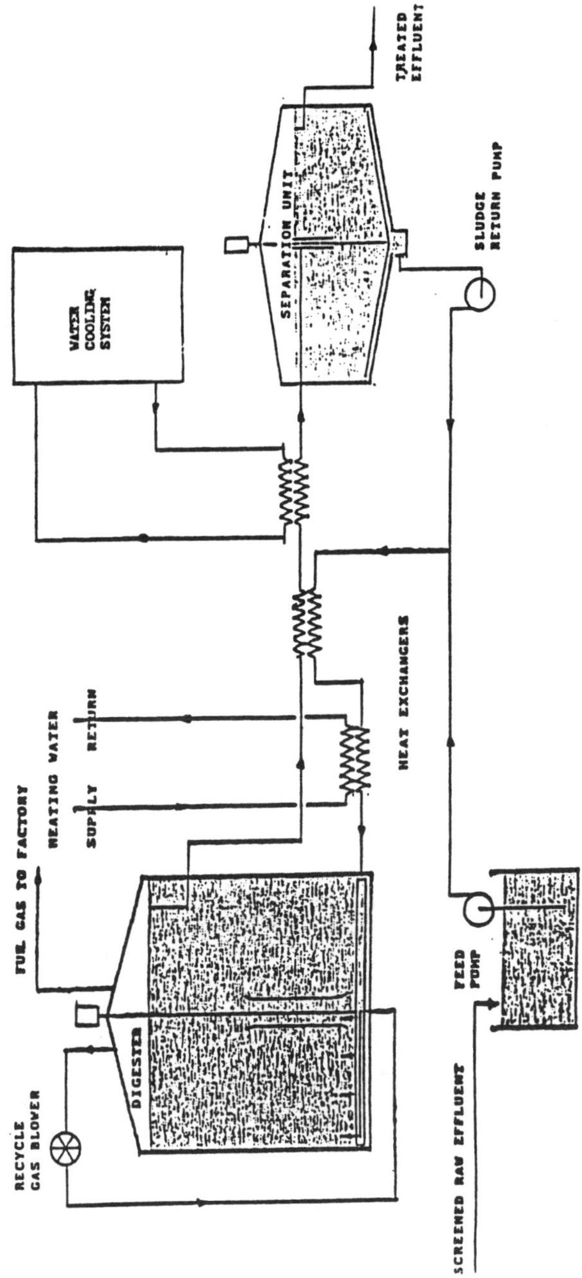

The feed to the anaerobic units was supplemented with nitrogen and phosphorus as well as the micronutrients iron, cobalt, and nickel at concentrations of 10, 5, and 1 mg/l, respectively. The feed was also supplemented for alkalinity at a level of 500 mg/l as $CaCO_3$. Soda ash was used as an alkalinity supplement. The major nutrients and micronutrients were not studied extensively or optimized during the course of this study. Past experience at Hercules has shown that introducing nutrients as a factor during a study can often needlessly complicate the study and obscure results from a study. Our philosophy has changed so that we now supplement all major nutrients and micronutrients at required levels and monitor the effluent carefully to assure that they are present in adequate quantities. One can then be confident that nutrients are not a factor to be considered in analyzing the results of these studies. From raw wastewater characteristics, one can determine whether nitrogen, phosphorus, iron, cobalt, and nickel will be required as supplements for successful anaerobic treatment. If they are required, they can always be optimized on a full-scale plant when the time comes.

The results of the anaerobic treatability study are typified by the plots of Reactor 2 performance. In Figure 6, the COD data for Reactor 2 are presented. In reviewing the soluble COD removal, it is immediately apparent that there was a very long acclimation period associated with this wastewater. It should be noted that Reactor 2 was originally started up with seed sludge from Elizabeth, NJ. That sludge proved to be highly unacceptable in terms of effluent TSS which will be shown in one of the later plots. The high effluent TSS was resulting in a massive washout situation, and it was clear that this was not an acceptable seed sludge. On day 38, a known good sludge was brought in from Goshen, PA, and it was used to reseed Reactor 2. As can be seen, it took approximately 50 days before there was any real improvement in the COD removal. Then, from approximately day 86 and on, the COD removal starts to advance upward in a very rapid fashion and reaches a plateau around day 130. This same pattern can be observed in the effluent soluble COD. Starting around day 50, the effluent soluble COD begins a steady trend downward and seems to reach its plateau value somewhere

Explosives Industry Wastewater 401

Figure 6: Bioenergy Reactor 2 COD Data

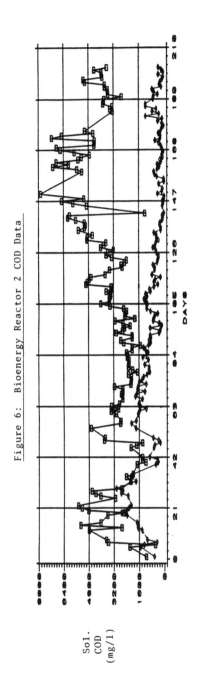
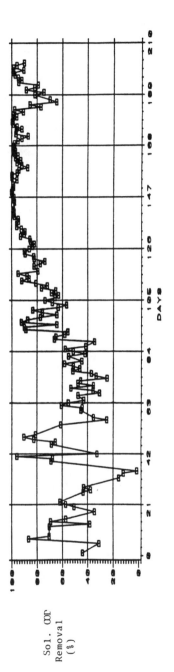

Sol. COD (mg/l)

Sol. COD Removal (%)

around day 130. This is an extremely long acclimation period, and it required great patience to stick with the study. However, previous work by Speece, et al (2) had shown that substantial acclimation periods can be expected with cyanide bearing wastewaters.

BOD_5 removal data is presented in Figure 7. Again, as with the COD data, the trends show a continuous slow but steady drop in effluent BOD_5 concentration until it seems to reach a plateau somewhere around day 130. The 5-day average of the gas yield is presented in Figure 8. Data is averaged over 5 days because of the metering system used. With small volumes of gas, the incremental counting meters can show wide variations from day to day which are not always true variations of process performance but, rather, are just variations in the partial increments that are counted from day to day. The 5-day gas yield confirms the same trend of slow but steady acclimation of the biomass to this wastewater. The gas yield seemed to reach its average value somewhere around day 120. It should be noted for the benefit of the reader that the little tail-off at the very end of the gas yield plot represents overloading and shock studies and it is not typical process performance. The volatile acids data is presented along with F/M data in Figure 9. The volatile acids were initially very high averaging 500 mg/l. This is normally considered the upset range for a suspended growth system. Around day 110, the volatile acids suddenly dropped and reached a very acceptable level of less than 100 mg/l. F/M loading on this particular reactor was kept at a very steady 0.1 on a COD to MLVSS basis until system performance appeared to reach a steady value. At that point, the loading was increased in steps which were generally held for 10-15 days. These steps were continued until a steady loading of 0.25 COD/D/MLVSS was achieved. Effluent TSS data is presented in Figure 10. As noted earlier, during the first 38 days, the seed sludge from Elizabeth, NJ was used. The very high values of effluent TSS during those days were simply not acceptable. The reactor was almost completely washed out by day 38 and so this was obviously not an acceptable seed

Figure 7: Bioenergy Reactor 2 BOD_5 Data

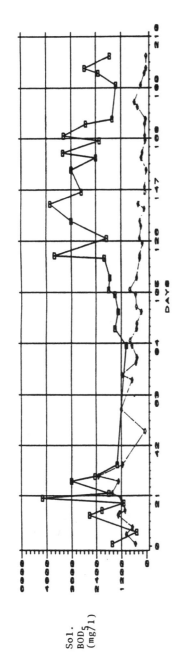
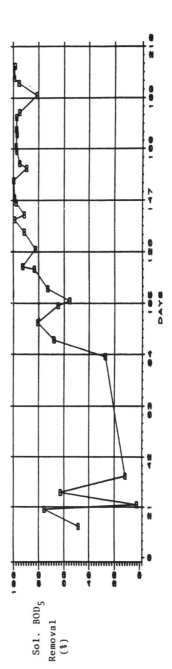

404 Biotechnology for Degradation of Toxic Chemicals

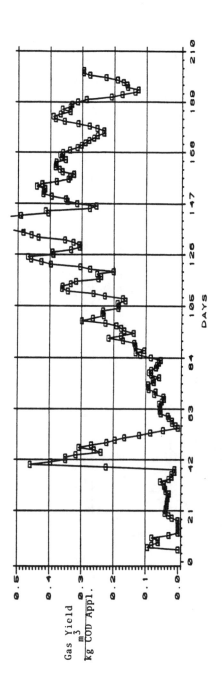

Figure 8: Bioenergy Reactor 2 - 5-Day Average Gas Yield

Explosives Industry Wastewater 405

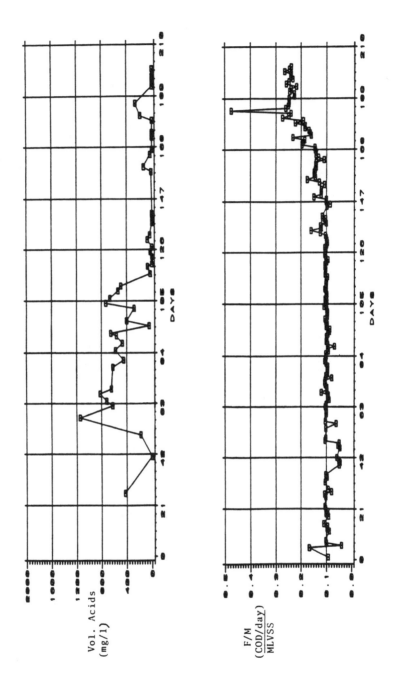

Figure 9: Bioenergy Reactor 2
Volatile Acids and F/M

406 Biotechnology for Degradation of Toxic Chemicals

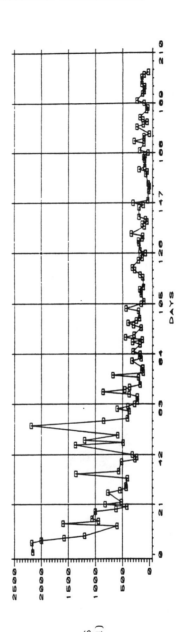

Figure 10: Bioenergy Reactor 2 Effluent TSS

sludge. When the West Goshen sludge was brought in, it too had some bioflocculation problems associated with acclimation to the new wastewater. However, the effluent TSS steadily decreases until it reaches a final plateau of approximately 100 mg/l.

The combined results of the Bioenergy bench-scale reactors are presented in Table 3. BOD_5 and COD removal are quite high for an anaerobic process. Although not quite as high as the aerobic process, this certainly represents good performance. The removal of nitroglycerine was almost equivalent to the aerobic units. Removal of cyanide was also essentially complete. The noted effluent cyanide concentration of less than 0.02 mg/l represents the limit of detection used in the lab analysis for cyanide. A question mark is next to the 100% removal because it was later discovered that this was really not complete removal but, rather, part of the cyanide was being complexed with the metal micronutrients, either iron or nickel. The analysis measured only free cyanide, so the complexed cyanide was not included in the effluent concentrations. The results of the Bioenergy study indicated that a conservative design F/M that would provide reliable performance with 0.20 COD applied per day per MLVSS. At this F/M, the net sludge yield was 0.02 kg of TSS per kg COD applied. The gas yield was 0.35 meters^3/kg COD applied, and the gas composition was 80.1% methane and 19.9% carbon dioxide. Design calculations based on the bench-scale results indicated a hydraulic retention time of 1.4 days and a sludge retention time of 180 days. This is an unusually high SRT even for an anaerobic system.

Once a steady anaerobic effluent was achieved, the aerobic activated sludge polishing units achieved 96% BOD_5 removal and 78% COD removal. Bioflocculation characteristics were good and supernatants with less than 30 mg/l TSS were typically achieved during SVI tests. Nitroglycerine removal could not be monitored due to analytical problems. Cyanide in the aerobic effluent was 0.25 mg/l which was higher than the feed to these units. The measurement was, however, as free cyanide and did not include the complexed cyanide. The effluent from the aerobic polishing stage was collected and stored by freezing over a course of approximately 3 weeks in order to develop a large enough volume to run a 96-hour fathead

TABLE 3

BIOENERGY BENCH-SCALE RESULTS

PARAMETER	TOTAL EFFLUENT CONC. (MG/L)	PERCENT REMOVAL
BOD_5	233 (MG/L)	89%
COD	396	85%
TSS	100	-
NG	1.7	92%
CN	<0.02	100% (?)

TABLE 4:

AEROBIC ACTIVATED SLUDGE POLISHING BENCH-SCALE RESULTS

PARAMETER	TOTAL EFFLUENT CONC. (MG/L)	PERCENT REMOVAL
T. BOD_s	19	94
S. BOD_s	16	96
T. COD	122	78
S. COD	86	90
TSS	30	70
NG	N/A - ANALYTICAL PROBLEMS	
CN	0.25	NEGATIVE

minnow acute toxicity test. The results of this test were highly successful. The LC_{50} could not be measured. There was, in fact, no measurable acute toxicity in any concentrations including the pure treated effluent.

COMPARISON OF STUDIES

The results of the two bench-scale studies are compared in Table 5. Effluent total BOD is essentially equivalent. The soluble BOD_5 is a little bit higher in the two-stage process. Total effluent COD was also very close. Nitroglycerine removal was essentially below the nominal detection limit of 1 part per million. Cyanide removal was also very high, and acute toxicity is greatly removed by both processes. It should be noted again that the LC_{50} developed for aerobic activated sludge process was developed under adverse operating conditions and represents only the minimum capacity for removing toxicity.

The basis of design for the full-scale wastewater treatment plant is presented in Table 5a. It is not a large plant in terms of flow, but it does have a very substantial BOD_5 loading of 2,500 lbs/day on average. Based on the Basis of Design Parameters in Table 5a, a conceptual design was developed for both treatment processes. These conceptual designs are compared in Table 6. From an engineering perspective, the two-stage system requires less total reactor volume. However, it is a more complicated system, involving 5 vs. 3 tanks and several more pumps, lines, control meters, etc. The energy balance on both processes comes out even. The anaerobic process runs at an elevated temperature and the gas produced by the anaerobic process is not quite enough to maintain that temperature through-out the whole system. Therefore, supplemental steam is required at most ambient temperature conditions. The aerobic process also requires supplemental steam to maintain removal efficiency under winter conditions and also requires approximately 34 horsepower to drive the blowers for aeration. The biggest difference between the two systems, in terms of operating costs, is the net sludge production. The sludge production from the aerobic activated sludge system is more than 4 times greater than that produced by the combined two-stage treatment system.

TABLE 5:

COMPARISON OF BENCH-SCALE STUDY RESULTS

PARAMETER	AAS EFFL. CONC. (MG/L)	TWO-STAGE EFFL. CONC. (MG/L)
T. BOD_5	14	19
S. BOD_5	6	16
T. COD	132	122
S. COD	111	86
NG	0.93	N.A.
CN	0.19	0.25
ACUTE TOXICITY	$LC_{50} \geq 68\%$	NONE

TABLE 5A
BASIS OF DESIGN

PARAMETER	UNITS	AVERAGE	MAXIMUM	MINIMUM
BOD_5	LBS/DAY	2,475	4,540	135
COD	LBS/DAY	2,890	4,870	450
TKN	LBS/DAY	10.1	-	-
TOTAL P	LBS/DAY	1.5	-	-
FLOW	GPD	135,000	175,000	82,000

TABLE 6:

COMPARISON OF CONCEPTUAL DESIGNS

PARAMETER	UNITS	AEROBIC ACTIVATED SLUDGE	TWO-STAGE ANAEROBIC/ AEROBIC
F/M	COD/D/MLVSS	0.30	0.20 ANAEROBIC 0.15 AEROBIC
MLSS	MG/L	4,500	8,000
TANKS	-	1-EQUALIZATION 1-REACTOR 1-SETTLER $\overline{3}$	1-EQUALIZATION 2-REACTOR 2-SETTLERS $\overline{5}$
REACTOR VOLUME	M GALS.	0.362	0.216 ANAEROBIC 0.035 AEROBIC
NET SLUDGE	LBS/DAY	364	90

In spite of the advantages in terms of operating costs, the two-stage system was not built. Time constraints precluded the design and construction of the two-stage plant. Rather, a single-stage conventional aerobic activated sludge system, with built-in capability for potential facile future conversion to a two-stage aerobic/anaerobic system, was constructed.

FULL-SCALE START UP

The results of the startup of the aerobic activated sludge system are presented in Figures 11 and 12. As with the bench-scale units, there was a 30 day acclimation period during which there was marginal treatment efficiency. As with the bench-scale units, there was also excessive foaming and other nuisance problems during this period. The results of the aerobic activated sludge system from May are presented in Table 7. The plant was started up in early January so this represents approximately 5 months after startup. Effluent BOD_5 and COD discharges are meeting permit conditions. During startup, there were some problems controlling the load and, consequently, a bloom of filamentous bacteria developed in this system. During the month of May, the filament content was high, and hence the high effluent TSS value of 160 mg/l. The nitroglycerine removal at this time is nearly in excess of 99%. The effluent is continuously less than 1 mg/l which is the nominal detection limit of the analytical method used. Cyanide removal is also very high, averaging 98.9 percent. The effluent cyanide is almost half of what was predicted from the bench-scale studies. One 96-hour static renewal acute toxicity test was run in May using fathead minnows. No measurable toxicity was detected, even in 100% effluent. This is much better than what was experienced on bench-scale, but again the bench-scale results for the aerobic activated sludge were under adverse conditions and represented minimal performance.

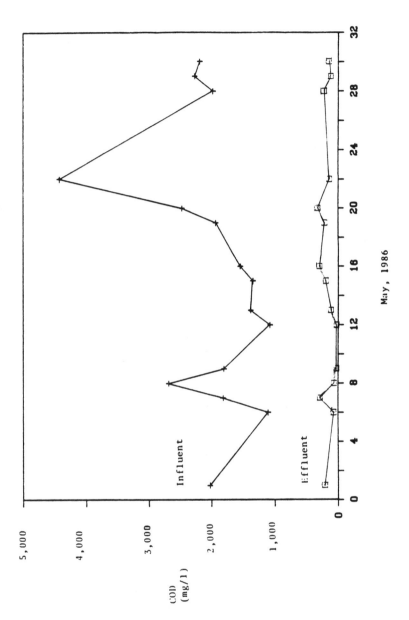

Figure 11: Full-Scale AAS COD

416 Biotechnology for Degradation of Toxic Chemicals

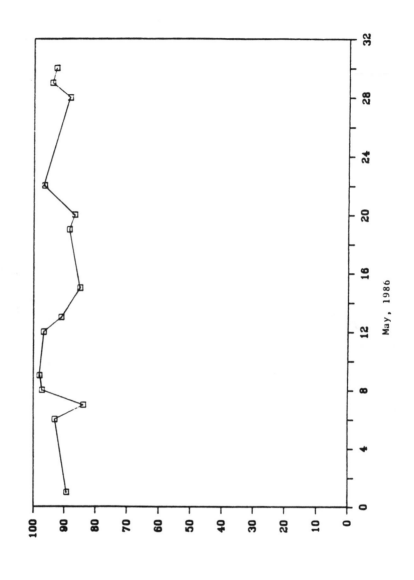

Figure 12: Full-Scale AAS COD Removal

TABLE 7:

FULL-SCALE RESULTS
AEROBIC ACTIVATED SLUDGE

PARAMETER	INFLUENT (MG/L)	EFFLUENT (MG/L)	PERCENT REMOVAL
BOD_5	1,350	42	96.9
COD	1,990	218	89.0
TSS	-	160	-
NG	156	< 1	99+%
CN	8.0	.095	98.9
ACUTE TOXICITY	-	NO MEASURABLE ACUTE TOXICITY	100%

SUMMARY

In sum, there were some very similar aspects to both treatment processes which were investigated and the one which was eventually built on full scale. No matter which process was used, there was a prolonged acclimation period associated with this wastewater. The raw wastewater was acutely toxic to fathead minnows at very low concentrations. Apparently, it takes a substantial period of time to develop bacteria which are able to degrade this wastewater. Also notable with both processes was the high sludge retention time that developed from successful bench-scale conditions. In the case of the strictly aerobic process, the SRT was 21 days. In the case of the combined two-stage anaerobic aerobic process, the SRT was 160 days. The long SRT is indicative of the recalcitrant nature of the organics in this wastewater. Once one has developed bacteria which are capable of degrading these organics, the bacteria must be maintained in the system for a long period of time in order to successfully treat the wastewater. Also characteristic of both systems investigated was complete toxicity removal. After treatment, a raw wastewater with an LC_{50} of 2% by volume is completely non-toxic. Perhaps most important, all of the results demonstrate that with patience and proper process control and acclimation, conventional biological treatment processes are capable of successfully treating acutely toxic wastewaters.

REFERENCES

[1] Anderson, G. K.; Donnelly, T.; Rippon, M.; and Duarte, A., "Pilot and Full-Scale Anaerobic Digestion for the Food and Drink Industry," <u>Proceedings of the 36th Industrial Waste Conference</u>, Purdue University, West Lafayette, IN (May 12-14, 1981), pp. 279-300.

[2] Yang, J.; Parkin, G. F.; and Speece, R. E., "The Response of Methane Fermentation to Cyanide and Chloroform," <u>Progress in Water Technology</u>, Vol. 12, p. 977 (1980).

UTILIZATION OF NITRITE OXIDATION INHIBITION TO IMPROVE THE NITROGEN ELIMINATION PROCESS

S. Suthersan and J.J. Ganczarczyk

Department of Civil Engineering
University of Toronto
Toronto, Canada

ABSTRACT

Elimination of the second stage of nitrification can make the process of biological removal of nitrogen from waste water much more economical. The basic factors affecting such controlled elimination have been investigated in batch type reactors.

INTRODUCTION

The removal of nitrogenous oxygen demand, and in some cases, complete nitrogen removal from wastewater has received increased attention in the water pollution control field in recent years. The most commonly encountered form of nitrogen in municipal and industrial wastewaters is ammonia-nitrogen. Even though there are several means for the elimination of ammonia from wastewater, nitrification is the most attractive method from a

* To whom all correspondence should be directed

technical and economic viewpoint. For the complete removal of nitrogen in all its forms from the wastewater, nitrification is followed by the denitrification process.

In the nitrification process, the ammonia present in wastewater is oxidised to nitrite and nitrate by the autotrophic nitrifying organisms <u>Nitrosomonas</u> and <u>Nitrobacter</u>, respectively. This process has been found to be more sensitive than other biological systems to inhibitory substances and various environmental factors such as temperature, pH, dissolved oxygen concentration and other factors. The inhibition of nitrification or, in other words the inhibition of the activity of nitrifying organisms, is understood here either as slowing down and/or temporarily stopping the bacterial metabolic activity. Inhibition, according to this definition, is a reversible phenomenon as opposed to toxicity. It should be noted that there has been a lot of confusion in the past on the usage of these two terms in the literature on the subject.

Ther are numerous reports on the inhibition of <u>Nitrobacter</u> which is responsible for the second stage of nitrification. Several factors such as pH (Martin et al. (1942), Duisberg et al. (1954), Morril (1959) and Court et al. (1964)), various organic compounds (Tomlinson et al. (1966), Hockenbury and Grady (1977),Sharma and Ahlert(1977) , low oxygen concentrations (Boon and Laudelaut (1962), Payne (1979), Mines (1983)) and heavy metals (Ganczarczyk (1979), Huang et.al (1982), Randall and Buth (1984)) were found to be inhibitory to the oxidation of nitrite to nitrate by <u>Nitrobacter</u>.

The substrates for <u>Nitrosomonas</u> and <u>Nitrobacter</u> themselves in the form of free ammonia (FA) and free nitrous acid (FNA) have been found to be inhibitory to these organisms in very low concentrations. Anthonisen (1974), in his extensive studies on the inhibition of <u>Nitrobacter</u> by FA, has reported that concentrations in a range as low as 0.1-1.0 mg/L inhibited **Nitrobacter**. Prakasam and Loehr (1972), Wong-chong and Loehr (1978), Alleman and Irvine (1980) and Alleman (1984) have also reported on the inhibitory effects of FA on **Nitrobacter**. Based upon results published by many authors, it appears that inhibition of **Nitrobacter** by FA takes place at significantly lower concentrations than required to inhibit **Nitrosomonas**.

Historically, all these reports on inhibition of nitrifying organisms have been considered as a hindrance to efficient nitrification process operation and emphasis has been put on finding remedial ways to overcome this inhibition. However, this paper deals with the efforts to suppress the second stage of nitrification, by which the nitrification-denitrification process can be modified so that it is the accumulated nitrite instead of nitrate which is denitrified. The pathway for the modified NITRITIFICATION - DENITRITIFICATION process will be:

$$NH_4^+ \longrightarrow NO_2^- \longrightarrow N_2$$

The benefits that will accrue by successfully implementing this modification will be the significant reduction in the oxygen requirements for the nitrification step, reduced organic carbon (e.g. methanol) requirements and increased reaction rates for the denitrification step, and reduced overall capital costs.

MATERIALS AND METHODS

Due to the ease of monitoring and controlling the parameters in a batch reactor in comparison to a continuous flow reactor, the reactor selected for these experiments was of the semi-batch mode. In a semi-batch or fed-batch operation (Yoshida et al. 1973; Yamane and Shimizu, 1984; Parulekar and Lim, 1985) the substrate is fed either intermittently or continually during the course of an otherwise batch operation. Unlike the batch and continuous reactors, the semi-batch bioreactors may be operated in a variety of ways, regulating the substrate feed rate in a pre-determined manner (feed forward control) or using feed back control. The mode selected for these experiments was a constant feed semi-batch reactor where the substrate (NH_4Cl) with the necessary nutrients, was fed at a constant rate to maintain a constant FA concentration in the reactor, under controlled pH conditions, at room temperature. This constant feed rate of NH_4Cl to the reactor was equal to the predetermined rate of oxidation of ammonia by the Nitrosomonas present in the reactor. The variation in the reactor volume due to the addition of substrate into the semi-batch reactor (dilution effect) was overcome by using the substrate feed with very high concentrations which made the volume change as small as possible. The growth rate of the nitrifying organisms was small and, also, no organic carbon was present in the substrate feed. Hence, the increase in biomass in the semi-batch nitrifying bio-reactors used in these experiments, either due to the nitrifying biomass or due to the heterotrophs present in the sludge, was negligible.

The biomass used in the experiments was obtained from a nitrifying activated sludge plant in Milton, Ontario, and was maintained in a stock culture for further enrichment of nitrifying organisms. The pH in the reactors was controlled automatically by a multichannel controller (Bach-Simpson, SAS1) with a RADIOMETER PHM62 pH meter and a TTT60 titrator, by dosing with an alkaline solution of the following composition: 25 g sodium bicarbonate, 10 g sodium carbonate and 25 g sodium hydroxide in a one litre solution. The dissolved oxygen concentration was maintained around 6 mg/L. The total ammonia concentration was determined using an ion selective electrode and an ORION 701 A ionalyser meter, while NO_2^- and NO_3^- were analysed on a Technicon Autoanalyser-I with sulfanilamide and a cadmium reduction column.

RESULTS AND DISCUSSION

Technical Concept of the Modification

The intended modification of the process should be able to achieve nitrite buildup, through selective inhibition of Nitrobacter present in the microbial population in the system. The FA inhibition to Nitrobacter was shown as occurring at concentrations substantially lower than the concentrations required to inhibit Nitrosomonas. It is this differential that will be utilised to inhibit Nitrobacter and thus accumulate nitrite without affecting the activity of Nitrosomonas.

The scheme for the modification is based on a novel concept of the RECOVERY TIME for the inhibited Nitrobacter (Suthersan and Ganczarczyk, 1986). When Nitrobacter is exposed to an inhibitory factor and then relieved from this inhibitory environment, there will be a specific time during which this organism's activity will remain suppressed. Therefore, recovery time is defined as " the time taken by the inhibited organism to recover its metabolic activity after being relieved from the inhibitory environment". Here, this factor was defined as the time required by the inhibited Nitrobacter to oxidise at least 5 percent of the accumulated nitrite after removal of the inhibitory conditions.

The Nitrobacter thus inhibited during its retention in the inhibition chamber which is located in the recycle line of an activated sludge process system (Fig. 1), will recover its activity to oxidize nitrite in the aeration tank unless steps are taken to control it. To effect such a control, the hydraulic retention time in the aeration tank has to be maintained at a value less than the recovery time of the inhibited Nitrobacter.

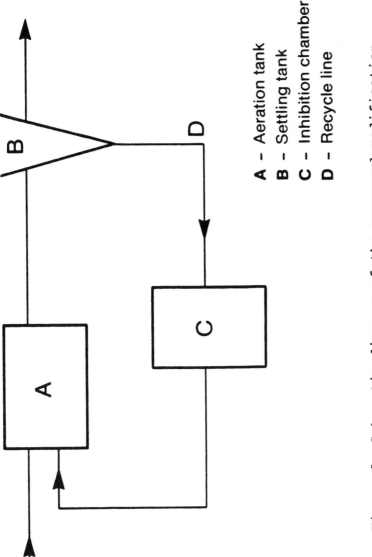

Figure 1. Schematic diagram of the proposed modification to activated sludge nitrifying system

A – Aeration tank
B – Settling tank
C – Inhibition chamber
D – Recycle line

Definition of Parameters

Quantitatively, recovery time is a function of the inhibitory free ammonia (FA) concentration, pH, time of exposure and the Nitrobacter population present. In the available reports on the inhibition of Nitrobacter by FA, no attempts were made to quantify the degree of inhibition. Usually an observed accumulation of nitrite was reported as inhibition. In this study, the degree of inhibition was measured as the percentage of the ratio between the accumulated nitrite concentration and the combined nitrite and nitrate concentrations.

$$\% \text{ of inhibition} = \frac{[NO_2^-]}{([NO_2^-] + [NO_3^-])} \times 100$$

In terms of biochemical definitions, inhibition is measured as a decrease in the enzymatically catalysed reaction rates. As an extension, it is appropriate to measure the degree of inhibition in terms of the relevant reaction rates. However for the purpose of this study, the above method, based on product formation seemed relevant because the objective of this study was to find out to what extent the Nitrobacter's ability to oxidise the available nitrite had been impaired in a given environment, with the aim to find whether complete cessation of this step could be achieved.

Adaptation to increasing FA concentrations

In most instances reported in the literature, the inhibition of Nitrobacter due to free ammonia (FA) concentrations was not the intended objective of any study with the exception of the work by Anthonisen (1974). Rather, when inhibition was observed during a study, it was often reported as an explanation of some other phenomenon. Hence, most data available on the values of FA concentrations inhibitory to Nitrobacter cannot be used for comparison since the environmental conditions under which these data were obtained were not consistent or controlled.

Anthonisen (1974) in his experiments designed to find out the FA concentrations inhibitory to Nitrobacter, selected the pH values in the optimum range for nitrification. The free ammonia (FA) concentrations of 0.1-1.0 mg/L which triggered an accumulation of nitrite were reported as the inhibitory range. In his study, the degree of inhibition caused by different FA concentrations was not investigated, but in our initial study (Suthersan and Ganczarczyk,1986) it was found that the inhibitory FA concentration values could substantially exceed the range suggested by Anthonisen (1974).

It was also found that the inhibitory concentration of FA which could be tolerated by Nitrobacter could be increased to significantly higher concentrations than the values suggested by Anthonisen (1974). This was achieved by continuously acclimatising the Nitrobacter population to increasing FA concentrations at a pH of 8.0. Starting with an FA concentration of 0.5 mg/L and increasing it by 0.25 mg/L per week, a continuously and gradually acclimatised nitrifying biomass completely oxidised NH_4^+ to NO_3^-. The importance of intermittent imposition of inhibition followed by recovery of the Nitrobacter for the modification becomes quite obvious from the above observations. In this continuation of the original study, it was also found that, during the initial stages, the stepwise increase required to maintain 80 percent inhibition of Nitrobacter was slower with respect to time than later stages, when the degree of acclimatisation was higher (Fig.2). After three days of acclimatisation at 2.5 mg/L of FA it was observed that complete nitratification was carried out by this Nitrobacter population.

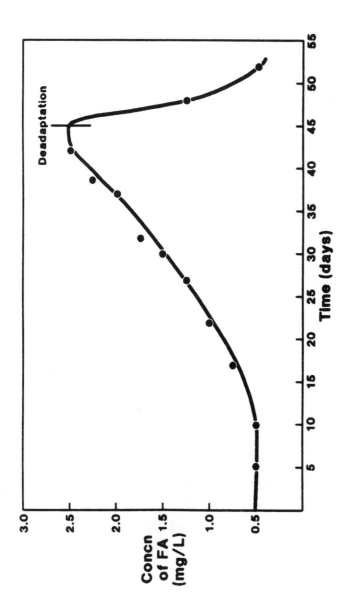

Figure 2. Effect of continuous adaptation and de-adaptation on the required FA concentration to cause 80 % inhibition of <u>Nitrobacter</u>

Deadaptation by reverting to favorable conditions

The same biomass which was continuously and gradually adapted for 45 days, was then provided with an environment with a pH value of 7.8 (which would be the pH in the aeration tank) and a total ammonium nitrogen concentration of about 10 mg/L which amounted to an FA concentration of 0.2 mg/L. After three days, the concentration of FA required to cause 80 percent inhibition of Nitrobacter decreased to 1.0 mg/L and, after one week at 0.5 mg/L of FA concentration, 80 percent inhibition of Nitrobacter was achieved.

A population may adapt to an adverse environment in several possible ways. Non-genetic modifications are temporary and can be readily reversed by a return to the non-inhibitory environment. These modifications are called phenotypic adaptations. Genotypic adaptations, by contrast, require a genetic modification and the change is inherited by daughter cells and may persist for some time even after the environment is restored to its original state. Upon the return to the favorable environment from the inhibited conditions, the inhibited population tries to de-adapt, but the length of time needed to regain the original characteristics is extremely variable. This period necessary for de-adaptation depends on the type of microorganism, the factor responsible for the initial inhibition and the mechanism of adaptation and de-adaptation. Restoration of the original behavioral pattern by a population following a phenotypic adaptation is rapid as a rule (Alexander, 1971). Since the effects due to adaptation were readily reversed by the return to the non-inhibitory environment of the Nitrobacter (Fig. 2) it could be assumed that the Nitrobacter had undergone only a phenotypic adaptation due to continuous and gradual adaptation under inhibitory conditions.

It would have been very interesting to compare the nitrite oxidation rates by Nitrobacter during the different stages of the adaptation phase and the deadaptation phase. But, since ammonia oxidation by Nitrosomonas is the rate limiting step in the nitrification process, it was not possible to measure the sequential nitrite oxidation rate in the reactor.

Role of pH

Anthonisen (1974) did not investigate the direct role of pH on the inhibition of Nitrobacter. It could be seen that the necessary information to ascertain whether pH itself has a direct role in inhibiting Nitrobacter or the role of pH is in influencing another inhibitory factor (eg: formation of free ammonia) is lacking in the literature. A series of experiments were conducted in semi-batch reactors, to investigate the direct role of pH on both recovery time and percentage of inhibition of Nitrobacter. It was found that the pH itself plays an additive role to the effects caused by FA on these two parameters (Suthersan and Ganczarczyk, 1986). Also it was apparent that the more adverse the environment was in terms of FA concentration and pH, the higher were the inhibitory effects in terms of both percentage of inhibition and recovery time. In the case of the recovery time, the "damage" caused by increasing FA concentrations was more apparent than that due to increasing the pH towards the adverse range.

Influence of Exposure Time

Suthersan and Ganczarczyk (1986), demonstrated the significance of exposure time on the recovery time of Nitrobacter. It was observed that, under similar inhibitory conditions a longer exposure time caused a longer recovery time. From Fig.3 it could also be seen that, for a biomass of MLSS 3000 mg/L inhibited at a pH of 8.8 and allowed to recover at a pH of 8.0, the increasing FA concentrations increased the available recovery time with longer exposure times. Exposure time seems to be more significant in yielding a longer recovery time at lower FA concentrations (Table.1).

Table 1. Percentage increase in Recovery Time (% RT) due to an increase in exposure time from 4 hrs to 8 hrs at different FA concentrations

FA concn. mg/L	RT (%)
1	71
2	63
3	33
4	27
5	19

Nitrobacter remains in a dormant state during the period which is defined here as the recovery time. The reason for this dormancy is not very clear and investigation of this phenomenon was beyond the scope of this study. Dormancy has been defined by Sussman and Halvorson (1968) as "any rest period or reversible interruption of the phenotypic development of an organism". It could be a condition in which development is delayed due to an innate property of the dormant stage such as a barrier to the penetration of nutrients, a metabolic block or the production of a self-inhibitor. The recovery time also could be hypothesised as the time taken for the decomposition of the enzyme-inhibitor or the enzyme-substrate-inhibitor complex formed under the inhibitory conditions. Thus recovery time may be the measure of the reversibility of the inhibition and the affinity between the enzyme and the inhibitor.

From Figs. 3 and 4 it can also be seen that high FA concentrations and higher exposure times

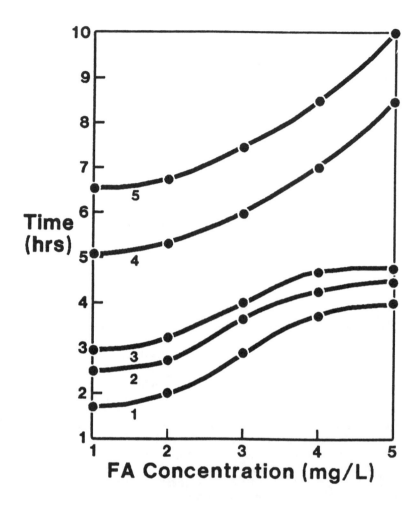

Figure 3. Influence of FA concentration and Exposure time on the short term and long term recovery of <u>Nitrobacter</u> ; 1- Exposure 4 hrs (5 % recovery), 2- Exposure 6 hrs (5 % recovery), 3- Exposure 8 hrs (5 % recovery), 4- Exposure 8 hrs (50 % recovery), 5- Exposure 8 hrs complete recovery

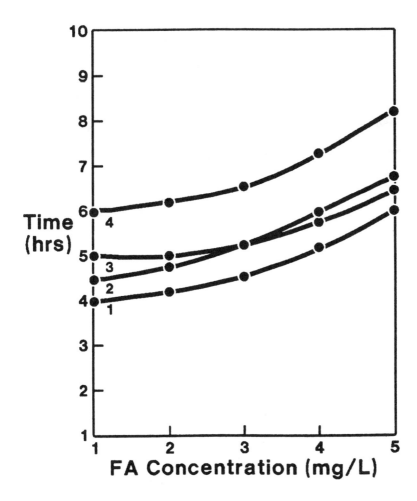

Figure 4. Long term recovery of the inhibited <u>Nitrobacter</u> ;
1- Exposure 4 hrs (50 % recovery),
2- Exposure 6 hrs (50 % recovery),
3- Exposure 4 hrs (complete recovery),
4- Exposure 6 hrs (complete recovery),

give significantly large values for the time taken by <u>Nitrobacter</u> for 100 percent recovery. Also, at the same exposure time , higher FA concentrations significantly increase the time taken for complete recovery. At all three ranges of exposure times, the time taken from 5 percent recovery to 50 percent recovery is higher than the time taken from 50 percent recovery to complete recovery. Particularly during the high FA and high exposure time conditions the ratio of the times taken for initial recovery to later recovery is larger (Figs. 3 & 4). This indicates that at longer exposures in a very adverse environment, the initial recovery is slower. This may be due to the initially slower decomposition of the enzyme - inhibitor complex causing the inhibition.

Inhibitor : Biomass Ratio

Suthersan and Ganczarczyk (1986) also reported that it would be more appropriate to evaluate the inhibitory effects indicating the FA / nitrifying biomass ratio, and not just the FA concentrations alone, since the reactors with higher biomass concentrations were found to be less sensitive to the same inhibitory conditions than those with lower biomass concentrations. But, these initial experiments were conducted only with biomass concentrations of less than 2000 mg/L as MLSS.

Since the inhibitory chamber in the proposed modified process is located in the recycle line, the biomass concentrations encountered will be significantly higher than 2000 mg/L. To study the effects on recovery time of <u>Nitrobacter</u> in higher biomass concentrations due to FA concentrations, a series of experiments were conducted. Nitrifying biomass with MLSS concentrations of 3000, 4000, 6000, 9000 mg/L was inhibited at pH of 8.8 and at an exposure time of 6.0 hrs under different FA concentrations. From Fig. 5 it can be seen that the increasing biomass concentrations decrease the obtainable recovery time at the same FA

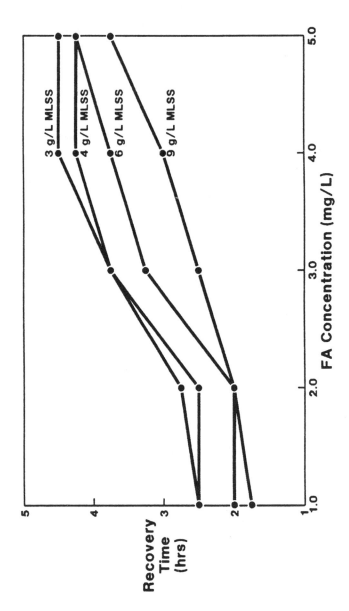

Figure 5. Recovery time vs FA concentrations at high biomass concentrations

concentration. This trend looks the same at low and high FA concentrations. At lower biomass concentrations the recovery time seems to reach a plateau around 4.5 hrs. But, with higher biomass concentrations, the recovery time seems to be increasing with higher FA concentrations. To get a recovery time of 4.5 hrs a significantly higher amount of FA will be needed.

Effects on Nitrosomonas due to the inhibition imposed on Nitrobacter

From our preliminary studies (Suthersan and Ganczarczyk, 1986) and in this extended study it was quite apparent that a combination of higher FA concentrations and higher pH values caused higher degrees of inhibition and longer recovery times of Nitrobacter. To provide a higher FA concentration with a particular total ammonium concentration, the highest possible pH value is preferrable. To find the effects of the nitrifying biomass retention in the inhibition chamber on the Nitrosomonas population, a series of experiments were again conducted.

Batch nitrification experiments were conducted with a nitrifying biomass of 6000 mg/L with an initial ammonia nitrogen concentration of 60 mg/L, and at pH values of 8.0, 8.8 and 9.2 without any exposure to inhibitory conditions. At the same time, a biomass of 6000 mg/L as MLSS taken from the same mother culture was inhibited at an FA concentration of 3.0 mg/L at 6.0 hrs of exposure time and at pH values of 8.4, 8.8 and 9.2. Then the supernatant was removed and the same experiment for batch nitrification was conducted again at a pH value of 8.0.

From Fig. 6 , it can be seen that, in the uninhibited experiments nitritification was faster at pH 8.0 than at 8.8 as expected. At pH 9.2 the nitritification rate was very small. The nitrite accumulation at pH 8.0 and 8.8 was due to the initially high FA concentrations present in the

Nitrite Oxidation Inhibition to Improve the Nitrogen Elimination Process 437

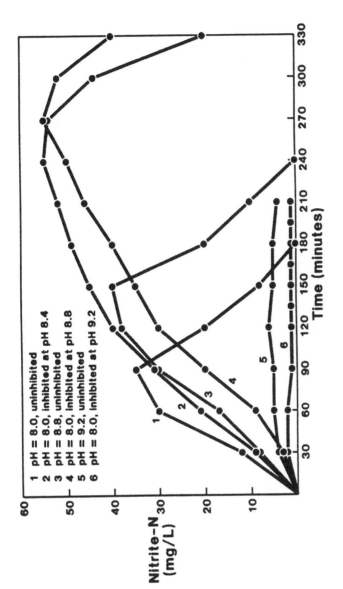

Figure 6. Batch nitrification results for inhibited and un-inhibited biomass

reactor. For the inhibited biomass, when returned to pH 8.0 and relieved from the inhibitory environment, during the first half an hour the nitrification rate is slower than that of the uninhibited biomass at pH 8.0. The reduction in the nitrification rate is greater for the biomass inhibited at pH 8.8 than the one inhibited at pH 8.4. After one hour of aeration, the differences are smaller. It can also be seen that providing an inhibitory environment in the chamber with a pH value of 9.2 is not possible because at such conditions <u>Nitrosomonas</u> seems to lose its activity permanently.

SUMMARY AND CONCLUSIONS

1). Under the usual circumstances in biological waste treatment systems, the phenomenon of inhibition is considered as an event deteriorating the performance of the system. But, for the purpose of modifying the nitrification - denitrification process, inhibition could be advantageous.

2). The concept of "Recovery Time" was introduced to study the inhibition of the second stage of nitrification.

3). It was found that, in addition to the inhibitory effects caused by FA, the pH itself played an additive role in the inhibition of <u>Nitrobacter</u>.

4). By continuously and gradually acclimatising <u>Nitrobacter</u> to increasing FA concentrations, it was found that adaptation to higher concentrations took place. But this adaptation was found to be only temporary and, by de-adapting, the <u>Nitrobacter</u> regained its original characteristics. Thus, by intermittently providing inhibitory and recovery environments in the inhibition chamber and aeration tank, respectively, the adaptation of <u>Nitrobacter</u>, to higher FA concentrations could be prevented. Hence, it will not be necessary to increase the required inhibitory FA concentration in the inhibition chamber due to adaptation of the biomass.

5). It was found that the more adverse the inhibitory environment was, the time taken by <u>Nitrobacter</u> for complete recovery was significantly longer. The initial recovery of <u>Nitrobacter</u> was found to be much slower than the later stage of this phenomenon.

6). Higher exposure times increased the recovery time obtained under the same conditions, particularly at lower FA concentrations.

7). Higher biomass concentrations were found to be less sensitive to the same inhibitory conditions than were as the more diluted mixed liquors. At lower biomass concentrations, the recovery time seemed to reach a constant value at around 4.5 hrs., but with higher biomass concentrations, a significantly higher FA concentration (> 4 mg/L) has to be provided to obtain the same recovery time.

8). During the initial period when brought back to favorable conditions, the activity of <u>Nitrosomonas</u> was reduced due to the inhibitory environment provided earlier to suppress <u>Nitrobacter</u> activity. The biomass inhibited at a pH of 8.8, was slower to recover initially than that inhibited at 8.4. Providing an inhibitory environment above the pH value of 8.8 was not possible, because <u>Nitrosomonas</u> lost its activity permanently.

9). A continuous flow modification of the activated sludge process to suppress the second stage of nitrification proved to be possible.

ACKNOWLEDGEMENTS: This study was supported by a

Natural Sciences and Engineering Research Council of Canada grant. Authors are indebted to Dr.Z. Lewandowski of Montana State University for his comments to our initial report on this research.

REFERENCES

Alexander, M, (1971), " Microbial Ecology ", John Wiley and Sons, Inc., New York.

Alleman, J.E. and Irvine, R.L, (1980), " Nitrification in the sequencing batch biological reactor", Jour. Water Poll. Contr. Fed., 52, 2747-2754.

Alleman, J.E, (1984), "Elevated nitrite occurrence in biological waste water treatment systems", Water Sci. and Tech., 17, 409-419.

Anthonisen, A.C, (1974), "The effects of free ammonia and free nitrous acid on the nitrification process", Ph.D. Thesis, Cornell Univ., Ithaca, NY.

Boon, B. and Laudelout, H, (1962), "Kinetics of nitrite oxidation by <u>Nitrobacter Winogradskyi</u>", Biochem. Jour., 85, 440-447.

Court, M.N., et al., (1964), "Toxicity as a cause of the inefficiency of urea as a fertilizer-I : Review", Jour. Soil Sci., 15, 42-48.

Duisberg,P.C.,et al., (1954), "Effect of ammonia and its oxidation products on rate of nitrification and plant growth", Jour. Soil Sci., 78, 37-49.

Ganczarczyk, J.J, (1979), "Nitrite in nitrification and denitrification - Preliminary findings" , Presented at the 14th Canadian symposium on WaterPollution Research, Toronto, Canada.

Hockenbury, M.R. and Grady, C.P.L, (1977), "Inhibition of nitrification effects of selected organic compounds", Jour. Water Poll. Contr.Fed., 49, 768-777.

Huang, J.Y.C., et al., (1982), "Metal inhibition of nitrification",Proc., 37th Ind. Waste Conf., Purdue Univ., Indiana, 85-98.

Martin, W.P.,et al., (1942), "Threshold pH value for the nitrification of ammonia in desert soils", Soil Sci. Soc. Amer. Proc., 7, 223-228.

Mines, R.O., (1983), Ph.D. Thesis, Virginia Polytechnic Institute and State University, VA, as cited by Alleman (1984).

Morril, E.G, (1959), "An explanation of the nitrification patterns observed when soils are perfused with ammonium sulfate", Ph.D. Thesis, Cornell Univ., Ithaca, NY.

Parulekar, S.J. and Lim, H.C, (1985), "Modeling optimization and control of semi-batch bioreactors",Adv.in Biochemical Engineering / Biotechnology, 32, 206-258.

Payne, W.J, (1979), "Reduction of nitrogenous oxides by micro organisms",Bacter.Reviews, 37, 409-452.

Randall, C.W. and Buth, D, (1984), "Nitrite build up in activated sludge resulting from combined temperature and toxicity effects", Jour. Water Poll. Contr. Fed., 56, 1045-1049.

Sharma,B.and Ahlert,R.C,(1977),"Nitrification and nitrogen removal", Water Res., 11, 897-925.

Suthersan, S. and Ganczarczyk, J.J, (1986), " Inhibition of nitrite oxidation during nitrification: some observations", Water Poll. Res. Jour. of Canada, Vol. 21, No.2, 257-266.

Tomlinson, T.G., et al., (1966), "Inhibition of nitrification in the activated process of sewage disposal", Jour.App.Bact., 29, 266-291.

Wong-Chong, G.M. and Loehr, R.C, (1978), "Kinetics of microbial nitrification - nitrite nitrogen oxidation", Water Res., 12, 605-609.

Yamane, T. and Shimizu, S, (1984), "Fed-batch techniques in microbial processes", Adv. in Biochemical engineering / Biotechnology ,30, 147-194.

Yoshida,Y , et al., (1973), "Fed-batch hydrocarbon fermentation with colloidal emulsion feed", Biotech. and Bioeng., 15, 257-270.

ANAEROBIC TREATMENT OF MOLASSE SUGAR CANE STILLAGE WITH HIGH MINERALS

Michel Henry, Emmanuel Michelot, and Jean Pierre Jover

SGN
Saint-Quentin-en Yvelines, France

SUMMARY

Mesophilic anaerobic digestion of molasse cane stillage containing high level of minerals has been studied with a 25 liters fixed film reactor.

At a constant loading rate of 16 kg $COD.M^{-3}.d^{-1}$, the reactor is fed with diluted molasse cane stillage ; this dilution is then progressively reduced, leading to an increase of soluble minerals in reactor form 2 up to 26,5 g/l without decreasing of the system performances, in terms of gas productivity and pollution abatement.

(Hydraulic detention time : 3,8 days ; loading rate : 16 kg $COD.m^{-3}\ d^{-1}$; biogas productivity : 5,8 $m^3.m^{-3}.d^{-1}$ and COD removal : 71 %).

INTRODUCTION

Since 1983, SGN has taken an active interest in the anaerobic digestion of molasse sugar cane stillage. An industrial pilot plan demonstration unit has been installed at the Société Industrielle de Sucrerie in the French west Indies (GUADELOUPE) which has led to the construction of a full scale plant treating 24 t COD/day ; this unit is now under start up. As the composition of the spent wash may vary from a distillery to another, especially concerning the level of minerals, SGN decided to carry out trials in collaboration with Electricité de France DTNE at Chatou on the digestion of molasse can stillage coming from India and Reunion Island.

These high salinity stillage are known as difficult substrates for treatment by anaerobic digestion.

MATERIAL AND METHODS

Reactor

Laboratory scale fixed film reactor of 25 liters containing a support matrix of PVC (Flocor R) with a liquid volume of 22,5 liters.

The reactor is fed and reactor liquid recycled on a continuous and dowflow mode (recycling flow = 12 $L.d^{-1}$).

The operation temperature is maintained at 35°C. Biogas output is monitored using a commercial wet gas meter.

Analytical Methods

All analyses were performed according to French standard methods.

Influents

Three types of influents were used for the trials. The values for main characteristics are given in table 1.

Data of molasse sugar cane stillage from Guadeloupe mentioned in the report are also given in this table.

Type of influent	Wine stillage	Molasse sugar cane stillage		
Origin	Cognac (FRANCE)	INDIA	REUNION ISLAND	GUADELOUPE (WEST INDIES)
pH	3,0-3,2	4,0-4,2	4,0-4,2	4,0-4,5
COD (mg $O_2.l^{-1}$)	26.000	97.105.000	110.115.000	45.55.000
Soluble ash content (g.l^{-1})	2	31	46,5	17
Potassium as K (g.l^{-1})	0,8	10	7	4
Calcium as Ca (g.l^{-1})	0,1	2,8	2,9	1,5

TABLE 1: COMPOSITION OF INFLUENTS

RESULTS AND DISCUSSION

The aim of these trials is to demonstrate that the methanigenic bacteria are able to keep their biological capacities in a high salty medium often considered as toxic.

Most of the time, the cations inhibition levels are studied for each element, not taking them into account the possible synergistic or antagonistic effects between cations and the ability of methanogenic biomass to adapt to the medium.

Seeding and start-up of reactor

The reactor was seeded with a sludge coming from a lavye industrial SGN anaerobic digester working in Cognac (France) on Cognac wine stillage. This sludge was well adapted to this spent wash, containing low levels of minerals.

Using Cognac wine stillage as substrate, the organic loading rate was then increased up to 16 kg $COD.m^{-3}.d^{-1}$ over a three months time period. The performances of the systems, determined under steady state conditions, are summurized in table N° 2.

Transition to molasse cane stillage

At same loading rate (16 kg $COD/m^{-3}.d^{-1}$), the reactor was then fed with diluted Indian molasse cane stillage, in order to have the same minerals level content as in Guadeloupe spent wash, for which we do not noticed any inhibitory effect (soluble ash content = 17 g/l).

The specific characteristics of these indian subtrates, especially their lower biodegredability explains the lower performances observed under steady state conditions as indicated in table N° 2.

Influence of salty level on the system stability

At the loading rate of 16 kg $DCO.m^{-3}.d^{-1}$, the dilution was stopped. This led to a quick loss of stability : the volatile acids content of the reactor increased rapidly showing a non-ability of the methanogenic bacteria to adapt themselves to rapid variation of minerals level of the medium.

Recovery of the system and progression of mineral load

Feeding of influent to the reactor was stopped for a few days to permit the degradation of excess of volatils acids accumulated before.

We decided then to begin feeding again with diluted indian molasse cane stillage and to decrease progressively the dilution factor to allow a slow increase of mineral content of the digester and to ensure the acclimatation of bacteria.

The monotoring of mineral load increase is performed by the synthesis of analytical results especially concerning volatile acids, alkalinity and biogas CO_2 contents.

Final results

The mineral content of reactor increase from 2 up to 26,5 g/l progressivly over a four months time period utilizing successivly indian and then Reunion molasse cane stillage without any change of performances of biogas productivity and COD removal.

Data are given in table 2.

Type of influent	Wine stillage	Molasse sugar cane stillage		
Origin	Cognac (FRANCE)	INDIA	INDIA	REUNION ISLAND
Dilution (spent wash:water) (V/V)	-	1 : 2	1 : 1/4	1 : 3/4
Soluble ash content (g/l)	2	10	25	26,5
Hydraulic detention time (d)	2,2	2,2	5,1	3,8
Volumetric loading rate (kg COD.m^{-3}.d^{-1})	16	16	16	16
Biogas productivity (m^3 biogas.m^{-3}.d^{-1})	8,6	5,5	5,5	5,8
% CH4	63	56	58	58
COD removal (%)	91	72	65	71

TABLE 2: ANAEROBIC DIGESTION PERFORMANCES ON STEADY STATE CONDITIONS

CONCLUSIONS

We demonstrate in this study the ability of the SGN fixed film anaerobic digestion process for the treatment of molasse sugar cane stillage containing up to 26,5 g/l soluble ash without anay dilution.

Analysis of results seems to show that inhibitions level were not reached : we had high and stable productivities and COD removal.

It also demonstrates the ability of methanogenic bacteria to adapt themselves to new environmental conditions provided these changes are progressive and controlled.

POTENTIAL FOR ANAEROBIC TREATMENT OF HIGH SULFUR WASTEWATER IN A UNIQUE UPFLOW—FIXED FILM—SUSPENDED GROWTH REACTOR

L. Syd Love

*Sydlo Inc.
578 Minette Circle
Mississauga, Ontario, Canada*

INTRODUCTION

We learn from McKinney (1) that three basic groups of bacteria are responsible for the anaerobic degradation of the organic component of wastewater.

First acid forming bacteria, (<u>acid formers</u>) breakdown the complex organic compounds into simple organic acids, aldehydes, ketones and alcohols, generally known as simple volatile fatty acids.

The <u>acid formers</u> are a very large, diverse group that include both faculative bacteria and strict anaerobes. Very little is known about them.

The <u>methane formers</u> are a second group of bacteria essential to anaerobic degradation, because they utilize the volatile fatty acids to produce a biogas containing about 70% methane (CH_4) and 30% carbon dioxide (CO_2).

The <u>methane formers</u> are strict anaerobes and include three distinct species of bacteria. <u>Methanobacterium</u>, <u>Methanosarcina</u> and <u>Methanococcus</u>.

A third group of bacteria (<u>Desulfovibrico</u>), are also important. These bacteria reduce sulfate compounds to produce hydrogen sulfide (H_2S) and therefore are the source of rather serious odour, corrosion and toxicity problems.

Little can be done to eliminate the odour and corrosion, however, the following Paper will describe a rather unique solution to sulfide toxicity in anaerobic degradation.

Sulfide Production in Anaerobic Degradation:

About twenty years ago Lawrence et al (2) completed an important laboratory study of "The Effects of Sulfides on Anaerobic Treatment". Not only did they quantify the problem, they also identified a solution.

Sulfides in anaerobic systems can result from:

(i) Introduction with the incoming wastewater
(ii) Biological reduction of sulfur containing compounds, i.e. Sulfates and proteins.

Sulfur reducing bacteria <u>Desulfovibrio Desulfuricans</u> use sulfates as their electron acceptor in metabolism of organic compounds to produce H_2S and CO_2. (2)

$$8H^+ + 8e^- + SO_4^= \xrightarrow{\text{sulfate reducing bacteria}} S^= + 4H_2O$$

The reduction of one mole of sulfur (32 gms), in this manner, corresponds to the oxidation of eight equivalents of organic matter, or about 64 gms on a C.O.D. basis.

Aulenbach and Heukelekian (3) have shown how the sulfur content of proteins can also result in the formation of H_2S during anaerobic degradation.

Therefore sulfur in wastewater has two distinct undesirable effects on anaerobic degradation.

(i) It results in the formation of H_2S which is odourous, corrosive and potentially seriously toxic.
(ii) It reacts with organic matter which otherwise would have been converted to valuable CH_4 gas.

Very little can be done to prevent the sulfate bacteria from reacting with these sulfur compounds. However, an anaerobic reactor can be designed to minimize, or completely eliminate, sulfide toxicity.

Sulfide Toxicity:

Rudolfs and Amberg (4) investigated the effects of soluble sulfides in the digestion of sewage sludge and found that 200 mg/L of sulfides (as S) decreased gas production by 70 percent and volatile matter by 50 percent.

In a separate study of acetate fermentation (4) they also concluded that methane fermentation was a linear function of the soluble sulfide concentration and that no CH_4 gas was formed with a sulfide addition above 165 mg/L (as S). However, additions of up to 300 mg/L (as S), of soluble sulfides, appeared to have little effect on the production of volatile fatty acids.

This would indicate that **methane formers** are much more sensitive to sulfide toxicity than the **acid formers** are.

This observation is supported by the laboratory study conducted by Lawrence et al (2) who have concluded that:

(i) "Concentrations of soluble sulfide higher than 200 mg/L (as S) produce severe toxic effects and complete cessation of gas production".

(ii) "Inhibitory concentration of sulfides affects gas production first, while significant volatile acid accumulation takes place much slower, and only after gas production has been severely retarded".

Because of the limited solubility of H_2S a certain portion will escape to form H_2S gas which will form a part of the biogas.

According to Henry's Law an equilibrium will exist between the H_2S in solution and the H_2S in the gas above the liquid.

$$[H_2S \text{ sol}] = K \cdot oc \, [H_2S \text{ gas}]$$

K = ionization constant
oc = absorption coefficient

Values for both K and oc are well documented in the literature. (5)

Boyles Law states that the amount of H_2S remaining in solution is a function of the concentration of H_2S in the gas phase above. Therefore if the H_2S in the gas phase is rapidly removed the H_2S solution will be greatly decreased.

This phenomenon may be the reason for McCarthy's (6) observation that:

"If the concentration of soluble sulfide percursors in a waste entering a digester were 800 mg/L (as S), the pH 7.0 and three cubic feet of gas were produced per gallon of waste added, only about 20%, or 160 mg/L of sulfides would remain in solution in the digester. The remainder, or 640 mg/L would escape with the other gases produced during treatment".

One Solution to Sulfide Toxicity: - (Ferric Chloride Addition)

Lawrence et al (2) found that by adding an iron salt, (ferric chloride $FeCl_3$) to the reactor, they were able to inactivate the sulfides by precipitating them as insoluble iron sulfide.

Since there appears to be no limit to the amount of insoluble iron sulfide that an anaerobic system can tolerate one merely has to add sufficient $FeCl_3$ to completely eliminate sulfide toxicity.

Today this is the only method to control sulfide toxicity in anaerobic degradation of high sulfur wastewaters. Apparently the process works well but is extremely costly because:

1. Ferric chloride is expensive
2. Increase sludge disposal cost
3. Precludes possibility of sulfur recovery.

An Alternative Solution to Sulfide Toxicity:

In the anaerobic degradation of high strength wastewaters it is not uncommon to generate ten to fifteen volumes of biogas for each volume of liquid treated.

If the anaerobic reactor operates in the upflow mode, the buoyant effect of this rising mass of biogas will result in a significant loss of biological solids and a corresponding reduction in treatment efficiency.

The Sydlo anaerobic reactor was designed specifically to solve this problem. Large quantities of supernatant are continuously removed from the reactor and directed to a gas separator where the biogas is removed. The supernatant, less the biogas, is then recycled back to the reactor.

By operating in this manner sixty to eighty percent (or more) of the biogas will be removed from the reactor. When treating a high sulfur wastewater this biogas will contain a substantial amount of the H_2S generated within the reactor.

According to the work completed by Lawrence et al (2) and McCarthy (6) the removal of large quantities of H_2S will greatly reduce the sulfide toxicity. In fact it will be possible to adjust the recirculation rate to a point where the soluble sulfide concentration is well below the toxicity level.

Figure No. 1 is a schematic showing the two points of recycle flow within the Sydlo anaerobic reactor:

1. Supernatant is first removed from the top of the reaction-mixing hood and directed to gas separator GS-1. Since this is basically in the acid phase of anaerobic degradation these gases will contain a substantial amount of H_2S.

2. The second recycle flow is taken from the upper section of the reactor, just below the fixed film media. These are directed to gas separator GS-2.

    The collection grid, which also serves as the media support, is designed to maximize gas collection.

Each biogas removal system is completely independent and can operate over a wide range of recycle flows. The actual flow rate can be regulated by a simple, adjustable weir on each of the gas separators.

Advantages of the Sydlo Reactor: (Lower operating cost)

By removing the sulfide as H_2S gas, rather than precipitating it as an insoluble sulfide, a substantial reduction in operating cost will be achieved as follows:

1. Elimination of need for ferric chloride:
   About three pounds of $FeCl_3$ is required to precipitate one pound of sulfide. $FeCl_3$ is a relatively expensive and rather corrosive chemical.

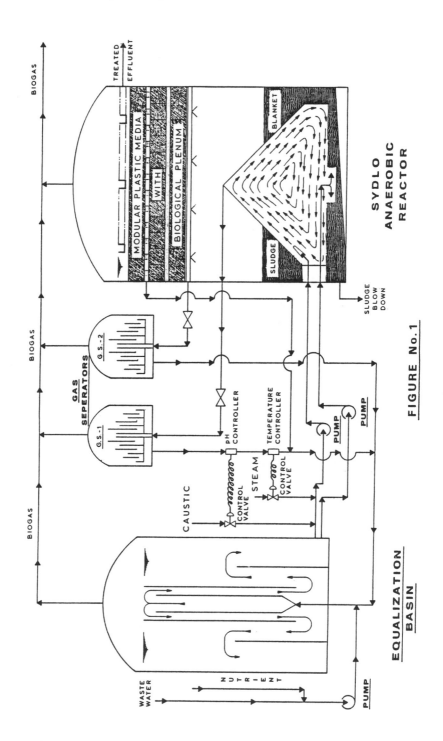

FIGURE No. 1

2. **Lower sludge disposal costs:**
 The precipitation of insoluble sulfides, by $FeCl_3$ addition will increase the volume of waste sludge and its disposal cost.

3. **Sulfur Recovery:**
 Sulfur is a potentially valuable by-product from the degradation of high sulfur wastewater. Presently its market value is excess of $150/ton (Canadian).

The Hiperion process, as described in Appendix "A", is an economical method for recovering H_2S from biogas.

The Sydlo Reactor - How It Operates:

The Sydlo anaerobic reactor is shown in Figure No. 2. It operates as follows:

A. If alkalinity or nutrients are required, they are added and thoroughly mixed with the wastewater prior to entering the reactor. This can best be accomplished in an equalization basin.

 Pre equalization of wastewater is desirable because anaerobic systems operate best when the loading rate (flow and C.O.D.) are as uniform as possible.

B. Wastewater, together with recycle supernatant (less biogas) from gas separator GS-1 enters the reactor through an eductor that mixes four volumes of biological solids with each volume of liquid.

C. Additional complete mixing of liquid and biological solids takes place under the reaction hood.

D. Supernatant (liquid, biogas and biological solids) is continuously removed from the top of the reaction hood and directed to, gas separator GS-1, where the biogas is removed.

 The supernatant, (less biogas), is recycled back to the reactor, usually via an equalization basin. The recycle rate, which can be extremely high, is

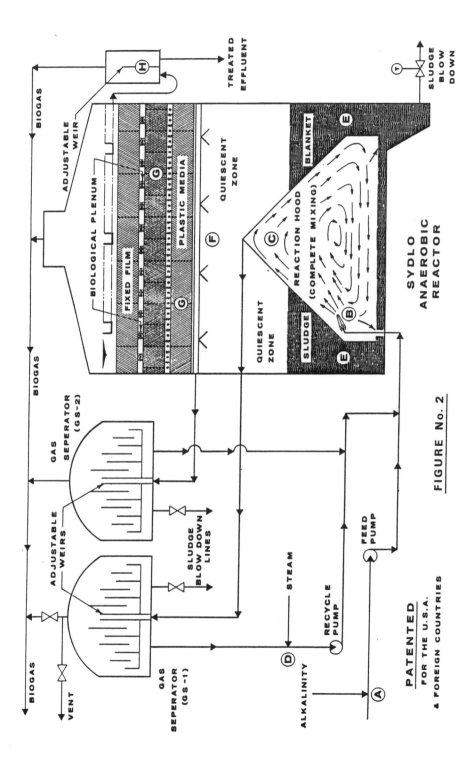

controlled by an adjustable weir on the gas separator GS-1.

E. The remaining liquid is laterally distributed, within the reactor, around the periphery of the reaction hood. A high degree of treatment takes place as this liquid slowly rises through the dense blanket of biological solids that forms, at the base of the reaction hood. In addition, upward filtration within this sludge blanket, will remove fine particles and thus increase the blanket density.

F. A supernatant removal system, designed to maximize biogas collection, is located in the quiescent zone, just below the fixed film media. Actually this system of pipes also supports the fixed film plastic media.

Supernatant (liquid and biogas) is continuously removed from the reactor and directed to, gas separator GS-2, where the biogas is removed. The liquid (less biogas) is returned to the reactor, usually via the equalization basin. The recycle rate is controlled by an adjustable weir on the gas separator GS-2.

G. The bed of plastic media normally consists of three, 2'0" deep, layers of cross flow, modular media each separated by a 6" biological plenum.

This unique bed configuration (patent pending) serves four important functions in achieving a high degree of efficiency and stability.

1. It provides a surface to which bacteria can attach, to further treat the passing liquid.

2. The numerous passageways, throughout the modular media, act as a multitude of tube settlers which greatly enhances the reactor's ability to retain biological solids.

3. These biological solids settle into the plenum below where additional treatment takes place in the suspended growth mode.

Therefore the media bed is actually several fixed film – suspended growth biological systems in series. An arrangement of this type will result in an extremely high degree of treatment efficiency.

    4. The cross flow plastic media provides uniform distribution of the slowly rising liquid.

H. Treated effluent is removed from the reactor by a system of surface launderers designed for uniform liquid removal over the entire reactor surface.

The liquid level, within the reactor, is controlled by an adjustable weir located in the effluent water seal chamber.

Scum (floating material) can be removed, from the reactor, while in operation, by adjusting this weir as follows:

    1. Raise liquid level about 4" – 6" above level of surface launderers. Scum will float above the open surface launderers.

    2. Return to normal operating level. Scum will flow into open surface launderers and out with effluent.

Conclusion:

The Sydlo anaerobic reactor has been designed specifically to remove large quantities of biogas as it is generated within the reactor. Because of this feature it is ideally suited for the treatment of high sulfur wastewater.

The degree of soluble sulfide removal can be controlled by selecting the appropriate supernatant recycle rate. Simple, adjustable weirs are provided, on the gas separators, for this purpose.

The advantage of this technology is that it will substantially reduce daily operating costs.

Although this reactor is particularly well suited for the treatment of high sulfur wastewaters, it may also be used, with considerable advantage, when treating other wastewaters where sulfide toxicity is not a problem.

REFERENCES

(1) McKinney, Ross E. Anaerobic Treatment Concepts - 1984 Environmental Conference TAPPI, pg. 163.

(2) Lawrence, A.W., McCarthy, P.L., and Guerin, F.J.A. "The Effects of Sulfides on Anaerobic Treatment". Proc. 19th Ind. Waste Conference Purdue University, 1964, pg. 343.

(3) Aulenbach, D.B., and Heukelekian, H. "Transformation and Effects of Reduced Sulfur Compounds in Sludge Digestion". Sewage and Industrial Waste 27, 1147 (1955).

(4) "White Water Treatment II Effects of Sulfides on Digestion" and "White Water Treatment III Factors Affecting Digestion Efficiency", Sewage and Ind. Waste 24 1278 and 1402 (1952).

(5) Hand Book of Chemistry and Physics, 39th Edition Chemical Rubber Publishing Co., Cleveland (1957).

(6) McCarthy, Perry L. Anaerobic Waste Treatment Fundamentals Part Three, Toxic Materials and Their Control, Public Works for November 1964, pg. 93.

APPENDIX A — HIPERION

ECONOMICAL H_2S REMOVAL

BY DIRECT CONVERSION TO ELEMENTAL SULFUR

PROCESS DESCRIPTION

The HIPERION Sulfur Removal Process is a wet desulfurization process that uses a proprietary naphthaquinone chelate redox-type catalyst to form elemental sulfur from sulfur bearing streams. With its capacity for rapid and essentially quantitative removal of H_2S, with little or no pickup of CO_2, this new process is ideally suited to the elimination of sulfur from a variety of gas streams; e.g. refineries, coke ovens, natural gas plants, pulp and paper mills, sewage treatment plants, chemical facilities and geothermal units.

The principal reaction in the removal of H_2S involves the redox transfer of the naphthaquinone chelate and H_2S to form naphthahydroquinone chelate and sulfur. The hydroquinone chelate is then air-oxidized back to the quinone chelate form and water. Absorption and reduction both take place at ambient temperature while pressure is determined by the upstream or downstream process conditions. Typical precursors used in forming the chelated catalyst include $FeSO_4$ and a metal salt of a dicarboxylic acid. Such compounds are relatively nontoxic, unlike the vanadium and arsenic employed in other processes for direct oxidation of H_2S to sulfur.

Illustrated is a simplified flow diagram of the HIPERION Sulfur Removal Process. The sour gas containing sulfurous compounds is fed into the bottom of an absorber column where the gas is contacted with an aqueous solution of the naphthaquinone chelate as it cascades downward in the absorber. Sweet gas is removed from the top of the absorber. The reduced solution is transferred from the bottom of the absorber to the top of the regeneration tower. Air is introduced through a sparger in the bottom of this tower to regenerate the spent catalyst. A slip stream

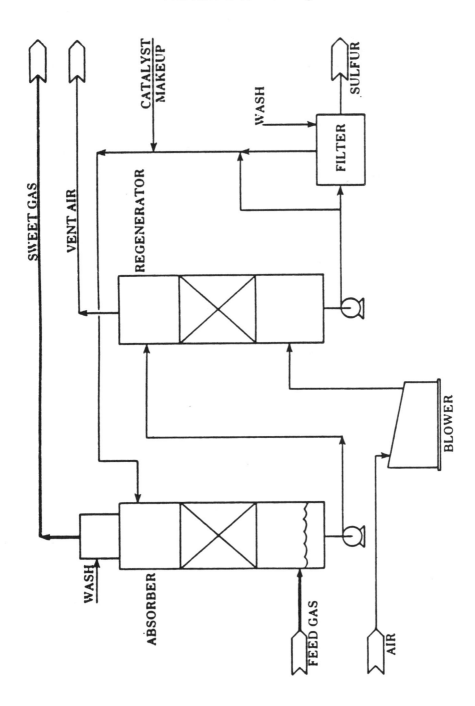

from the bottom of the regenerator is fed to a filter where the sulfur is removed. The regenerated naphthaquinone is then recycled to the absorption tower. Both towers have several packed sections; the patented packing is designed for minimum holdup of sulfur and low pressure drop.

The filter cake is washed with water prior to its release from the filter. The filtrate and wash liquor are used in the preparation of the make-up solution of fresh naphthaquinone. This process required low catalyst make-up and minimum utilities.

Advantage of the HIPERION Sulfur Removal Process:

- High efficiency - high throughput rates
- Adaptable to a variety of feed compositions
- Nontoxic catalyst
- Operable over a wide range of temperatures and pressures
- Low capital - low operating costs
- Skid mounted and mobile units available
- Catalyst not subject to biodegradation
- Operable over a wide range of CO_2
- No chemical consumption in reaction

For further information on the HIPERION H_2S Removal Process, please contact:

>Mr. Jackson Yu
>Ultrasystems Inc.
>16, 845 Von Kariman Avenue
>Irvine, California 92714
>Tel: (714) 863-7000

ANAEROBIC DIGESTION OF CHEMICAL INDUSTRY WASTEWATERS CONTAINING TOXIC COMPOUNDS BY DOWNFLOW FIXED FILM TECHNOLOGY

Michel Henry, Yves Thelier, and Jean Pierre Jover

SGN
Saint-Quentin-en-Yvelines, France

SUMMARY

Treatment of high-strength acidic wastewater containing cyclic acetals in a 20 m^3 industrial-scale downflow mesophilic, SGN anaerobic fixed film process occured without any problem after reaching a loading rate of 11 kg COD/m^3.d within 7 months.

These compounds showed reversible inhibitory effects against methanogenic microbial biomass during start-up phase.

Specific probes for anaerobic digestion process control in industrial full-scale plant should allow a very quick reaction in case of abnormal concentrations of some toxic compounds in digester influent to prevent biological treatment capacity upsets.

INTRODUCTION

Anaerobic digestion gives us today an extremely advantageous solution to reduce organic pollution with concomittant energy recovery.

SGN developed a high-rate process, based on active biomass retention on a plastic media called Flocor R, in a downflow anaerobic fixed film reactor. This process is applied in several agro-industrial plants, at full scale, to treat wastewaters with low suspended solids content at high organic loading rates (up to 25 kg COD/m^3.d) with soluble COD reduction in order of 90 % (1).

SGN has recently developed this process for application in the field of chemical and pharmaceutical industry. Wastewaters from these plants often contain low molecular organics degradable by anaerobic digestion. One of the treatment problems, beside minerals implementation, is the presence of toxic molecules associated with the biodegradable organic compounds (2). For investigating the technical and economic feasability of treating a high-strength acidic wastewater coming from a chemical plant, in a downflow anaerobic fixed film reactor, experimental studies were conducted in a 20 m^3 industrial pilot-plant unit.

Results of these trials will be discussed here after.

MATERIAL AND METHODS

Reactor

A schematic drawing of the pilot plant is given in Figure 1.

This 20 m^3 reactor contains Flocor R plastic media (30 x 34 mm PVC rings, specific area = 230 m^2/m^3, porosity = 96 %).

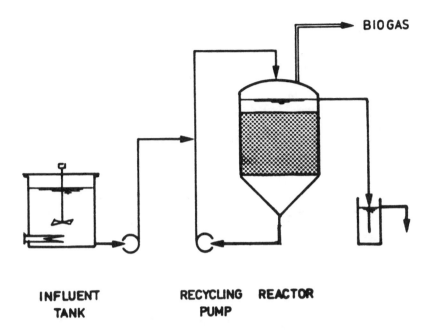

FIGURE NR 1 : SGN ANAEROBIC FIXED FILM PILOT PLANT

Influents

Two types of influents were treated during trials. Both of them were totally demineralized and contained no suspended solids. Mean values for main characteristics of influents are given in table NR 1.

TABLE NR 1 : COMPOSITION OF WASTEWATERS

		INFLUENT NR 1	INFLUENT NR 2
pH		2.1	2.1
Soluble COD	(kg/m^3)	55.0	51.0
Acetate	(kg/m^3)	49.0	46.3
Formate	(kg/m^3)	16.0	13.5
Cyclic acetals (CA)	(kg/m^3)	0.1	1.3

Before anaerobic digestion, pH of influent was corrected to 3.5 and nutrients were added.

Analytical methods

- pH, total alkalinity, soluble COD, total suspended solids were performed according to the French standard methods (3).
- Volatile acids spectra (C1 to C5) were measured by high pressure liquid chromatography (HPLC).
- Cyclic acetals (CA), CH_4 and CO_2 content were measured by gas chromatography (GC).

Frequency of sampling

One liquid sample of influent and effluent was taken daily for above mentionned analysis.
One biogas sample was taken daily for CH_4 and CO_2 determination.

RESULTS

Start-up phase - Loading rate increase - Influent NR 1

Anaerobic sludge, coming from a distillery effluent digester, served as seeding source (VSS = 39.0 kg/m^3).
A 7 months time period was necessary to reach a volumetric loading rate (VLR) of 11 kg COD/m^3.d.

Toxicity of CA - Effect of CA shock during start-up phase

$$\text{Ca (general formula : } R_1\!\!>\!\!C\!\!<\!\!\genfrac{}{}{0pt}{}{O-CH-O}{O-CH-O}\!\!>\!\!C\!\!<\!\!R_3 \text{)}$$

are by products issued during synthesis of disinfectants. Normal concentration in influent NR 1 is 0.1 kg/m^3. A process problem led to a sudden increase in water of CA up to 13 kg/m^3. Due to dilution of influent by continuous recycling, the instantaneous CA concentration applied to the first zone of fixed film biomass was about 0.5 kg/m^3 for a 4 hours time period. This led to a dramatic decrease of methane content in gas (table NR 2).

TABLE NR 2 : EFFECT OF CA SHOCK ON METHANE CONTENT OF BIOGAS

Volumetric loading-rate	(kg COD/m^3.d)	3.5
Cyclic acetals applied to biofilm	(kg/m^3)	0.5
Methane content of biogas before application of CA	(% V)	45.0
Methane content of biogas after 4 hours of CA application	(% V)	22.0

Recovery of the system

After 4 hours of feeding this water with abnormal CA concentration, we decided to stop the feeding of the plant for 24 hours to prevent biological activity upset. Recovery of system back to a loading rate of 3.5 kg COD/m^3.d was achieved within 4 days after feed stop.

Performances obtained with influent NR 1

Performances observed during a 20 days period, under steady-state conditions, are summarized in table NR 3.

TABLE NR 3 : PERFORMANCE of SGN ANAEROBIC FIXED FILM PROCESS WITH INFLUENT NR 1

Volumetric loading-rate	(kg COD/m^3.d)	11.0
Reaction temperature	(°C)	36
Volumetric loading rate	(Kg Acetate/m^3.d)	9.6
CA concentration	(kg/m^3)	0.1
Soluble COD reduction	(%)	95.6
Volatible acids reduction	(%)	95.0
Specific methane production at 30°C - 1 bar	(m^3 CH_4/m^3 dig.d)	4.37

Transition to influent NR 2

Due to the CA concentration of water NR 2, the anaerobic treatment should lead to a CA concentration in reactor about two times higher of the concentration which showed acute toxicity effect on methanogenic biomass (see point 3.2.).

We decide, after 7 months work on influent NR 1 (reached loading rate = 11 kg COD/m^3.d) to make the transition to influent NR 2.

We reached a stable CA concentration of 1.0 kg/m^3 in reactor after ten days of feeding influent NR 2 (see figure NR 2) and then measured performances of the anaerobic digestion with influent NR 2, under steady-state conditions (table NR 4).

TABLE NR 4 : PERFORMANCES OF SGN ANAEROBIC FIXED FILM PROCESS WITH INFLUENT NR 2

Volumetric loading-rate	(kg COD/m^3.d)	11.0
Reaction temperature	(°C)	36
Volumetric loading rate	(Kg C2/m^3.d)	9.8
CA concentration	(kg/m^3)	1.0
Soluble COD reduction	(%)	90.5
Volatile acids reduction	(%)	96.0
Specific methane production at 30°C - 1 bar	(m^3 CH_4/m^3 dig.d)	4.0

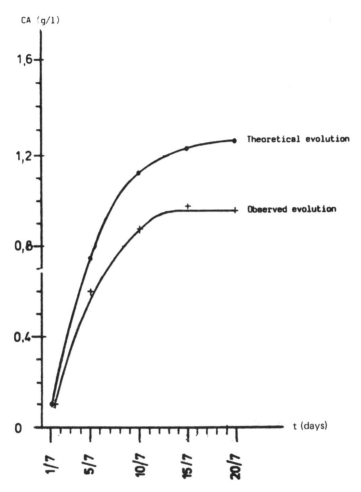

FIGURE NR 2
EVOLUTION OF CA IN REACTOR AFTER TRANSITION TO INFLUENT NR 2

The theoretical CA concentration is calculated as follows :

$CA_{th} = ((CA_0 - k) \times (1 - EXP(-t/HRT)) + k$

with : CA_{th} = theoretical cyclic acetals concentration (kg/m^3)
CA_0 = CA concentration of influent NR 2 (kg/m^3)
k = CA concentration in reactor at $t = 0$ (kg/m^3)
HRT = Hydraulic retention time (days).

DISCUSSION

- CA showed acute toxicity at 0.5 kg/m^3 against methanogenic biomass while under start-up (loading rate = 3.5 kg COD/m^3.d).
- The effect of these compounds in the fixed film digester is reversible.
- Two times higher concentration, observed while treating influent NR 2, are tolerated without problem, after reaching a loading rate of 11 kg COD/m^3.d, and consequently, after finishing methanogenic biomass accumulation in the fixed film reactor. The toxicity of CA seems to depend of the biomass quantity.
- The CA concentration observed in reactor (1.0 kg/m^3) with influent NR 2 is 25 % lower than the theoretical concentration, as the influent contains 1.3 kg/m^3). This should indicate a percentage of fixation or degradation of these compounds by the fixed biofilm.
- The biodegradation hypothesis seems the more realistic, because we observed, after stopping the feed of influent to the pilot, a 50 % percent decrease of CA concentration in reactor, as soon as the residual volatile acids in reactor became lower (1.5 kg/m^3).
- Increasing CA concentration from 0.1 to 1.0 kg/m^3 did not affect volatile acids reduction by biological methanogenic activity.

CONCLUSION

SGN anaerobic fixed film process can be utilized for digestion of high strengh acidic wastewater containing up to 1.3 kg/m^3 of cyclic acetals after biomass accumulation.

Utilizing this process in the chemical or pharmaceutical industry, where the wastewaters often contains toxic molecules, requires careful attention during start-up phases. Installation of specific probes (methane analyser or total oxygen demand meter for instance) in industrial full-scale anaerobic digestion plants allows to react very quickly in case of digestion failure due to abnormal concentrations of toxic compounds in water. This permits to protect the anaerobic biomass against these compounds.

This is a very important fact since the methanogenic bacteria doubling time is rather long compared to aerobic population.

REFERENCES

1. HENRY M. "Industrial performance of a fixed film anaerobic digestion process for methane production and stabilization of sugar distillery and piggery wastes" in IGT Symposium Papers Energy from Biomass and Wastes IX, Lake Buena Vista, Florida, Jan. 28 Feb. 1, 1985 - pp. 829-852.

2. SPEECE R., "Anaerobic biotechnology for industrial wastewater treatment" Environmental Science Technology, Vol. 17, N° 19, 1983 - pp. 416A-427A.

3. AFNOR, "Recueil des normes françaises des eaux - Méthodes d'essais" - 2ème édition (1983). Published by AFNOR - Tour Europe - Cedex 7 - 92080 PARIS La Défence Cedex - France.

TREATMENT OF PROCESS WASTEWATER FROM PETROCHEMICAL PLANT USING A ROTATING BIOLOGICAL CONTACTOR—A CASE STUDY

Warren C. Davis, Jr. and Tom M. Pankratz
Royce Process Equipment Company, Inc.
Pearland, Texas

GENERAL

A Rotating Biological Contactor (RBC) process can be utilized in virtually any industrial or municipal application where the liquid wastewater characteristics are amenable to biological oxidation. Biological oxidation, in this fixed film process, uses innocuous bacteria, protozoa, and algae to convert the soluble (dissolved) organic pollutants into new cell mass and gaseous by-products.

Conventional RBC technology consists of large diameter plastic media discs mounted on a horizontal shaft that slowly rotates through a tank partially filled with screened sewage. The rotating media is a support surface for the biofilm and provides contact between the biofilm, nutrients, and the organic constitutuents, and atmospheric oxygen.

SUMMARY

A six month field pilot study, using a Rotating Biological Contactor (RBC), was conducted at a southern petrochemical plant. The purpose of this study was to determine the feasibility of the RBC process as a means of treating a unique industrial wastewater for direct discharge.

This pilot study utilized industrial wastewater containing blowdown from an ethane and propane gas cracking furnace quench tower co-mingled with storm water runoff. The polluted wastewater must be treated prior to reuse or discharge.

Following startup, and after the establishment of a biofilm on the contactor media, the operating parameters were varied in an effort to determine the optimum operating conditions. Variations included increasing and decreasing the plant flow rates, varying the speed of shaft rotation, pre-aeration, and the addition of powdered activated carbon.

The pilot test results indicated that the RBC can provide effective biological treatment of the industrial wastewater within the requirements of the permit limits. Discussion of the experiments is brief and not intended to be rigorous.

INTRODUCTION

Figure 1 illustrates a typical process scheme at the petrochemical gas processing plant. Refinery gas, rich in ethane and propane, enters the cracking furnace and is separated, by steam heat, into the lighter products of propylene and ethylene. The gaseous products flow to the quench tower to be cooled by recycled quench water. Any water entrained in the product gas is removed by a proprietary separation process and recycled back to the quench tower. Blowdown from the quench tower drains to the plant treatment facilities.

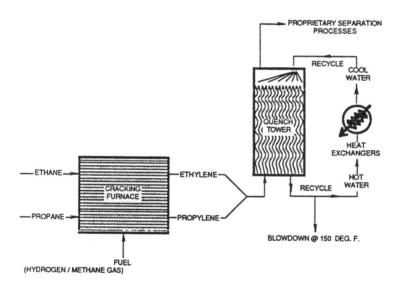

Figure 1. Typical petroleum gas processing flow schematic.

During the steam cracking process, several hundred organic compounds may be synthesized in the part per billion (ppb) range. The compounds flow to the quench tower and are concentrated in the quench tower blowdown. These compounds are collectively refered to as polynuclear aromatics.

Many of the compounds in the blowdown may be considered toxic at sufficiently high doses. Such polynuclear aromatics generated include:

*trichloroethylene
*tetrachloroethylene
*carbon tetrachloride
*vinyl chloride
*1,2 dichloroethane
*1,1,1 trichloroethane
*chloroform

Trichloroethylene and related solvents do not occur naturally, and because of their volatility, they are seldom detected in concentrations greater than a few micrograms per liter in the blowdown. However, due to their proliferation in groundwater, the Environmental Protection Agency (EPA) is considering maximum contamination levels for waste discharges containing these solvents. These same regulations may require special treatment of the storm water runoff which can collect trace amounts of chromium, mercury, zinc, phenols, benzene, and other chlorinated solvents used in the plant.

To reduce the organic content, the quench tower blowdown and plant runoff are collected and treated as shown in Figure 2. Existing treatment consists of oil skimming, equalization coagulation/flocculation, and clarification. From the clarifier, water is pumped through a sand filter and supplied to the cooling tower as make-up. The blowdown and plant runoff either evaporate in the cooling tower system or become part of the cooling tower blowdown discharge.

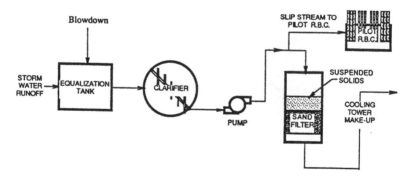

Figure 2. Blowdown / runoff treatment system.

The process make-up poses several problems in the cooling tower system. The make-up to the cooling tower is rich in organics. This creates an ideal biological reactor in the cooling tower with a consistent source of nutrients, oxygen, and controlled (warm) water temperature. As a result, chemicals must be added to the cooling water to prevent both chemical and biological corrosion and fouling.

Biological fouling can be controlled through the use of industrial biocides. Chemical fouling, due primarily to the precipitation of calcium and magnesium salts, can be inhibited with phosphonates. Corrosion can be controlled by the addition of soluble hexavalent chrome and zinc. The corrosion inhibitors establish thin protective films on the metal surfaces. Due to these chemical additions and a high organic content, cooling water discharge may be considered toxic to the receiving stream.

In an effort to reduce potential problems and due to the pressures of increasing EPA regulations, this petrochemical plant initiated a water management study to determine the need for additional wastewater treatment facilities.

One scheme considered was to upgrade the existing facility, reducing the volume of make-up to the cooling tower system. A portion of the make-up water could be treated to reduce the contaminant levels and discharged into a receiving stream.

A plant review of the wastewater treatment options determined that biological treatment would be the best alternative. The main advantage would be the removal of the organic pollutants in the quench blowdown and storm water runoff. The reduced flow to the cooling tower would also reduce the required chemical additions and cooling water discharge.

Due to the fluctuations in organic loading, and the hydraulic surges associated with storm occurances, a fixed film process was thought to be preferable to suspended growth systems. An RBC pilot study was selected because RBCs generally are more reliable than other fixed film processes and land area was at a premium (1). The RBC process was selected to reduce a major portion of the dissolved organics in the process make-up. Solids handling and disposal were not part of the pilot study.

Table 1 lists the influent chemical oxygen demand (COD), total organic carbon (TOC), and phenol concentrations as well as the current discharge limitations imposed for the petrochemical plant. The design flow is 282 gallons per minute (gpm).

TABLE I. Permit Discharge Limits

Influent COD	950 lbs/day (282 mg/l)
COD Permit Limit	350 lbs/day (104 mg/l)
Influent Phenol	4 lbs/day (1100 ppb)
Phenol Permit Limit	0.4 lbs/day (120 ppb)

Influent and effluent COD, TOC, and phenol concentrations were chosen as the RBC process monitoring parameters.

COD represents the total organic carbon with the exception of certain aromatics, such as benzene (2). Since TOC also represents total organic content, it would act as a reference.

PRODUCT (SYSTEM) DESCRIPTION

The RBC pilot unit was arranged in four stages in a self-supporting steel reactor tank as shown in Figure 3. The reactor tank was divided into four baffled compartments to provide series flow. The use of consecutive compartments, or stages, maximizes the treatment efficiency of the RBC system by allowing different microbial fauna and flora a chance to proliferate as the organic strength is reduced in each successive stage.

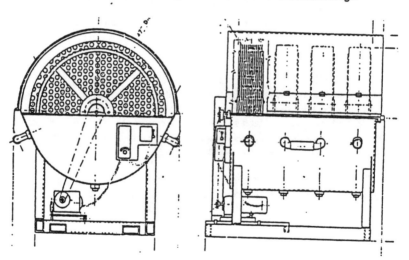

Figure 3. RBC pilot unit.

Each stage contained 18 segmented polyethylene discs, four feet in diameter, all mounted on a common shaft. Discs were arranged with adjacent discs having alternate vertical and horizontal corrugations to serve as spacers and provide passageways to prevent sloughed biomass from bridging and clogging between adjacent discs.

Pilot Plant Description:

Overall dimensions, feet.	7.5'x4.5'x5.5'
Total Disk Area, square feet	2200
Sq. ft./Cu. ft. Ratio	32
Disk diameter, feet	4
Disk thickness, inches	0.06
Submergence, %	40
Eff. liquid volume, gallons	530
Speed Range, rpm	1-6
Maximum flow, gpd	5000
Number of stages	4
Number of discs	115

PILOT TEST CHRONOLOGY

NOVEMBER	1984	- start-up phase of study began
DECEMBER	1984	- actual data collection began determined flowrate
JANUARY	1985	- varied disc rpm
FEBRUARY	1985	- added powered activated carbon
MARCH-APRIL	1985	- installed pre-aeration

Table II shows the monthly average influent and effluent COD, phenols, and TOC data collected from the pilot RBC wastewater stream.

TABLE II. MONTHLY AVERAGES FROM DECEMBER TO APRIL

	TI	TE	CI	CE	PI	PE
DEC	81	48				
JAN	83	50	270	134	2070	91
FEB	78	48	263	156	1507	35
MAR	86	44	258	113	1775	24
APR	74	41	274	89	2186	63

(TI - TOC Influent, TE - TOC Effluent, CI - COD Influent, CE - COD Effluent, PI - Phenols Influent, PE - Phenols Efffluent, TOC and COD data given as mg/l, Phenol data given as ppb)

TEST METHODS

All lab testing was conducted at the petrochemical plant. Tests for the COD, TOC, and phenols concentration in the RBC influent and effluent were conducted three times a week. Lab test procedures were according to "Standard Methods-15th edition".

To carry out the study, the RBC pilot unit was fed from the pre-clarified pressure filter influent line. The feed water to the RBC pilot unit was a mixture of quench tower blowdown and plant runoff.

RESULTS AND DISCUSSION

Start-up - November

For the start-up period, the RBC pilot unit was seeded with biological mixed liquor from a nearby municipal treatment plant and provided a nutrient addition of ammonia and phosphorous to maintain a ratio of 100:5:1 of $BOD5:N:P$. After approximately three days, the discs began to develop a biological slime coating. The slime became a smooth, even, tan-colored coating in the first two compartments.

The third and fourth compartments took longer to develop their coatings, presumably because the reduction of organics in the first two compartments made less substrate available for microbial consumption. Following a 4 week acclimation period, the unit was operated continuously for five months.

Microscopic examination of the RBC biofilm showed diverse and healthy heterogeneous cultures.

The temperature of the process make-up averaged 104 F throughout the pilot study.

Flowrate - December

In December, the flowrate was adjusted to maintain an organic loading rate between 2.5 to 4.0 lbs soluble BOD5/1000 square feet/day. Assuming the majority of the wastewater organic constituents were readily biodegradable, COD and BOD concentrations would be equivalent. It was determined that a 4 gallon/minute (gpm) average flow was best suited for this application.

Indications from literature reviews and field observations are that loadings in the higher end of the organic loading range will increase the likelihood of developing problems such as dissolved oxygen (DO) limitations, heavier than normal biofilm thickness, nuisance organisms, and deterioration of overall process performance (3).

A steady flow rate was difficult to maintain due to variable head pressures in the sand filter. When there was a suspended solids build-up in the filter, the back pressure would increase, in turn, increasing the flow to the pilot RBC. When the sand filter was backwashed to remove the suspended solids, the flow to the pilot RBC would decrease. Flow through the sand filters ranged between 187 to 281 gallons per minute.

Disc Rotation - January

Disc rotation provides the mechanism of soluble organic transport from the wastewater to the attached biomass on the disc while the speed of rotation determines the rate of oxygenation.

Until January, the pilot unit disc speed had been operating between 2.5 to 3.5 rpm (peripheral velocity was 35 to 45 feet/min) to allow for the establishment of a biofilm on the disc surface. The biofilm was dark in color and produced an odor that indicated anaerobic conditions. To increase the DO levels in the basin and to simulate the peripheral velocity of a full-scale RBC unit, the disc speed was increased and maintain at 4.75 rpm (60 feet/min) for the rest of the pilot study.

Steady-state performance in the pilot RBC was difficult to determine because of the organic variations in the feedwater. The average percent removal rate was COD, 47% and phenols, 89%. During January, the average removal rate was below the necessary permit removal rates of COD, 63% and phenols, 90%.

Based on average influent TOC levels of 83 mg/l, the RBC effluent TOC levels should be 3.5 times less than effluent COD levels or 30 mg/l. This gives an estimated permit TOC percentage removal rate as 64%. For January, the TOC removal rate averaged 38%.

PAC Addition - February

During February, powdered activated carbon (PAC) was added to the unit to investigate whether increased effluent quality could be achieved and if so, what was a cost effective dosage.

Figure 4 shows that the COD and TOC reduction levels were not acceptable. The use of PAC carbon did not significantly improve operations. Biological utilization of adsorbed organic constituents was not evident.

The average percent removal rate was TOC, 37%; COD, 40%; and phenols, 98%. It should be noted that the organic loading and the COD percent removal rate may have been adversely affected by PAC additions. The PAC appeared to have covered the biofilm and inhibited the adsorption of dissolved organics and oxygen.

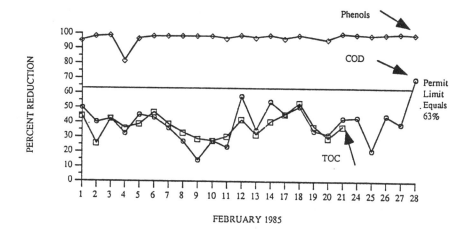

Figure 4. RBC COD, TOC, phenol reductions in February.

Phenols removal level exceeded the required permit level of 90%.

Pre-aeration - March & April

Pre-aeration was added to the system in March and April. Atmospheric oxygen was added by introducing plant air in a mixing tank upstream of the RBC.

Pre-aeration was selected as an alternative to increasing the disc speed or adding air in the RBC tank. Either additonal disc speed or air sparging in the tank may result in increased biomass sloughing from the RBC media discs.

Pre-aeration improved effluent quality as shown by the graphs.

During March, the average percent removal rate was TOC, 52%; COD, 56%; and phenols, 98%. In April, the percent removal rate was TOC, 45%; COD, 67% and phenols, 98%. Figures 5 & 6 show the March and April percent reduction data. The COD and phenols percent removal met the permit requirements at an average flow of 4 gpm and organic loading of 3.38 lbs. COD/1000 square feet/day as shown in Figure 7.

Process Wastewater from a Petrochemical Plant 481

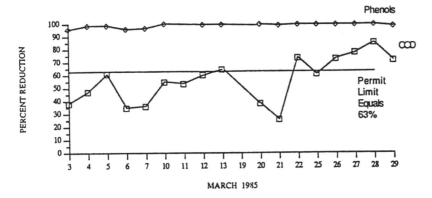

Figure 5. RBC COD and phenol reductions in March.

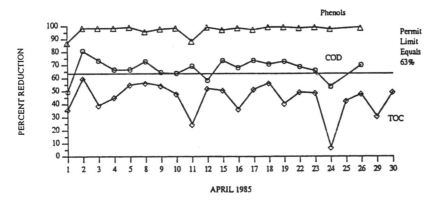

Figure 6. RBC COD, TOC, phenol reductions in April.

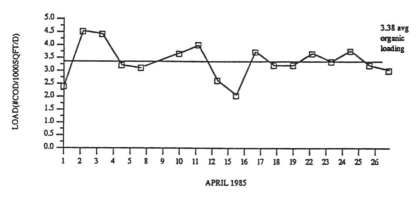

Figure 7. RBC #COD organic loading in April.

TOC percent removals did not meet expected levels. This may be attributed to refractory organics which require longer exposure to the biofilm for biodegradation. There may also be inorganic or non-biogradable constituents in the wastewater stream.

CONCLUSIONS

1. The field pilot study using this wastewater demonstrated the feasibility of the RBC process in applications with diverse/unique wastes with varying strengths and flows.

2. Effluent from the pilot plant complied with all requirements of the permit restrictions for COD and TOC at a loading rate of 3.38 lbs. COD/1000 square feet/day.

3. PAC did not show any apparent COD reduction for this particular wastewater.

4. Discs rotation may not supply sufficient oxygen for aerobic respiration in all operating conditions. Supplemental air may be required.

ACKNOWLEDGEMENTS

The RBC pilot unit used in this case study was manufactured by Royce Process Equipment Company, Inc., Pearland, Texas. The pilot scale RBC was operated by Royce Process and plant personnel.

REFERENCES

1. Metcalf & Eddy, Inc., Wastewater Engineering: Treatment/Disposal.Reuse. McGraw-Hill Book Company, 1979, Chapter 9

2. Eckenfelder, W.W., Principles of Water Quality Management, CBI Publishing Company, Inc.,1980, Chapter 2

3. Opatken, E.J., Rotating Biological Contactors - Second Order Kinetics In: Proceedings of Second International Conference on Fixed Film Biological Processes, Washington, D.C., July 1984.

LAND TREATMENT OF NITROGUANIDINE WASTEWATER

Richard T. Williams
A. Ronald MacGillivray

David E. Renard

Roy F. Weston, Inc.
West Chester, Pennsylvania

U.S. Army Toxic and Hazardous
Materials Agency
Aberdeen Proving Ground, Maryland

ABSTRACT

Nitroguanidine (NQ) wastewaters contain nitroguanidine, guanidine nitrate (GN), ammonia, nitrate, and sulfate. Simulated NQ wastewater is being applied to continuous and perfusion soil columns, with continuous flow column influent and effluent samples being analyzed for wastewater components and transformation products (nitrosoguanidine, guanidine, cyanamide, melamine, and cyanoguanidine). Whey, molasses, and glucose are being tested as carbon supplements. Mineralization rate experiments are being conducted using ^{14}C-NQ and ^{14}C-GN as test substrates. The number of microbes capable of degrading NQ and GN is being determined, as is microflora acclimation.

Preliminary data indicate that carbon supplements facilitate NG degradation after 70 days of application in continuous flow soil columns. Batch mineralization experiments generally support these findings. To date, cyanamide is the only transformation product detected in significant quantities.

INTRODUCTION

Nitroguanidine (NQ) is a water-soluble nitroamino compound used as an ingredient in munitions propellants. A number of wastestreams are produced during NQ manufacture, and these wastewaters may contain NQ (0.4 - 2,500 mg/L), guanidine nitrate (1.0 - 200 mg/L), ammonia, (1.0 - 200 mg/L), nitrate (1.0 - 11,800 mg/L), and sulfate (0 - 5,500 mg/L). Land application has been proposed as a treatment method for wastewater. WESTON is under U.S. Army Toxic and Hazardous Materials Agency (USATHAMA) contract to investigate the land treatability of simulated NQ wastewater. The objective of this study is to develop sufficient data to recommend either for or against land application.

Attempts to demonstrate the biological mineralization of NQ in aerobic, aqueous systems have failed (6). NQ was, however, cometabolically reduced to nitrosoguanidine (NSQ) in anaerobic continuous culture using an acclimated sewage treatment plant digest as an inoculum (6). Glucose was utilized as the cometabolic substrate. Further studies (4) revealed that NSQ was not biologically degraded under aearobic or anaerobic conditions in aqueous systems.

The chemically unstable NSQ that formed in the above study was transformed into cyanamide, cyanoguanidine, melamine, guanidine, and nitrosamide. The nitrosamide further decomposed to nitrogen gas and water (6).

Studies (5) also have been conducted on the biodegradability of NQ in columns packed with garden soil. NQ (150 mg/L) was degraded when a glucose supplemented (0.5 to 1.0 percent) solution was applied at 4 mL/hour. A C:N ratio between 34 to 1 and 68 to 1 was required for biodegradation of NQ to ammonia. No significant concentrations of potentially hazardous organic intermediates were detected.

The fate of actual NQ wastewater has been studied under land application, field conditions (1, 9), but the results have been inconclusive regarding the fate of wastewater components.

MATERIALS AND METHODS

Diluted, simulated NQ wastewater containing NQ (129 mg/L), guanidine nitrate (10.5 mg/L), ammonia (12.5 mg/L), nitrate (12.5 mg/L), and sulfate (166 mg/L) is being applied to six continuous flow soil columns (5) which contain 1 kg of soil. The columns were established with the following experimental conditions.

No.	Medium	Microorganisms	Supplement
1	Tap water	Active	None
2	Wastewater	Active	None
3	Wastewater	Sterile	Glucose, 1 g/L
4	Wastewater	Active	Glucose, 1 g/L
5	Wastewater	Sterile	Whey, 1 g/L
6	Wastewater	Active	Whey, 1 g/L

Perfusion columns (7), also containing 1 kg of soil, are being tested under similar conditions. NQ manufacturing site soil is being utilized. Sterile columns are maintained with 0.75% mercuric chloride.

Batch mineralization studies (3) are conducted in erlenmeyer flasks containing 30 g of NQ production site soil. Mineralization of 400,000 DPM of ^{14}C-NQ or ^{14}C-guanidine nitrate to $^{14}CO_2$ is being monitored under various test conditions. Flasks are inoculated with an acclimated flora.

Enumeration of extracted (2) soil microorganisms is by plate count and modified ^{14}C-MPN (8).

RESULTS

As of 1 May 1986, data for the continuous flow soil columns were available through day 90. Data from a screening, batch mineralization study using ^{14}C-NQ and ^{14}C-GN were also available, as were physiological background, soil characterization data for the test soils.

Wastestream components were absent in the control column (#1) influent and effluent. NQ in the effluent of Column #2 remained at a low level through day 39, then increased until reaching the influent concentration at day 84. TOC in the effluents of columns #3 to 6 remained at a low level until day 60, then increased sharply. NQ in the effluent of columns #4 and 6 decreased after day 70. The only transformation product that increased as NQ decreased was cyanamide. GN and other transformation products were not detected at a significant level in any of the column effluents. Tentative evidence indicates that microbial acclimation increased significantly after approximately 70 days of wastewater application, and that both whey and glucose stimulated NQ transformation.

Most of the ^{14}C-GN added in the batch mineralization experiment was trapped from the displaced headspace gas after two weeks of incubation. Tests are underway to verify that biological mineralization is the basis for this observation. ^{14}C-NQ was slowly converted to ^{14}CO$_2$ in a preliminary experiment, and no acclimation was observed through day 44. The presence of whey, glucose, or phosphorus and trace metal nutrients did not enhance ^{14}C-NQ mineralization through day 44. Aerobic versus anoxic conditions also did not effect ^{14}C-NQ transformation during this time period.

SUMMARY

The organic components of NQ wastewater appear to be transformed within soil to a degree which indicates land application may be feasible.

Carbon supplementation, however, is required, and a long acclimation period appears to be needed before NQ is significantly mineralized. A significant concern is nitrogen loading, which may result in groundwater contamination after repeated wastewater application.

REFERENCES

1. Boldt, R.E., W.J. Critz, P. Fiancu and M.D. Nickelson. 1982. Land Treatment Feasibility Study No. 32-24-0410-83 Sunflower Army Ammunition Plant. United States Army Environmental Hygiene Agency, Aberdeen Proving Grounds, Maryland.

2. Ellis, W.R., G.E. Ham, and E.L. Schmidt. 1984. Persistence and Recovery of Rhizobium japonicum inoculum in a field soil. Agronomy Journal 76:573-576.

3. Howard, P.H., J. Saxena, P.R. Durkin, and L.T. Ou. 1975. Review and evaluation of available techniques for determining persistence and routes of degradation of chemical substances in the environment. NTIS PB243825 (Also, 1981 update of 1975 Report, NTIS NO. PB84-168731.)

4. Kaplan, D.L. 1983. Biodegradation of Nitrosoguanidine. U.S. Army Natick Research and Development Laboratories (unpublished).

5. Kaplan, D.L. and A.M. Kaplan. 1985. Degradation of Nitroguanidine in Soils. U.S. Army Natick Research and Development Laboratories (unpublished).

6. Kaplan, D.L., J.H. Cornell, and A.M. Kaplan. 1982. Decomposition of Nitroguanidine. Environmental Science and Technology. 16:488-492.

7. Kaufman, D.D. 1966. An inexpensive, positive pressure, soil perfusion system. Weeds 90:90-91.

8. Lehmicke, L.G., R.T. Williams, and R.L. Crawford. 1979. ^{14}C-Most-probable-number method for enumeration of active heterotrophic microorganisms in natural waters. Appl. Environ. Microbiol. 38: 644-649.

9. Nickelson, M.K. 1984. Land Treatment Feasibility Study No. 32-24-0419-84 Sunflower Army Ammunition Plant. United States Army Environmental Hygiene Agency, Aberdeen Proving Ground, Maryland.

COMBINED FIXED BIOLOGICAL FILM MEDIA AND EVAPORATIVE COOLING MEDIA TO SOLIDIFY HAZARDOUS WASTES FOR ENCAPSULATION AND EFFICIENT DISPOSAL

Sheldon F. Roe, Jr.

Independent Consultant
5375 Coral Ave.
Cape Coral, FL

1.0 INTRODUCTION

Hazardous wastes are a problem; not only because they are toxic, but also because they often occur in dilute solution. Although the waste may have been concentrated when it was dumped or used; rain water, surface water, ground water, or waste water frequently dilutes the waste. The premise of this paper is that the process problem is not only to detoxify the waste, but to concentrate it. The emphasis herein is on treating dilute solutions. Indeed, if the waste can be concentrated economically, detoxification may not be necessary--the waste may be stored in its toxic form, or it may be practical to incinerate it whereas incineration of dilute waste is impractical.

This paper will outline a particular kind of evaporation process common in evaporative cooling for agricultural and industrial uses. The emphasis here, however, is not cooling but evaporation of water in combination with biotechnology, volatile stripping, and other conventional unit operations.

Further, the evaporative process described herein emphasizes purposely scaling or fouling the evaporative cooling media, encapsulating it, and disposing of it as a solid waste. A second possibility is to incinerate the evaporative cooling media containing the hazardous waste.

Existing technology will be discussed in terms of conventional evaporation, utility cooling tower evaporation, evaporative cooling, VOC Stripping, and fixed film biotechnology. Additional discussions will include a process description and a process example as well as the various combinations of aerobic--anaerobic digestion and evaporation.

2.0 DISCUSSION

2.1 Existing Technology

2.1.1 Conventional Evaporation

Evaporation technology ranges from modern multistage evaporators such as those used in landfills or industries (1) to evaporative ponds (2).

Evaporative ponds are difficult to beat economically--especially in the days before liners. Land availability and time to do the evaporative job are disadvantages. Operating costs for the gas phase are particularly advantageous as compared with the fans of the evaporative cooler or the compressors of the multistage evaporators. More discussion on this point will be in the Section 2.3 on gas phase.

Conventional evaporation for hazardous wastes is perhaps best summarized by Woodland (3). Costs are quoted at $0.40 to $4.40/1000 lbs. of liquid evaporated for steam alone at a steam cost of $4/$10^6$ BTU. Other costs are quoted at approximately $0.35/1000 gal., but capital costs are greater here in large multiple effect systems.

2.1.2 Utility Cooling Tower Evaporation

Evaporation, of course, occurs in utility cooling towers. Reference 4 is particularly interesting because it does deal with hazardous wastes and with the gas stripping phenomenon in utility cooling towers. However, operating conditions are controlled to prevent scaling (fouling), and the towers are designed for cooling--not evaporation and fouling.

2.1.3 Description Of Evaporative Cooling With Structured Packing

The differences between biological fixed film and evaporative cooling media are as follows:
1. Evaporative cooling media is used only in crossflow towers (do not confuse with the term crossflow media used in biotechnology as explained in Reference 5) as a trickle bed.
2. In evaporative cooling media, the flute angles are 15° and 45° to promote liquid drainage against the crossflow gas stream.
3. The evaporative cooling media contains a finer structure (more ft^2/ft^3).
4. The evaporative cooling media is composed of cellulose or glass fibers bound with plastic resins. The fiber enhances the evaporative ability of the sheets. This media will pick up four times its weight in water, for instance.

Figures 1 and 2 show evaporative cooling media performance and configuration as used in greenhouses and chicken houses (1). Also of interest is evaporatively assisted dry cooling or peak shaving on dry cooling as reported in Reference 6. Carryover of mists is often a concern here, and this concern should also apply to concentration of hazardous waste. Very little mist carryover occurs or it would foul the large and fine surface area of the dry cooler directly behind the evaporative cooling tower. Supple (7) gives an idea of the performance in various geographies. However, remember that

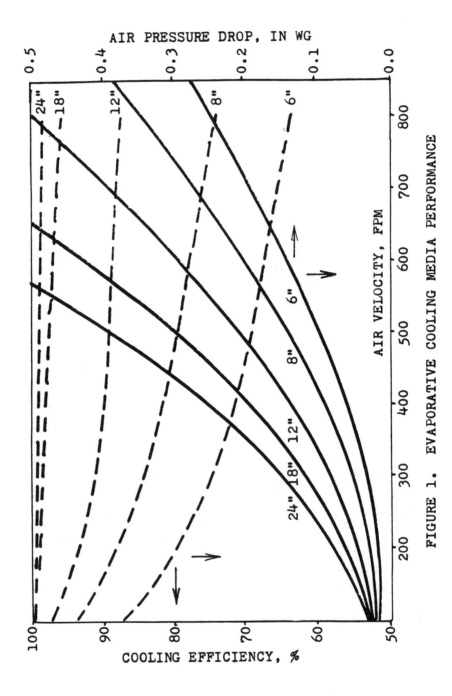

FIGURE 1. EVAPORATIVE COOLING MEDIA PERFORMANCE

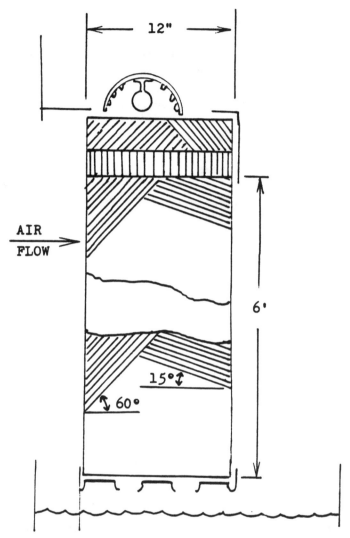

FIGURE 2. EVAPORATIVE COOLING MEDIA WITH SUPPORTS AND LIQUID DISTRIBUTION SYSTEM

comfort cooling or peak shaving do not necessarily consider low humidities at night.

Figure 3 illustrates the finer structure of the evaporative cooling media. Structures described are the conventional 60° angle for counterflow towers. The finest structure (6560) is usually utilized for evaporative cooling, while the coarser structures (19060 and 27060) are conventionally used for trickling filters and anaerobic digesters.

2.1.4 C.L.E.A.R.S. Process

The closed loop evaporative atmospheric recovery system (C.L.E.A.R.S.) by Heil Process Equipment Company was invented several years ago under U.S. Patent No. 3,661,732. This process uses heated water in a spray or packed counterflow tower for evaporation of water from metal plating wastes. Steam usage is quoted at 1000 lbs./100 gallons of rinse water. Disadvantages of this process are: the steam-boiler requirement, lack of site adaptability because of the boiler requirement, and the solids handling problems characteristic of the metal plating industry. This process is interesting because it utilizes dropwise heat and mass transfer as compared to evaporative cooling which utilizes film transfer.

2.1.5 VOC Stripping

The use of packed bed or other stripping towers in water cleanup has been described in References 5, 8-10. If hazardous chemicals are lower boiling point than water (see Henry's Law Coefficient), they will evaporate preferentially to the water in an evaporative process. Indeed, because large percentages of water are being evaporated, smaller percentages of compounds with higher boiling points than water may also volatilize and disappear before the water does.

2.1.6 Biotechnology

Fixed film biotechnology exists in several forms as outlined in Reference 5, and details will not be discussed--

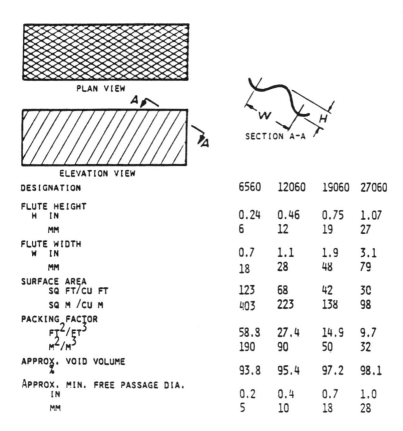

DESIGNATION	6560	12060	19060	27060
FLUTE HEIGHT				
H IN	0.24	0.46	0.75	1.07
MM	6	12	19	27
FLUTE WIDTH				
W IN	0.7	1.1	1.9	3.1
MM	18	28	48	79
SURFACE AREA				
SQ FT/CU FT	123	68	42	30
SQ M /CU M	403	223	138	98
PACKING FACTOR				
FT^2/FT^3	58.3	27.4	14.9	9.7
M^2/M^3	190	90	50	32
APPROX. VOID VOLUME %	93.8	95.4	97.2	98.1
APPROX. MIN. FREE PASSAGE DIA.				
IN	0.2	0.4	0.7	1.0
MM	5	10	18	28

FIGURE 3. CONSTRUCTION OF 60° SHEET PACKING

particularly at this conference. VOC stripping in biotowers has also been discussed elsewhere. The common point as far as heat and mass transfer is concerned is that the evaporative cooling media is similar for cooling towers, biotechnology, strippers, and evaporation. Indeed, the similarity extends to liquid-liquid separation and solid-liquid separation media. The general description of the media is corrugated sheets arranged at an angle as is illustrated in Figure 3 showing various sizes related to fouling potential. So much for existing technology, now the process of this paper will be illustrated with an example.

2.2 Process Description

A typical all inclusive process flow design is shown in Figure 4. Much of the equipment such as plate separators or belt filters might be unnecessary or small because a major portion of the solids will be removed with the evaporative cooling media.

It should be explained that the evaporative cooling media will be operated as a batch type process. The media will be run until it plugs with solids. Then the media will be removed and disposed of. A typical installation to be removed might be 6 ft. high x 1 ft. wide x 100 ft. long or 600 ft^3 of media. This 600 ft^3 would either be directly disposed of as a solid waste, encapsulated, or incinerated.

An analysis of a typical cubic foot is illustrated below:

Weight of media	3 lbs./ft^3
Plus Water	6 lbs./ft^3
Plus Hazardous Chemical for $\rho=1$, (biomass or organic) 50% of a ft^3	30 lbs./ft^3
Total Max Weight @$\rho=1$	39 lbs./ft^3
Total Max Weight @$\rho=2.5$ (inorganic)	87 lbs./ft^3
Total Max Weight @$\rho=7$ (metallic)	227 lbs./ft^3

It is assumed that 50% of the cubic foot would be filled with waste at the time of disposal. The weights are illus-

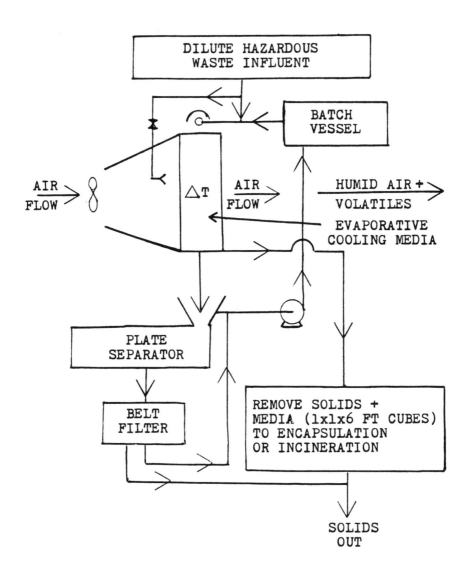

FIGURE 4.
PROCESS FLOW DIAGRAM

trated for three specific gravities of hazardous waste--organic, inorganic, and metal. Remember that if these originally occurred in dilutions of 100ppm, then 10,000 times weight of water would have been processed or 300,000 pounds of water to obtain the 30 pounds of waste in the organic case.

The media is operated as a batch type process, but the liquid should also be operated as a batch type process as in an evaporative pond--not as in multistage evaporators. The reason is that it is expected that the evaporative cooling media will become less efficient as it becomes fouled. Thus, the clean media will be operated in dilute solutions while the more fouled media will be operated in the more concentrated solutions. At the end of a cycle, the fouled media will be changed out, and a new batch of dilute liquid will be put in the vessel.

2.3 A Process Example

The evaporative capacity of evaporative cooling media is illustrated in Figure 1 for various air flow path lengths of from 6 inches to 24 inches. Twelve inches has been selected for this example. Figure 1 is utilized in conjunction with the psychometric chart for calculations.

Two arbitrary psychometric conditions have been chosen for illustrative purposes. Condition No. 1 is most advantageous as might be found in arid climates (80°F, 10%RH). Condition No. 2 is perhaps more characteristic of the remainder of the U.S.A. (50°F, 50%RH). Of course, the process will not run at 100%RH because relative humidity is the driving force.

The flow rate of water was selected at 70,000 lbs./hr. (140 gpm) at 10 ft. pumping head. The crossflow tower would have 6 ft. of media height and be wetted at a rate of 3 gpm/ft^2. Therefore, the tower size would be 6 ft. high x 1 ft. wide x 47 ft. long. Air flow would be 84,600 ACFM (280 FPM) at 0.1 in WG pressure drop.

The tower at Condition No. 1 would evaporate 34 lbs. of water/min. while at Condition No. 2, evaporation would be 10 lbs./min. Temperature drops of the air would be 24°F and 6°F respectively.

If the liquid batch were 70,000 lbs., then it would take 34 hrs. at Condition No. 1 and 117 hrs. at Condition No. 2 to evaporate the batch.

If the liquid contained 100ppm solids, and if all the solids deposited on the media; then it may be calculated how long it would take for the media to fill with solids as described in Section 2.2 with a solid of specific gravity (P) of 1.0. The solids would be 30 lbs. of hazardous waste for 282 ft^3 of tower or 8460 lbs. of hazardous waste per tower. At Condition No. 1 of 34 hrs. to produce 7 lbs. of hazardous waste from 70,000 lbs. of water, the media would be 10% full at 340 days. Of course, this time could be reduced by other operations such as those in Figure 4, and it is doubtful that the tower will be run until zero percent water. At this point the 8460 lbs. (282 ft^3) of hazardous waste would be encapsulated and disposed of as a solid waste, or incinerated. Power requirements for the tower are estimated at 13 HP for fans and 1½ HP for liquid pumping for a total of 14½ HP.

As was mentioned in 2.1.1 the gas handling cost of an evaporative pond is low while vapor compressors of multistage evaporators require power. A major cost of this process is the fans to handle air. However, this process was developed to provide cool air--not evaporate water.

When the purpose is evaporation, lower gas velocities and lower pressure drops may be possible resulting in lower power requirements. Indeed, the ultimate would be no fan with some evaporation occurring from the high surface area in combination with natural draft or diffusion. The optimum gas side condition is one of the parameters that needs to be studied for this process.

The preceding example has hopefully provided an understanding of the evaporative (cooling) process. Now, biotechnology will be considered.

2.4 Evaporation--Biotechnology Combination

The process sequence of unit operations is significant. Should the evaporation or the anaerobic--aerobic process be performed first? What are the advantages and disadvantages of each sequence?

2.4.1 Evaporation--Anaerobic

If the evaporation is performed before anaerobic digestion, the advantage is that the soluable organic concentration will be greater in the anaerobic process. Higher concentrations are desirable to prevent washout in a fixed film or mixed phase anaerobic process. The disadvantage is that the fine structure of the evaporative cooling media may act as filter for solids. Indeed, it may produce solids if it runs long enough to become a trickling filter. Air evaporation will, of course, always lead to an aerobic biological process if allowed to grow.

If nutrients, bugs, or other organics are not in the influent, these may be added intentionally before or after (preferably after) the evaporation process.

2.4.2 Anaerobic--Evaporation

If the anaerobic digestion is performed first, ammonia stripping will occur in the evaporation step because nitrogen compounds are converted to ammonia. Of course, CO_2, CH_4, H_2S, or odorous compounds will also be stripped. A disadvantage of this sequence is that the particulate biomass produced in the anaerobic step cannot be put directly on the fine evaporative media. An intermediate solids separation step will be necessary.

2.4.3 Aerobic--Evaporation

A trickling filter, activated sludge, or super bug process followed by evaporation is perhaps desirable; but the large volume of solids produced by the aerobic biological process should probably be removed before the evaporation step--depending on the scale of the operation. Simultaneous evaporation will probably be the choice here.

2.4.4 Evaporation--Aerobic

Since it usually takes several weeks for biomass to grow on a trickling filter, this should be long enough for the evaporation-concentration to be performed. Soluable organics or inorganic toxics could then be fed to a trickling filter or activated sludge plant. Partial aerobic destruction of toxics may occur on the media, and subsequently be encapsulated or incinerated.

2.5 Ambient Temperature Evaporation

Discussions of the aerobic, anaerobic, evaporation combination should include the advantages and disadvantages of ambient temperature evaporation. In cases where the evaporation step precedes aerobic or anaerobic digestion, the low temperature does not destroy the biogrowth. Indeed, if the object is to harvest (or dispose of) the biomass, low temperature evaporation is essential. Further, if the object is to promote the growth of biomass, the evaporative cooling media could be preimpregnated with cells, enzymes, or nutrients. The media could be predried and then when the evaporation step starts up, the biogrowth will start also.

3.0 RECOMMENDATIONS

3.1 The evaporative cooling process described herein is perhaps intermediate between evaporative ponds and multistage evaporators. This process shares the low energy and capital costs of evaporative ponds without the disadvantage of rain water dilution suffered by evaporative ponds. Yet, this process shares the chemical plant heritage of multistage evaporators in that it is a predictable, reliable crossflow tower without the fouling problems of multistage evaporators.

3.2 The evaporative process for hazardous waste concentration and disposal is certainly not a universal solution to all hazardous waste problems, but must be tailored to the properties of the waste. Then it must be evaluated on a pilot or full scale basis.

3.3 The advantages of ambient temperature evaporation in combination with conventional fixed film biological processes should be explored further--particularly, simultaneous fixed film aerobic digestion and evaporation.

3.4 Specific research areas recommended for the future include:
1. The use of super bugs in combination with aerobic digestion.
2. Since the dissolved oxygen content of the water will immediately be saturated or supersaturated, oxidation effects on toxic materials should be evaluated.
3. Introduction of cement or a polymer into the water distribution system at the final stages of solids buildup on the media might perform a partial encapsulation.
4. Operation of the media at .01 inches air flow pressure drop or less and 80-90% evaporation efficiency for economical operation.

References

1. Karoly, J. A., et.al., "Evaporation/Incineration: An Energy Efficient Option," Chem. Eng. Prog., Oct., 1981, pp. 42-47.
2. Grabacki, E., "Designing an Evaporation Spray Pond for Pollutants," Pollution Engineering, January, 1986, p. 38.
3. Woodland, Lawrence R., "Evaporation," <u>Unit Operations for Treatment of Hazardous Wastes</u>, DeRenzo, (Ed.), Noyes Data Corp., Park Ridge, N.J., 1978, pp. 445-451.

4. Neufeld, R. D., et.al., "Cooling Tower Evaporation of Treated Coal Gasification Wastewaters," Journal WPCF, Vol. 57, No. 9, Sept., 1985, pp. 955-963.
5. Roe, S. F., "Packed Bed Reactors for Wastewater Treatment," Chapter to be published in the book Bioenvironmental Systems, D. Weiss, (Ed.), CRC Corp., Boca Raton, FL, 1987.
6. Adler, J. W., "Different Cooling Techniques to Reduce Water Consumption," Seventh Annual Industrial Energy Technology Conference, Houston, TX, May, 1985.
7. Supple, R. G., "Evaporative Cooling for Comfort," ASHRE Journal, Aug., 1982, pp. 36-42.
8. Mumford, R. L., et.al., "Mass Transfer of Volatile Organics in a Packed Bed Stripper," presented at AWWA Annual Conference, Washington, DC, June, 1985.
9. Thibodeaux, L. J., et.al., "Measurement of Volatile Chemical Emissions from Wastewater Basins," EPA-600/S2-82-095, Aug., 1983.
10. Truong, K. N., et.al., "The Stripping of Organic Chemicals in Biological Treatment Processes," Environmental Progress, Vol. 3, No. 3, Aug., 1984, pp. 143-151.

FATE OF COD IN AN ANAEROBIC SYSTEM TREATING HIGH SULPHATE BEARING WASTEWATER

G.K. Anderson
J.A. Sanderson
C.B. Saw
Department of Civil Engineering
University of Newcastle upon Tyne
United Kingdom

T. Donnelly
Unilever Research
Port Sunlight Laboratory
Merseyside, United Kingdom

INTRODUCTION

Two pilot plants, an anaerobic contact process and an anaerobic packed bed reactor were operated on site, treating a high sulphate bearing wastewater from an edible oil refinery. The objectives of the research project were as follows:
- to test the amenability of wastewater from the acid water fat trap to treatment in continuously operated anaerobic processes;
- to compare the relative performances of two anaerobic process configurations.

This work has shown that anaerobic processes operating on sulphate-bearing acid water from edible oil refining, result in the development of a microbial system with sulphate reducing bacteria, rather than methanogenic bacteria as the main terminal group responsible for the majority of the COD removal capacity. This paper presents the data obtained from the pilot plant study to date and discusses the ability of the anaerobic systems to degrade fat and dissolved organics in the presence of sulphate.

ROLE OF SULPHATE REDUCING BACTERIA IN ANAEROBIC DIGESTION

Sulphate is found in high concentrations in the wastewaters from many industrial processes, such as molasses fermentation (bakers yeast and ethanol), pulp and paper, pectin, wine distillery, palm oil, petroleum refineries, edible oil production and a variety of other chemical industries. The presence of sulphur-containing chemicals in wastewaters is of major concern in anaerobic wastewater treatment. In sulphate reduction, organic matter is diverted from methane production to sulphide generation. Sulphur-reducing bacteria utilize sulphate as an electron acceptor with hydrogen sulphide being the end product of reduction. The kinetics of competition for the available electron donors between sulphate reducing bacteria and methane producing bacteria have received considerable attention (Abram & Nedwell, 1978a; Kristjansson et al., 1982; Schonheit et al., 1982; Valcke and Verstraete, 1983; Ingvorsen et al., 1984; Robinson and Tiedje, 1984). The SRBs apparently have a higher affinity for hydrogen and acetate which are the major methane precursors. As shown in Table 1 (Archer, 1983), the utilisation of hydrogen and acetic acid by sulphate reducing bacteria is favoured thermodynamically over its use by the methanogens.

TABLE 1: COMPARISON BETWEEN ENERGY YIELD FOR SULPHATE REDUCTION AND BIOMETHANATION

Reaction	kJ/reaction
$4H_2 + SO_4^= + H^+ \rightarrow HS^- + 4H_2O$	-152.6
$4H_2 + HCO_3^- + H^+ \rightarrow CH_4 + 3H_2O$	-135.9
$CH_3COO^- + SO_4^= \rightarrow HS^- + 2HCO_3^-$	-71.7
$CH_3COO^- + H_2O \rightarrow CH_4 + HCO_3^-$	-31.0

There may be at least two stages of methanogenesis inhibition
resulting from hydrogen sulphide formation:
- primary inhibition of methanogenesis due to competition for
  hydrogen and acetate by sulphate reducing bacteria, and
- secondary inhibition of methanogenesis resulting from the
  decline of the methanogenic population due to direct
  inhibition of the cell's functions by a high concentration
  of soluble sulphides.

Considerable research has been carried out to determine which
form of sulphur compound causes inhibition, and whether pH has
an effect on the threshold concentration through ionization.
It has been suggested that sulphite, thiosulphate and
trithionate are intermediates in the bacterial sulphate
reduction to sulphide (Akagi et al., 1974; Chambers &
Trudinger, 1975; Lee et al., 1973). In the anaerobic
degradation of cellulose to methane, the toxic effect of
different sulphur compounds has been reported to increase as
follows (Khan & Trottier, 1978):

$$\text{sulphate} < \text{thiosulphate} < \text{sulphite} < \text{sulphide} < H_2S$$

Puhakka at al. (1985) also reported that there are differences
in the degree of inhibition of methanogenesis between the
various oxidation states of inorganic sulphur compounds.
Studying the effects of different sulphur compounds on gas
production from a simulated evaporator condensate wastewater,
they found that dithionite inhibition was the most
conspicuous.

The toxicity level of sulphide is closely related to the free
hydrogen sulphide concentration. This means that low pH (<6.5)
increases toxicity, whereas the presence of iron reduces
toxicity due to precipitation of ferrous sulphide. Capone et
al. (1983) studied the interaction between sulphate reduction
and methanogenesis, and found that different metals could
markedly alter the flow of carbon in sediments. Kroiss and
Wabnegg (1983) have related methanogenesis inhibition to the
level of free H_2S in solution. The sulphide produced by the
microbial reduction of sulphate is distributed between H_2S, HS^-
and S^{2-} in solution, and H_2S in the biogas. At pH 7.5, about
20% of the total sulphide (H_2S, HS^-, S^{2-}) present in solution
exists as free H_2S. They found that a free H_2S level of 50
mg/l inhibits acetoclastic methane producing bacteria by about
50 per cent, while complete inhibition occurred at a free H_2S
level of about 200 mg/l. Winifrey and Zeikus (1977) report a

complete cessasation of methane production at a sulphide concentration of 340 mg/l. Others report different values but in most cases there is no information about the pH in the reactor which makes comparison difficult.

Much of the research in this field has been to attempt to identify a threshold sulphide concentration below which methanogenesis is relatively unaffected. The interactions between methanogenic and sulphate reducing bacteria have also been well established in freshwater and marine sediments (Cappenberg, 1974; Martens and Berner, 1977; Abram & Nedwell, 1978 a, b; Oremland & Taylor, 1978; Mountfort and Asher, 1981; Oremland & Polcin, 1982; Lovley & Klug, 1983). The general conclusion is that SRBs inhibit methanogenic bacteria by outcompeting them for hydrogen and acetate. Lovley et al. (1982) further concluded that methanogenic bacteria and SRBs can coexist in the presence of sulphate, and that the outcome of competition at any time was a function of the rate of hydrogen production, the relative population sizes and sulphate availability. Thus, methanogenic bacteria and SRBs should coexist in environments where the rate of uptake of hydrogen and acetate by the SRBs is lower than the rate of hydrogen and acetate production.

In addressing the possibility of a toxicity effect of sulphide on methanogenesis, the significance of competition can easily be missed. As shown in the sediment work, while high levels of sulphate relative to organics can inhibit methane production completely, sulphate reducing and methane producing bacteria can also co-exist over a wide range of sulphate concentrations. A simple description of the competition which takes place between these two groups is shown schematically in Figure 1. Since both the methanogenic bacteria and SRBs have the ability to utilize acetate and hydrogen, it would seem reasonable to assume that in a carbon limited system, rich in sulphate, the SRBs would dominate and should replace the methanogens as the terminal group responsible for the majority of the COD removal capacity. An important objective of this study was, therefore, to identify the capability of a SRB-dominated system to provide efficient COD removal from a fat-bearing wastewater.

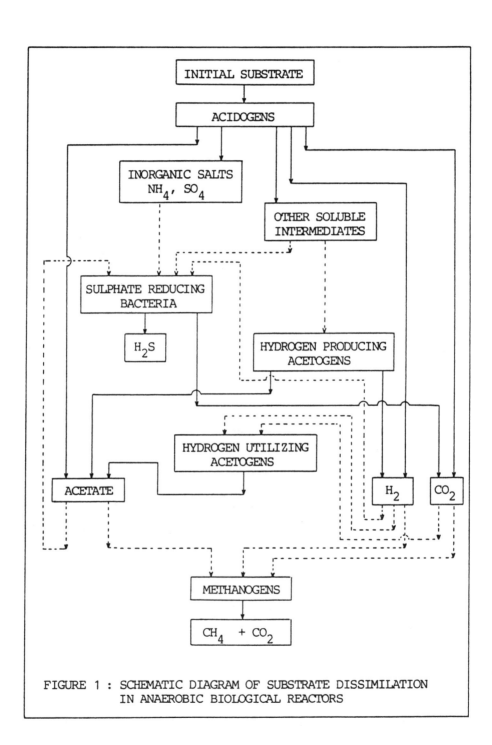

FIGURE 1 : SCHEMATIC DIAGRAM OF SUBSTRATE DISSIMILATION IN ANAEROBIC BIOLOGICAL REACTORS

SOURCE OF SULPHATE BEARING EFFLUENT IN EDIBLE OIL REFINING

The process of neutralization, which is carried out during the refining of certain edible oils, produces a soapstock from which fatty acids are recovered by acid splitting. Acid splitting is often carried out by the addition of sulphuric acid to the soapstock which causes the free fatty acids to separate from the acidic water phase. Once this separation has occurred, the water phase is drained to a fat separator where residual free fats are recovered. The final acid water represents the effluent from the process, and contains predominantly glycerol, fat and sodium sulphate.

Conventionally, the effluent from the acid water fat trap (AWFT) would be treated by chemical treatment, sedimentation and activated sludge. This combination of treatment is known to be effective but involves high capital and operating costs. Anaerobic digestion is a new effluent treatment technology which is currently emerging in certain sectors of the food and drinks industry and is attractive due to its significantly lower operating costs and lower land requirements when compared with the conventional effluent treatment systems. Although proven for carbohydrate-rich effluents, it has yet to be applied, in the U.K., to the treatment of effluent from edible oil processing. Following successful laboratory trials, the aim of this present study was to investigate the process further by carrying out a pilot plant study.

DESCRIPTION OF PILOT PLANTS

Due to fears of the possible formation of a fatty scum in upflow processes, two pilot plants, the completely mixed contact process, and the upflow anaerobic filter, were operated in parallel to establish a comparison of the effects of process configuration. Flow diagrams of each process are shown in Figures 2 and 3.

Each plant had a reactor volume of 0.5 m^3, and as may be seen from the flow diagrams, raw wastewater, following correction of pH and nutrient addition, was fed continuously, by means of separate feed pumps, to each unit.

In the case of the contact process, inflow of the wastewater caused an overflow of the reactor contents to a sedimentation

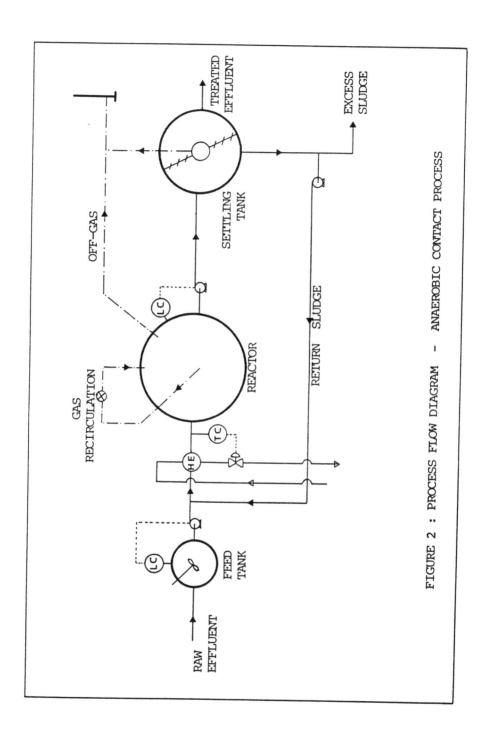

FIGURE 2 : PROCESS FLOW DIAGRAM — ANAEROBIC CONTACT PROCESS

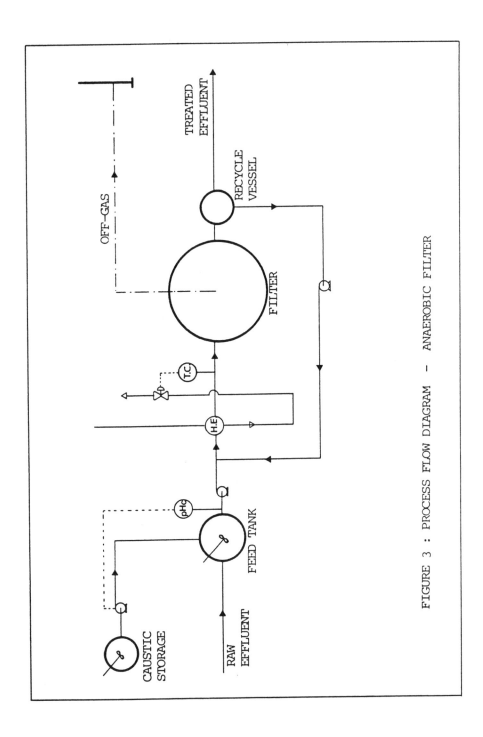

FIGURE 3 : PROCESS FLOW DIAGRAM — ANAEROBIC FILTER

tank, and the settled biomass was then recycled continuously back into the reactor.

The upflow packed bed process consisted of a reactor 3.0 m high, 0.5 m diameter, partly filled with 50 mm polypropylene packing media. The wastewater was pumped into the bottom of the reactor along with the recycled flow of treated effluent in order to give a net upflow velocity of 17 m/d. Treated effluent was displaced from the top of the reactor into a small recycle vessel from where the flow was divided into the recycled flow and final effluent.

Both plants were suitably lagged and thermostatically controlled in order to maintain a temperature of 35°C within each reactor. Off-gases from each reactor were monitored for flow rate using dry-type gas meters, and for composition by means of an in-line analyser for methane and by manual sampling for carbon dioxide and hydrogen sulphide.

The wastewater discharged into the acid water fat trap (AWFT) varied, as dictated by upstream operation, between a low COD "acid water" and a "margarine waste". The margarine waste was typically stronger than the acid water with higher levels of total fatty matter (TFM), thus higher COD and higher pH, although sulphate levels were lower, as might be expected. Hence the wastewaters in the AWFT itself consisted of a mixture of the two extreme types, as shown in Table 2.

The two pilot plants were operated on site, in parallel, to facilitate comparison of loading responses and treatment capacity. Raw feed was periodically drawn from the AWFT into a 2.5 m^3 storage tank in order to provide a composite and representative feedstock. The pH of the wastewater was adjusted with sodium hydroxide to 6.5 - 7.2, and the nutrient source maintained to give a COD:N:P ratio of 100:1.6:0.9.

TABLE 2: CHARACTERISTICS OF THE RAW WASTEWATERS

Parameter	Range
COD (mg/l)	1010 - 8200
pH	1.5 - 2.0
SS (mg/l)	380 - 1420
TKN (mg/l)	55 - 65
PO_4-P (mg/l)	90 - 100
TFM (mg/l)	140 - 1510
SO_4 (mg/l)	3100 - 7400
Na (mg/l)	2940 - 3000

RESULTS AND DISCUSSION

A comparison of the relative performances of the two anaerobic process configurations has been discussed in detail by Anderson et.al. (1986), and will not be repeated here. Average performance data of the two anaerobic systems for each loading rate period are shown in Tables 3 to 6, from which it may be seen to date that the average loading rate on the contact process has ranged from 0.42 to 1.12 kg $COD/m^3/d$, whilst that on the filter has ranged from 0.41 to 3.39 kg $COD/m^3/d$. The hydraulic retention time ranged from 9.26 to 1.98 days and 8.06 to 0.67 days, respectively for the contact process and the anaerobic filter.

Start-up and Development of the SRB-Dominated System

Sludge from a heated municipal sludge digester was screened through a size 80 mesh and used as the initial seed material for both reactors. An initial organic loading rate of 0.42 and

0.41 kg COD/m³/d was applied to the contact process (CP) and anaerobic filter (AF) respectively. This relatively low loading was maintained for a period of 25 days as shown in Table 3(a), in order to permit gradual acclimatization of the biomass to the industrial wastewater (acid water).

Both methane and hydrogen sulphide were evolved within 24 hours of applying the acid water. The initial methane yield was, however, only 0.23 m³/kg COD applied and fell steadily over the 25 day period to approximately 0.02 m³/kg COD. Gas production fell to negligible levels during this period, at the end of which the gas composition in the headspace of both reactors averaged 75 percent methane, 20 percent carbon dioxide and 5 percent hydrogen sulphide on a volume to volume basis, reflecting the relative solubilities of each gas.

Development of the SRB-dominated system was clearly rapid during this first phase of loading, but methanogenesis never completely stopped.

Inhibition of Sulphate Reducing Bacteria

The inhibition of SRBs which then encourages the growth of methanogenic bacteria has been tested both in the laboratory and at pilot plant scale. It was found that maintaining a suitable concentration of inhibitor in both the contact and the packed bed reactors effectively controlled the SRBs, as evidenced by low levels (<1% v/v) of hydrogen sulphide in the methane rich off-gas. However, periodic addition of the inhibitor has proved to be ineffective for this particular wastewater, and unless the dosage rate was maintained, the SRBs soon became re-established. Due to the high cost of the inhibitor, the fact that initial stabilization of the SRBs had no lasting effect and the discovery that SRBs may grow successfully in preference to mathanogenic bacteria, the addition of this chemical has no longer been continued during this study. The digesters are thus operated as binary microbiological systems, i.e with no inhibition of the sulphate reducing bacteria (SRBs).

Gross COD Removal Efficiency

Wastewaters from edible oil processing, as characterised by its high sodium, sulphate and fat concentrations, represent

TABLE 3 : AVERAGE PERFORMANCE DATA FOR THE ANAEROBIC CONTACT AND ANAEROBIC FILTER PROCESSES

a) Operating Conditions and TFM-Related Performance

PROCESS	PERIOD days	FLOW l/d	H.R.T. days	O.L.R. kg.COD/m^3/d	M.L.S.S. mg/l	EFF. S.S mg/l	INF. TFM mg/l	EFF. TFM mg/l	TFM REMOVAL %
CONTACT	0-26	54	9.26	0.42	7400	1065	582	67	88
CONTACT	27-54	93	5.38	0.51	10430	245	548	56	90
CONTACT	55-143	252	1.98	1.12	11360	390	600	61	90
FILTER	0-26	62	8.06	0.41	-	608	691	45	93
FILTER	27-49	146	3.42	0.67	-	295	282	49	83
FILTER	50-95	272	1.84	1.70	-	580	594	70	88
FILTER	96-143	471	1.06	1.95	-	335	285	63	78
FILTER	201-221	747	0.67	3.39	-	300	-	54	-

b) Gross COD-Related Performance

PROCESS	PERIOD	O.L.R.	INF. COD	EFF. COD	FILTERED EFF. COD	REMOVAL (WHOLE)	REMOVAL (FILTERED)
	days	kg.COD/m^3/d	mg/l	mg/l	mg/l	%	%
CONTACT	0-26	0.42	3865	1940	1515	50	61
CONTACT	27-54	0.51	2734	1250	981	54	64
CONTACT	55-143	1.12	2213	1741	1324	21	40
FILTER	0-26	0.41	3306	1620	1092	51	67
FILTER	27-49	0.67	2295	1174	978	49	57
FILTER	50-95	1.70	3125	1795	1381	43	56
FILTER	96-143	1.95	2068	1564	1153	24	44
FILTER	201-221	3.39	2269	1590	1028	30	55

c) Sulphate-Related Performance

PROCESS	PERIOD days	O.L.R. kg.COD/m³/d	SO₄ LOADING RATE kg.SO₄/m³/d	INF. SO₄ mg/l	EFF. SO₄ mg/l	SO₄ REDUCTION %
CONTACT	0-26	0.42	0.49	4574	2093	54
CONTACT	27-54	0.51	1.16	6258	3526	44
CONTACT	55-143	1.12	1.83	3631	2067	43
FILTER	0-26	0.41	0.50	4000	2290	43
FILTER	27-49	0.67	1.22	4164	1710	59
FILTER	50-95	1.70	2.44	4482	2544	43
FILTER	96-143	1.95	3.25	3454	2423	30

one of the most difficult challenges for anaerobic treatment. The moderate COD removal efficiencies in anaerobic processes treating high sulphate wastewaters have for years caused some concern resulting in a delay in the application of anaerobic digestion to the treatment of such wastewaters. In order to determine the true performance of the anaerobic processes, the components which make up the inlet and outlet COD concentrations must be clearly understood.

In the raw wastewater, the polluting load (COD) is made up predominantly of dissolved organics and emulsified fats. It has been found that both anaerobic processes remove virtually all of this organic material, including the fatty matter, with removal efficiencies greater than 80 per cent, as shown in Table 3(a). However, gross COD removal efficiencies, as shown in Table 3(b), appear to be only moderate, ranging from 21 to 54 percent and 24 to 51 percent for the CP and AF processes, respectively. Volatile fatty acids concentrations (VFA) in the treated effluent were consistently low and ranged from 124 to 332 mg/l for both processes as shown in Table 4. Combined with the low TFM concentrations in the treated effluent, these low levels of VFAs seem inconsistent with the apparently low COD removal efficiencies. As an additional check, the concentration of organic carbon was measured in the effluent from each process in order to determine whether or not the VFAs and TFM represented the sole source of carbon in the effluent. As shown in Table 4, the organic carbon from the VFAs and TFM combined, accounted for most of the organic carbon in the treated effluent. This proved that, not only was hydrolysis of the raw effluent proceeding efficiently, but also that the removal of carbonaceous matter was proceeding efficiently. Clearly then, the COD in the treated effluent was largely due to inorganic material, and the obvious source was sulphides formed by the SRB´s.

Relative Contribution of Carbonaceous COD and Sulphides to the Treated Effluent COD

Sodium sulphide is oxidized in the COD test, as shown below:

$$Na_2S + 2O_2 \longrightarrow Na_2SO_4$$

Thus 2 g oxygen is required to oxidize 1 g sulphide. By subtracting the carbonaceous COD (estimated from organic carbon measurements, or from VFA and TFM concentrations) from

TABLE 4 : CARBONACEOUS CONTENT OF TREATED EFFLUENT

PROCESS	PERIOD	COD	VOLATILE ACIDS			FATTY MATTER		ORGANIC CARBON
			TOTAL	ACETIC	COD EQUIV.	TFM	COD EQUIV.	
	days	mg/l	mg/l	mg/l	mg/l	mg/l	mg/l	mg/l
CONTACT	0-26	1515	125	88	150	67	191	–
CONTACT	27-54	981	125	90	150	56	160	–
CONTACT	55-143	1324	231	173	277	61	174	179
FILTER	0-26	1092	124	88	149	45	128	–
FILTER	27-49	978	124	86	149	49	140	–
FILTER	50-95	1381	332	250	398	70	200	–
FILTER	96-143	1153	260	190	312	63	180	172
FILTER	201-221	1028	299	229	348	54	154	–

the measured filtered effluent COD, it is possible to obtain
an estimate of the COD exerted by dissolved sulphides. As
shown in Table 5 and Figure 4, sulphide-related COD accounted
for between 66 to 77 percent of the total COD in the case of
the contact process, and between 51 to 75 percent of the COD
in the case of the anaerobic filter.

Removal of Carbonaceous COD

Taking into account the effect of the dissolved sulphides on
effluent COD, the true COD removal characteristics of the SRB
system may be estimated. As may be seen from Table 6 and
Figure 5, the removal efficiency of carbonaceous COD was very
high in both processes with efficiencies ranging from 76 to 92
percent and carbonaceous COD removal rates ranging from 0.38
to 2.64 kg COD/m³/d for equivalent organic loading rates of
0.41 to 3.39 kg COD/m³/d. The effect of organic loading rates
on the substrate utilization rate is shown in Figure 6, from
which it is evident that at the lower organic loading rates
(<1.5 kg COD/m³/d) at which the CP was operated, the
performances of the CP and the AF were very similar. In
addition, at the higher loading rates (3.39 kg COD/m³/d),
which the AF was able to accomodate, efficiency of treatment
hardly diminished, indicating that the maximum treatment
capacity of the system had not been reached.

COD:Sulphate Utilization Ratio

Sulphate reducing bacteria (SRB) such as <u>Desulfovibrio
desulfuricans</u> derive energy for synthesis and maintenance from
the metabolism of organic matter, and use sulphate as their
terminal electron acceptor. The reduction of sulphates in this
process may be expressed by the following equation (Abram &
Nedwell, 1978 a):

$$8H^+ + 8e^- + SO_4^= \xrightarrow{\text{Sulphate Reducing Bacteria}} S^= + 4H_2O$$

Thus, the reduction of one mole of sulphate (96 g) corresponds
to the oxidation of eight equivalents of organic matter, or
about 64 g COD. The theoretical COD:SO_4 ratio required for the
reduction of sulphate is thus 0.67:1.0. However, this excludes
the removal of COD for cell growth (synthesis) and the effects
of COD removal by other microorganisms such as methanogenic

TABLE 5 : DISTRIBUTION OF COD COMPONENTS IN TREATED EFFLUENT

PROCESS	PERIOD	TOTAL COD (FILTERED)	CARBONACEOUS COD		SULPHIDE RELATED COD	
	days	mg/l	mg/l	% Total	mg/l	% Total
CONTACT	0-26	1515	341	23	1174	77
CONTACT	27-54	981	310	32	671	68
CONTACT	55-143	1324	451	34	873	66
FILTER	0-26	1092	277	25	815	75
FILTER	27-49	978	289	30	689	70
FILTER	50-95	1381	598	43	783	57
FILTER	96-143	1153	492	43	661	57
FILTER	201-221	1028	502	49	526	51

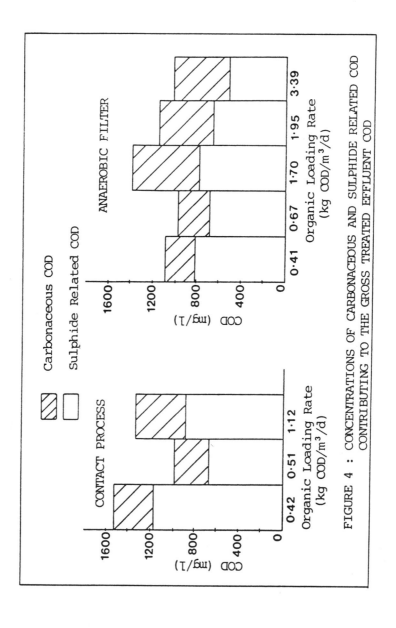

FIGURE 4 : CONCENTRATIONS OF CARBONACEOUS AND SULPHIDE RELATED COD CONTRIBUTING TO THE GROSS TREATED EFFLUENT COD

TABLE 6 : AVERAGED CARBONACEOUS COD REMOVAL CHARACTERISTICS OF THE SRB-DOMINATED DIGESTERS

PROCESS	PERIOD	O.L.R.	CARBONACEOUS COD UTILIZATION RATE	CARBONACEOUS COD REMOVAL EFFICIENCY	COD:SO$_4$ RATIO APPLIED	UTILIZED
	days	kg.COD/m^3/d	kg.COD/m^3/d	%	kg/kg	kg/kg
CONTACT	0-26	0.42	0.38	91	0.84	1.42
CONTACT	27-54	0.51	0.45	89	0.44	0.89
CONTACT	55-143	1.12	0.89	80	0.61	1.13
FILTER	0-26	0.41	0.38	92	0.83	1.77
FILTER	27-49	0.67	0.59	87	0.55	0.82
FILTER	50-95	1.70	1.37	81	0.70	1.30
FILTER	96-143	1.95	1.48	76	0.60	1.53
FILTER	201-221	3.39	2.64	78	-	-

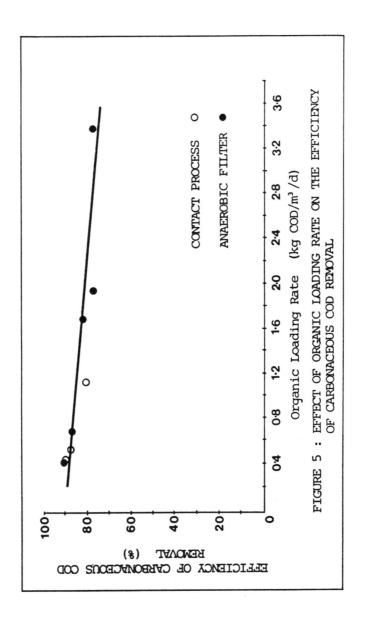

FIGURE 5 : EFFECT OF ORGANIC LOADING RATE ON THE EFFICIENCY OF CARBONACEOUS COD REMOVAL

Treating High Sulfate Bearing Wastewater 525

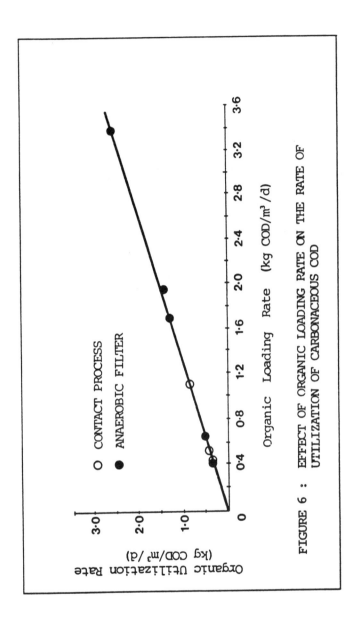

FIGURE 6 : EFFECT OF ORGANIC LOADING RATE ON THE RATE OF UTILIZATION OF CARBONACEOUS COD

bacteria operating within the same system, which would tend to result in a higher $COD:SO_4$ ratio.

In this study, the applied $COD:SO_4$ ratios ranged from 0.44 to 0.84, but the $COD:SO_4$ utilization ratios ranged from 0.82 to 1.53 as shown in Table 6. The higher utilization ratios observed in the anaerobic filter were believed to be due to greater methanogenic activity which may have been encouraged by the spatial separation of the species throughout the height of the reactor. However, the fact that methane continued to be produced at all in both processes was surprising since the limiting substrate was clearly carbon, with sulphate being very definitely in excess as shown in Table 3(c). Both systems demonstrated higher $COD:SO_4$ ratio in the initial period after start-up, reflecting the greater methanogenic activity which occurred during that period when the SRB population was not fully established. Excluding these initial loading phases, the average $COD:SO_4$ utilization ratios were 1.01 kg/kg for the contact process and 1.22 kg/kg for the anaerobic filter.

Downstream Treatment

This finding regarding the fate of COD in anaerobic systems treating high sulphate wastewater is of great importance to the successful treatment of such wastewaters and reverses the findings of much of the scientific work which has been carried out. In the past, it has been thought that a combination of low gas production and poor COD removal for such wastewaters has been the result of hydrogen sulphide toxicity. In fact, as shown in this study, the bacteria responsible for reduction of the sulphate, namely SRBs, may provide a more robust system than the methanogenic bacteria, and for each kilogram of sulphate reduced, 0.67 kilograms of organic COD is removed. However for every 1 kg of sulphide formed in the effluent, 2 kg of inorganic COD is exerted.

The species of COD change in such a binary microbial system. Biologically the system does not suffer from toxicity, but is in fact perfectly healthy. The implications for those industries producing high sulphate wastewaters could be significant since, by simply stripping out the dissolved sulphide from the reactor liquor, COD removal efficiencies in excess of 80 per cent may be attainable. The biological system, based on sulphate reducing bacteria, would then be capable of giving stable and effective treatment of high sulphate bearing wastewaters.

In order to realise the full treatment potential of the process, it is therefore necessary to remove hydrogen sulphide from the reactor liquor. This would have two effects, firstly the baseline COD in the treated effluent, which is attributed to the sulphide, would be removed and would thus automatically raise the COD removal eficiency. Secondly, any inhibition caused by excessive hydrogen sulphide would be reduced and thus the higher treatment potential of a biological system dominated by sulphate reducing bacteria may be achieved.

Removal of the hydrogen sulphide could be effected simply by recirculation of the reactor head space gas through the reactor, in the case of the contact process, or through the recycle tank in the case of the anaerobic filter. The hydrogen sulphide would be stripped from solution and then removed by passing the recirculated gas through a small caustic scrubber prior to the gas being returned to the reactor. The newly developed hydrogen sulphide absorption process used by Olthof et al. (1985) which selectively scrubs the hydrogen sulphide out of the gas stream may be applicable. The use of oxygen for sulphide control is also another possibility (Kite and Stringer, 1986). Published and practical work on sulphide oxidation is minimal, so it is important to study the best solution to the sulphide problem. This is currently being investigated by the authors.

It is envisaged that the SRB-dominated bioreactor would have the following potentials:

i) Possible higher treatment capacity:
Due to the growth rate of SRBs being greater than methanogens, it is likely that the SRB dominated system would operate at a higher organic loading rate, which in turn would result in a smaller full-scale plant.

ii) Increased flexibility:
Sulphate reducing bacteria are known to be tolerant to a wide range of pH and temperature variations. It has been found in this study that SRBs naturally dominate the methanogenic bacteria in the treatment of acid water. The tolerance of the SRB dominated system to pH, treatment conditions and load variations should provide a more robust system than a methanogenic digester.

CONCLUSIONS

Results to date show that both anaerobic systems are able to provide stable and reliable treatment of combined acid and margarine water as received from the outlet of the acid water fat trap without the need for chemical treatment to remove fat.

Although inhibition of sulphate reducing bacteria (SRB) is technically feasible, the anaerobic reactors can also be operated as binary microbiological systems. This work has shown that anaerobic processes operating on sulphate-bearing acid water from edible oil refining, result in the development of a microbial system with sulphate reducing bacteria, rather than methanogenic bacteria as the main terminal group.

In order to realise the full treatment potential of the anaerobic process, the sulphides must be stripped from the treated effluent. The resultant biological sulphate reducing systems may then provide a high efficiency of carbonaceous COD removal at rates at least comparable with methanogenic systems.

REFERENCES

1. Abram,J.W. and Nedwell,D.B. (1978 a), "Inhibition of Methanogenesis by Sulphate Reducing Bacteria Competing for Transferred Hydrogen", Arch. Microbiol., 117, p89.

2. Abram,J.W. and Nedwell,D.B. (1978 b), "Hydrogen as a Substrate for Methanogenesis and Sulphate Reduction in Anaerobic Saltmarsh Sediments", Arch. Microbl., 117, p93.

3. Akagi,J.M., Chan,M. and Adams,V. (1974), "Observations on the Bisulfite Reductase (P 582) Isolated from Desulfotomaculum nigrificaus", J. Bact., 120, p240-244.

4. Anderson,G.K., Donnelly,T., Sanderson,A.J. and Saw,C.B. (1986), "Comparison of the Anaerobic Contact and Packed Bed Processes for the Treatment of Edible Oil Wastewaters", Presented at 41st Annual Purdue Conference on Industrial Waste, U.S.A.

5. Archer,D.B. (1983), "The Microbiological Basis of Process Control in Methanogenic Fermentation of Soluble Wastes", Enzyme Microb. Technol., 5, p162-170.

6. Capone,D.G., Reese,D.D. and Kiene,R.P. (1983), "Effects of Metals on Methanogenesis, Sulphate Reduction, Carbon Dioxide Evolution and Microbial Biomass in Anoxic Salt Marsh Sediments", Appl. Environ. Microbiol., 45:5, p1586-1591.

7. Cappenberg,T.E. (1974), "Interrelation Between Sulfate-reducing Bacteria and Methane-producing Bacteria in Bottom Deposits of a Freshwater Lake. 1.Field Observation", Antonie van Leeuwenhoek J. Microbiol. Serol., 40, p285-295.

8. Chambers,L.A. and Trudinger,P.A. (1975), "Are Thiosulfate and Trithionate Intermediates in Dissimilatory Sulfate Reduction", J. Bact., 123, p36-40.

9. Ingvorsen,K., Zehnder,A.J.B. and Jorgensen,B.B. (1984), "Kinetics of Sulphate and Acetate Uptake by Desulfobacter postgatei", Appl. Environ. Microbial., 47, p403-408.

10. Khan,A.W. and Trottier,T.M. (1978), "Effect of Sulfur Containing Compounds on Anaerobic Degradation of Cellulose to Methane by Mixed Cultures Obtained from Sewage Sludge", Appl. Env. Microbiol., 35, p1027-1034.

11. Kite,O.A. and Stringer,P.R. (1986), "The Use of Oxygen for Sulphide Control", PIRA Conference on Cost Effective Treatment of Papermill Effluents Using Anaerobic Technologies, Leatherhead, U.K., January 1986.

12. Kristjansson,J.K., Schonheit,P. and Thauer,R.K. (1982), "Different K_s Values for Hydrogen of Methanogenic Bacteria and Sulphate Reducing Bacteria: An Explanation for the Apparent Inhibition of Methanogenesis by Sulphate", Arch. Microbiol., 131, p278-282.

13. Kroiss,H. and Plahl-Wabnegg,F. (1983), "Sulphide Toxicity with Anaerobic Waste Water Treatment", Proc. Anaerobic Waste Water Treatment, Symposium held in Noordwijkerhout, The Netherlands, Nov. 1983, p72 - 85.

14. Lee,J.P., LeGall,J. and Dech,H.D.Jr. (1973), "Isolation of Assimilatory- and Dissimilatory-type Sulfite Reductases from Desulfovibrio vulgaris", J. Bact., 115, 529-542.

15. Lovley,D.R., Dwyer,D.F. and Klug,M.J. (1982), "Kinetic Analysis of Competition Between Sulphate Reducers and Methanogens in Sediments", Appl. Environ. Microbiol., 43, p1373.

16. Lovley,D.R. and Klug,M.J. (1983), "Sulfate Reducers can Outcompete Methanogens at Freshwater Sulfate Concentrations", Appl. Environ., Microbiol., 45, p187-192.

17. Martens,C.S. and Berner,R.A. (1977), "Interstitial Water Chemistry of Anoxic Long Island Sound Sediments. 1. Dissolved Gases", Limnol. Oceanogr, 22, p10.

18. Mountfort,D.O. and Asher,R.A. 1981), "Role of Sulphate Reduction Versus Methanogenesis in Terminal Carbon Flow in Polluted Intertidal Sediment of Waimea Inlet, Nelson, New Zealand", Appl. Environ. Microbiol., 42, p252.

19. Olthof,M., Kelly,W.R., Oleszkiewicz,J. and Weinreb,H. (1985), "Development of Anaerobic Treatment Process for Wastewaters Containing High Sulphates", 40th Annual Purdue Conference on Industrial Waste, U.S.A.

20. Oremland,R. and Polcin,S. (1982), "Methanogenesis and Sulfate Reduction: Competitive and Non-competitive Substrates in Estuarine Sediments", Appl. Environ. Microbiol., 44, p1270-1276.

21. Oremland,R.S. and Taylor,B.F. (1978), "Sulfate Reduction and Methanogenesis in Marine Sediments", Geochim. Cosmochim. Acta., 42, p209-214.

22. Puhakka,J.A., Rintala,J.A. and Vuoriranta,P. (1985), "Influence of Sulfur Compounds on Biogas Production From Forest Industry Wastewater", Wat. Sci. Tech., 17, No.1, p281-288.

23. Robinson,J.A. and Tiedje,J.M. (1984), "Competition Between Sulphate Reducing and Methanogenic Bacteria for H_2 Under Resting and Growing Conditions", Arch. Microbial., 137, p26-32.

24. Schonheit,P., Kristjansson,J.K. and Thauer,R.K. (1982), "Kinetic Mechanism for the Ability of Sulphate Reducers to Outcompete Methanogens for Acetate", Arch. Microbial., 132, p285-288.

25. Valcke,D. and Verstraete,W. (1983), "A Practical Method to Estimate the Acetoclastic Methanogenic Biomass in Anaerobic Sludge", J. Wat. Pollut. Control Fed., 55, p1191-1195.

26. Winfrey,M.R. and Zeikus,J.G. (1977), "Effect of Sulfate on Carbon and Electron Flow During Microbial Methanogenesis in Freshwater Sediments", Appl. Environ. Microbiol., 33, p275-281.

THE FATE OF 4,6-DINITRO-o-CRESOL IN MUNICIPAL ACTIVATED SLUDGE SYSTEMS

Henryk Melcer and Wayne K. Bedford

Environment Canada
Wastewater Technology Centre
Burlington, Ontario

INTRODUCTION

To improve the knowledge base required for the development and implementation of Canadian regulations governing industrial discharges to municipal sewers, bench-scale experiments have been conducted at Environment Canada's Wastewater Technology Centre (WTC) to assess the fate of specific toxic and inhibitory organic contaminants in municipal wastewater treatment plants.

This paper describes work carried out to calibrate and test experimental protocols used in these experiments. 4,6-dinitro-o-cresol (DNOC) was selected to serve as a tracer organic compound in view of its low volatility (1), potential for adsorption (2) and reported resistance to biodegradation (3,4). It was expected that sufficient quantities of DNOC would be found in the effluent and sludge to verify experimental protocols.

DNOC has been used extensively as an insecticide and a herbicide (5). Its vapour pressure is low (5.2×10^{-5} mm Hg at 25°C) and volatilization was not considered to be a significant mechanism of removal from activated sludge systems (1). DNOC has a log octanol/water partition coefficient of

2.85 which suggests that adsorption by organic matter can occur (2). In Warburg respirometer tests (3) and static shake flask studies (4), Tabak et al. concluded that DNOC had not undergone significant degradation and that it was not possible to develop an accumulated biomass capable of satisfactorily degrading DNOC under conditions expected at wastewater treatment plants.

Contrary to expectations, DNOC was degraded. Experiments were, therefore, extended to determine any limitations on the degradation of DNOC.

METHODS AND MATERIALS

A flow schematic of the bench-scale activated sludge systems used in these experiments is shown in Figure 1. All reactor and clarifier components were made of glass and all tubing which came into contact with DNOC was made of Teflon except for the silicone tubing used in the peristaltic pumps. The working volume of the reactor was 15 L. Air was supplied at 3.85 L/min (0.26 volume of air/volume of reactor·minute) through a fritted glass diffuser. This level of aeration provided adequate mixing. Dissolved oxygen concentration was maintained between 1.0 and 4.0 mg/L. Three systems were operated at a nominal hydraulic retention time (HRT) of 6 hours and sludge retention times (SRTs) of 5, 10 and 15 days. The 15-day system was converted to a 2-day SRT near the end of the study period. Primary effluent was pumped continuously to the apparatus from a nearby municipal wastewater treatment plant.

DNOC dosing solutions were prepared in 15 L volumes by dissolving DNOC crystals in 200 mL of methanol. The remaining volume was made up with distilled water. Methanol was used as the solvent because of its biodegradability and availability. DNOC was pumped at a controlled rate into the primary effluent which was then mixed with return sludge prior to entering the reactor. The apparatus was air-tight so that the only DNOC losses were via exhaust vapours, clarifier effluent and waste sludge. The exhaust vapours were passed through a flask containing 0.5 L of a saturated solution of sodium hydroxide (NaOH). The quantity of DNOC measured in the NaOH solution was that mass which had been adsorbed between sampling periods.

DNOC concentrations were determined in grab samples of the dosing solution and the mixed liquor at weekly intervals. The DNOC concentration in the clarifier effluent was also

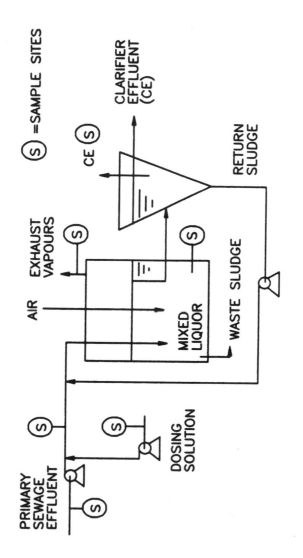

FIGURE 1. BENCH–SCALE ACTIVATED SLUDGE SYSTEM

measured at weekly intervals in 24-hr refrigerated composite samples collected in acid solution. All samples containing biomass were submitted for DNOC analysis in acidified 40 mL sampling vials. The total volume of NaOH solution in the exhaust vapour sampling bottle was submitted weekly for DNOC analysis. High pressure liquid chromatography (HPLC) procedures were used for DNOC analysis.

Twice weekly routine sampling for BOD_5 and TSS was performed on primary and secondary effluents and for physical parameters such as mixed liquor suspended solids (MLSS), specific oxygen uptake rate (SOUR) and sludge volume index (SVI). These analyses were performed according to procedures defined by Standard Methods (6).

Dosing of the reactors with DNOC commenced after equilibrium operating conditions had been established. An initial DNOC dose was selected to give a primary effluent concentration of approximately 300 µg/L. Incremental increases of two, three and four times this value were effected when analysis showed that clarifier effluent DNOC values had reached steady-state. This gave nominal feed DNOC concentrations of 600, 900 and 1200 µg/L.

The experimental operating conditions that were evaluated are summarized in Table I. At the nominal initial feed DNOC

TABLE I. Experimental Operating Conditions

Nominal Feed DNOC Concentration (µg/L)	Days of Operation			
	SRT (days)			
	2	5	10	15
300	46	47	89	89
600	-	11	-	11
900	-	21	-	21
1200	-	28	31	15

- Runs not part of experimental design.

concentration of 300 µg/L the 5, 10 and 15-day SRT systems required 3 to 4 SRTs before equilibrium DNOC values were observed in the clarifier effluent and the sludge. Operation was maintained for an additional 3 to 4 SRTs at this dosing level to observe any change in the performance of the activated sludge system. After a change in the feed DNOC level, the systems were operated for 1 to 3 SRTs, wherever possible, to establish steady-state conditions before sampling for DNOC analysis was resumed. In the case of the 15-day SRT system, pseudo-equilibrium conditions would have been achieved at the higher DNOC feed levels since time constraints prevented operation for the full 3 SRTs.

The 10-day SRT system was not operated at the 600 and 900 µg/L feed DNOC levels. Bulking sludge conditions were experienced due to excessive filamentous bacterial growth. Upon reseeding and acclimation at the 300 µg/L level, the 600 and 900 µg/L levels were omitted as the other systems had by that time shown no deterioration in performance nor a shift from equilibrium DNOC conditions at these dosing levels.

To investigate the effect of low SRTs on DNOC degradation, the 15-day SRT system was changed to a 2-day SRT and operated for a 46-day period at a nominal feed DNOC concentration of 300 µg/L.

RESULTS AND DISCUSSION

Conventional Parameters

A summary of the conventional and physical parameters monitored during the study is presented in Table II. The mean primary effluent BOD_5 concentration for the whole study period was 71 mg/L. This BOD_5 is exclusive of the DNOC dose. It does not include the contribution of methanol to the overall oxygen demand. It was calculated that the BOD_5 of the methanol was approximately four times that of the primary effluent. This was the least amount of methanol that could be used because of the limited solubility of DNOC in water. This use of methanol influenced the F/M of the four systems. Corresponding F/M values for the 2, 5, 10 and 15-day SRT systems were 1.39, 0.57, 0.39 and 0.28 g BOD_5/g VSS·d, respectively. The 10-day and 15-day SRT system organic loading rates were comparable to that of a conventional activated sludge system. The other reactors were classed as high-rate systems.

The mean effluent BOD_5 and TSS values for the systems were in the range 10 to 17 and 18 to 25 mg/L, respectively. Mean SVI and SOUR data for the 5, 10 and 15-day SRT reactors were within acceptable ranges for activated sludge systems treating primary sewage effluent. Although the mean SVI of the 2-day SRT reactor was higher than that of the other systems, its settling properties did not adversely affect its

Table II. Response of Activated Sludge Systems to DNOC

Parameter	Primary Effluent		Clarifier Effluent SRT (day)							
			2		5		10		15	
	x̄	s	x̄	s	x̄	s	x̄	s	x̄	s
BOD_5 (mg/L)	71	25	17	12	13	10	10	7	12	5
TSS (mg/L)	—	—	18	22	25	26	23	25	23	15
MLVSS (mg/L)	—	—	931	398	2286	429	3343	602	4629	1150
SVI (mL/g)	—	—	137	53	85	63	78	44	60	23
SOUR (mg O_2/ mg VSS·d)	—	—	0.544	0.217	0.274	0.069	0.205	0.055	0.178	0.049

performance. Filamentous growth occurred in the 5, 10 and 15-day SRT reactors during the latter portion of the 300 µg/L DNOC dosing period. It was postulated that this was due to some component of the primary effluent other than the DNOC as all the reactors exhibited the filamentous characteristics simultaneously. The 5 and 15-day SRT reactors recovered while the 10-day SRT system was reseeded. No further problems with filamentous growth were experienced for the remainder of the study.

It was concluded that the DNOC had no significant effect on the operation of the activated sludge systems at the influent DNOC concentrations investigated.

Fate of 4,6-Dinitro-o-cresol

DNOC data for feed, effluent, sludge and air for the 4 SRT test conditions are shown in Table III. They represent equilibrium values measured just before DNOC dose levels were increased. Clarifier effluent DNOC values for the 5, 10 and 15-day SRT reactors ranged from 19 to 51 µg/L for the entire range of influents tested. Except for the 5-day SRT reactor, when operated at a feed DNOC concentration of 1176 µg/L, all clarifier effluent DNOC values were close to the detection limit for DNOC in aqueous samples. There did not appear to be a trend demonstrated either with change in SRT or with feed DNOC concentration. Equilibrium sludge DNOC data ranged from 7 to 25 µg/g on a dry weight basis. These are low values and it would appear that, if DNOC adsorption was occurring, the adsorbed DNOC was being degraded as quickly as adsorption was taking place. Cumulative air DNOC values indicated an insignificant loss of DNOC by volatilization for all cases. This confirmed earlier findings by Fuentes (7).

Mass balance closures were calculated for each experiment by comparing the mass of DNOC fed to each system with the mass measured in the exhaust air, clarifier effluent and waste sludge samples. The difference in values was attributed to loss by degradation since very small losses were observed in the exhaust air and in the biomass. The values of DNOC in the biomass were considered to be equilibrium values given the number of SRTs each reactor cycled through before the data in Table III were collected. DNOC removal by degradation ranged from 93 to 98% for the 5, 10 and 15-day SRT reactors. This may have included some loss by adsorption. Only 52% DNOC loss by degradation was registered by the 2-day SRT system.

Table III. Steady-State DNOC Data

SRT (day)	DNOC Concentration			Cumulative DNOC in Exhaust Vapour (µg/d)	Per Cent Removal of DNOC
	Feed (µg/L)	Effluent (µg/L)	Sludge (µg/g)*		
2	269	128	19	4	52
5	348	23	10	2	93
	671	19	7	2	97
	1124	25	25	2	98
	1176	51	14	2	96
10	387	21	23	2	95
	1255	24	12	5	98
15	348	23	7	2	93
	440	23	8	2	95
	1252	21	12	2	98
	1321	24	18	4	98

– Except for exhaust vapour data, which represent cumulative DNOC adsorbed between sample times, all values are actual measured values just before dose levels were increased.

* All values are less than the indicated values based on a dry weight basis.

The mean values of all equilibrium DNOC clarifier effluent and sludge data were pooled by SRT in Table IV. For low SRT systems, 3 or 4 data points were available whereas for high SRT systems only 1 or 2 data points were available. The data show that there was no significant difference in mean clarifier effluent DNOC values for the 5, 10 and 15-day SRT reactors for all influent conditions tested. The 2-day SRT clarifier effluent value of 82 µg/L DNOC was approximately four times the level in the other systems. The corresponding high standard deviation of approximately 37 µg/L indicated that equilibrium conditions were never achieved in this reactor. Sludge DNOC values were relatively consistent for all systems. The sludge DNOC values for the 5-day SRT reactor showed the greatest variability. This was attributed to the period of operation following an increase in the influent DNOC dose level from 900 to 1200 µg/L. Although the clarifier effluent DNOC values remained constant, the sludge DNOC values increased to approximately 50 µg/g for the following three weeks. Subsequently, this returned to a level consistent with the other reactors.

Table IV. Comparison of Overall Equilibrium DNOC Data

SRT (day)	DNOC Concentration					
	Feed (µg/L)		Effluent (µg/L)		Sludge (µg/g)	
	Range		$\bar{x}$	s	$\bar{x}$	s
2	265		82.2	36.7	20.0	1.4
5	337 - 1105		26.4	8.7	29.9	17.8
10	294 - 1196		21.9	3.7	19.2	8.3
15	298 - 1388		22.2	2.3	12.2	4.6

- Data are based on all equilibrium data collected at each feed DNOC concentration.

This work appears to contradict the findings of Tabak et al. (3,4). This difference may be attributed to the fact that Tabak et al. worked with batch systems with high initial DNOC concentrations of 5 and 10 mg/L. More recent work on DNOC degradation was reported by Garcia-Orozco et al. (8). In this case, continuous flow systems with powdered activated carbon were used albeit with a synthetic substrate. The feed DNOC concentration was varied from 0 to 27 mg/L over a range of SRTs from 4 to 12 days. Very limited degradation of DNOC was observed in the control reactors (without powdered activated carbon) which may be attributed to the high feed DNOC values or to the rapidity with which the feed DNOC concentration was changed.

Adequate data were collected to validate this experimental procedure. It will be used in further work to evaluate the fate of pentachlorophenol and other compounds in municipal activated sludge systems.

CONCLUSIONS

Municipal activated sludge systems operated at a 6-hr HRT and SRTs of 5 to 15 days can degrade 4,6-dinitro-o-cresol from levels in the range 0.3 to 1.2 mg/L to detection limits. At SRTs of less than 5 days, DNOC removal is incomplete.

Insignificant losses of DNOC by volatilization were observed.

The major mechanism of DNOC removal appeared to be by biodegradation accounting for 93 to 98% removal of the DNOC fed to the systems.

ACKNOWLEDGEMENTS

This work was sponsored by Environment Canada and the Great Lakes Water Quality Toxics Subcommittee. The support of the WTC Organics Laboratory is gratefully acknowledged.

REFERENCES

1. Melnikov, N.N., "Chemistry of Pesticides," Residue Review, Vol. 38, 1971, pp. 97-98
2. U.S. Environmental Protection Agency, Treatability Manual, Vol. 1, Treatability Data, Washington, D.C., 1980, p. 1.8.13-1.
3. Tabak, H.N., Chambers, C.W., and Kabler, P.W., "Microbial Metabolism of Aromatic Compounds," Journal of Bacteriology, Vol. 87, No. 4, 1964, pp. 910-919.
4. Tabak, H.N. et al., "Biodegradability Studies With Organic Priority Pollutant Compounds," Journal WPCF, Vol. 53, No. 10, 1981, pp. 1503-1518
5. Ware, G.W., The Pesticide Book, W.H. Freeman and Co., San Francisco, 1978, pp. 42, 84.
6. American Public Health Association, Standard Methods for the Examination of Water and Wastewater, 15th Edition, Washington, DC, 1980.
7. Fuentes, H.R., "The Fate of 2,4-dichlorophenol and 4,6-dinitro-o-cresol in Continuous Flow Complete Mix Aerobic Activated Sludge Systems," Ph.D. Dissertation, Vanderbilt University, 1982.
8. Garcia-Orozco, J.H., Fuentes, H.R., and Eckenfelder, W.W., "Modelling and Performance of the Activated Sludge - Powdered Activated Carbon Process in the Presence of 4,6-dinitro-o-cresol," Journal WPCF, Vol. 58, No. 4, 1986, pp. 320-325.

PILOT-SCALE ANAEROBIC BIOMASS ACCLIMATION STUDIES WITH A COAL LIQUEFACTION WASTEWATER

David N. Young
Eric B. Vale

*Dearborn Environmental Consulting
Services
Dearborn Chemical Company Limited
Mississauga, Ontario, Canada*

Eric R. Hall

*Environment Canada
Wastewater Technology Centre
Burlington
Ontario, Canada*

INTRODUCTION

Anaerobic biological treatment is currently being investigated as an alternative to aerobic treatment and solvent extraction methods for bulk organics removal from high-strength petroleum refining, coal coking, coal conversion and petrochemical process wastewaters. Phenolic compounds generally comprise the largest fraction of organic pollutants in these wastewaters and have consequently been the focus of many single-substrate anaerobic biological toxicity and treatability studies employing batch culture enrichment as well as continuous-flow pilot scale reactor techniques

(1-9). Several synthetic and actual phenolic wastewater treatability studies have also been conducted using the above-mentioned techniques (10-15).

These studies have demonstrated that many of the phenolic and other toxic organic compounds commonly found in hydrocarbon processing effluents are fermentable at relatively high mass loading rates provided the biomass is acclimated to their presence over a period of months to years. Thus, the most serious impediment to use of anaerobic biological treatment as a first-stage bulk organics removal and detoxification technique will likely be the requirement of a long acclimation phase. Biomass acclimation to high organic loading rates will be necessary to reduce the size and cost of industrial scale reactors.

The purpose of the study described in this paper was to assess the influence of pretreatment, reactor design and operational strategies on the rate of anaerobic biomass acclimation to a toxic phenolic wastewater. Pilot scale upflow anhybrid and granular activated carbon (GAC) fluidized bed reactors were acclimated to dilutions of steam-stripped H-Coal liquefaction process condensate using the four following strategies:

1) addition of a readily-biodegradable organic co-substrate
2) wastewater toxicity reduction by low severity MIBK solvent extraction pretreatment
3) increase reactor organic loading rate by decreasing wastewater dilution rate at a constant hydraulic residence time (HRT)
4) increase organic loading rate by decreasing HRT at constant wastewater dilution rate

Acclimation strategies 2, 3 and 4 were evaluated in the anhybrid and GAC fluidized bed reactors while the co-substrate strategy was tested in the anhybrid reactor only. A total of 3 anhybrid and 2 fluidized bed reactors were used in the study. Each reactor was started by seeding with biomass obtained from a large pilot scale sludge blanket reactor treating starch plant wastewater. The study was conducted over a $2\frac{1}{2}$ year period at Environment Canada's Wastewater Technology Centre, Burlington, Ontario, Canada.

MATERIALS AND METHODS

The upflow anhybrid reactors used in the study consisted of PVC columns of 12.7 cm. internal diameter and 210 cm. height. A granular sludge blanket zone was maintained in the lower 102 cm. of the reactor topped by a fixed film filter zone of 86 cm. height containing 2.45 cm. dumped Norton bioring plastic media supported on a perforated plate. The initial void volume was 23.5 L in each anhybrid reactor. Effluent was recirculated to a perforated plate inlet diffuser in the bottom of the reactor at 46 L/day to promote complete mixing once every 12 hrs. Heat tape and insulation wrapping around the main column and an automatic sensor/controller maintained a constant temperature of 35 deg. C in the reactor. The GAC fluidized bed reactors consisted of a clear PVC main column 142 cm. in height and 7.6 cm. internal diameter (with a tapered inlet), topped by a disengaging section 25.4 cm. in height and 10 cm. internal diameter. The effluent recycle system included a clear PVC carbon trap 41 cm. in height and 10 cm. internal diameter connected to a heat traced stainless steel recycle line which used a sensor/controller to maintain a constant temperature of 35 deg. C in the reactor. The fluidized bed reactors contained 3.3 L (dry settled volume) of 16 x 40 mesh Filtrasorb 400 granular activated carbon (GAC). Effluent was recycled at a sufficient rate to expand and fluidize the GAC media to a height in the column of roughly two-thirds the total height. The initial reactor void volume was 7.8 L. All of the reactors used in the study employed a siphon break gas collection system with gas evolution monitored daily by Triton-W.R.C. Model p.181 liquid displacement type gas meters. Neutral pH was maintained in each reactor by manual daily measurement and adjustment. Schematics of the pilot scale reactors are shown in Figure I.

Two large sequential samples of undiluted steam-stripped H-Coal process condensate were obtained from the Ashland Synthetic Fuels Catlettsburg, Kentucky pilot plant and stored in a stirred tank. For solvent extraction pretreatment studies, a large aliquot was processed in a pilot scale rotating disk extraction unit with MIBK (methyl isobutyl ketone) solvent at a solvent-to-feed ratio of 0.05:1 v/v. The extracted sample was then air stripped for 24 hrs. to remove residual MIBK and stored in a stirred tank. Start-up of the anhybrid reactors and the co-substrate studies were

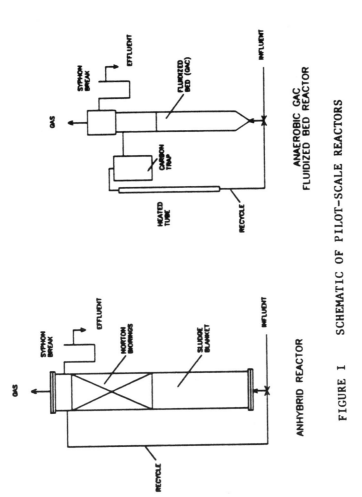

FIGURE I SCHEMATIC OF PILOT-SCALE REACTORS

conducted with starch processing plant effluent (supplemented with ammonium hydroxide) obtained approximately once per month and stored in a stirred tank. The fluidized bed reactors were started on diluted H-Coal wastewater. All H-Coal wastewater feeds were supplemented with synthetic growth medium containing trace salts and vitamins based on the formula used in the initial serum bottle studies as shown in Table I. Biomass seeding of the anhybrid reactors was conducted over a 3 week period. The fluidized bed reactors were seeded over a 3 month period.

Table I   Growth Medium Characteristics

Compound	Concentration in Growth Medium (mg/L)
$NaCl$	500
$CaCl_2 \cdot 2H_2O$	100
NH_4Cl	500
$MgCl_2 \cdot 6H_2O$	100
KH_2PO_4	33
K_2SO_4	43
$(NH_4)_6Mo_7O_{24} \cdot 4H_2O$	10
$ZnSO_4 \cdot 7H_2O$	0.1
H_3BO_3	0.3
$FeCl_2 \cdot 4H_2O$	1.5
$CoCl_2 \cdot 6H_2O$	10
$MnCl_2 \cdot 4H_2O$	0.03
$NiCl_2 \cdot 6H_2O$	0.03
$AlK(SO_4)_2 \cdot 12H_2O$	0.1
Nicotinic Acid	0.1
Cyanocobalamine	0.1
Thiamine	0.05
p-Aminobenzoic acid	0.05
Pyridoxine	0.25
Pantothemic acid	0.025

All analyses except individual phenols, gas composition and volatile organic acids (VOA's) were performed according to Standard Methods (16). The 4AAP colorimetric method was used for determination of total phenolics per Standard Methods (16). Individual phenols in the H-Coal condensate were analyzed by GC during initial studies conducted by Fedorak and Hrudey (15) at the University of Alberta, Edmonton, Alberta, Canada. Feed and recycle rates, temperature, GAC bed expansion and fines carryover, gas production and influent/effluent pH (Canlab Digital pH Meter Model 607) were manually checked daily. Influent and effluent samples were obtained from each reactor twice per week and analyzed for soluble COD, total phenolics (4AAP method) and VOA's. These samples were acidified with sulphuric acid to pH 2 then centrifuged and filtered through a glass fiber filter. VOA's were analyzed by a method outlined by Boone (17) in a Carle Instruments Model 211 Gas Chromatograph. Total volatile organic acids (TVA) were then reported as the sum of acetic, proprionic, butyric and isobutyric acids. Reactor product gas was sampled at random intervals into a 1 L Tedlar bag and analyzed by a GC method described by Fedorak and Hrudey (18). One set of reactor influent and effluent samples were analyzed for individual phenolics by a reverse phase direct injection HPLC method using a BAS electrochemical detector and an elutant consisting of acetonitrile and water mixture (22:78) modified with 1 percent acetic acid. All analyses except individual phenolics by GC were performed by the Analytical Services Section of the Wastewater Technology Centre.

EXPERIMENTAL DESIGN

Anhybrid reactors 1, 2, and 3 were started on starch plant effluent to establish an anaerobic granular sludge blanket zone in the lower portion and a fixed-film filter zone in the upper portion of the reactors. Anhybrid reactor 1 was utilized to examine the co-substrate acclimation strategy after operating day 72. Initially solvent extraction pretreated H-Coal wastewater was fed with the starch waste until day 326 when it was converted to non-pretreated H-Coal waste at 5% v/v strength plus starch waste. Performance of the co-substrate strategy is reported over days 362 to 430.

Anhybrid reactor 2 was utilized over days 73 to 325 to treat solvent pretreated H-Coal waste plus growth medium. Reactor 2 was then converted to non-pretreated H-Coal feed plus growth medium over the remainder of the study (days 326 to 705) to examine acclimation to increasing organic loading rates by reducing reactor hydraulic retention time (HRT).

Anhybrid reactor 3 was converted to non-pretreated H-coal waste plus growth medium after start-up for the remainder of the study. Acclimation performance data from days 73 to 340 is used for comparison with reactor 2 performance when treating solvent extracted H-Coal waste. Over the whole study period, reactor 3 was acclimated to increasing organic loading rate by increasing the H-Coal waste feed strength. The performance data can be compared with the latter operating period data of reactor 2.

GAC fluidized bed reactor 4 was started on 5% v/v strength H-Coal waste plus growth medium and was subsequently operated over days 154 to 260 to generate data for comparison with start-up and acclimation on solvent extraction pretreated wastewater. Over the remainder of the study (days 261 to 654), reactor 4 was operated to assess acclimation to increasing organic loading rate by HRT reduction. GAC fluidized bed reactor 5 was started and operated initially on solvent extraction pretreated waste plus growth medium until day 260. Reactor 5 was then converted to non-pretreated H-Coal waste plus growth medium and acclimated to increasing organic loading rate by increasing wastewater feed strength.

RESULTS AND DISCUSSION

Wastewater Characteristics

An analysis of the undiluted steam-stripped H-Coal process condensate used in this study is shown in Table II. The breakdown of individual phenolics is shown in Table III. The analyses indicate that phenolic compounds account for the majority of the wastewater organic carbon with pure phenol and cresol isomers constituting 86% of the identified phenolics. The initial batch serum bottle culture enrichment studies showed that unacclimated biomass methanogenesis was inhibited at greater than 6% v/v H-Coal wastewater strength and that inhibition was caused by a non-phenolic organic component(s) (14).

Table II. Steam-Stripped H-Coal Process Condensate Characteristics

Parameter	Concentration (mg/L)
COD (soluble)	21,100
Total Phenolics (colorimetric)	7,600
Organic Carbon (sol.)	7,870
TKN	267
Nitrite-Nitrogen	0.2
Nitrate-Nitrogen	0.8
Ammonia-Nitrogen	210
Total Phosphorus	5.0
Total Cynanide	0.21
pH	6.8
Aluminum	0.18
Calcium	4.41
Cadmium	0.003
Chromium	0.10
Copper	0.03
Iron	18
Manganese	0.10
Nickel	0.03
Lead	0.03
Zinc	0.15

Table III. Analysis of Phenolic Compounds in H-Coal Process Condensate *

Compound	Concentration (mg/L)
phenol	4,900
o-cresol	586
m/p-cresol	1,650
m-cresol **	(1,230)
p-cresol **	(420)
2,4/2,5-dimethyphenol	63
3,5-dimethylphenol	213
3,4-dimethylphenol	44
Total Phenolics (sum)	7,456
Total Phenolics (colorimetric)	7,600

* analysis reported by Fedorak and Hrudey (15)
** based on m/p ratio of 2.9/1

Co-Substrate Addition

A readily-biodegradable starch plant wastewater co-substrate was added to anhybrid reactor 1 for comparison to anhybrid reactors 2 and 3 to assess the effect of greater biomass activity on acclimation to H-Coal waste. H-Coal waste at 5% v/v strength was initially fed with co-substrate to reactor 1 on day 72 and acceptable phenolics removal (>75%) was established after 100 days of operation. Similarily, reactor 3 required 110 days to acclimate to 5% v/v strength H-Coal waste plus growth medium after start-up on starch plant wastewater. Thus, co-substrate addition did not affect the initial acclimation of biomass to the presence of phenolics in the feed.

Between days 362 to 430, reactors 1 and 2 were subjected to a three-times organic loading rate increase by HRT reduction only with each reactor receiving sub-inhibitory levels of H-Coal waste. The performance of reactor 1, treating 5% v/v H-Coal waste plus co-substrate, and reactor 2, treating 5% v/v H-Coal waste plus growth medium, is compared over this period in Table IV.

The operating data show that reactor 2, which did not receive co-substrate, was capable of slightly greater phenolics removal during similar phenolics loading rate increases in both reactors despite poorer COD removal performance. In addition, relative gas production doubled in both reactors over this period. Thus, there was no apparent enhancement to phenolics fermentation caused by co-substrate addition during acclimation to organic loading increases by HRT reduction.

The phenolics and other organic compounds in the H-Coal wastewater were present a sub-inhibitory levels throughout the above studies. The overall results indicate that specific "phenol-degrading" anaerobic biomass is established in an anaerobic reactor treating even low levels of phenolics in a diverse mixture of organic compounds. Recent studies, as summarized by Young and Rivera (7), have established that very specific multi-step metabolic pathways are employed by anaerobic microorganisms to degrade phenolics to volatile acids which strongly indicates a need for specialized or acclimated biomass for this purpose.

Low Severity Solvent Extraction Pretreatment

Low severity solvent extraction pretreatment has been shown to render H-Coal wastewater biodegradable by aerobic cultures with minimal dilution while significantly reducing the volume of by-product phenols generated (19). This pretreatment method, using MIBK at a solvent-to-feed ratio of 0.05:1, was applied to H-Coal wastewater fed (plus growth medium) to anhybrid reactor 2 and GAC fluidized bed reactor 5. Table V shows the phenolics removals during the pretreatment, however, the primary intention in applying the pretreatment was to reduce the level of an unidentified non-phenolics organic component(s) which caused inhibition to batch anaerobic cultures at > 6% v/v wastewater strength (14). The pretreatment was shown to preferentially remove phenol, cresols and resorcinol producing a higher proportion of dimethyl phenols and catechol in the pretreated feed.

Table IV. Comparative Acclimation Performance of Anhybrid Reactors Treating H-Coal Wastewater with and without Co-Substrate (Operating Days 362 to 430)

Duration (d)	HRT (d)	Organic Loading Rate (kg COD/m³/d)		Phenolic Loading Rate (kg/m³/d)		COD Removal (%)		Total Phenolics Removal (%)		Gas Production (L/L/d)	
		R1*	R2*	R1	R2	R1	R2	R1	R2	R1	R2
13	3.4	0.77	0.29	0.09	0.08	77	59	75	79	0.34	0.11
10	2.4	1.8	0.37	0.12	0.11	85	50	73	79	0.68	0.13
22	1.8	1.6	0.58	0.14	0.16	82	60	76	71	0.52	0.19
17	1.6	1.3	0.54	0.16	0.18	79	67	78	86	0.53	0.20
6	1.3	2.3	0.71	0.19	0.20	77	54	77	85	0.70	0.19

* R1: Anhybrid Reactor 1 treating 5% v/v H-Coal & Starch Wastewater
  R2: Anhybrid Reactor 2 treating 5% v/v H-Coal & Growth Medium Wastewater

Table V. Removal of Phenolic Compounds by Low Severity MIBK Solvent Extraction Treatment of H-Coal Condensate (Solvent-to-feed ratio 0.05:1)

Compound	Removal (%)
phenol	59
m & p-cresol	77
o-cresol	83
2,4 dimethyl phenol	nil
3,4 dimethyl phenol	nil
3,5 dimethyl phenol	nil
Resorcinol	77
Catechol	nil
Total Phenols (4AAP)	60

Anhybrid reactor 2 was fed pretreated H-Coal waste (plus growth medium) at 5% v/v strength for 63 days, 7% strength for the next 74 days and 9% strength for the next 113 days at constant HRT following start-up on starch plant wastewater. Anhybrid reactor 3 was acclimated to unpretreated H-Coal waste plus growth medium in an identical fashion. Operating and performance data for the reactors, taken from periods where the organic loading rate was similar, is shown in Table VI.

Table VI. Comparative Performance of Anhybrid Reactors with and without Solvent Extraction Pretreatment of the Feed

	REACTOR 3 (Unextracted)	REACTOR 2 (Extracted)	
Duration (d)	74	128	74
Wastewater Strength (% v/v)	7	9	7
HRT (d)	5	5	5
Influent COD (mg/L)	1360	1200	840
Influent Phenolics (mg/L)	440	220	160
Organic Loading Rate (kg COD/m^3/d)	0.25	0.23	0.16
Organic Removal Rate (kg COD/m^3/d)	0.17	0.06	0.06
COD Removal (%)	69	25	37
Phenolics Removal (%)	75	30	40
Gas Production (L/d)	1.78	0.84	1.22

* - mean values over reporting period

These data show that reactor 2 was inhibited in terms of organics removal at one-half the feed phenolics concentration of reactor 3. Since the methane content of the gas produced from reactor 2 was not measured, it is not clear whether methanogenesis was also inhibited during treatment of the pretreated waste. Reactor 2 was also inhibited, although less severely, during feed of 7% v/v strength pretreated waste. Addition of pretreated feed to both the anhybrid and GAC fluidized bed reactor was discontinued at this point. Since the activated carbon had not yet been saturated in reactor 5, there is no comparative data available for the fluidized bed reactors treating extracted H-Coal waste.

The results of this evaluation indicate that low severity solvent extraction pretreatment with MIBK does not remove the inhibitory organic component in the H-Coal waste nor was the biomass able to acclimate to this component over a three month period. In addition, while the overall strength of the waste was lowered by pretreatment, its toxicity appears to have been enhanced indirectly by the preferential removal of the more biodegradable phenolics which likely inhibited the establishment of active phenolics-degrading biomass in the reactor.

Anhybrid Reactor Acclimation to Increasing Organic Loading Rate

Reactor 2 was operated in two separate phases after the pretreatment evaluation was discontinued: an HRT reduction phase followed by a feed strength increase phase. Commencing on day 326, reactor 2 was converted to 5% v/v H-Coal waste plus growth medium feed and acclimated over the next 215 days (to day 542) to step-wise organic loading rate increases from 0.21 to 2.0 kg $COD/m^3/d$ by HRT reduction from 4.7 days to 0.47 days. The feed strength was then increased step-wise from 5 to 15% v/v H-Coal waste at constant HRT of 0.78 days over a 65 day period which raised the organic loading rate from 1.4 to 3.6 kg $COD/m^3/d$. Phenolics removal performance over the entire operation of reactor 2 is reported as phenolics loading and removal rates in Figure II. The organic loading and removal rates as well as volumetric gas production for reactor 2 are shown in Figure III.

Initial acclimation to the phenolics at sub-inhibitory H-Coal waste strength was rapid in reactor 2 due to its previous contact with pretreated waste. Approximately 80% phenolics removal was maintained over the HRT reduction phase. Influent phenolics and COD concentrations during this phase averaged 330 to 1080 mg/L respectively. Effluent total volatile acids (TVA) were consistently below 50 mg/L while the influent TVA were generally nil. Gas production, as shown in Figure III, increased proportionately with increasing organic loading rate during the HRT reduction phase. Gas methane content was consistent and averaged 72%

556 Biotechnology for Degradation of Toxic Chemicals

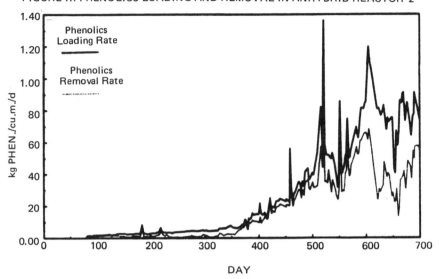

FIGURE II: PHENOLICS LOADING AND REMOVAL IN ANHYBRID REACTOR 2

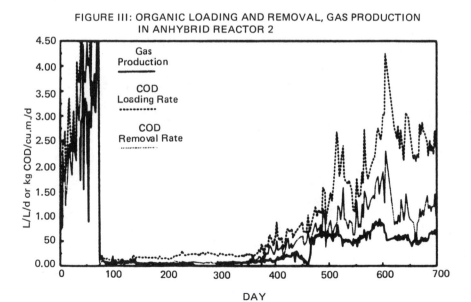

FIGURE III: ORGANIC LOADING AND REMOVAL, GAS PRODUCTION IN ANHYBRID REACTOR 2

over the period. Cessation of gas production on day 458 was caused by an influent phenolics spike (Figure II) with full recovery requiring 7 days. In terms of COD conversion, gas production ranged between 400 to 1000 L/kg COD removed which is primarily a reflection of fluctuations in loading rates during the study.

During the feed strength increase phase, reactor 2 acclimated to a maximum H-Coal waste strength of 13% v/v, however, attempts to acclimate the reactor to 15% strength caused phenolics-degrading biomass inhibition although, significantly, gas production and methane content remained constant. Similar inhibition was caused by a feed phenolics spike on day 524 with recovery in 14 days and no effect on gas production. Towards the end of the study, reactor 2 appeared to be recovering phenolics removal capability at 15% strength waste and an organic loading rate of 2.4 kg COD/m/d. During this period, influent COD and phenolics levels were the same as previous at 13% feed strength. The operating conditions and performance of reactor 2 at the maximum feed strength and organic loading rate achieved in the study is shown in Table VII.

Commencing on day 73, reactor 3 was acclimated to step-wise increases in H-Coal feed strength from 5 to 22% v/v (plus growth medium) over a total period of 578 days at a constant HRT of 4.7 days which raised the organic loading rate from 0.20 to 0.78 kg COD/m^3/d. Phenolics removal performance over the entire operation of reactor 3 is reported as phenolics loading and removal rates in Figure IV. Reactor 3 organic loading and removal rates as well as gas production are shown in Figure V.

The initial acclimation to the influent H-Coal waste phenolics at sub-inhibitory (<6% v/v) levels required a 63 day period in reactor 3 following start-up on starch plant wastewater. Acclimation to 13% strength waste required an additional 332 day acclimation period (to day 468) at which time consistent gas production was re-established as shown in Figure V. A period of phenolics removal inhibition followed almost every step increase in feed strength during this phase of operation as indicated in Figure IV. Reactor 3 was then further acclimated up to a maximum 20% feed strength over a 116 day period (to day 605). Severe inhibition to both phenolics-degrading and methanogenic biomass occurred after the feed strength was increased to 22%. The feed strength was reduced to 15% and the organic loading rate set a 2.4 kg COD/m/d as per the reactor 2 recovery operating conditions.

558 Biotechnology for Degradation of Toxic Chemicals

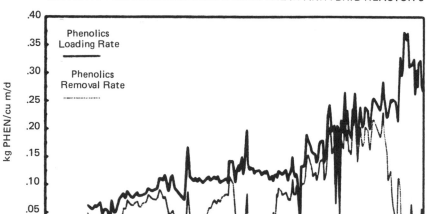

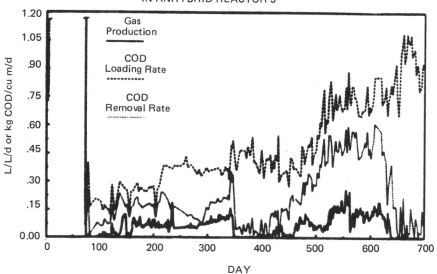

However, after 53 days of operation under these conditions, neither phenolics removal nor methanogenesis recovery appeared to be imminent. The operating conditions and performance of reactor 3 at its maximum feed strength and organic loading rate achieved in the study is compared with the reactor 2 results under maximum conditions in Table VII.

The comparative results of anhybrid reactor acclimation by the two different operating methods clearly indicates the superiority of the HRT reduction strategy applied to reactor 2 to achieve maximum organic loading rates over a minimum period of time. Reactor 2 also exhibited more stable operation and greater resistance to upsets since the influent toxic component(s) concentration was lower, allowing more rapid build-up of acclimated biomass, and toxic spikes were more rapidly washed out of the reactor due to the lower HRT. An important observation was that methane production activity was unaffected during overloading upsets in reactor 2. Reactor 2 also generated roughly double the gas production per kg COD removed compared to reactor 3, however, reactor 3 gas nonetheless contained consistently high methane levels. These results suggest that the phenolics-degrading anaerobic biomass is more sensitive to the H-Coal wastewater constituents than the methanogens. From an operational point of view, therefore, monitoring of gas production or gas methane content may not give a clear indication of the overall activity of the biomass when treating H-Coal or similar phenolic wastewaters in anhybrid reactors. In their recent work with H-Coal wastewater, Fedorak and Hrudey (15) showed that monitoring the emergence of m-cresol provides early indication of reduced phenolics degrading capability by an anaerobic biomass.

The comparative results in Table VII also show that the actual volume of H-Coal waste treated in reactor 2 was 3.9 L/d versus 1.0 L/d in reactor 3, which offsets the higher dilution volume requirement in reactor 2. Reduction of HRT at a reduced feed strength is, therefore, the preferred acclimation strategy to high organic loading rates in anhybrid reactors treating H-Coal or similar complex phenolic wastewater due to the significantly reduced reactor sizing requirement, higher volumetric gas production and greater overall process stability.

Table VII. Comparative Maximum Performance of Anhybrid Reactors Following Acclimation to H-Coal Process Condensate*

	Reactor 2**	Reactor 3***
Duration (days)	13	84
Acclimation Period (days)	323 ****	449
Feed Strength (% v/v)	13	20
HRT (days)	0.78	4.7
H-Coal Waste Feedrate (L/d)	3.9	1.0
Influent COD (mg/L)	2395	3840
Influent Phenolics (mg/L)	720	1100
Organic Loading Rate (kg COD/m^3/d)	2.7	0.72
Phenolics Loading Rate (kg/m^3/d)	0.81	0.21
Organic Removal Rate (kg COD/m^3/d)	1.7	0.49
Phenolics Removal Rate (kg/m^3/d)	0.64	0.17
COD Removal (%)	64	68
Phenolics Removal (%)	78	82
Gas Production (L/L/d)	0.86	0.14
Specific Gas Production (L/kg COD remv'd)	490	280
Gas Methane Content (%)	74	68
Effluent Volatile Acids (mg/L)	13	35

*   - mean values over duration
**   - acclimated by HRT reduction followed by feed strength increase
*** - acclimated by feed strength increase only
**** - includes an additional 63 days to account for initial acclimation to H-Coal waste phenolics

GAC Fluidized Bed Reactor Acclimation to Increasing Organic Loading Rates

Reactor 4 was acclimated in a similar fashion to anhybrid reactor 2 in two phases: HRT reduction followed by feed strength increase. Greater than 99.5% and 95% of the influent H-Coal waste phenolics and COD were removed respectively by adsorption on the activated carbon media over the first 280 days of operation. GAC media saturation was indicated by a drop in phenolics and COD removal and sustained gas production. Phenolics removal performance over the entire operation of reactor 4 is reported as phenolics loading and removal rates as shown in Figure VI. The organic loading and removal rates as well as gas production for reactor 4 are shown in Figure VII.

Between days 281 and 406, the organic loading rate was increased step-wise from 5.4 to 13.9 kg COD/m^3/day primarily by HRT reduction at a feed strength of 15% v/v H-Coal waste plus growth medium. On day 407, the feed strength was increased to 25% which caused immediate gas production inhibition and eventual phenolics removal inhibition. After a 57 day period of inhibition, the feed strength was reduced to 15% and the HRT increased to 1.6 days, whereupon recovery occurred in 10 days. Reactor 4 was then reacclimated to an organic loading rate of 13.9 kg COD/m^3/d at 15% feed strength and 0.39 days HRT over the next 124 days (to day 589). Over the reacclimation period, phenolics removal averaged 86 percent, COD removal averaged 72 percent and gas production increased from 1.18 to 5.13 L/L/d. Specific gas production averaged 550 L/kg COD removed while the gas methane content was consistent and averaged 70%.

On day 582, influent and effluent samples from reactor 4 were analyzed for individual phenolics with the results shown in Table VIII. The samples were obtained 14 days after a modest increase in organic loading rate by HRT reduction and no increase in feed strength, therefore, it is assumed that the GAC media was at equilibrium with the liquid phase concentration of adsorbable compounds.

The results show that the dimethyl phenol isomers and catechol were not degraded during a stable and active period of reactor operation. The m and p-cresols were partially degraded. Fedorak and Hrudey (15) have observed preferential degradation of p-cresol over m-cresol during H-Coal waste treatment in semi-continuous anaerobic culture experiments, however, the presence of activated carbon media may have

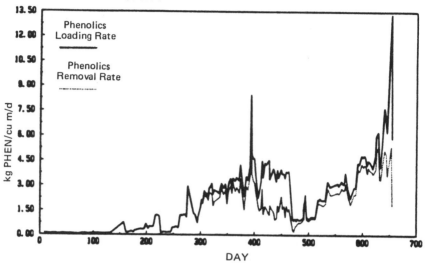

FIGURE VI: PHENOLICS LOADING AND REMOVAL IN GAC FLUIDIZED BED REACTOR 4

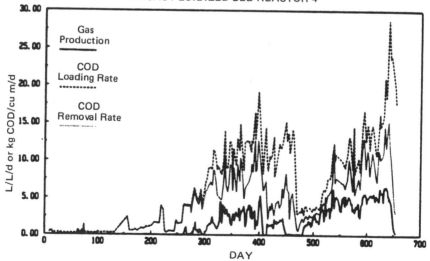

FIGURE VII: ORGANIC LOADING AND REMOVAL, GAS PRODUCTION IN GAC FLUIDIZED BED REACTOR 4

altered the pattern of degradation of these compounds. The undegraded phenolics in Table VIII appeared to account for all of the effluent phenolics and greater than 50 percent of the effluent COD from reactor 4.

Table VIII. Removal of Individual Phenolics in Reactor 4 on Day 582 *

	Feed (mg/L)	Effluent (mg/L)
Phenol	398	15
m & p cresol	152	69
o-cresol	70	8
Catechol	10	15
Resorcinol	18	2
2,4 dimethylphenol	18	14
3,4 dimethylphenol	3.4	2.5
3,5 dimethylphenol	6.1	5.4
Total Phenolics (sum)	676	131
Total Phenolics (4AAP)	717	93
COD	3010	625

* operation at 15% v/v H-Coal feed strength plus growth medium, organic loading rate ~ 10.6 kg $COD/m^3/d$.

Over the next 50 days of operation (to day 639), the feed strength was increased step-wise at constant HRT (0.39 days) from 15% to 25% H-Coal waste increasing the organic loading rate from 13.9 to 21 kg $COD/m^3/d$. A further increase to 30% feed strength caused gas production and phenolics removal inhibition to the end of the study. The operating conditions and performance of GAC fluidized bed reactor 4 at the maximum feed strength and organic loading rate achieved in the study is shown in Table IX.

GAC fluidized bed reactor 5 was acclimated to increased organic loadings by feed strength increases at constant HRT in a similar fashion to anhybrid reactor 3. Reactor 5 was fed solvent extraction pretreated H-Coal waste over the initial 260 day start-up period. From day 261 on, reactor 5 was fed unpretreated H-Coal waste plus growth medium with activated carbon media saturation occurring around day 307 as evidenced by consistant gas production and reduced COD and phenolics removal. Phenolics removal performance over the entire operation of GAC fluidized bed reactor 5 is shown in Figure VIII. The organic loading and removal rates as well as gas production for reactor 5 are shown in Figure IX.

Between days 261 and 325, the feed strength was increased from 15% to 35% v/v H-Coal waste at a constant HRT of 3 days which caused gas production inhibition only. The slight increase in effluent phenolics may have been caused by the lowering of the feed strength which led to phenolics desorption from the GAC media. Gas production recovered in 14 days and the feed strength was then increased step-wise from 15% to 45% (constant HRT of 5 days) between days 374 and 440. Severe inhibition occurred to both gas production and phenolics removal during this period. Recovery occurred after 37 days of operation at 5% feed strength and 2 days HRT. Between days 477 and 654 (end of study), the feed strength was again increased from 15% to 40% H-Coal waste at 5 days HRT although in smaller incremental steps. The feed strength increases raised the organic loading rate from 1.5 to 9.0 kg $COD/m^3/d$. Over the last 5 days of the study, the feed strength was lowered by mistake to 15% strength after a phenolics spike was introduced into the reactor which caused an apparent upset due in part to desorption of organics from the GAC media.

During the final acclimation period phenolics and COD removals averaged 91% and 82% respectively which strongly suggests partial removal by activated carbon absorption as a new equilibrium with the increased liquid phase organics concentration was being established. The average time between feed strength increases was 14 days during this period. Gas production increased from 0.54 to 2.45 L/L/day, while specific gas production averaged 400 L/kg COD removed. Gas methane content was consistent and averaged 68%. The operating conditions and performance of GAC fluidized bed reactor 5 at the maximum organic loading rate and feed strength achieved in the study is compared with reactor 4 results under its maximum conditions in Table IX. The

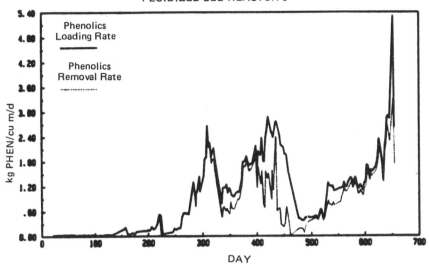

FIGURE VIII: PHENOLICS LOADING AND REMOVAL IN GAC FLUIDIZED BED REACTOR 5

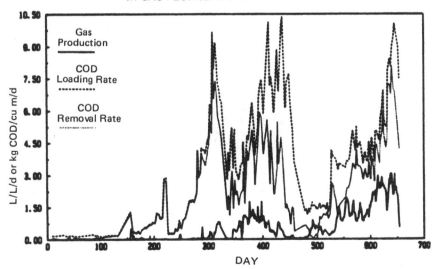

FIGURE IX: ORGANIC LOADING AND REMOVAL, GAS PRODUCTION IN GAC FLUIDIZED BED REACTOR 5

Table IX. Comparative Maximum Performance of GAC Fluidized Bed Reactors Following Acclimation to H-Coal Process Condensate *

	Reactor 4 **	Reactor 5 ***
Duration (days)	6	10
Acclimation Period (days)****	358	350
Feed Strength (% v/v)	25	40
HRT (days)	0.39	1.6
H-Coal Waste Feedrate (L/d)	5	2
Influent COD (mg/L)	4350	6160
Influent Phenolics (mg/L)	975	2010
Organic Loading Rate (kg COD/m^3/d)	21	8.7
Phenolics Loading Rate (kg/m^3/d)	4.8	3.1
Organic Removal Rate (kg COD/m^3/d)	13.4	6.0
Phenolics Removal Rate (kg/m^3/d)	3.8	2.5
COD Removal (%)	63	76
Phenolics Removal (%)	80	88
Gas Production (L/L/d)	6.39	2.45
Specific Gas Production (L/kg COD remv'd)	480	380
Gas Methane Content (%)	72	65
Effluent Volatile Acids (mg/L)	101	ND

* - mean values over duration
** - acclimated by HRT reduction followed by feed strength increase
*** - acclimated by feed strength increase only, reactor in non-steady state condition
**** - initial period to saturate activated carbon media not included

reactor 5 results should be viewed as non steady-state since it is likely that activated carbon absorption accounted for some of the COD and phenolics removal during the period following the increase in feed strength from 35% to 40%.

The comparative results show that reactor 4 was capable of acclimating to a significantly higher organic loading rate by the HRT reduction, reduced feed strength strategy compared to reactor 5. The reduced feed concentration and residence time of inhibitory H-Coal wastewater components promoted greater resistance to upset during acclimation and faster build-up of phenolics-degrading biomass. Very high rate operation was achieved in reactor 4, typical of anaerobic reactor operation with readily-biodegradable high organic strength wastewaters. While reactor 5 was capable of treating 40% v/v strength feed versus 25% in reactor 4, the actual H-Coal wastewater feed rate in reactor 4 was 5 L/d versus 2 L/d in reactor 5. Thus, acclimation to the higher organic loading rate significantly reduced reactor size requirements despite a higher dilution rate requirement. The greater stability and volumetric gas production rates achieved in reactor 4 are additional benefits of the HRT reduction acclimation strategy. The results for reactor 5, while not indicative of steady-state operation, show that anaerobic biomass acclimation to high influent phenolic compound concentrations can be achieved in a GAC fluidized bed reactor at relatively high organic loading rates.

The effect of carbon adsorption on the levels of individual phenolics in the reactor liquid phase was not determined. Adsorption data for phenolic compounds other than phenol, which is strongly adsorbed (20), is not available in the general literature, however, based on their similarity in molecular weight and solubility, it is unlikely that the relative levels of individual phenolics in the liquid phase were significantly altered by carbon absorption. The GAC media reduced the level of the unidentified inhibitory component in the H-Coal waste such that the reactors could be started on 15% strength waste as compared to 5% strength in the anhybrid reactors.

Roughly one-third of the GAC media was replaced in reactors 4 and 5 over the entire study. Media replacement was conducted only as a precautionary measure during a major upset rather than on a routine basis. It is very likely that bioregeneration of the media occurred during periods of feed strength reduction as was observed in GAC expanded-bed reactor studies by Kim et al (8). Build-up of biorefractory or compounds on the GAC media, although not believed to be responsible for inhibition observed during this study, will likely necessitate media replacement on a routine basis in GAC fluidized bed and similar reactors treating toxic wastewaters over a sustained period.

Comparative Performance of Anhybrid and GAC Fluidized Bed Reactors

The beneficial effect of the adsorptive capacity of the GAC media is clearly demonstrated in this study. Roughly order-of-magnitude greater organic loading rates were achieved in reactor 4 versus reactor 2 and reactor 5 versus reactor 3 where the acclimation strategies employed were similar. Adsorption of inhibitory feed components by the GAC media allowed fluidized bed reactor treatment at double the feed strength and feedrate used the anhybrid reactors. Concentration in reactor 4 was only 35% greater than in reactor 2. The presence of the GAC media, however, did not enhance the specific gas production rate or methane content of the gas. Operating data for reactors 4 and 5 during periods of operation at organic loading rates near 3 kg COD/m^3/d compared with reactor 2 data indicate that the volumetric gas production rate was also very similar in both the anhybrid and GAC fluidized bed reactors over the study. These results coupled with the anhybrid reactor acclimation data strongly indicate that the phenolics-degrading biomass is more sensitive to the H-Coal waste inhibitory components than the methanogens.

Based on the maximum organic loading rates attained and the respective reactor void volumes, the size requirement for GAC fluidized bed reactor is one-quarter that of an anhybrid reactor treating an equivalent volume of H-Coal wastewater. In addition, the dilution requirement for the GAC fluidized bed reactor was significantly lower during the start-up and acclimation phase. In terms of full-scale implementation, therefore, the availability of dilution water becomes a critical issue, particularly where it is economically feasible to recycle low strength process wastewaters in the plant with minimal or no pretreatment as an alternative use.

The long acclimation period of the GAC fluidized bed reactors may be reduced significantly since in this study, gas production was primarily observed for signs of reactor upset. Based on the previous arguments, this method indicated upset some time after the phenoics-degrading

biomass was inhibited. Thus, monitoring of a surrogate
phenolic compound such as m-cresol, as described earlier in
this paper, should allow better control of the acclimation
process to prevent major upsets, particularly where the waste
strength and flow is variable. In addition, the GAC media
saturation and reactor start-up period of roughly 300 days in
this study could be reduced to several weeks by forced media
saturation with pure phenol or a mixture of waste phenolics
and seeding with acclimated biomass. Wang et. al. (9)
describe rapid start-up of a phenol-treating GAC expanded-bed
reactor in 12 days by this method.

SUMMARY AND CONCLUSIONS

Neither readily-biodegradable organic co-substrate
addition nor low severity solvent extraction pretreatment
improved the acclimation of anaerobic biomass in anhybrid
reactors to H-Coal process wastewater. Acclimation to
increasing organic loading rates by HRT reduction at
sub-inhibitory feed strengths of the H-Coal wastewater
resulted in higher loading rates attained in both the
anhybrid and GAC fluidized bed reactors than the alternate
strategy of increasing feed strength at long HRT's. The
study results show the beneficial effect of GAC media on
anaerobic treatment of a toxic wastewater. Maximum organic
loading rates of 2.6 kg $COD/m^3/day$ at 0.78 days HRT and 13%
v/v feed strength and 21 kg $COD/m^3/d$ at 0.39 days and 25%
feed strength were attained in the anhybrid and GAC fluidized
bed reactor respectively over an acclimation period of
approximately one year. A maximum feed strength of 40% v/v
was treated in the GAC fluidized bed reactor during non
steady-state conditions at an organic loading rate of 8.7 kg
$COD/m^3/d$ and influent total phenolics concentration of 2000
mg/L. Gas was produced at similar rates in both reactor
types with a consistent methane content of 70%. These
results and observations of the behaviour during upset
conditions agree with independently generated semi-continuous
culture data that indicate the phenolics degrading biomass is
more sensitive than the methanogens to the H-Coal wastewater.

Careful acclimation of the biomass by monitoring
phenolics-degrading activity coupled with pre-saturation of
the activated carbon media by phenolic compounds prior to
start-up should allow a significant reduction in the
acclimation period of GAC fluidized bed reactors treating
high strength phenolic wastewaters. The overall results

indicate that anaerobic treatment of a toxic high-strength phenolic wastewater can be carried out at organic loading rates similar to those used with a readily biodegradable wastewater following a carefully executed acclimation phase. This treatment method should, therefore, be suitable for use as a first-stage bulk organics and detoxification process in a multi-stage treatment system.

ACKNOWLEDGEMENTS

Funds for this study were provided through the Interdepartmental Panel on Energy Research and Development, the Unsolicited Proposal Program of the Department of Supply and Services, and the Environmental Protection Service of Environment Canada. The contributions of Mrs. Rashne Baetz, U.S. EPA, and the Analytical Services Section of the Wastewater Technology Centre, Environment Canada, are gratefully acknowledged. This paper has not been subjected to the Environment Canada's required peer review and as such does not necessarily reflect their views.

REFERENCES

1. Chmielowski, J., Grossman, A. and Labuzek, S., "Biochemical Degradation of Some Phenols during Methane Fermentation," Zesz. nauk. Politech. Slask, Inz. Sanit., 8, 1965, pp. 97-122.
2. Healy, J.B. Jr., and Young, L.Y., "Anaerobic Biodegradation of Eleven Aromatic Compounds to Methane," Appl. Envir. Microbiol., 38, 1979, pp. 84-89.
3. Suidan, M.E., Cross, W.H., and Fong, M., "Continuous Bioregeneration of Granular Activated Carbon during the Anaerobic Degradation of Catechol," Prog. Wat. Technol., 12, 1980, pp. 203-214.
4. Khan, K.A., Suidan, M.T., and Cross, W.H., "Anaerobic Activated Carbon Filter for the Treatment of Phenol-bearing Wastewater," J. Wat. Pollut. Control Fed., Vol. 53, 1981, pp. 1519-1532.
5. Boyd, S.A., Shelton, D.R., Berry D., et. al., "Anaerobic Biodegradation of Phenolic Compounds in Digested Sludge," Appl. Envir. Microbiol., Vol. 46, 1983, pp. 50-54.
6. Fedorak, P.M. and Hrudey, S.E., "The Effects of Phenol and Some Alkyl Phenolics on Batch Anaerobic Methanogenesis," Water Res., Vol. 18, 1984, pp. 361-367.
7. Young, L.Y. and Rivera, M.D., "Methanogenic Degradation of Four Phenolic Compounds," Water Res., Vol. 19, No. 10, 1985, pp. 1325-1332.
8. Kim, B.Y., Chian, E.S.K., Cross, W.H. et. al., "Adsorption, Desorption and Bioregeneration in an Anaerobic Granular Activated Carbon Reactor for the Removal of Phenol," J. Wat. Pollut. Control Fed., Vol. 58, No. 1, 1986, pp. 35-40.
9. Wang, Y.T., Suidan, M.T., and Rittman, B.E., "Anaerobic Treatment of Phenol by an Expanded-Bed Reactor", J. Wat. Pollut. Control Fed., Vol. 58, No. 3, 1986, pp. 227-233.
10. Chou, W.J., Speece, R.E., and Siddiqi, R.H., "Acclimation and Degradation of Petrochemical Wastewater Components by Methane Fermentation," Biotechnol. Bioengng Symp., Vol. 8, 1978, pp.391-414.
11. Cross, W.H., Chian, E.S.K., Pohland, F.G., et. al., "Anaerobic Biological Treatment of Coal Gasifier Effluent," Biotechnol. Bioengng Symp., Vol. 12, 1982, pp. 49-363.

12. Suidan, M.T., Strubler, C.E., Kao, S.W., et. al., "Treatment of Coal Gasification Wastewater with Anaerobic Filter Technology," J. Wat. Pollut. Control Fed., Vol. 55, 1983, pp. 1263-1270.

13. Harper, S.R., Cross, W.H., Pohland, F.G., et. al., "Adsorption-enhanced Biogasification of Coal Conversion Wastewater," Biotechnol. Bioengng Symp., Vol. 12, 1983, pp. 401-420.

14. Fedorak, P.M. and Hrudey, S.E., "Batch Anaerobic methanogenesis of Phenolic Coal Conversion Wastewater," Wat. Sci. Technol., Vol. 17, 1985, pp. 143-154.

15. Fedorak, P.M. and Hrudey, S.E., "Anaerobic Treatment of Phenolic Coal Conversion Wastewater in Semicontinuous Cultures," Water Res., Vol. 20, No. 1, 1986, pp. 113-122.

16. "Standard Methods for the Examination of Water and Wastewater," 15th Ed., Am. Public Health Assoc., Washington, D.C., 1980.

17. Boone, D.R., "Terminal Reactions in the Anaerobic Digestion of Animal Waste," Appl. Envir. Microbiol., Vol. 43, 1982, pp. 57-64.

18. Fedorak, P.M. and Hrudey, S.E., "A Simple Apparatus for Measuring Gas Production by Methanogenic Cultures in Serum Bottles," Envir. Technol. Lett., Vol. 4, 1983, pp. 425-432.

19. Young, D.N., "Environmental Implications of Coal Gasification/Liquefaction Technologies in Canada," unpublished report prepared by Dearborn Environmental Consulting Services for the Industrial Programs Branch, Environmental Protection Service, Environment Canada, Ottawa, Ontario, DSS File No. KE 204-1-0375, April, 1983.

20. Perrich, J.R., "Activated Carbon Adsorption for Wastewater Treatment", CRC Press, Cleveland, Ohio, 1981.

ANOXIC/OXIC ACTIVATED SLUDGE TREATMENT OF CYANOGENS AND AMMONIA IN THE PRESENCE OF PHENOLS

Deanna J. Richards and Wen K. Shieh

Department of Civil Engineering
University of Pennsylvania
Philadelphia, Pennsylvania

INTRODUCTION

Petrochemical, steel manufacturing, mining and synthetic fuel processing are among several industries that generate wastewaters containing relatively high concentrations of phenols, cyanide, thiocyanate and ammonia. Individually these compounds have severe environmental consequences (1,2,3). Also well documented is the toxicity of cyanide and the adverse effects on health of both phenolic compounds and nitrates (4).

Despite the fact that wastewaters with this composition are inhibitory to biodegradation, biological oxidation through activated sludge processes is utilized widely to achieve required treatment (5,6,7,8,9,10). Biological treatment usually follows pretreatment methods such as equilization and storage, ammonia stripping, chlorination and air floatation which render the wastewater more amenable to biodegradation.

The literature provides evidence that there are a few microorganisms in nature which are capable of utilizing cyanide and thiocyanate as carbon and energy sources in their metabolic and respiration processes (11,12,13). Source of organic carbon other than the cyanide -ion is sometimes required for cell systhesis (14). The main end products from these microbial reactions are ammonia and nitrate which can be converted biochemically through nitrification and denitrification to innocuous nitrogen gas (15). Therefore, the biodegradation of cyanogens together with the removal of ammonia via nitrification and denitrification in an anoxic/oxic activated sludge system represents a more promising and economical alternative to handling ammonia, cyanogen laden phenolic wastewater.

This paper primarily describes and discusses an on-going laboratory investigation which evaluates the efficacy of an anoxic/oxic activated sludge system for treating this type of waste water, with the following objectives:

- to compare the effect of cyanide concentration increase on an activated sludge system and an anoxic/oxic activated sludge system
- to report on the removal of cyanide and thiocyanate from both systems.

MATERIAL AND METHODS

Experimental Units

Two completely mixed continuous flow activated sludge reactors with internal recycle were used as experimental units. One unit was run with an anoxic chamber in series with an oxic chamber. The anoxic unit was provided with continuous stirring. Made of plexiglas, the activated sludge units had an aeration volume of 4.8L and a settling chamber volume of 1.3L. The anoxic unit of the anoxic/oxic system had a working volume of 1.5L. Schematics of the activated sludge unit and the anoxic/oxic unit are shown in Figure 1.

Diffused air was introduced to the aeration chambers through aquarium aerators and air diffusers to maintain adequate mixing and aeration. The mixed liquor dissolved

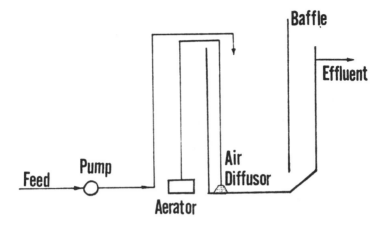

Figure 1(a) Schematic of the Experimental Activated Sludge Set Up

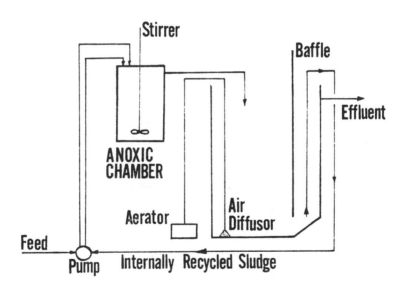

Figure 1(b) Schematic of the Anoxic/Oxic Set-Up

oxygen (DO) concentration in the activated sludge unit was always above 2.0 mg/L and the DO in the aeration (oxic) chamber of the anoxic/oxic system was maintained at 1.5 to 2.0 mg/L.

Synthetic wastewater (or feed) of the composition shown in Table 1 was made up daily and stored in a refrigerator in a covered container.

Table 1
Composition of the Synthetic Feed

Component	mg/L
Phenol	200
m-cresol	50
o-cresol	50
p-cresol	50
catechol	50
NH_4-N	150
KH_2PO_4	100
$NaHCO_3$	2,500
$FeCl_3$	5
$MgSO4$	10
$CaCl_2$	10
Yeast	10
CN^-	10-45
SCN^-	100

Variable-speed peristaltic pumps were used to deliver the stored feed as influent to the units. The pumping was maintained at rates to achieve a hydraulic retention time (HRT) of 30 hours. For the anoxic/oxic system, both feed to the anoxic chamber and sludge recycle from the oxic chamber to the anoxic chamber were transferred continuously through the same peristaltic pump as shown in Figure 1(b). The internally recycled sludge flow was maintained at 3.5 times the flow rate of the influent.

The experimental units were operated at room temperatures ranging from 20 to 23°C. The pH in the reactors were in the 7.8 - 8.6 range.

The initial seed organisms were obtained from the Southwest Philadelphia Water Pollution Control Plant.

The mixed liquor collected was allowed to settle for 45 minutes. After decanting, 60 mL concentrated mixed liquour was transferred to the activated sludge unit and acclimated to the phenolic feed (as in Table 1) but without the cyanide and thiocyanate. After three weeks, half the mixed liquor in the activated sludge unit was transferred to the anxoic/oxic unit. At this point, the cyanide and thiocyanate component of the feed was added. The HRT and sludge retention time (SRT) of each unit were maintained at 30 hours and 30 days, respectively.

Experimental Design

The SRT was maintained by taking into account unintentional sludge wasting in the effluent, and daily deliberate slude wasting.

Under the operating conditions of HRT (30 hours) and SRT (30 days and 45 days), steady state data were collected for three changes in the cyanide concentration (10 mg/L, 30 mg/L and 45 mg/L). Nitrogen balances were also done using the data obtained from the activated sludge unit to evaluate if the cyanogens accumulated in the sludge.

Analytical Procedure

For each change in cyanide (CN^-) concentration, samples were collected and analyzed daily for total organic carbon (TOC), ammonia-nitrogen, total suspended solids (TSS) and mixed liquor volatile suspended solids (MLVSS). At steady-state, total inorganic nitrogen, cyanide and thiocyanate concentrations were also determined daily over a period of one week to ensure that representative results were obtained. Analytical task were performed in accordance to **Standard Methods** (16).

RESULTS AND DISCUSSION

TOC Removal

There was no appreciable difference in the performance of both systems under SRT of 30 days and 45 days with regards to overall TOC removal, when CN^- concentration was increased from 10 mg/L to 45 mg/L. Both systems removed 90% of the influent TOC (Table 2).

Table 2
TOC Removal

SRT	CN⁻ Influent Concentration mg/L	Activated Sludge Unit %	Anoxic/oxic Sludge Unit %
30 days	10	91	92
	30	92	91
	45	89	92
45 days	10	92	93
	30	94	92
	45	91	91

The variation of TOC of 200 mg/L, 300 mg/L and 500 mg/L during the SRT = 45 days and CN⁻ = 30 mg/L run did not alter TOC removal efficiency.

Cyanide and Thiocyanate Removal

Ludzack and Schafer (17) made the observation that after two to three weeks of acclimation, cyanide, cyanate and thiocyanate were effectively degraded without loss of suspended solids from the activated sludge system. It has been noted that the microbial oxidation of cyanide and thiocyanate by aerobic autotrophes may be given by:

$$HCN + 0.5O_2 + 2H_2O \longrightarrow HCO^- + NH_4^+ \qquad [1]$$

and

$$SCN^- + 2O_2 + 2H_2O \longrightarrow CO_2 + NH_4^+ + SO_4^= \qquad [2]$$

A typical nitrogen balance carried out (Table 3) seemed to validate this. Total nitrogen leaving the system accounted for 88.8% of the influent nitrogen. The influent nitrogen was made up of 84% ammonia-nitrogen with the remaining nitrogen content resulting from cyanide and thiocyanate.

Table 3
Nitrogen Balance on Activated
Sludge Unit (SRT: 30 days CN^- 10 mg/L

Influent		Effluent	
NH_4-N	150.0 mg/L	NH_4-N	2.2 mg/L
CN^--N	5.3 mg/L	NO_3-N, NO_2-N	140.0 mg/L
SCN^--N	24.0 mg/L	CN^--N	0.3 mg/L
		SCN - N	1.3 mg/L
Feed-N	179.3 mg/L	VSS - N	15.5 mg/L
		Effluent - N	159.3 mg/L

Biomass deliberated wasted = 0.0 mL
∴ Nitrogen leaving system = 88.8% Nitrogen entering system
% CN^-- N + SCN^-- N in influent = 16%

In both the activated sludge system and the anoxic/oxic system cyanide concentration was reduced consistently from the 10, 30 and 45 mg/L to around 0.35 mg/L CN^- when the SRT was 30 days. At the SRT of 45 days CN^- concentration was reduced to 0.1 to 0 mg/L CN^-. The increase in concentration of CN^- did not affect the removal of SCN^-. Both systems, at both SRTs of 30 and 45 days, achieved 95% removal of the thiocyanate.

Formation and Removal of Ammonia-Nitrogen

In both the activated sludge unit and the anoxic/oxic systems, ammonia-nitrogen removal steadily decreased with increase in cyanide concentration. Results of SRT 30 days are shown in Figure 2. The trend was similar for SRT of 45 days.

Although further study is required, the anoxic/oxic system demonstrated advantages over the activated sludge system in treating phenolic wastewater containing cyanide, thiocyanate and ammonia.

In the anoxic/oxic system, the oxidation of the organic components of the feed and the degradation of cyanide and thiocyanate to ammonia-nitrogen (Equation 1 and 2) takes place in this oxic zone. Here the feed

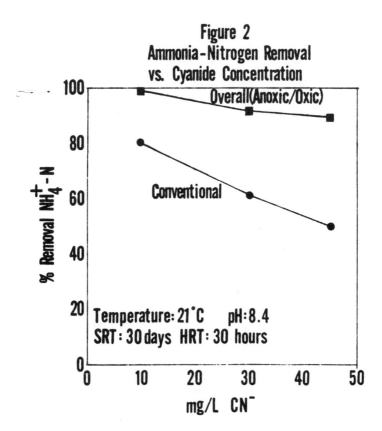

Figure 2
Ammonia-Nitrogen Removal vs. Cyanide Concentration

ammonia-nitrogen and ammonia-nitrogen from the breakdown of the cyanogens is nitrified according to:

$$2NH_4^+ + 3O_2 \longrightarrow 2NO_2^- + 4H^+ + 2H_2O \qquad [3]$$

$$2NO_2^- + O_2 \longrightarrow 2NO_3^- \qquad [4]$$

The mixed liquor from the oxic zone and settled sludge in the secondary clarifier are recycled back to the anoxic zone. Together with the influent feed, in the absence of DO, with the feed organics serving as the organic carbon source, the nitrates present are denitrified to nitrogen gas. Residual organics from this zone are further oxidized in the oxic zone.

CONCLUSION

While further investigation is in progress, the results from this study, to date, indicate that there are several advantages to the anoxic/oxic system over the activiated sludge system in handling cyanide, thiocyanate laden phenolic wastewater.

Specific conclusions are:

- Cyanides and thiocyanates are effectively removed in both the activated sludge system and the anoxic/oxic system in the presence of up to 45 mg/L CN^-.

- Overall TOC removal is consistent in both systems.

- The anoxic/oxic system reacts better to changes in cyanide concentration changes than does the activated sludge unit. This is probably due primarily to the dilution factor afforded in recycling from the oxic to the anoxic chamber.

- The anoxic/oxic system also has an added advantage over the activated sludge system in terms of ammonia-nitrogen removal. It is possible to reduce ammonia-nitrogen to nitrogen gas.

ACKNOWLEDGEMENT

This project is financially supported in part by the University of Pennsylvania Research Foundation. The authors gratefully acknowledge the seed money provided by the Foundation.

REFERENCES

(1) Baughman, G.L. and Paris, D.L., "Microbial Bioconcentration of Organic Pollutants from Aquatic Systems - A Critical Review", <u>CRC Critical Review of Microbiology</u>, Vol. 8, 1981, p.205.

(2) Huff, J.E. and Bigger, J.M., <u>Cyanide Removal From Refinery Wastewater Using Powdered Activated Carbon</u>, 1980 EPA 600/2-8-125.

(3) Luthy, R.G. "Treatment of Coal Coking and Coal Gasification", <u>JWPCF</u>, Vol. 53, p. 325.

(4) National Research Council, Commission on Life Sciences, Board on Toxicology and Environmental Health Hazards. <u>Toxicity Testing: Strategies to Determine Needs and Priorities.</u> National Academy Press, Washington, D.C. 1984.

(5) Fisher, C.W., Hepner, R.D.amd Tallon, G.R., "Coke Plant Effluent Treatment Investigations", <u>Blast Furnace and Steel Plants</u>, Vol. 58, 1970, p.315.

(6) Ganczarczyk, J.J., "Second-Stage Activated Sludge Treatment of Coke-Plant Water Research, Vol. 13, 1979, p. 337.

(7) Kostenbader, P.D. and Flecksteiner, J.W., "Biological Oxidation of Coke Plant Weak Ammonia Liquor", <u>JWPCF</u>, Vol. 41, Part 1, 1969, p. 199

(8) Luthy, R.G. and Jones, L.D., "Biological Oxidation of Coke Plant Effluent", <u>J. Environ. Eng. Div., Proc. Am. Soc. Civ. Eng.</u>, Vol. 106, 1980, p. 847.

(9) Melor, H. et al, "Combined Treatment of Coke Plant Wastewater and Blast Furnace Blowdown Water in A Coupled Biological Fluidized", JWPCF, Vol. 56, 1984, p. 192.

(10) Nutt, S., Melcer, H. and Pries, J.H., "Two-Stage Biological Fluidized Treatment of Coke Plant Wastewater for Nitrogen Control", JWPCF, Vol. 56, 1984, p. 851.

(11) Atkinson, A., "Bacterial Cyanide Detoxification", Biotech. Bioengr., Vol. 17, 1975, p. 457.

(12) Howe, R.H.L., "Biodestruction of Cyanide Wastes: Advantages and Disadvantages" in International J. Air and Water Pollution, Vol. 9, 1965, p. 473.

(13) Knowles, C.J., "Microorganisms and Cyanide", Bacteriological Reviews, Vol. 40, 1976, p. 652.

(14) Murphy, R.S. and Nesbitt, J.B., "Biological Treatment of Cyanide Wastes", Engineering Research Bulletin B-88, Pennsylvania State University, 1964.

(15) Ekama, G.A. and Marais, G.V.R., "Biological Nitrogen Removal" in Theory, Design and Operation of Nutrient Removal Activated Sludge Processes, Water Research Commission, South Africa, 1983, p. 6-1

(16) Standard Methods for the Examination of Water and Wastewater (15th ed), American Public Health Association, 1980.

(17) Ludzack, F.J. and Schaffer, R.B., "Activated Sludge Treatment of Cyanide, Cyanate and Thiocyanate" in Purdue University Engineering Bulletin Extension Service, Vol. 106, 1960, p. 439.

PARTITIONING OF TOXIC ORGANIC COMPOUNDS ON MUNICIPAL WASTEWATER TREATMENT PLANT SOLIDS

Richard A. Dobbs
Michael Jelus

Wastewater Research Division
Water Engineering Research Laboratory
U.S. Environmental Protection Agency
Cincinnati, Ohio

Kuang-Ye Cheng

Department of Civil and Environmental
Engineering
University of Cincinnati
Cincinnati, Ohio

INTRODUCTION

Partitioning on solids is one of the fundamental processes controlling the removal of toxic organic compounds in municipal wastewater treatment plants. The literature provides ample evidence that organics may accumulate in sludges at concentrations several orders of magnitude greater than the influent concentrations. Compounds which have a special affinity for sorption include pesticides, phthalates, and polynuclear aromatic hydrocarbons which have shown concentration factors of 13-13,000 (1,2,3). The fate of hydrophobic organic pollutants in water systems is highly dependent upon their sorptive behavior. The degree of sorption not only affects a chemical's mobility, but also is a dominant factor in fate processes such as volatilization and biodegradation.

Since interphase partitioning phenomena are one of the major steps in the removal of organics from the bulk aqueous flow in wastewater treatment plants, it is important that both quantitative and qualitative information about the sorption process be obtained. These results will enable the prediction of removal by wastewater solids, provide a basis for modeling the process, and provide a data base to correlate sorption on wastewater solids with other sorption processes and with molecular properties.

In dilute systems typical of most environmental situations, the partition coefficient Kp for sorption on wastewater solids can be defined as follows:

$$K_p = C_s/C_L \qquad (1)$$

where: C_s = concentration of pollutant in the solid phase and C_L = concentration of pollutant in the liquid phase.

The usual method for measuring partition coefficients is to determine a sorption isotherm. Slurries of the wastewater solids are prepared in different initial concentrations of the toxic organic chemical being studied. After the systems reach equilibrium, the concentration in both the solution and solid phases are measured (solid phase concentration can be calculated based on the difference between the initial concentration added and the solution concentration at equilibrium). Data are fitted to the Freundlich equation which can be written:

$$X/M = KC_e^n \qquad (2)$$

where: X = $C_o - C_e$ which is the amount of solute sorbed from a given volume of solution
C_o = initial concentration of solute added
C_e = concentration of solute at equilibrium
M = weight of solids added to the solution
K and n = empirical constants

Data are fitted to the logarithmic form of Equation (2) which can be written as follows:

$$\log X/M = \log K + n \log C_e \qquad (3)$$

For dilute solutions this equation yields a straight line with a slope of n and an intercept equal to the value of K (X/M at C_e = 1.0) when X/M is plotted as a function of C_e on logarithmic paper. The intercept is an indicator of sorption capacity and the slope of sorption intensity. K is often referred to as the sorption or adsorption coefficient. To obtain the partition coefficient K_p) when X/M is in mg/gm and C_e is in mg/l, the following equation is used:

$$K_p = \frac{X/M}{C_e/1000} \tag{4}$$

(Equation (4) also applies to the case where X/M is in µg/gm and C_e is in µg/l)

The Freundlich equation is a useful method for the treatment of sorption data. Since $X = C_o - C_e$, substitution of $C_o - C_e$ for X in Equation (2) yields:

$$(C_o - C_e)/M = KC_e^n \tag{5}$$

which gives removal of toxic organic compound as a function of solids concentration.

MATERIALS AND METHODS

Reagents

Organic compounds were obtained from Eastman Kodak Co., Rochester, NY; Aldrich Chemical Company, Inc., Milwaukee, WI; and Chem Service Inc., West Chester, PA.* All chemicals were the purest available and were used as received from the supplier.

Analytical Methods

Analytical methods for analysis of the selected organic compounds used in the present study are summarized in Table I.

Volatile chlorinated organics were analyzed using a Tracor LSC-2 sample concentrator and a Tracor 222 gas chromatograph equipped with a Tracor 700 Hall electrolytic conductivity detector. Pesticides and the cresol were analyzed using a Varian Model 3700 gas chromatograph equipped with a Varian CDS-111 chromatographic data system. Dyes were determined with Beckman Model 25 spectrophotometer equipped with a clinical sipper system.

---

* Mention of trade names or commercial products does not constitute endorsement or recommendation for use.

Table I. Summary of Analytical Methods Used for Analysis of Toxic Organic Compounds

Compounds	Analytical Methods	Conditions
Methylene chloride Chloroform 1,1-dichloroethylene Carbon tetrachloride Tetrachloroethylene Chlorobenzene	purge and trap gas chromatography	Texas trap; 3% Carbowax 1500 on chromosorb column; Hall detector
Lindane Heptachlor epoxide Dieldrin	hexane extraction gas chromatography	1.5% SP-2250/ 1.9% SP-2401 on Supelcoport column election capture detector
2,4-dinitro-o-cresol	freon extraction gas chromatography	1% SP-1240 DA on Supelcoport column flame ionization detector
Methylene blue dye Azo dye 88 Azo dye 151	visible spectroscopy	665 nm 505 nm 512 nm

Wastewater Solids Preparation

Wastewater solids were collected from municipal wastewater treatment plants within a fifty mile radius of the Water Engineering Research Laboratory, Cincinnati, Ohio. Wastewater plants were selected to include a wide range of domestic/industrial contribution to the total plant flow. Samples of primary sludge, mixed liquor solids, and digested sludge were collected in a series of five gallon carboys. Solids were removed by centrifugation in a perforate bowl centrifuge. Analysis of the sludge or mixed liquor and the centrate showed solids capture of approximately 95±3% by centrifugation. The collected solids (approximately 10% on a dry weight basis) were stored at 4°C until used. No visible change was observed during the cold storage period of 30-90 days required for completion of experimental testing for a given wastewater solid system.

EXPERIMENTAL PROCEDURE

Experimental Protocol Development

Preliminary Studies

A standard experimental protocol for determination of the partition coefficient or sorption capacity of wastewater solids for toxic organic compounds has not been developed. Reports in the literature by different investigators are in direct conflict with regard to several fundamental concepts related to the partitioning process. In order to resolve these issues and establish a basis for partition measurements, several preliminary studies were conducted. Key issues to be resolved include: (1) effect of solids-to-water ratio on the partition measurement; (2) rate of attainment of equilibrium; and (3) use of viable versus non-viable solids in the isotherm procedure.

Several reports in the literature state that the equilibrium partition coefficient (K_p) decreases as the ratio of solids-to-water increases (4,5,6). These observations challenge the classical view of the sorption process. The two procedures available for sorption capacity measurements involve the use of a constant concentration of compound with varied amounts of solids or the use of a varied concentration of compound with a constant amount of solids. If solids-to-water ratio affect the measurement of partition coefficients, the first method described should not be used. In order to resolve this issue, a preliminary set of isotherms were measured using both methods. The test involved determination of the sorption or partition coefficient for methylene blue dye using mixed-liquor solids.

In the first of the two procedures, a constant methylene blue dye concentration of 10.0 mg/l was used with mixed-liquor solids concentrations of 0, 0.05, 0.10, 0.25, 0.30, and 0.5 g/l on a dry weight basis. These solid doses selected gave a 10-fold variation in solids-to-liquid ratio and are typical of the conditions used to measure partition coefficients. In the second procedure, a constant mixed-liquor solids concentration of 0.1 g/l dry weight was used in conjunction with methylene blue dye concentrations of 0, 2.5, 5.0, 7.5, 10.0, and 20.0 mg/l. Methylene blue dye was chosen for the preliminary test because it is non-biodegradable, non-volatile, and easily determined by visible spectroscopy at 665 nm. Test samples containing the specified solids and dye concentration were prepared and equilibrated for 8 hours. Solids were separated by centrifugation and the concentration of dye measured. The isotherms obtained by both methods are shown in Figure 1 where X/M is plotted as a function of C_e according to the logarithmic form of the Freundlich equation.

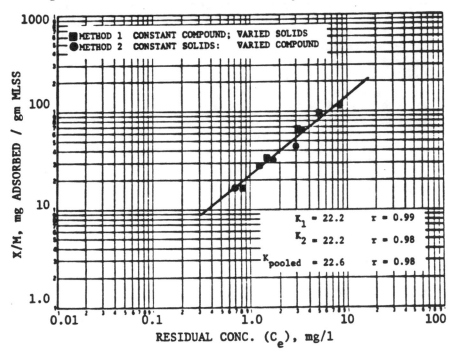

Figure 1. Comparison of two isotherm methods for sorption capacity measurements.

The sorption coefficients (K values) calculated from the least squares treatment of the separate isotherm data were the same for both methods (22.2 mg/gm). The pooled data yielded a K value of 22.6 mg/gm in excellent agreement with the two individual isotherm values. These data indicate that the sorption coefficient and the partition coefficient remain constant over the range of solid-to-water ratios used in this study. It has been emphasized that if precautions are taken to eliminate or account for non-settling (or non-filtrable) microparticles which remain in the aqueous phase during sorption tests, the partition coefficient does remain constant (7) in agreement with the present study. Even though either method can be used to experimentally determine sorption and partition coefficients, the second method discussed was selected for use in the present study. The choice was based on the desire to maintain a constant solids level in the samples during the solids' separation step by centrifugation or filtration prior to analysis. A second reason for the use of constant solids in the sorption test is that frequently measurable levels of a given test compound will be found in the wastewater·solids as a natural background. If this occurs, it is easier to correct analytical values of the equilibrium concentrations in the samples by a constant amount.

A great deal of confusion exists in the literature regarding the rate of attainment of sorption equilibrium. Sorption kinetics for phenol and 1,4-dichlorobenzene were reported to be exceedingly fast (equilibrium in <4 minutes) (8). In the case of polynuclear aromatic hydrocarbons on natural sediments 35 to 60% of the equilibrium sorption capacity was achieved in minutes (mixing limited) with <u>apparent</u> equilibrium after a few hours (9).

Based on the conflicting reports in the literature on the rate of attainment of equilibrium, kinetic measurements on the sorption of methylene blue dye on mixed-liquor solids were made to determine the proper contact time for experimental purposes. Mixed-liquor solids from three diferent municipal wastewater treatment plants were collected. Rate of sorption of methylene blue dye from a 10 mg/L solution was measured at 0.1, 0.2 and 0.5 g/L dry weight solids. All three mixed-liquor samples produced similar results. Data for the mixed-liquor solids from the same wastewater treatment plant used in the isotherm runs are presented in Figure 2.

The kinetic data for the sorption of methylene blue dye are in agreement with the two box model proposed for hydrophobic pollutants in water (10). The two box model described sorption dynamics in terms of a rapid or "labile" exchange and a highly retarded or "non-labile" sorption requiring days to weeks to occur. In the case of methylene blue dye and 0.5 g/l mixed-liquor solids dose, 92% of the compound was sorbed in approximately 30 minutes, while the remainder required several hours. In order to minimize biodegradation during the sorption test, the remaining organic compounds were equilibrated for 6 hours. This avoided overnight contact and permitted samples to be separated from the solids the same day.

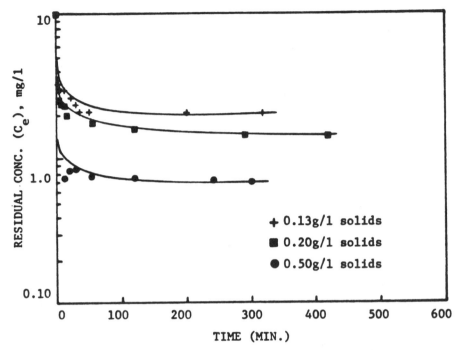

Figure 2. Kinetics of sorption of methylene blue dye on mixed-liquor solids.

The third issue to be resolved in preliminary studies involved methods to eliminate biological degradation of test compounds during the partitioning process. It is obvious that other removal mechanisms such as biodegradation and volatilization would have to be controlled if accurate sorption data were to be obtained. Otherwise, measurements of changing aqueous phase compound concentrations would include the effects of all removal mechanisms resulting in apparent sorption capacities greater than actually achieved. Volatilization was eliminated from sorption measurements by using completely filled containers with zero head space. The impact of biodegradation on the sorption measurements was more difficult to control. Lyophilization followed by dry heat treatment has been reported to produce biological solids which had flocculation and settling properties which were indistinguishable from live biomass (8). These authors claimed the batch experimental technique with nonviable biomass appeared to give realistic results compared to tests with viable biomass. Based on the reported successful use of lyophilization for control of biological activity, this technique was evaluated. A batch of mixed-liquor solids was collected and separated from the supernatant in a perforate bowl centrifuge. The moist solids were divided into two portions. One portion was freeze-dried and the other was stored at 4°C. Sorption isotherms were measured using 2,4-dinitro-o-cresol. The isotherms were conducted at pH 3.0 for two reasons. First, the low pH would inhibit or eliminate biological activity in the test with viable biomass. Secondly, the decrease in pH would convert the 2,4-dinitro-o-cresol to the undissociated (sorbable) form (pKa = 4.35). Results of the isotherm comparison are shown in Figure 3.

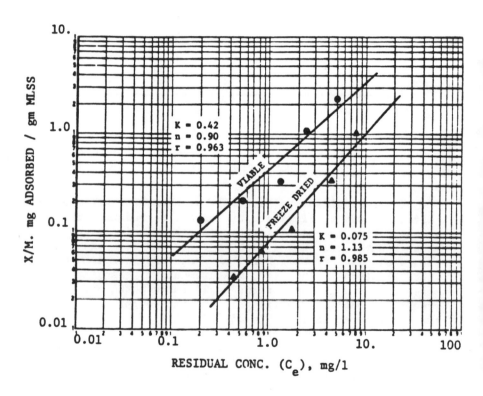

Figure 3. Comparison of freeze-dried and viable biomass for sorption of 2,4-dinitro-o-cresol at pH 3.0.

The use of freeze-dried solids in the isotherm procedure was not acceptable based on the results shown in Figure 3. The viable biomass clearly has a significantly greater sorption capacity than the freeze-dried material. A similar result has been reported for sorption of azo and triphenylmethane dyes on microbial populations (11).

Additional studies are under way to develop methods to control biological activity during the partitioning process. Preliminary data suggest that pH adjustment to 3.0 may effectively eliminate biological activity without significantly affecting the partitioning process for neutral and acidic compounds.

Sorption or Partitioning of Selected Toxic Organics

Based on the results of the preliminary tests, additional sorption isotherms were measured using viable mixed-liquor solids with a 6-hour contact time. Compounds selected for experimental use were either resistant to biodegradation or were shown not to significantly degrade during the course of the isotherm test by specific analysis on whole samples (both aqueous and solid phases). Chemical structures for the toxic organic compounds investigated to date are shown in Figure 4.

DISCUSSION OF RESULTS

The isotherms obtained for the toxic organic compounds studied showed a wide range of sorption capacities. Relevant Freundlich parameters for sorption on mixed-liquor solids are summarized in Table II.

Sorption coefficients (K values) ranged from 0.062 mg/gm for methylene chloride to an extrapolated value of 38.9 mg/gm for dieldrin at equilibrium concentrations of 1.0 mg/l. Slopes (n values) were generally close to 1.0 and ranged from 0.31 to 1.15 for sorption on mixed-liquor solids. Slopes for sorption on soils have been reported to range from a low of 0.3 to a high of 1.7 (12). Compiled data on 26 chemicals (mostly pesticides) had a mean value of n of 0.87 with a coefficient of variation of ±15% (13). (If a measured value is not available, it is frequently assumed to be equal to 1.0.) Excellent correlation coefficients (r values) were obtained for all the isotherms measured.

In soil and sediment sorption, the extent to which an organic chemical partitions itself between the solid and solution phases is expressed in terms of a parameter K_{oc}, which is largely independent of the properties of the soil or sediment.

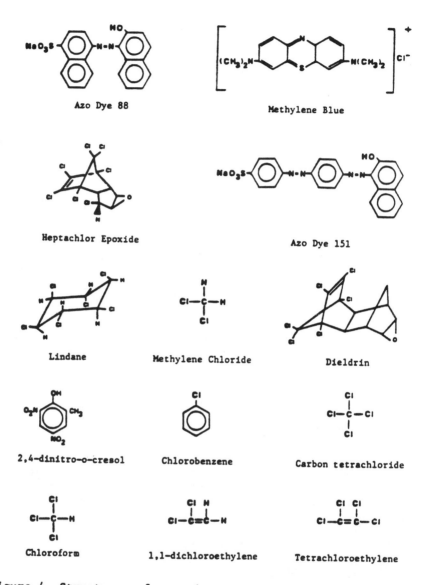

Figure 4. Structures of organic compounds used in sorption studies.

Table II. Summary of Freundlich Parameters For Sorption of Toxic Organics on Mixed-Liquor Solids

Compound	K, (mg/gm)[a]	n[b]	r[c]
Methylene chloride	0.062	1.15	0.92
Chloroform	0.094	0.90	0.98
1,1-dichloroethylene	0.150	0.71	0.96
Carbon tetrachloride	0.597	0.31	0.97
Chlorobenzene	0.330	0.47	0.91
2,4-dinitro-o-cresol	0.420	0.90	0.96
Tetrachloroethylene	0.970	1.16	0.98
Lindane[d]	0.762	1.00	0.99
Heptachlor epoxide[d]	17.9	1.09	0.98
Dieldrin[d]	38.9	1.11	0.99
Methylene blue dye	22.6	0.82	0.98
Azo dye 151	8.30	1.00	0.99
Azo dye 88	1.35	0.65	0.99

[a] K = sorption coefficient = X/M at C_e = 1.0 mg/l
[b] n = slope of isotherm plot
[c] r = correlation coefficient
[d] K values extrapolated to C_e = 1.0 mg/l for comparison purposes.

Once K has been determined from the Freundlich equation, it is divided by % organic carbon (oc) contained in the soil or sediment and multiplied by 100 to obtain K_{oc}. In the case of wastewater solids used in the present study, organic carbon was based on weight loss on ignition at 600°C and is defined by the following equation:

$$K_{om} = K \cdot \frac{100}{\% \text{ VSS}} \qquad (6)$$

where K_{om} = amount of solute or pollutant sorbed per unit weight of organic matter in solids.
VSS = volatile suspended solids.

Sorption data are usually correlated based on organic carbon or organic matter. Therefore, the sorption coefficient, K, must be converted to K_{om} for correlation purposes. Thus, we can define a corrected partition coefficient by the following equation:

$$K_p' = \frac{K_{om}}{C_e/1000} \qquad (7)$$

Emphasis on the environmental impact of toxic organic chemicals has resulted in increased reliance on the physical-chemical properties of these compounds to assess their environmental behavior. One such parameter, the octanol/water partition coefficient, (K_{ow}) has proved useful as a means for predicting soil adsorption (14), biological uptake (15), lipophilic storage (16), and biomagnification (17, 18, 19, 20). K_{ow} is defined by the following equation:

$$K_{ow} = C_o/C_w \qquad (8)$$

where C_o = equilibrium concentration of organic compound in octanol layer
C_w = equilibrium concentration of organic compound in water layer

Based on the soil-sediment literature, an attempt was made to correlate sorption on wastewater solids with the octanol/water partition coefficient for the compounds used in the present study. Sorption coefficients in Table II were converted to K_{om} values using Equation 6. The K_{om} values obtained were used to calculate corrected partition coefficients (K_p' values) using Equation 7. The values obtained are summarized in Table III along with the corresponding octanol/water partition coefficients.

A simple linear regression of log K_p' on log K_{ow} yielded a correlation coefficient (r) of 0.985. A plot of log K_p' versus K_{ow} is shown in Figure 5. The following equation was obtained for the relationship between partitioning on wastewater solids and octanol/water partition coefficient:

$$\log K_p' = 1.06 + 0.60 \log K_{ow} \qquad (8)$$

If K_{ow} is given in numerical form, the equation can be written as follows:

$$K_p' = 11.6(K_{ow})^{0.6} \qquad (9)$$

Table III. Summary of Values Used to Correlate Sorption on Mixed-Liquor Solids with Octanol/Water Partition Coefficient

Compound	K_{om} [a]	K_p'	log K_p'	log K_{ow} [b]
Methylene chloride	0.067	67	1.83	1.26
Chloroform	0.117	117	2.07	1.97
1,1-dichloroethylene	0.185	185	2.27	2.13
Carbon tetrachloride	0.735	735	2.87	2.64
Chlorobenzene	0.407	407	2.61	2.84
2,4-dinitro-o-cresol	0.590	590	2.77	2.85
Tetrachloroethylene	1.11	1110	3.04	2.88
Lindane	2.57	2570	3.41	3.85
Heptachlor epoxide	15.33	15,330	4.19	5.40
Dieldrin	25.16	25,160	4.40	5.48

[a] K_{om}: units are mg/gm volatile suspended solids at $C_e = 1.0$ mg/l for first seven compounds listed and μg/gm volatile suspended solids at $C_e = 1.0$ μg/l for the three pesticides.

[b] K_{ow}: values were taken from reference (21) or were calculated by fragment method (22).

Limited tests have been completed on the sorption of toxic organics on primary sludge and digested sludge. Data for two compounds are presented in Table IV.

Table IV. Comparison of Partition Coefficients (K_p [a] Values) for Sorption of Toxic Organics on Wastewater Solids

Compound	Primary Sludge	Mixed-Liquor	Digested Sludge
1,1-dichloroethylene	0.20	0.15	0.17
Chlorobenzene	0.35	0.33	0.38

[a] K_p: units are mg/gm solids

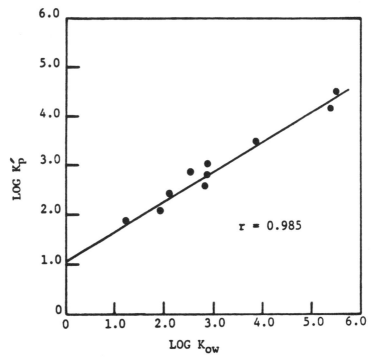

Figure 5. Correlation of partition coefficients for sorption on mixed-liquor solids with octanol-water partition coefficients.

Based on limited data, it appears that the relationship between K_p' and K_{ow} established for mixed-liquor solids is also valid for other municipal wastewater treatment plant solids. Additional isotherms are being measured to expand the data base and validate the relationship for other solids.

SUMMARY AND CONCLUSIONS

Preliminary studies have shown that partitioning on municipal wastewater treatment plant solids was not affected by solids-to-liquid ratio. Kinetic data on sorption of toxic organic compounds on wastewater treatment plant solids showed an initial rapid uptake followed by a slower rate over an extended period of time. Freeze-dried solids did not exhibit the same sorption characteristics as viable biomass.

A correlation between sorption of toxic organics on wastewater treatment plant solids and octanol/water partition coefficient has been established. The relationship should be useful for estimating the removal of toxic organic compounds in municipal and industrial wastewater treatment plants by the sorption mechanism. The correlation also provides a basis for predicting concentrations of toxic compounds in various sludges provided the equilibrium concentration in the aqueous phase is known (or assumed).

REFERENCES

1. Fate of Priority Pollutants in Publicly-Owned Treatment Works, EPA-440/1-79-300, Oct., 1979.
2. Hannah, S.A., and Rossman, L., Monitoring and Analysis of Hazardous Organics in Municipal Wastewater - A Study of Twenty-Five Treatment Plants, proceedings of the Seminar on Hazardous Substances in Wastewaters, Pollution Control Association of Ontario, and the Ontario Ministry of the Environment, Nov. 3, 1982.
3. Hovious, J.C., Waggy, G.T., and Conway, R.A., Identification and Control of Petrochemical Pollutants Inhibitory to Anaerobic Processes, EPA-R2-73-194, Office of Research and Monitoring, Washington, D.C., 20460.
4. O'Connor, D.J., and Connolly, J.P., Water Research, Vol.14, 1517, (1980).
5. Voice, T.C., Rice, C.P., and Weber, W.J., Environmental Science and Technology, Vol. 17, 513, (1983).
6. Voice, T.C., and Weber, W.J., Sorbent Concentration Effects in Liquid/Solid Partitioning, Environmental Science & Technology, Vol. 19, No. 9, 7890, (1985).
7. Gschwend, P.M., and Wu, S., On the Constancy of Sediment-Water Partition Coefficients of Hydrophobic Organic Pollutants, Environmental Science & Technology, Vol. 19, No. 1, 90, (1985).
8. Blackburn, J.W., and Troxler, W.L., Prediction of the Fates of Organic Chemicals in a Biological Treatment Process, Environmental Progress, Vol. 3, No. 3, 163, (1984).
9. Karickhoff, S.W., Sorption Kinetics of Hydrophobic Pollutants in Natural Sediments, Contaminants and Sediments, R.A. Baker, ed., Vol. II, 193, (1980), Ann Arbor Science, Ann Arbor, MI.
10. Karickhoff, S.W., and Morris, K.R., Sorption Dynamics of Hydrophobic Pollutants in Sediment Suspensions, Environmental Toxicology and Chemistry, Vol. 4, 469, (1985).

11. Michaels, G.B., and Lewis, D.L., Sorption and Toxicity of Azo and Triphenylmethane Dyes to Aquatic Microbial Populations, Environmental Toxicology and Chemistry, Vol. 4, 45, (1985).
12. Hamaker, J.W., and Thompson, J.M., Adsorption, In: Organic Chemicals in the Soil Environment, Vol. 1, pp 49-143, edited by C.A.I. Goring, and J.W. Hamaker, Marcel Dekker, Inc., New York, (1972).
13. Rao, P.S.C., and Davidson, J.M., "Estimation of Pesticide Retention and Transformation Parameters Required in Non-Point Source Pollution Models," Environmental Impact of Non-Point Source Pollution, Ann Arbor Science Publishers, Inc., Ann Arbor, MI, (in press, 1979/1980).
14. Briggs, G.G., A Simple Relationship Between Soil Adsorption of Organic Chemicals and Their Octanol/Water Partition Coefficients, proceedings of the 7th British Insecticide and Fungicide Conference, (1973).
15. Kenaga, E.E., Res., Rev., Vol. 44, 73, (1972).
16. Davies, J.E., et.al., Environmental Health, Vol. 30, 608, (1975).
17. Lu, P.Y., and Metcalf, R.L., Environmental Health Perspect., Vol. 10, 269, (1975).
18. Metcalf, R.L., et.al, Environmental Health Perspect., Vol. 4, 35, (1973).
19. Metcalf, R.L., et.al., Arch. Environ. Contam. Toxicol., Vol. 3, 151, (1975).
20. Neely, W.B., Branson, D.R., and Blau, G.E., Environmental Science & Technology, Vol. 8, No. 13, 1113, (1974).
21. Water-Related Environmental Fate of 129 Priority Pollutants, EPA-440/4-79-029a and 029b, December, 1979.
22. Hansch, C., and Leo, A., Substituent Constants for Correlation Analysis in Chemistry and Biology, pp. 18-43, John Wiley & Sons, Inc., New York, New York, 1979.

PATAPSCO WASTEWATER TREATMENT PLANT TOXICITY REDUCTION EVALUATION

John A. Botts, Jonathan W. Braswell and Elizabeth C. Sullivan

Engineering-Science, Inc.
Fairfax, Virginia

William Goodfellow

EA Engineering, Science and Technology
Sparks, Maryland

Burton D. Sklar

City of Baltimore
Patapsco Waste Water Treatment Plant
Baltimore, Maryland

Dolloff F. Bishop

Water Engineering Research Laboratory
Environmental Protection Agency
Cincinnati, Ohio

INTRODUCTION

The U.S. Environmental Protection Agency (EPA) has established a policy to develop water quality-based permit limitations to control the discharge of toxic materials through the nation's wastewater treatment systems. A toxics management process is evolving to support this policy and one step in this process involves conducting Toxicity Reduction Evaluations (TREs) at wastewater treatment facilities identified as having toxic influent wastewaters. The objectives of these TREs are to assess the sources of toxicity, the impact of treatment, and the pass-through following treatment of wastewater toxicity.

The EPA and the City of Baltimore (City) are conducting a TRE at the City's Patapsco Waste Water Treatment Plant (Patapsco WWTP). The Patapsco TRE was initiated in April 1986 and will provide the first case history of a toxics management program at a municipal wastewater treatment plant. The overall approach and specific tasks that have been developed since the TRE was proposed are described herein. Initial results of the study are also presented and discussed.

Background

The Patapsco WWTP was chosen for a TRE because there was evidence of influent wastewater toxicity and toxicity pass-through to the receiving waters. The City had implemented a toxics monitoring program in 1980 through the use of Microtox and a modified respirometry test. As a result of this program, the City has developed considerable experience in in-plant toxicity monitoring and the toxicity data collected to date provides a basis for the more definitive work being conducted in this TRE.

Wastewater Sources

The Patapsco WWTP is fed by two main sewage streams. The major influent wastewater stream, the Southwest Diversion (SWD) sewer, was designed for a flow of 210 MGD but its present wastewater flow is approximately 25 to 30 MGD. This wastewater flow is primarily domestic wastewater and the toxicity, as measured by Microtox, has generally been minimal; however, during 1981 through 1983 there was a significant increase in the Microtox toxicity of the SWD influent wastewater during the summer months.

The other influent wastestream, the IPI sewer, presently has a wastewater flow of approximately 10 to 15 MGD. The IPI wastewater is primarily industrial wastewater and has a consistently high Microtox toxicity throughout the year. Additionally, toxic spills of short duration (usually less than 6 hours) are noted regularly in the IPI sewer. High concentrations (1,000 ppb or greater) of xylenes, benzenes, and toluenes are consistently present in these wastewater spills. (1)

Wastewater Treatment Facilities

The Patapsco wastewater treatment facilities consist of primary settling followed by oxygen activated sludge secondary treatment. Sludge is dewatered using plate and frame filter presses and incinerated in multiple hearth furnaces. A schematic of the Patapsco WWTP including the wastestream sampling locations for the TRE is shown in Figure 1.

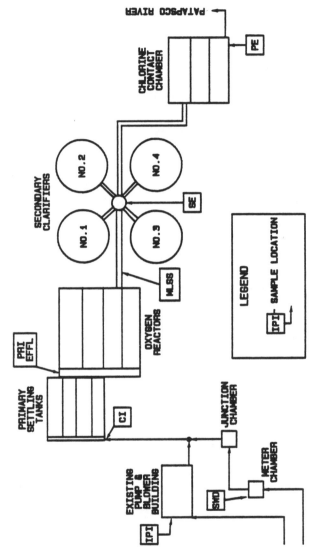

Figure 1. Patapsco WWTP schematic and wastewater sampling locations. The sampling locations are the industrial sewer (IPI), the Southwest Diversion (SWD), the combined influent (CI), the settled primary effluent (PRIM EFFL), the unchlorinated secondary effluent (SE) and the chlorinated plant effluent (PE).

The plant was designed in the early 1970's and construction was completed in September 1982. Prior to 1982, treatment was limited to primary treatment and high Microtox toxicity levels were consistently measured in the plant effluent. Start-up problems were experienced with the solids treatment facilities in 1982 and the first half of 1983 and an inability to dispose of accumulated process solids led to a shutdown of the secondary process in mid 1983. During this period the plant was unable to meet its effluent limitations and significant Microtox toxicity was present in the plant effluent.

Since the beginning of 1984, plant performance has stabilized and the activated sludge biomass appears to have acclimated to the influent. At the start of 1984 effluent BOD averaged between 40 and 50 mg/l and effluent suspended solids ranged from 40-50 mg/l. By the second half of 1984 however, average BOD had decreased to 15 to 20 mg/l and suspended solids had decreased to 20 mg/l. Specific oxygen uptake by the plant biomass also increased through 1984. At the start of 1984 the specific oxygen uptake averaged 20 mg O_2/l/hr/g MLSS, but by August the average specific oxygen uptake had increased to 50 mg O_2/l/hr/g MLSS. Although plant performance has improved, the City is concerned about the impact of transient toxic loads on the plant's biological treatment system, and the continued pass-through of toxicity in the plant effluent.

Key Project Tasks

The tasks to be implemented in the TRE are:

o Evaluate the fate and impact of toxicity during treatment using in-plant toxicity monitors, i.e., Microtox (2), respirometry (3), and ATP (4), and analyses for key plant performance parameters;

o Evaluate the correlation between the in-plant toxicity monitors and standard EPA bioassays;

o Identify the specific chemical inventory entering and leaving the plant;

o Assess the treatability of industrial wastewaters entering the plant; and o   Develop techniques to minimize toxicity impacts on plant operations and toxicity pass-through to receiving waters.

RELATIONSHIP OF PLANT PERFORMANCE TO EFFLUENT TOXICITY

A preliminary evaluation of the historical plant data was made for the period August, 1984 to February, 1986. The purpose of this evaluation was to provide an understanding of the variations in influent wastewater toxicity over time and how these variations are related to plant performance and the occurrence of effluent toxicity.

Plant performance and effluent toxicity were compared by (1) sorting key performance data into classes (e.g., 501-600 mg/l COD, 601-700 mg/l COD, etc.) and recording the number of parameter values occurring in each of these classes (e.g. 5 days had COD values from 501-600 mg/l); (2) relating the occurrence of effluent toxicity to the parameter values for each of the classes (e.g. of the 5 days with COD values 501-600 mg/l, 3 days had toxic effluent); and then (3) comparing the percent occurrence of effluent toxicity between the parameter classes. The key parameters that were compared to effluent toxicity were influent BOD and COD, influent toxicity, influent suspended solids, plant flow, food to mass (F/M) ratios, mean cell retention times (MCRT), mixed liquor suspended solids (MLSS) and percent BOD and COD removals. For purposes of this evaluation, effluent toxicity was defined as a daily Microtox EC_{50} value of less than 100% for the plant secondary (unchlorinated) effluent.

In addition, the data were reviewed to examine the relationship between plant performance and effluent toxicity during periods of sustained high influent toxicity (i.e., Microtox EC_{50} values less than 10 for at least one week).

Influent Parameters versus Effluent Toxicity

Plant influent BOD and COD and plant influent flow appeared to have no direct relationship to effluent toxicity. The occurrence of effluent toxicity also appeared to be independent of influent toxicity as shown in Table I below.

Table I. Comparison of Influent Toxicity to Effluent Toxicity

Combined Influent Microtox EC_{50}	No. of Toxic Effluent Days	No. of Values in Range	% of Occurrence of Effluent Toxicity	% of Total Influent Values in Range
51-100	4	22	18.2	3.9
21-50	36	106	34.0	18.6
11-20	58	163	35.6	28.7
6-10	54	168	32.1	29.5
1-5	38	110	34.5	19.3
Total	190	569	Average 33.4	

The percent occurrence of effluent toxicity when the influent toxicity was high (34.5% when influent EC_{50} = 1 to 5, and 32.1% when influent EC_{50} = 6 to 10) was no greater than the average percent occurrence of effluent toxicity overall (33.4%). There was a lower percent occurrence of effluent toxicity when influent toxicity was low (18.2% when EC_{50} = 51 to 100); however, the statistical significance of this value was not determined.

The lack of correlation between influent parameters and effluent toxicity may be due, in part, to the influent and effluent wastewater samples not being lagged. The data used for this preliminary evaluation are analytical results of samples of plant influent and effluent collected on the same 24-hour composite schedule, and thus do not take into account the hydraulic retention time (HRT) of wastewater in the plant. Since the start of the TRE, effluent samples are being collected on a schedule that lags the influent sample collection by the average HRT of wastewater in the plant secondary system.

Operational Parameters versus Effluent Toxicity

As shown in Table II, effluent toxicity occurred less frequently when MLSS concentrations were above 5,000 mg/l. Likewise, the occurrence of the effluent toxicity decreased at lower F/M ratios. While the occurrence of effluent toxicity was 30% at the design F/M ratio of 0.55, at an F/M ratio of

0.7 or greater the occurrence of effluent toxicity increased to 50% and at an F/M ratio of 0.3 or less the occurrence of toxicity decreased to 22%.

Table II. Comparison of MLSS Concentrations To Effluent Toxicity

MLSS (mg/l)	No. of Toxic Effluent Days	No. of Values in Range	% Occurrence of Effluent Toxicity	% of Total MLSS Values in Range
>5,500	21	121	17.3	22.6
5,001-5,500	21	70	30.0	13.1
4,501-5,000	61	149	40.9	27.8
4,001-4,500	48	126	38.1	23.5
3,501-4,000	21	54	38.9	10.1
3,000-3,500	5	12	41.7	2.2
<3,000	4	4	100.0	0.7
Total	181	536	Average 33.8	

BOD and COD Removals versus Effluent Toxicity

To approximate the HRT of wastewater in the plant, BOD and COD removal were obtained by comparing a daily influent value with the following day's effluent value. Decreased plant performance as measured by low BOD and COD removal (i.e., BOD removal <80% and COD removal <75%) appeared to be associated with events of effluent toxicity. Although these events did not necessarily result in permit violations, effluent toxicity was frequently associated with the pass-through of high concentrations of residual COD as shown below in Table III.

Table III. Comparison of Effluent COD to Effluent Toxicity

Plant Effluent COD (mg/l)	No. of Toxic Effluent Days	No. of Values in Range	% Occurrence of Effluent Toxicity	% of Total COD Values in Range
>150	39	60	65.0	11.5
100-150	79	182	43.4	34.7
50-100	61	282	21.6	53.8
Total	179	524	Average 34.2	

Performance During Toxic Events

The effect of long-term influent toxicity on plant performance was evaluated by studying the data on periods of seven days or more with influent Microtox toxicity less than an EC_{50} of 10. Two examples of toxic events are described as follows.

One extended period of high influent toxicity occurred from September 9 to 20, 1984. As shown in Figure 2, as influent toxicity increased from the 9th through the 13th, there was a corresponding increase in effluent toxicity. This toxicity pass-through was associated with a slight decrease in plant performance (i.e., lower MLSS concentration and COD removal). Later in the period, although influent toxicity continued to be high, effluent toxicity abated and the MLSS concentration and COD removal increased. The improvement in plant performance and the decrease in effluent toxicity may be due to operational changes (e.g., increased solids recycle) and/or the acclimation of the biomass to the influent toxicity.

Another event of sustained high influent toxicity occurred from June 20 to July 7, 1985 (Figure 3). As in the previous example, the increase in influent toxicity is reflected in an increase in effluent toxicity. However, in this case, the apparent effect of toxicity on plant performance lagged the occurrence of influent and effluent toxicity. The MLSS concentration and COD removal decreased three to four days following the onset of high influent toxicity. Possible reasons

Patapsco Wastewater Treatment Plant 609

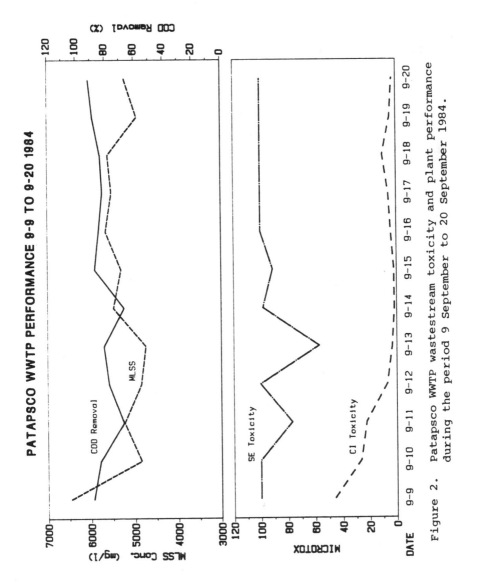

Figure 2. Patapsco WWTP wastestream toxicity and plant performance during the period 9 September to 20 September 1984.

610 Biotechnology for Degradation of Toxic Chemicals

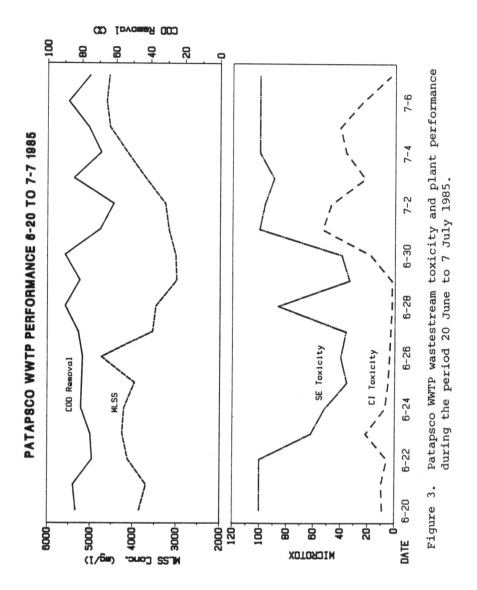

Figure 3. Patapsco WWTP wastestream toxicity and plant performance during the period 20 June to 7 July 1985.

for the delayed effect of toxicity on plant performance include: (1) an increase in recycle solids flow from 16 to 25 mgd on the 26th and 27th temporarily mitigated toxicity effects, and (2) the toxic compounds exhibited largely chronic toxicity to the plant biomass.

SUMMARY

The evaluations performed to date of the historical data demonstrate that the pass-through of toxicity at the Patapsco WWTP can be related in part to plant performance and operating conditions. Effluent toxicity was correlated with reduced plant performance (e.g., BOD and COD removal) and operation of the plant outside of the design F/M ratio. Although the historical influent data (i.e., BOD, COD, SS and toxicity) were not found to be related to effluent toxicity, recent data that account for the HRT of the wastewater will be used to re-evaluate this relationship.

The evaluation of plant performance during toxic events found additional evidence for the relationship of plant performance to toxicity pass-through. Effluent toxicity appeared to be related to influent toxicity, MLSS concentration and COD removal. However, it was not determined whether wastewater toxicity was a result of poor performance or vice versa.

INITIAL STUDY RESULTS

Wastewater Toxicity Characterization

The unknown relationship between toxicity as measured by the in-plant monitors (Microtox, respirometry and ATP assays) and the standard EPA bioassays for NPDES toxicity regulation are being evaluated by comparing the plant's toxicity monitoring tools with the EPA 48-hour Ceriodaphnia dubia and 96-hour Mysidopsis bahia acute tests and the 7-day C. dubia chronic test. The bioassays are being performed on 24-hour composite samples of the plant's primary settling tank effluent (PRIM EFFL) and secondary unchlorinated effluent (SE) approximately twice each week during the TRE. Each bioassay in a test series requires appropriate aliquots of the same composite samples used for in-plant toxicity tests and conventional characterization analyses. The bioassay procedures involve static renewal steps using a portion of the original composite sample for each static renewal addition.

Less than 8 hour old C. dubia neonates (young) and 1 to 5 day old post-release M. bahia juveniles are being tested according to procedures outlined in Peltier and Weber (6). The C. dubia 7-day survival and reproduction testing is following procedures described in Horning and Weber (7). The culture and dilution water used for testing C. dubia is Patapsco River water being collected from Henryton Bridge, Howard County, Maryland; for M. bahia the water is 20 ppt synthetic sea water (Forty Fathoms, Marine Enterprises, Towson, Maryland).

Initial TRE toxicity data are presented in Tables IV and V. To date, C. dubia has been the most sensitive indicator of acute toxicity for both the primary effluent and the secondary effluent wastewaters (Table IV). The acute responses for M. bahia and Microtox have been considerably less sensitive than C. dubia. The Ceriodaphnia were approximately 8 and 9.9 times more sensitive than M. bahia and Microtox, respectively; while M. bahia were approximately 1.5 times more sensitive than Microtox. The acute response of C. dubia to secondary effluent does not appear to follow the trend of responses of M. bahia or Microtox. For example, from 9 to 23 April the 96-hour LC_{50}s of secondary effluent for M. bahia steadily increased from 34.8 to 63.7 percent effluent, while during the same time C. dubia LC_{50}s only ranged from 2.9 to 5.5 percent effluent. The EC_{50} values determined from Microtox appeared more similar to the acute responses of M. bahia than C. dubia.

Results of all three toxicity tests show that significant toxicity reduction occurs across the plant's secondary treatment system. The primary effluent was approximately 3.3 times (range of 1.9 to 7.9) more acutely toxic to C. dubia and M. bahia than the secondary effluent (Table IV). Microtox showed a greater difference than the other two species, i.e., the primary effluent was approximately 9.6 times (range of 1.8 to 28.6) more toxic than the secondary effluent.

The 7-day chronic values (MATCs) for the primary and the secondary effluent were only slightly lower than the 48-hour LC_{50}s (Table V). The acute/chronic ratios for Ceriodaphnia, calculated as the 48-hour LC_{50} divided by the 7-day MATC,

Table IV. Acute Toxicity Data for Three Aquatic Organisms Exposed to Influent and Effluent Wastewater from Patapsco WWTP

Sample[a]	Sample Date	Microtox 5-min. EC_{50}	Mysidopsis bahia 96-hour LC_{50} (95% Conf. Int.)	Ceriodaphnia dubia 48-hour LC_{50} (95% Conf. Inf.)	Ceriodaphnia dubia 7-day LC_{50} (95% Conf. Inf.)
SE	4-9-86	61.3	34.8 (30-100)	5.5 (3-10)	5.5 (3-10)
SE	4-10-86	52.9	34.7 (26.0-47.0)	5.5 (3-10)	5.5 (3-10)
Prim Effl	4-16-86	34.7	b	b	b
SE	4-16-86	81.5	44.1 (36.1-67.2)	5.5 (3-10)	5.5 (3-10)
Prim Effl	4-17-86	53.2	b	b	b
SE	4-17-86	93.2	49.0 (35.3-77.4)	5.5 (3-10)	2.1 (1-3)
Prim Effl	4-23-86	19.5	21.1 (16.2-27.2)	2.9 (1.6-3.2)	1.1 (0.8-1.4)
SE	4-23-86	100	63.7 (47.9-77.4)	5.2 (3.2-6.5)	4.2 (3.2-6.5)
Prim Effl	4-30-86	9.3	27.9 (20.0-33.2)	2.2 (1.6-3.2)	1.1 (0.8-1.4)
SE	4-30-86	100	62.5 (39.3-80.4)	9.4 (6.5-13.5)	5.9 (3.2-6.5)
Prim Effl	5-8-86	24.0	b	b	b
SE	5-8-86	88.9	70.7 (50-100)	4.6 (3.2-6.5)	1.8 (0.8-3.2)
Prim Effl	5-14-86	19.2	33.3 (25-50)	4.6 (3.2-6.5)	b
SE	5-14-86	57.2	80.6 (50-100)	9.4 (6.5-13.5)	2.3 (1.6-3.4)
Prim Effl	5-28-86	13.6	20.6 (12.5-50)	2.2 (1.6-3.2)	1.2 (0.8-3.2)
SE	5-28-86	100	56.7 (25-100)	17.3 (6.5-40.5)	5.5 (3.9-8.1)
Prim Effl	6-4-86	4.3	b	b	b
SE	6-4-86	100	53.5 (25-100)	23.4 (13.5-40.5)	9.4 (6.5-13.5)
Prim Effl	6-5-86	3.5	b	b	b
SE	6-5-86	100	57.0 (50-100)	10.7 (6.5-40.5)	8.1 (3.2-13.5)

[a] SE = Secondary effluent
Prim Effl = Primary effluent
[b] Acute tests not performed on this sample

Table V. Chronic Toxicity Data for *Ceriodaphnia dubia* Exposed to Influent and Effluent Wastewater from Patapsco WWTP

Sample[a]	Sample Data	7-Day Chronic End Points[b] (Expressed as Percent Sample)		
		LOEC[c]	NOEC[d]	MATC[e]
SE	4-9-86	10.0	3.0	5.5
SE	4-10-86	10.0	3.0	5.5
SE	4-16-86	10.0	3.0	5.5
SE	4-17-86	10.0	3.0	5.5
Prim Effl	4-23-86	3.2	1.5	2.2
SE	4-23-86	6.5	3.2	4.6
Prim Effl	4-30-86	1.5	0.7	1.0
SE	4-30-86	6.5	3.2	4.6
SE	5-8-86	3.2	1.5	2.2
SE	5-14-86	6.5	3.2	4.6
Prim Effl	5-28-86	1.5	0.7	1.0
SE	5-28-86	13.5	6.5	9.4
SE	6-4-86	13.5	6.5	9.4
SE	6-5-86	13.5	6.5	9.4

[a] SE = secondary effluent, Prim Effl = primary effluent
[b] Chronic values based on *Ceriodaphnia* reproductive potential expressed as neonates per surviving female.
[c] LOEC = Lowest Observable Effect Concentration
[d] NOEC = No Observable Effect Concentration
[e] MATC = Maximum Acceptable Toxic Concentration

ranges from 1 to 2.5 which is not an uncommon acute/chronic ratio for municipal effluents (8). C. dubia appear to either survive and produce normal numbers of young when exposed to secondary effluent, or die; very little supression of young production was observed. This trend also appears to be the same for primary effluent for the four samples tested. In fact a slight stimulation of young production was observed at low effluent test concentrations when compared to the river water controls. This reproductive enhancement is probably due to added nutrients and/or bacterial food sources contributed by the effluent which are beneficial to Ceriodaphnia at very low effluent concentrations. However, this does not necessarily mean that the effluent is beneficial to the receiving water system, because the increase in reproduction may be an artifact of the testing procedures.

It is difficult to determine why the relative sensitivity of C. dubia and M. bahia differ in Patapsco effluent because only a small data base exists in the literature on a few chemicals. No studies could be found that compared C. dubia with other species, however, a few studies have compared C. reticulata, which is the same genus as C. dubia, to other organisms. Mount and Norberg (9) suggested that C. reticulata, Daphnia pulex, and D. magna have similar sensitivities to effluents and single chemicals, and generally did not differ in toxic response by more than a factor or two when compared on 13 substances tested. LeBlanc (10) compared the acute sensitivity of several species, including D. magna and M. bahia, to nonpesticide organics. For the 19 different chemicals investigated, M. bahia was more sensitive to 13 compounds, in four cases D. magna was more sensitive, and for two chemicals, comparisons could not be made.

Based on other studies it can be concluded that C. dubia and M. bahia are similar in their responses to many substances that are toxic. However, it is difficult to determine the reason or reasons that M. bahia are less sensitive to the Patapsco WWTP wastewaters in this study. One possible reason may be that the toxicity of the effluent is physically and/or chemically modified by the addition of the synthetic sea salts. The salts added to the test samples may coagulate or bind toxic organics and metals making them unavailable to the test organisms, and thereby reducing the toxicity to M. bahia.

FUTURE PROJECT TASKS

Characterization of Plant Performance

In-plant toxicity is currently being monitored by Microtox and a modified respirometer test. In addition the TRE is evaluating an ATP test as a potential toxicity indicator method. The ATP test involves performing a bench-scale batch treatment test of a wastewater using the plant's activated sludge, and measuring the ATP content of the batch test activated sludge over time. To simulate the plant's biological treatment process, the wastewater and return activated sludge are mixed in a ratio consistent with plant F/M, and the mixed liquors are aerated for a period of time equivalent to the average detention time of wastewater in the plant reactors. The ATP, Microtox, respirometry data will be used to assess toxicity reduction across each treatment process. The fate and impact of toxicity during wastewater treatment will be evaluated by correlating the toxicity data with the conventional data used to characterize plant performance.

Identification of Specific Toxics

The existing data on specific priority pollutants and non-priority organic pollutants are being evaluated to identify specific toxic compounds in the Patapsco WWTP wastestreams. In addition, a new toxics identification approach developed by Don Mount (EPA Diluth, MI) will be applied to plant influent and effluent wastewaters in an attempt to identify specific components of toxicity entering and leaving Patapsco WWTP. This test involves fractionation of the wastewater and testing the fractions for toxicity. Wastewater fractions found to have high toxicity will be analyzed by GC/MS in an effort to determine the specific contributors to toxicity. The wastewater sampling and fractionation for this approach is scheduled for one warm weather month (July) and one cold weather month (November) during the TRE.

A schematic of the toxics identification procedure is presented in Figure 4. Initially the whole wastewater and fractions of wastewater treated by aeration, filtration, and pH adjustment will be tested for toxicity using acute or time lethality _Ceriodaphnia_ tests. The toxicity test on aerated wastewaster will indicate if toxicity is associated with volatile compounds. The filtration step is designed to determine whether toxicity is found in the suspended particulate

Patapsco Wastewater Treatment Plant 617

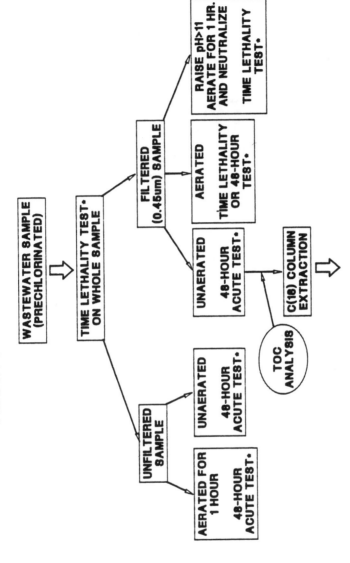

Figure 4. Toxicity identification procedure schematic. (continued on next page)

618 Biotechnology for Degradation of Toxic Chemicals

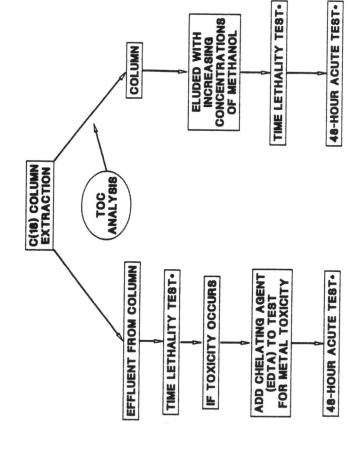

Figure 4. Toxicity identification procedure schematic.

phase or the soluble phase of the wastewater. Aeration in conjunction with pH adjustment will be used to test for ammonia toxicity. If the filtered unaerated fraction is found to be toxic, further fractionation will be conducted to determine whether toxicity is associated with non-polar organic compounds or metals that may be present in the wastewater. Fractionation of non-polar organics will be accomplished by eluting the wastewater with varying concentrations of methanol through a C_{18} column. Fractionation for metals partitioning will be achieved by adding varying concentrations of a chelating agent (EDTA) to aliquots of the eluted wastewater. The fractions of wastewater generated by the Mount procedure will be tested for toxicity using acute and time lethality <u>Ceriodaphnia</u> tests and wastewater fractions found to have high toxicity will be analyzed by GC/MS.

<u>Wastewater Treatability</u>

Although the Patapsco WWTP has stabilized since start-up problems in 1981 to 1983, the City is concerned that the current influent wastewater toxicity may be having an impact on the efficiency of operation of the biological treatment system. Specifically, the removal of BOD per pound of activated biomass may be lower than normal for domestic sewage. A second concern relates to the pass-through of wastewater toxicity following treatment. There is a need to demonstrate that the plant, in addition to having the capability of achieving BOD/SS permit concentrations, can achieve effective detoxification of toxic influents at the current F/M. To address these concerns, wastewater treatability studies will be conducted to assess the capability of the plant biomass for substrate (BOD and COD) removal and for detoxification of industrial wastewaters currently entering the plant. These studies will involve batch treatment testing of toxic industrial effluent samples from each of eight candidate industries using the plant's return activated sludge as the test biomass. Batch tests are scheduled in a warm weather month (July) and a cold weather month (December) to coincide with the toxics identification task.

The treatability test procedure will utilize 24-hour composite samples of selected industrial wastewaters. For each industrial sample, a range of three sample dilutions will be made that includes the dilution (or a dilution factor) of industrial wastewater expected in the Patapsco WWTP influent. The sample dilutions will be mixed with plant biomass in a ratio consistent with current plant F/M and a batch test will be performed in the Arthur Techline respirometer (3). The test length will be equivalent to the detention time of the plant reactors (2 hours). During this test period, measurements of oxygen uptake and substrate (BOD and COD) utilization will be made. At the completion of the test, the original wastewater and the settled supernatant (effluent) of the batch reactor will be analyzed for BOD, COD and Microtox toxicity. Selected effluents will also be analyzed for toxicity by the Ceriodaphnia time lethality test.

Industrial wastewaters found to have high pass-through toxicity following batch treatment will be re-sampled and tested again. In addition, the new toxics identification procedure will be performed on a batch test effluent found to have high Microtox or Ceriodaphnia toxicity in an attempt to identify specific toxic components. The treatability tests will provide perspective on the effect of influent wastewater toxicity on the activity of the plant biomass, and the pass-through of BOD, COD and toxicity in the plant effluent.

PROJECT BENEFITS

Considerable progress has already been achieved on a TRE at the Patapsco WWTP with regard to the City's historical toxicity database and the development of a TRE approach. As the TRE proceeds, important information will be gained for used in the development of an effective toxics management program at Patapsco. Aspects of the TRE that will support the establishment of such a program is the relationship of the pragmatic in-plant toxicity monitors to the EPA's regulatory bioassays and the results of the toxics identification and treatability assessments.

The documentation of toxicity management methodology developing at Baltimore offers the EPA important technical support for the Water Quality-Based Toxics Control Policy. Another key benefit of the TRE is the information that will be provided on potential in-plant operational changes and pretreatment alternatives needed to minimize plant upset and toxicity pass-through.

REFERENCES

1. Engineering-Science, <u>Evaluation of Alternatives for the Reduction of Toxic Inhibition at the Patapsco Waste Water Treatment Plant,</u> Prepared for the City of Baltimore, Baltimore, MD, 1985.

2. Beckman Instruments, <u>Microtox™ Operating System Manual,</u> Beckman Instruments, Inc., Carlsbad, CA, 1982.

3. Techline Instruments, <u>Techline Laboratory Respirometer Operating Manual,</u> Techline Instruments, Inc., Fond du Lac, WI, 1984.

4. 3M-Lumac, <u>Operations Manual: Lumac/3M Biocounter,</u> 3M-Lumac, Minneapolis, MN, 1981.

5. Peltier, W. and C.I. Weber, Methods for Measuring the Acute Toxicity of Effluents to Freshwater and Marine Organisms, U.S. Environmental Protection Agency, Cincinnati, OH. EPA 600-4-85-013, 1985.

6. Horning, W.B., II and C.I. Weber, Short-term Methods for Estimating the Chronic Toxicity of Effluents and Receiving Waters to Freshwater Organisms, U.S. Environmental Protection Agency, Cincinnati, OH. EPA-600-4-85-014, 1985.

7. EA Engineering, Science, and Technology, Inc. (EA), <u>Ceriodaphnia</u> sp. Reproductive Potential Tests on Ambient Stations and Selected Effluents, Naugatuck River, Connecticut, Prepared for U.S. Environmental Agency, Monitoring and Support Division, Washington, D.C. EA Report RTI41A, 1984.

8. Mount, D.I. and T.J. Norberg, A Seven-day Life-cycle Cladoceran Toxicity Test, Environmental Toxicology and Chemistry 3:425-434, 1984.

9. LeBlanc, G.A., Interspecies Relationships in Acute Toxicity of Chemicals to Aquatic Organisms. Environmental Toxicology and Chemistry 3:47-60, 1984.

MICROTOX ASSESSMENT OF ANAEROBIC BACTERIAL TOXICITY

Doris S. Atkinson and Michael S. Switzenbaum

Environmental Engineering Program
Dept. of Civil Engineering
University of Massachusetts
Amherst, Massachusetts

INTRODUCTION

Recently anaerobic treatment processes have been receiving increased attention owing to their advantages over aerobic biological treatment processes. These advantages include the production of smaller amounts of waste sludge and the production of methane gas as a useful by-product. One concern about the use of anaerobic treatment methods is the reliability of processes for wastes which may contain substances toxic to the methanogenic microorganisms. In response to this concern, bioassay methods have been developed for measuring the presence of inhibitory substances in wastewaters being treated by anaerobic processes.

The anaerobic toxicty assay (ATA) was initially developed by Owen et al. (1) and is based on techniques developed by Hungate (2) which were modified by Miller and Wolin (3). It is currently one of the more widely used methods of determining anaerobic toxicity. The ATA is a batch method which measures the adverse effect of a substance or mixture on the rate of methane production from an easily degraded methanogenic substrate. While this method is simple and relatively inexpensive, it is time consuming, requiring up to two weeks before results are available.

For improved process monitoring of anaerobic treatment systems which may receive influents of variable composition, it would be useful to have a quick, inexpensive, screening test which could be used as a surrogate to the ATA test. The Microtox ^{TM} toxicity analyzer, developed by Beckman Instruments and currently marketed by Microbics Corporation of Carlsbad, CA., provides a means for rapid, inexpensive assessment of toxicity of aqueous samples. The Microtox system is a relatively inexpensive test which employs aerobic bioluminescent marine bacteria (Photobacterium phosphoreum) and can yield reproducible results within one hour. Good correlations between Microtox and rat, fish, daphnid, shrimp, algal, and other aerobic bacterial bioassays have been reported in the literature (4-14).

The primary objective of this research has been to determine whether the Microtox system can be used as a suitable surrogate test for the longer ATA test. The study was conducted in two parts. The available literature on both anaerobic toxicity and Microtox testing was reviewed and the reported toxcity levels for chemicals tested by both sytems were compared. Further laboratory tests using both toxicty methods were also performed. The combined literature and laboratory data have been analyzed, and the results of these studies are discussed in terms of the applicability of the Microtox system as a prescreening tool for anaerobic treatment systems.

BACKGROUND

Anaerobic Toxicity Testing

Over the last twenty or so years a number of different approaches to anaerobic toxicity testing have been used. Tests may be conducted under batch (1,15-25) or continuous flow conditions(17-19,21-23,26-28). The organisms may be in suspended culture or attached (27). The culture used may be a mixed cuture from a reactor, may be a washed cell suspension prepared from reactor contents (25), or may be a pure culture (29) isolated in the laboratory. Tests may be conducted with non-acclimated organisms or organisms which have had some previous exposure to the toxicant being evaluated (15,18,21-23,27,28) and a variety of different substrates may be used as the food source. These test parameters as well as other important factors such as solids retention time, temperature, pH, osmolarity, the presence or absence of antagonistic or synergistic substances will all bear on the test results.

The anaerobic toxicity assay (ATA) is a batch test, which uses suspended, mixed methanogenic cultures obtained from reseach reactors

(1). The organisms are fed acetate and propionate as an easily degraded methanogenic substrate. The toxic substance to be tested may be either a pure substance or a complex mixture. While the test was designed to be used with non-acclimated organisms, it can be adapted to aclimation studies. The test is usually run under quiescent conditions, but continuous stirring may be used by adding magnetic stirrer bars to individual serum bottles. The pH of the test environment is held at or near 7.1 by means of a bicarbonate buffer. The incubation temperature is 35° C. The parameter measured in the ATA test is either the rate of total gas production or the rate of methane production. The volume of gas produced is measured using a lubricated glass syringe. The suggested means of reporting data is to present the maximum rate ratio (MRR) which is the maximum rate of gas production of the toxified sample normalized to that of a non-toxified control.

The ATA is a simple, inexpensive test, which has advantages over more complicated continuous toxicity tests. The ATA test is a practical test which has been used sucessfully by a number of investigators to test a variety of substances (1,15,17,18,20,21,23). However, it is a time consuming test (requiring up to two weeks before tests are available) and is therefore, not as practical for applications such as influent monitoring.

Microtox Testing

The Microtox test was originally developed by Beckman Instruments, but is now marketed by Microbics Corporation of Carlsbad, California. The Microtox test is a quick, relatively inexpensive bioassay which uses the aerobic, bioluminescent marine bacterium Photobacterium phosphoreum. The light emitting biochemical pathway used by the Microtox organism is known to be an electron transport pathway, providing the organism with an alternative to cytochrome electron transport. This pathway is thought to provide the organism with an adaptive advantage under microaerophilic conditions (30-32).

The principle of the Microtox test is that the intensity of bioluminescence is diminished in response to exposure to toxicants. This response is generally linear over some range of toxicant concentration. The light output at specified temperature, pH, and salinity is easily measured with the Microtox Analyzer, which contains incubation wells, temperature controls and a photomultiplier tube connected to a digital output display. The analyzer can also be connected to a strip chart recorder for a permanent record of the test data. Once the bacteria have been prepared for the test procedure, the initial light level is measured, a toxic challenge added, and the light level measured again after a specified period of exposure.

Generally the concentration of toxicant causing a 50% decrease in light output relative to a reagent blank is the reported result (33).

The Microtox test is a versatile test, which may be usd to evaluate almost any aqueous sample. The test is short (one to two hours) and requires no laboratory culturing of test organisms (the manufacturer supplies standard cultures of lyophilized bacteria in easy to handle single test reagent vials). It also has good reproducibility, with a coefficient of variation of 15 to 20% (4). The Microtox system previously has been used for effluent monitoring (4,10,13,14) as well as wastewater treatability monitoring (34), studies of toxicity removal in activated sludge treatment plants (35), evaluation of fossil fuel process waters (9), landfill leachate studies (36) and sediment toxicity studies(37).

Good correlations between Microtox tests and a number of other bioassays using organisms from a wide variety of phylogenetic groups have been found (4-14). Chang et al. found correlation coefficients (r) of 0.9 and 1.0 between Microtox results and rat and fish LD50's respectivily for typical organic toxicants (5). Curtis et al. and Indorato et al. found correlation coeficients between Microtox and different species of fish between 0.80 and 0.95 (6,8). Looking at 20 chlorophenols, 12 chlorobenzenes and 13 para-substituted phenols, Ribo and Kaiser compared results of 7 different bioassays with Microtox results (12). Two bacterial assays, three fish assays, and Daphnia and shrimp assays were included. Correlation coefficients for these studies all fell between 0.82 and 0.96. No examples of Microtox comparisons to anaerobic systems were found in the literature.

METHODS

Literature Study

The literature on anaerobic toxicity and Microtox testing was reviewed. Reported toxicity values for chemicals tested both by Microtox and anaerobic methods were compared. Initial classification of the data included dividing chemicals into three categories: more toxic to Microtox, same order of magnitude toxicity, and more toxic to anaerobes. For chemicals for which published reports contained detailed enough information, either estimates of the concentation, or the reported concentration causing 50% inhibition of the rate of total gas or methane production was plotted against the concentration causing 50% inhibition in the Microtox assay.

Anaerobic Toxicity Assays (ATA's)

Anaerobic toxicity assays (ATA's) were performed following the method developed by Owen et al. (1). The assay was conducted in 125 mL serum bottles which contained defined media, anaerobic seed inoculum, toxicant, and a methanogenic substrate. Total gas production or methane production was monitored over a two week incubation period, and the rates of gas production in toxified bottles were compared to non-toxified controls.

The defined medium, containing vitamins, nutrient salts, bicarbonate, sodium sulfide as a reducing agent, and resazurin to detect oxygen contamination, was transferred anaerobically to 70% nitrogen - 30% carbon dioxide purged serum bottles. A 30% by volume seed inoculum was then added. Seed was obtained from a 15L mixed culture anaerobic research reactor operated at 20 day solids retention time, fed a sucrose based feed solution. Varying concentrations of toxicants were added via syringe in volumes up to 2 mL. Similarly, a 2 mL substrate "spike" containing acetate and propionate was added, bringing the final liqid volume to 50 - 52 mL. Bottles were then equilibrated to the incubation temperature of 35° C, and the gas volume of each zeroed to ambient atmospheric pressure using a lubricated glass syringe. Total gas production, measured by the glass syringe method, as well as percent methane, determined by gas chromatography, were measured periodically over a two week incubation period. At each measurement, bottles were zeroed to ambient pressures. The maximum rate ratio (MRR) for each toxicant concentration was determined as the maximum rate of gas production for the toxified sample normalized to the average maximum rate of gas production of the non-toxified controls. The concentration of toxicant causing a 50% reduction in gas production rates was then determined by interpolation of the data.

Microtox Toxicity Assays

Microtox toxicity assays were performed following the procedures described by the manufacturers (33). The Microtox system employs lyophilized aerobic marine bacteria which, upon reconstitution, emit light. The initial light level is measured; a toxic challenge is added and the percent light loss due to the toxicant is determined. Up to four concentrations may be tested at a time.

Microtox reagent (lyophilized bacteria) supplied by the manufacturer, was reconstituted with a buffered saline reconstitution solution (also supplied by the manufacturer). The reconstituted bacteria were allowed to equilibrate for at least 15 minutes at the test temperature (15° C), after which, 10 microliter aliquots were pipetted into 10 vials, each containing 0.5 mL of 2.2% saline dilution fluid. After 15-20 minutes the initial light level of each vial was measured and recorded. Then either 0.5 mL dilution fluid (blank) or 0.5 mL of one of four serial dilutions of toxicant in dilution fluid was added to each of 2 of the 10 test vials. After five minutes exposure time the light level of each vial was again measured, and the percent light loss relative to the average of the reagent blanks was calculated. The percent light loss was plotted against toxicant concentration, and the concentration resulting in 50% light loss determined by interpolation. This value is referred to as the five minute, 50% effective concentration or 5EC50.

RESULTS

Literature Review

Table 1 presents the results of the initial literature review. Literature data for both Microtox and anaerobe toxicity was found for a total of 39 chemicals. Of these, 24 (62%) were found to be more toxic to Microtox. Eleven (28%) were of the same order of magnitude toxicity to both Microtox and methanogens, and the remaining 4 (10%) were more toxic to methanogens. Considerably more data are available for Microtox testing than for anaerobe testing, and are reported in a more consistant manner. Many different types of anaerobic toxicity data were included in the initial review. Some of the data are based on mixed culture studies while others are based on enriched or pure cultures of methanogens. Many procedural differences were found in the anaerobe literature. For some chemicals, reports were found simply stating that a certain chemical was not toxic at a given concentration. When this value was appreciably above the reported Microtox 5EC50, the chemical was listed as more toxic to Microtox.

For chemicals for which there were more detailed literature data, the concentration of toxicant causing 50% anaerobic inhibition was determined. The reported values, or in some cases estimates, were plotted against the Microtox 5EC50 values on logarithmic axes. This information is presented graphically in Figure 1and in tabular form in Table 2. In cases where more than one Microtox value was available, the average value was plotted. Statistical analysis of the data yielded a correlation coefficient of 0.77. The solid line shown on Figure 1 is the line of equal value. Ten out of 14 data

628 Biotechnology for Degradation of Toxic Chemicals

Table 1 Relative Toxicity - Literature Survey

More Toxic to Microtox	Equally Toxic (Same Order of Magnitude)	More Toxic to Methanogens
Acrolein 16; 38	Ammonia (Total) 21,47; 11,38	Acrylonitrile 16; 38
Ammonia (Free) 22; 4,8,11,38	Cadmium 26; 38,48	Chloroform 25-27,39;11,38
Arochor 1242 20; 4	Carbon tetrachloride 25,49,50;38	1,2-Dichloroethane 14;43
Benzyl Alcohol 39; 38	Copper 16,26,44,51; 4,38	Methanol 27; 6
Catechol 16,39,45; 9,38	m-Cresol 39,40; 5,38	
4-Chloro-3-methylphenol 20; 38	Ethylacetate 16,39; 38	
2-Chlorophenol 20; 6	Nickel Chloride 21,44; 7,38,48	
p-Cresol 39,40; 4,9	Nitrite 52; 4,38	
Cyanide 27,41; 4,5,8,11,13	Nitrobenzene 16,20; 38	
2,4-Dicholorophenol 20; 6	4-Nitrophenol 20,39,40; 8	
2,4-Dimethylphenol 20; 6	2-Propanol 16; 6,38	

Table 1 - Relative Toxicity - Literature Survey (Continued)

More Toxic to Microtox

Dimethylphthalate 20,39; 38
2,4-Dinitrophenol 16,42; 6,8
Formaldehyde 15,16,21,2224, 27; 4,5,38
Mercuric Chloride 43,44; 4,7, 11,38
m-Methoxyphenol 39,40; 38
Naphthalene 20; 13
1-Octanol 39; 6
Pentachlorophenol 18,39; 6
Phenol 16,20,22,24,39,45; 4-8, 9,11,13,38
Resorcinol 16; 9
1,1,2,2-Tetrachloroethane 46; 6, 38
2,4,6-Trichlorophenol 20; 6,38
Zinc Sulfate 22,44; 7,11,38

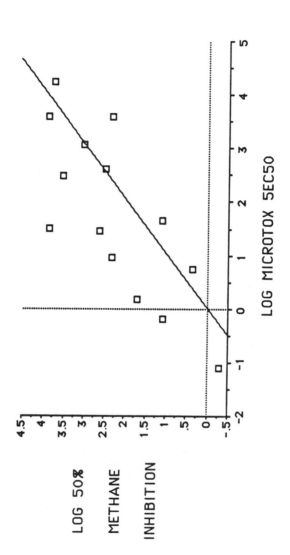

FIGURE 1 – MICROTOX 5EC50 vs METHANE INHIBITION LITERATURE DATA

Table 2 - Literature Toxicity Values

Chemical	Anaerobe 50% Inhibition	Microtox 5EC50
Acroleine	11.2 mg/L [16]	0.67 mg/L [38]
Acrylonitrile	212 mg/L [16]	3910 mg/L [38]
Ammonia (Free)	59 mg/L [22]	1.56 mg/L [8,11,38]
Ammonium (Total)	750 mg/L [21]	4833 mg/L [11,38]
Carbon Tetrachloride	2.2 mg/L [22]	5.6 mg/L [38]
Catechol	6966 mg/L [16]	32 mg/L [9,38]
Ethyl acetate	969 mg/L [16]	1180 mg/L [38]
Formaldehyde	200 mg/L [22]	9.1 mg/L [5,38]
Nickel Chloride	300 mg/L [21]	410 mg/L [38]
Nitrobenzene	12.3 mg/L [16]	46.2 mg/L [38]
Pentachlorophenol	0.5 mg/L [18]	0.08 mg/L [6]
2-Propanol	5408 mg/L [16]	17700 mg/L [38]
Resorcinol	3193 mg/L [16]	310 mg/L [9]
Zinc Sulfate	400 mg/L [22]	29.7 mg/L [7,11,38]

Table 3 - Laboratory Results vs. Literature Values

		LABORATORY RESULTS	LITERATURE VALUES
MICROTOX			
PHENOL	5EC50 (ave.)	29.3 mg/L	31.1 mg/L
	(st. dev.)	4.5 mg/L	7.8 mg/L
	(n)	4	9
$HgCl_2$	5EC50 (ave.)	0.050 mg/L	0.069 mg/L
	(st. dev.)	0.004 mg/L	0.006 mg/L
	(n)	3	4
ATA	(Sample Concentration Causing 50% Inhibition)		
CHLOROFORM		0.91 mg/L	0.96 mg/L Thiel et al. (25)
PHENOL		400 mg/L	500 mg/L Pearson et al. (22)

points (71%) lie on or above this line. These represent chemicals which are either more toxic or equally toxic to Microtox.

Laboratory Results

In order to assure that our laboratory results could be reasonably compared to results reported in the literature, Microtox and ATA assays were performed using chemicals for which there were previously reported literature data. Table 3 compares results generated in our laboratory with literature values.

Fourteen chemicals were tested by Microtox and/or ATA methods. Where good literature values were available for one test, but not the other, literature values were included. Laboratory results are presented graphically in Figure 2 and in tabular form in Table 4. The correlation coefficient for the laboratory data is 0.10. Eight (57%) of the chemicals tested were of equal toxicity or more toxic to Microtox.

Combined Results

Figure 3 represents the combined data from the literature and laboratory studies. For the whole set of data, (28 points) the correlation coefficient is 0.43. A total of 18 (64%) chemicals were of equal toxicity or more toxic to Microtox.

The combined data set was broken into three subsets to determine whether particular classes of chemicals showed stronger correlations between the Microtox 5EC50 values and the concentration causing 50% inhibition of methanogenesis. The subsets evaluated were organic and inorganic chemicals, and priority pollutants (53). The results of the analysis are presented in Table 5, along with summary information for literature, laboratory and combined results.

DISCUSSION

A toxicity test may be useful as a surrogate for another test if it can be shown to satisfy at least one of two criteria. The first criterion is that a good surrogate test should have some predictable, quantifiable relationship to the test it is to replace. The second criterion is that the test should be as sensitive or more sensitive to toxicants and act as a conservative estimator of toxicity. The first criterion is best judged by a statistical relationship while the second may be judged by the percentage of toxic cases the test sucessfully identifies, or screens out.

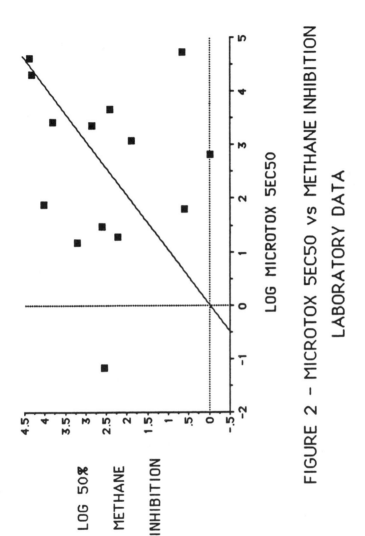

FIGURE 2 – MICROTOX 5EC50 vs METHANE INHIBITION LABORATORY DATA

Table 4 - Laboratory Results

Chemical	Anaerobe 50% Inhibition	Microtox 5EC50
Acetone	21000 mg/L	21700 mg/L
Arsenate	4.0 mg/L	64.5 mg/L [5,11]
Bacitracin	245 mg/L [19]	4780
Beryllium Sulfate	1500 mg/L	15 mg/L [38]
2-Bromoethanesulfonic Acid	4.35 mg/L [42]	55500 mg/L
n-Butyl Alcohol	6075 mg/L	2690 mg/L [4,6,8,38]
Chloroform	0.91 mg/L	677 mg/L [38,11]
1,2-Dichloroethane	75 mg/L [23]	1220 mg/L
Isopropyl Alcohol	23500 mg/L	42000 mg/L [4]
Mercuric Chloride	333 mg/L	0.050 mg/L
Methyl Isobutyl Ketone	10000 mg/L	80 mg/L [6]
Phenol	400 mg/L	29.3 mg/L
Potassium Chromate	160 mg/L	22 mg/L [38]
Vinyl Acetate	689 mg/L [16]	2081 mg/L

Table 5 - Summary Analysis

CATEGORY	n	r	% Equally Toxic to or More Toxic to Microtox
INITIAL SURVEY	--	--	90
LITERATURE DATA	14	0.77	71
LABORATORY DATA	14	0.10	57
COMBINED DATA	28	0.43	64
ORGANIC CHEMICALS	20	0.49	55
INORGANIC CHEMICALS	8	0.30	87
PRIORITY POLLUTANTS	17	0.27	65

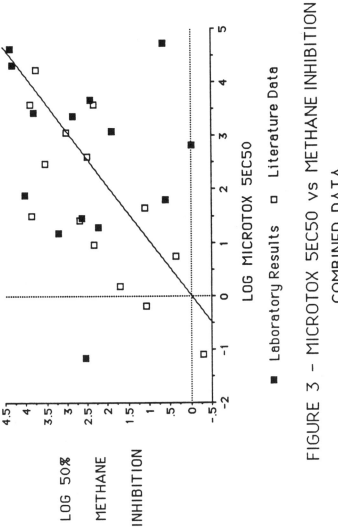

FIGURE 3 - MICROTOX 5EC50 vs METHANE INHIBITION COMBINED DATA

The initial literature survey looked promising in meeting the second criterion. Based on the initial survey, it would seem that the Microtox test should be able to identify methanogenic toxicants with approximately 90% reliability. (Only 10% of the chemicals were more toxic to methanogens.) Further study of literature values through regression analysis also looked promising, yielding a correlation coefficient of 0.77. This value is close to some of the values reported in studies comparing Microtox to other bioassays.

The laboratory data, however, did not yield such positive support for the use of Microtox as a surrogate for the ATA test. The correlation coeffient for this set of data was too low to reject the null hypothesis at any level of significance (54). The laboratory data also did not indicate Microtox as a good screening tool (43 % of chemicals were more toxic to methanogens). One reason for such a discrepancy between literature and laboratory results may be the method of choosing chemicals to test. In the literature study, data were gathered for each test without presupposition of the toxicity of chemicals to the other test. In the laboratory study, an attempt was made to look at some chemicals known to be specifically toxic to each system, some chemicals with very little toxicity to one of the tests, and at some which were thought likely to fall in some middle range. This deliberate testing of extremes may account for the lack of correlation of the data. The combined data show a slight correlation, and taken as a whole, 64 % of the chemicals studies were of equal or greater toxicity to Microtox.

Of the three subsets of the combined data which were evaluated, the organic chemicals showed only a slight correlation, and inorganic and priority pollutants insignificant correlation. However, Microtox seemed to be more suitable as a screening tool for inorganic toxicants than for other the other two groups. Only 1 of 8 inorganics (13%) was more toxic to methanogens. When considering the types of toxicants most likely to enter an anaerobic treatment unit, this observation is of interest. The 1980 EPA study of priority pollutants in publicly owned treatment works (55) lists the occurrence of priority pollutants in raw sludge samples as a percentage of times detected. Of the inorganic toxicants reported on in both this study and the EPA study, the range of percent times detected was from 32 to 98 % with an average detection rate of 82%. Organic toxicants, on the other hand, had a range of 2-62% with an average detection rate of 24%.

SUMMARY AND CONCLUSIONS

Overall, the results of this study do not indicate that Microtox would be expected to serve as a particularly good surrogate for the ATA to be used for monitoring potentially toxic wastes entering an anaerobic treatment unit. The Microtox may have an application for monitoring waste streams which may be subject to inorganic toxicants but are unlikely to be contaminated with organic toxicants. The work presented in this paper has been directed at evaluating the toxicity of pure compounds. Studies on toxified sludges would help in further evaluating the Microtox as a surrogate for the ATA.

ACKNOWLEDGEMENT

This research has been funded by the Massachusetts Division of Water Pollution Control (contract no. 83-31). The authors would also like to thank R. Hickey, J. Robins and D. Wagner for their technical assistance.

REFERENCES

1. Owen, W.F., Stuckey, D.C., Healy, J.B.Jr., Young, L.Y., and McCarty, P.L., "Bioassay for Monitoring Biochemical Methane Potential and Anaerobic Toxicity," *Water Research*, Vol. 13, 1979, pp.485-492.
2. Hungate, R.E., " A Role Tube Method for Cultivation of Strict Anaerobes," *Meth. Microbiol.*, Vol. 3B, 1969, pp. 117-132.
3. Miller,T.L., and Wolin, M.J., " A Serum Bottle Modification of the Hungate Technique for Cultivating Obligate Anaerobes," *Appl. Microbiol.*, Vol. 27, 1974, pp. 985-987.
4. Bulich, A.A., Greene, M.W., and Isenberg, D.L., " Reliability of the Bacterial Bioluminescence Bioassay for the Determination of the Toxicity of Pure and Complex Effluents," In *Aquatic Toxicology and Hazard Assessment: Fourth Conference*, D.R. Branson and K.L. Dickson (eds.), Amer. Soc. Testing and Materials, Spec. Techn. Publ. 737, 1981, pp. 338-347.
5. Chang, J.C., Taylor, P.B., and Leach, F.R., " Use of the Microtox Assay System for Environmental Samples," *Bull. Environm. Contam. Toxicol.*, Vol. 26, 1981, pp. 150-156.
6. Curtis, C., Lima, A., Lozano, S.J., and Veith, G.D., " Evaluation of a Bacterial Bioluminescence Bioassay as a Method for Predicting Acute Toxicity of Organic Chemicals to Fish," In *Aquatic Toxicology and Hazard Assessment: Fifth Conference*, J.G. Pearson, R.B. Foster and W.E. Bishop (eds.), Amer. Soc. Testing and Materials, Spec. Techn. Publ. 737, 1982, pp. 170-178.
7. Dutka, B.J., and Kwan, K.K., " Comparison of Three Microbial Toxicity Screening Tests with the Microtox Test," *Bull. Environ. Contam. Toxicol.* Vol. 27, 1981, pp. 753-757.
8. Indorato, A.M., Snyder, K.B., and Usinowicz, P.J., " Toxicity Screening Using Microtox," First International Symposium on Toxicity Testing Using Bacteria, at The National Water Research Institute, Burlington, Ontario, Canada May 1983.
9. Lebsack, M.E., Anderson, A.D., DeGraeve, G.M., and Bergman, H.L., "Comparison of Bacterial Luminescence and Fish Bioassay Results for Fossil-Fuel Process Waters and Phenolic Constituents," in *Aquatic Toxicology and Hazzard Assessment: Fourth Conference*, D.R. Branson and K.L. Dickson (eds.), Amer. Soc. for Testing and Materials, Spec. Techn. Publ. 737, 1981, pp. 348-356.
10. Peltier, W., and Weber, C.I., " Comparison of the Toxicity of Effluents to Fish, Invertebrates and Microtox," Aquatic Biology Section Biological Methods Branch Environmental Monitoring and Support Laboratory U.S. Environmental Protection Agency Cincinnati, Ohio July, 1980.
11. Qureshi, A.A., Flood, K.W., Thomson, S.R., Janhurst, S.M., Inniss, C.S., and Rokosh, D.A., " Comparison of a Luminescent Bacterial Test with Other Bioassays for Determining Toxicity of Pure Compounds and Complex Effluents," In *Aquatic Toxicology and Hazard Assessment: Fifth Conference*,

J.G. Pearson, R.B. Foster and W.E. Bishop (eds.), Amer. Soc. Testing and Materials, Spec. Techn. Publ. 737, 1982, pp. 179-195.

12. Ribo, J.M., and Kaiser, K.L.E., " Effects of Selected Chemicals to Photoluminescent Bacteria and their Correlations with Acute and Sublethal Effects on Other Organisms," *Chemosphere*, Vol. 12, 1983, pp. 1421-1442.

13. Samak, Q.M., and Noiseux, R., " Acute Toxicity Measurement by the Beckman Microtox," Presented at the Seventh Annual Aquatic Toxicity Workshop in Montreal, Canada, November, 1980.

14. Vasseur, P., Ferard, J.F., Vail, J., and Larbaight, G., "Comparison of Bacterial Luminescence and Daphnia Bioassay Results for Industrial Wastewaters Toxicity Assessment, " *Environmental Pollution*, Vol. 34, 1984, pp. 225-235.

15. Benjamin, M.M., Woods, S.L., and Ferguson, J.F., " Anaerobic Toxicity and Biodegradability of Pulp Mill Waste Constituents," *Water Res.*, Vol. 18, 1984, pp.601-607.

16. Chou, W.L., Speece, R.E., Siddiqi, R.H., and McKeon, K., " The Effect of Petrochemical Structure on Methane Fermentation Toxicity," *Prog. Wat. Tech.*, Vol. 10, 1978, pp. 545-558.

17. Eis, B.J., Ferguson, J.F., and Benjamin, M.M., " The Fate and Effect of Bisulfate in Anaerobic Treatment," *Journal WPCF*, Vol. 55, 1983, pp.1355-1365.

18. Guthrie, M.A., Kirsch, E.J., Wukasch,R.F., and Grady, C.P.L., " Fate of Pentachlorophenol in Anaerobic Digestion," in 1981 National Conference on Environmental Engineering, F.M. Saunders, (Ed.), Proceedings of the ASCE Environmental Engineering Division Specialty Conference Atlanta, Georgia, July 8-10, 1981, published by ASCE, New York.

19. Hilpert, R., Winter, J., and Kandler, O., " Agricultural Feed Additives and Disinfectants as Inhibitory Factors in Anaerobic Digestion," *Agricultural Wastes*, Vol. 10, 1984, pp. 103-116.

20. Johnson, L.D. and Young, J.C., " Inhibition of Anaerobic Digestion by Organic Priority Pollutants ," *Journal WPCF*,Vol. 55, 1983, pp.1441-1449.

21. Parkin, G.F., Speece, R.E., Yang, C.H.J., and Kocher, W.M., " Response of Methane Fermentation Systems to Industrial Toxicants ," *Journal WPCF.*, Vol. 55, 1983, pp.44-53.

22. Pearson, F., Shiun-Chang, C., and Gautier, M., " Toxic Inhibition of Anaerobic Biodegradation," *Journal WPCF.*, Vol. 52, 1980, pp. 472-482 .

23. Stuckey, D.C., Owen, W.F., McCarty, P.L., and Parkin, G.F., " Anaerobic Toxicity Evaluation by Batch and Semi-Continuous Assays," *Journal WPCF*, Vol. 52, 1980, pp.720-729.

24. Sullivan, D.E., and Lambert, W.P., " Development of a Short Term Anaerobic Digestion Effects Test," Presented at the WPCF 56th Conference, Industrial Waste Symposia, Atlanta, Georgia, Oct. 3-7, 1983.

25. Thiel, P.G., " The Effect of Methane Analogues on Methanogenesis in Anaerobic Digestion ," *Water Res.*, Vol. 3, 1969, pp. 215-223 .

26. Mosey, F.E., and Hughes, D.A., " The Toxicity of Heavy Metal Ions to

Anaerobic Digestion," *Wat. Pollut. Control,* Vol. 74, 1975, pp. 18-33.
27. Parkin, G.F., and Speece, R.E., " Attached Verses Susppended Growth Anarobic Reactors: Response to Toxic Substances ," *Wat. Sci. Tech.*, Vol.15 ,1983, pp. 261-289.
28. Varel, V.H., and Hashimoto, A.G., " Methane Production by Fermentor Cultures Acclimated to Waste from Cattle Fed Monensin, Lasalocid, Salinomycin, or Avoparcin ," *Appl. Environ. Microbiol.*, Vol. 44, 1982, pp. 1415-1420.
29. Jarrell, K.F., and Hamilton, E.A., " Effect of Gramicidin on Methanogenesis by Various Methanogenic Bacteria," *Appl. Environmental Microbiology,* Vol. 50, 1985, pp. 179-182.
30. Hastings, J.W., and Nealson, K.H., " Bacterial Bioluminescence," *Annu. Rev. Microbiol.*, Vol. 31, 1977, pp. 549-595.
31. Hastings, J.W., Potrikus, C.J., Gupta, S.C., Kurfurst, M., and Makemson, J.C., " Biochemistry and Physiology of Bioluminescent Bacteria," *Adv. Microbial Physiol.,* Vol. 26, 1985, pp. 235-291.
32. Nealson, K.H., and Hastings, J.W., " Bacterial Bioluminescence: Its Control and Ecological Significance," *Microbiol. Rev.*, Vol. 43, 1979, 496-518.
33. Beckman Inc., " Microtox System Opperating Manual," Beckman Instruments, Inc., 1982, pp. 1-59.
34. Slattery, G.H., " Biomonitoring and Toxicity Testing : Toxicant Treatability Definition and Measurement," Presented at the WPCF Analytical Techniques in Water Pollution Control Conference, Cininnati, Ohio, May 2-3, 1985.
35. Neiheisel, T.W., Horning, W.B., Petrasek, A.C., Asberry, V.R., Jones, D.A., Marcum, R.L., and Hall, C.T., " Effects on Toxicity of Volatile Priority Pollutants Added to a Conventional Wastewater Treatment System, " NTIS 1982.
36. Sheehan, K.C., Sellers, K.E., and Ram, N.M., " Establishment of a Microtox Laboratory and Presentation of Several Case Studies Using Microtox Data," Technical Report to the Massachusetts Dept. of Environmental Quality Engineering, University of Massachusetts, Engineering Reopt No. 77-83-8, 1984, pp. 1-76.
37. Atkinson, D.S., Ram, N.M., and Switzenbaum, M.S., "Evaluation of the Microtox^{TM} Analyzer for Assessment of SEdiment Toxicity," Technical Report to the Massachusetts Dept. of Environmental Quality Engineering, University of Massachusetts Environmental Engineering Reopt No. 86-85-3, 1985, pp. 1-63.
38. Beckman Inc., "Microtox Application Notes: Microtox EC-50 Values," Beckman Instruments Inc., 1983, pp.1-6.
39. Shelton, D.R., and Tiedje, J.M., " General Method for Determining Anaerobic Biodegradation Potential," *Appl. Environ. Microbiol.*, Vol. 48, 1984, pp. 840-848.
40. Boyd, S.A., Shelton, D.R., Berry, D., and Tiedje, J.M., " Anaerobic Biodegradation of Phenolic Compounds in Digested Sludge ," *Appl. Environ. Microbiol.*, Vol. 46, 1983, pp. 50-54.

41. Yang, J., Speece, R.E., Parkin, G.F., Gossett, J., and Kocher, W., " The Response of Methane Fermentation to Cyanide and Chloroform," *Prog. Wat. Technol.*, Vol. 12, 1980, pp. 977-989.
42. Smith, M.R., and Mah, R.A., " Growth and Methanogenesis by Methanosarcina Strain 227 on Acetate and Methanol ," *Appl. Environ. Microbiol.*,Vol. 36, 1978, pp. 870-879 .
43. Lingle, J.W., and Hermann, E.R., " Mercury in Anaerobic Sludge Digestion ," *Journal WPCF.*,Vol. 47, 1975, pp. 466-471.
44. Capone, D.G., Reese, D.D., and Kiene, R.P., " Effects of Metals on Methanogenesis, Sulfate Reduction, Carbon Dioxide Evolution, and Microbial Biomass in Anoxic Salt Marsh Sediments ," *Appl. Environ. Microbiol.*,Vol. 45, 1983, pp. 1586-1591.
45. Healy, J.B., and Young, L.Y., " Anaerobic Biodegradation of Eleven Aromatic Compounds to Methane ," *Appl. Environ. Microbiol.*, Vol.38, 1979, pp. 84-89.
46. Jackson, S., and Brown, V.M., " Effect of Toxic Wastes on Treatment Processes and Watercourses," *Water Poll. Cont.*, Vol. 69, 1970, p. 292.
47. McCarty, P.L., " Anaerobic Waste Treatment Fundamentals Part Three: Toxic Materials and their Control," *Public Works*, Vol. 95, 1964, pp. 91-94 .
48. Sellers, K.E., <u>Studies on the Actions and Interactions of Heavy Metals on Bioluminescent Bacteria</u>. Thesis, Dept. Civil Engineering, Univ. Mass., Amherst, 1985.
49. Sykes, R.M., and Kirsch, E.J., " Accumulation of Methanogenic Substrates in CCl_4 Inhibited Anaerobic Sewage Sludge Digester Cultures ," *Water Res.*, Vol. 6, 1972, pp. 41-55 .
50. Russell, L.L., Cain, C.B., and Jenkins, D.I., " Impact of Priority Pollutants on Publicly Owned Treated Works Processes: A Literature Review ," Proceedings of the 37th Industrial Waste Conference, Purdue University, 1982, pp. 871-883.
51. McDermott, G.N., Moore, W.A., Post, M.A., and Ettinger, M.B., " Copper and Anaerobic Sludge Digestion ," *Journal WPCF.*, Vol. 35, 1963, pp. 655-662.
52. Balderston, W.L., and Payne, W.J., " Inhibition of Methanogenesis in Salt Marsh Sediments and Whole-Cell Suspensions of Methanogenic Bacteria by Nitrogen Oxides," *Appl. Environ. Microbiol.*, Vol. 32, 1976, pp. 264-269.
53. Keith, L.H., and Telliard, W.A., " Priority Pollutants I - a Perspective View," *Environmental Science and Technology*, Vol. 13, 1979, pp. 416-423.
54. Sharp, V.F., <u>Statistics for the Social Sciences</u>, Little, Brown and Co., Boston, 1979, pp. 381.
55. United States Environmental Protection Agency, "Fate of Priority Pollutants in Publicly Owned Treatment Works : Interim Report, " EPA - 440/1-80-301, 1980, pp. 1-180.

RESPIRATION-BASED EVALUATION OF NITRIFICATION INHIBITION USING ENRICHED *NITROSOMONAS* CULTURES

James E. Alleman

School of Civil Engineering
Purdue University
West Lafayette, IN

INTRODUCTION

This paper will present an expedient and cost-effective bioassay mechanism for the evaluation of potential wastewater toxicity. The procedure is based on a batch-wise analysis of biomass respiration using an enriched *Nitrosomonas* culture blended into raw or primary wastewater samples. An alternative use of this enriched culture would be to conduct bioassays based on a comparative analysis of colorimetrically-determined ammonium-nitrogen oxidation rates.

Both of the aforementioned techniques are presently being employed by the City of Indianapolis, Indiana in an attempt to characterize, and possibly identify, industrially generated toxicants which periodically upset their WWTP facility's nitrification capability. These bioassay techniques have been employed for nearly one year and have become an expedient mechanism for investigative analysis of suspected industrial waste generators.

BACKGROUND

Wastewater bioassays may be completed using a variety of analytical procedures, including: freshwater fish assays, freshwater invertebrate assays, freshwater algal assays, and soil microcosm studies.[1] However, all of these methods tend to be technical complicated, time consuming and inherently expensive.

Within the past few years, several authors have referenced the use of a simplified bacterial-based bioassay generally referred to by its proprietary name, *Microtox*. (Casseri, et al.,[2] Kurz, et al.,[3] and Slattery[4,5]). This bioassay technique employs a marine photo-luminescent bacterium (e.g. *Photobacterium phosphoreum*[2]) whose light emission characteristics provide an expedient measure of microbial toxicity. Aside from the benefit of obtaining real-time bioassay results, the *Microtox* technique is believed to be approximately an order-of-magnitude less expensive than the traditional procedures mentioned earlier (e.g. approximately $10 per test *versus* $100's per test).[1,3]

In conjunction with the use of this *Microtox* technique, Slattery[4,5] employed a supplemental analysis of biomass respiration (referred to as the Patapsco Modified Respiration procedure) as a means of augmenting the information obtained with the *Microtox* test. This respiration analysis was completed with existing mixed liquor samples dosed with either influent wastewater or a standard feed (i.e. a prepared glucose, glutamic acid, etc. solution used for control purposes). Information obtained with both tests was then used to assess wastewater treatability.

In contrast to the use of a photo-bacterium, Williamson and Johnson[1] employed an enriched *Nitrobacter* culture to identify wastewater toxicity on the basis of observed reductions in the relative metabolic rate. Nitrite-nitrogen oxidation was accordingly monitored with a colorimetric test on a batchwise basis. Interestingly, the *nitritifying* biomass used with this test was previously cultured, enriched, and then freeze-dried for subsequent use as required for individual tests. Excluding labor, the cost for this freeze-dried *Nitrobacter* bioassay was estimated at $20→50.[1]

The technique presented within this paper of using a respiration-based bioassay involving nitrifying microorganisms, therefore, represents a hybrid procedure coupling both oxygen-uptake measurements (i.e. as with the Slattery[4] approach) and nitrification biomass (i.e. as with the Williamson and Johnson[1] approach). In this case, however, an enriched *Nitrosomonas* population was used due to the relative ease of maintaining a stock culture within the lab. The apparent benefits to be derived from this bioassay approach are as follows:

- Based on their autotrophic characteristics and recognized environmental sensitivity, the employed *Nitrosomonas* bacteria can be expected to yield a prompt indication of the presence of wastewater toxicants,

- In the context of similitude between the employed bioassay and real-world process behavior, the response associated with the *Nitrosomonas* is expected to be as good, and possibly better, than that afforded by the *Microtox* test, particularly with regard to nitrification systems,

- The employed *Nitrosomonas* stock biomass can be maintained with nominal daily care (i.e. preparation of substrate, cleaning of the aeration stones, etc.....in essence, much the same effort as required to maintain a large aquarium),

- The oxygen uptake rate (OUR) test can be completed within minutes, and requires minimal operator training,

- Aside from the stock cultivation tank, the required hardware amounts to a standard respirometer (~ $4500) and shake table (~ $800),
- The recurring non-labor costs associated with this procedure are limited to inexpensive biomass feedstocks (i.e. primarily ammonium chloride and sodium bicarbonate),
- The overall cost of this technique should be considerably less than that of the *Microtox* test (at approximately $2→4 per test).

METHODS AND MATERIALS

Enriched Culture Development and Maintenance

Enriched *Nitrosomonas* biomass used throughout this study was progressively cultured from a nitrifying activated sludge over a period of four months and subsequently maintained for one year under the conditions given in Table 1:

TABLE 1 Growth Conditions for Experimental Nitrosomonas Culture	
Reactor Mode	CFSTR
Reactor Volume	20 l
Nitrogen loading rate	60 mg N l^{-1} hr^{-1}
Hydraulic retention time	24 hrs
Solids retention time	55 to 75 dys
MLSS	900 to 1000 mg l^{-1}
Effluent pH	7.2 to 7.4
Effluent ammonium-nitrogen	0.1 to 0.5 mg l^{-1}

Ammonium chloride and sodium bicarbonate were used as the two primary feedstocks at a respective weight ratio of 1:3, along with sufficient K_2HPO_4 to avoid any phosphate limitation. Dilution water (i.e. tap groundwater) used for preparation of the feed also received nutrient supplementation equivalent to that required for the standard BOD_5 test.

Batch Bioassay Procedure

During each test, 25 ml aliquots of either a buffered tap water (i.e. for the control) or wastewater sample were first added to 125 ml Erlenmyer flasks and then placed on a shake table maintained at approximately 100 rpm. Concentrated industrial wastewater samples were previously diluted with tap water to an extent commensurate with their expected sewer dilution levels. Next, 10 ml aliquots of the enriched *Nitrosomonas* culture were quickly added to each such test flask using an 'Eppendorf' syringe. These flasks were then immediately dosed with a concentrated NH_4-N solution (1 ml of a 0.08 M NH_4Cl solution) using another 'Eppendorf' syringe to insure that all of the flasks had sufficient ammonium-nitrogen substrate to last for at least 75 minutes of batchwise contact.

Immediately after initiating the exposure of the *Nitrosomonas* to the given wastewater samples on the shake table, testing of either the OUR or NH_4-N oxidation rates was begun. This information was then reviewed in comparison with the control results to ascertain apparent metabolic inhibition.

Test samples whose OUR levels or ammonium-nitrogen oxidation rates showed substantial variation from the control were subjected to an additional set of tests for pH and residual NH_4-N to verify that these parameters were not responsible for the aberrant performance. It must also be emphasized that these tests should either be completed under low-light conditions or with 'red' room light (i.e. darkroom conditions) in order to obviate potential problems with the sensitivity of dilute *Nitrosomonas* suspension to visible/near-UV light.[6]

In the event that the *Nitrosomonas* concentration within the stock solution were to vary on a daily basis (e.g. due to excessive biomass removal for bioassay testing) the obtained metabolic data (i.e. OUR or NH_4-N oxidation rate) should likely be be expressed on a specific basis (i.e. mg O_2 or N-oxidized hr^{-1} mg $MLSS^{-1}$). This change will facilitate comparisons of the raw metabolic data over time.

Respirometric Analysis

Oxygen uptake rate was employed as the primary bench-mark parameter for comparison of *Nitrosomonas* activity. These tests were conducted with a Gilson Model 5/6 Oxygraph and Clark electrode system. Biomass taken directly from the enriched culture reservoir normally exhibited a maximal respiration of 270 mg O_2 l^{-1} hr^{-1} when spiked with excess substrate (via a direct 30 ul injection of 0.1 M NH_4Cl). After dilution into the test flasks, uninhibited test samples exhibited maximal OUR values of 75 to 80 mg O_2 l^{-1} hr^{-1}.

After only a few minutes of contact on the shake table, respiration tests were completed on the biomass within each flask to establish an initial OUR level. A second respiration test was then completed after 60 minutes of contact on the shake table. This initial and 60 min OUR data was then used to ascertain percentage variations from the control tests as an indicator of biomass toxicity.

Ammonium-Nitrogen Oxidation Rate Analysis

Bioassays completed on the basis of relative ammonium-nitrogen oxidation rates involved successive hourly sampling from the shaking test flasks over a four (4) hr period. Colorimetric analyses of the residual NH_4-N within these samples were completed according to the Nesslerization Method.[7]

RESULTS AND DISCUSSIONS

Figures 1→5 provide a representative sampling of the chronological bioassay data obtained with five (5) regional wastewater streams, including two adjacent interceptor streams (undiluted) and three industrial wastewater streams (which were diluted prior to bioassay testing). This information is presented as percentage values

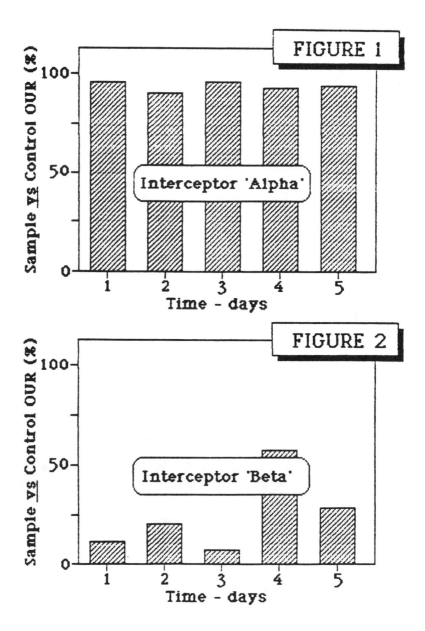

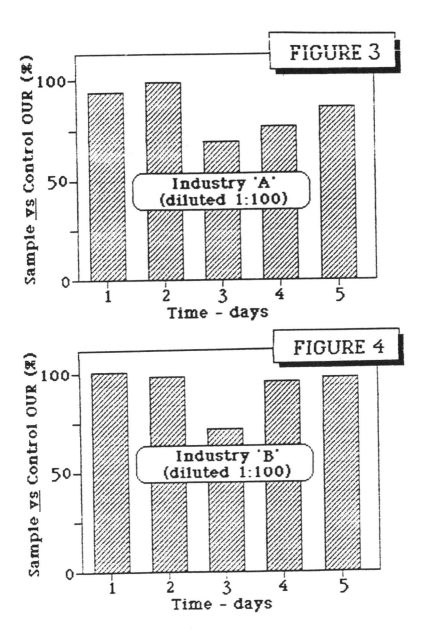

FIGURE 3. Industry 'A' (diluted 1:100)

FIGURE 4. Industry 'B' (diluted 1:100)

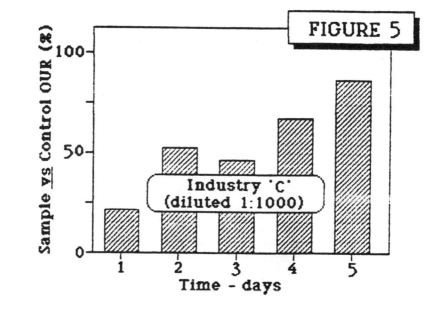

on OUR for the tested wastewater streams *versus* the control.

The data obtained on the targetted interceptors (i.e. covertly identified as 'alpha' and beta') is given in Figures 1 and 2, respectively, over a five (5) day period of observation. These results indicated that whereas the 'alpha' interceptor line had only a negligible impact on the employed *Nitrosomonas* biomass the 'beta' line had a sizable decline in OUR.

Industries A→C were all located along the 'beta' interceptor. The results given on industries A and B by Figures 3 and 4 revealed only a nominal decrease in biomass respiration for samples diluted at a 1:100 level.

However, as shown in Figure 5, industry C showed strong evidence of a persistent inhibitory effect on the *Nitrosomonas* in spite of dilution levels of 1:1000. This information corresponded with the toxic effects observed in conjunction with the 'beta' interceptor line into which this industry discharged.

SUMMARY

The enriched *Nitrosomonas* bioassay procedure described within this paper has proven to be an extremely useful, expedient, and cost-effective mechanism for rapid evaluation of wastewater toxicity. Extending beyond this immediate application, the procedure has been employed for investigative evaluations of suspected industrial waste generators. Furthermore, this test has been successfully used to diagnostically determine appropriate dilution levels for industrial waste streams at their respective sewer outfalls.

Although this bioassay may be completed with either respirometric (based on OUR) or colorimetric (based on NH_4-N oxidation rates) analyses, the OUR strategy appears preferable due to its simplicity. Monitoring of the ammonium-nitrogen oxidation rate with a colorimetric NH_4-N test requires tedious sample clarification and treatment. Within an eight (8) hr workday, two (2) laboratory technicians can routinely complete between 100 and 120 bioassay tests based on the OUR procedure; this number of samples is approximately four (4) times larger than that which can be managed with the colorimetric procedure.

As with any bioassay procedure, questions may be raised about the inherent reliability and veracity of the tests. In the particular case of its use by the City of Indianapolis, however, the enriched *Nitrosomonas* bioassay technique has proven to be an excellent 'indicator' procedure.

REFERENCES

1. Williamson, K.J. and Johnson, D.G., "A Bacterial Bioassay for Assessment of Wastewater Toxicity." *Proceedings of the 34th Purdue Industrial Waste Conference,* 1979, pgs. 264-273.

2. Casseri, N.A., et al., "Use of a Rapid Bioassay for Assessment of Industrial Wastewater Treatment Effectiveness." *Proceedings of the 38th Purdue Industrial Waste Conference*, 1983, pgs. 867-878.

3. Kurz, G.E., et al., "A Rapid Industrial Waste Screening Method." *Proceedings of the 39th Purdue Industrial Waste Conference*, 1984, pgs. 395-406.

4. Slattery, G.H., "Biomonitoring and Toxicity Testing (Toxicant Treatability Definition & Measurement)." Paper Presented at the WPCF Analytical Techniques, Cincinnati, OH, May, 1985, 6 pp.

5. Slattery, G..H., "Effects of Toxic Influent on Patapsco Waste Water Treatment Plant Operations." Paper Presented at the Annual WPCF Conference, New Orleans, LA, October, 1984, 20 pp.

6. Alleman, J.E. and Keramida, V., "Light Induced *Nitrosomonas* Inhibition During Endogenous Respiration." *Water Research*, 1986 (Paper currently under review).

7. "Standard Methods for the Examination of Water and Wastewater." APHA-AWWA-WPCF, 16th Edtn., 1985.

ASSESSMENT OF THE DEGREE OF TREATMENT REQUIRED FOR TOXIC WASTEWATER EFFLUENTS

G. Fred Lee and R. Anne Jones

*Department of Civil & Environmental Engineering
New Jersey Institute of Technology
Newark, New Jersey*

INTRODUCTION

Increasing attention is being given by regulatory agencies to controlling aquatic life toxicity in the surface waters of the US. The US EPA and other pollution control agencies have indicated for a number of years that they would be using biomonitoring (bioassays/toxicity tests) to assess the toxicity of effluents and to develop the basis for effluent limitations, but the implementation of these control programs is only now beginning to be aggressively pursued at the federal and state levels. While water quality standards of many states have included statements to the effect that toxic chemicals shall not be discharged in toxic amounts, the procedures for implementation of these provisions have generally not been outlined. Some states such as New Jersey (1) have adopted water quality standards which include specific limitations on the amount of aquatic life "toxicity" (as defined by the test and conditions used) that may be present in a wastewater treatment plant effluent. These effluent limitations are, however, arbitrarily established without sufficient regard to the receiving water characteristics and can thus lead to treatment requirements which are excessive or inadequate for aquatic life-related beneficial uses of the receiving waters.

The US EPA has recently released several guidance documents pertinent to evaluating and limiting the "toxicity" of effluents. These include: its recommended approach for establishing the allowable "toxicity" in an effluent (2), a guidance manual for determining the acute toxicity of effluents to aquatic life (3), and a draft guidance manual for determining the chronic toxicity of an effluent to aquatic life (4). The approach being used to implement "toxicity" control is to limit the "toxicity" of wastewater treatment plant effluents and other discharges; the premise is that if the effluent discharged is non-toxic there is limited likelihood of there being toxicity in the receiving water. As discussed below, there are significant problems with this approach; it can lead to requiring excessive treatment of the wastewaters compared to what is needed to protect aquatic life from the effects of the toxicants, or to inadequate treatment for aquatic life protection.

With these new regulatory approaches, the design and operation of wastewater treatment works will soon be prescribed based on the degree of treatment needed to protect beneficial uses of receiving waters rather than on readily available technology. Therefore, it is appropriate in a conference concerned with innovative biological treatment of toxic wastewaters to review approaches being used to establish the degree of treatment that will be required of these processes, that is, the approaches that are being used to evaluate the potential aquatic life toxicity associated with wastewater effluents and to develop the control/limitation of components of the wastewaters responsible for the toxicity. This paper presents a critical overview of these topics.

CHEMICAL CONCENTRATION-BASED EFFLUENT LIMITATIONS

The overall water quality objective in any aquatic life toxicity control program should be to achieve the designated aquatic life-related beneficial uses of the receiving waters. Traditionally this has been translated to mean not exceeding a set of single-value, numeric concentration criteria or standards in the receiving waters. Generally, the design engineer has used the state water quality standards as a basis for determining the degree of treatment necessary to limit toxicity in receiving water from specific effluent contaminants. However, as discussed below, a critical evaluation of these criteria and standards and of the characteristics of the receiving waters can significantly influence what may be an acceptable load for many toxic contaminants.

The September 1985 US EPA guidance document (2) specifies two approaches for setting limitations on the toxicity of an effluent to aquatic life. One of these is a chemical-specific approach based on water quality criteria and standards. The other is based on effluent toxicity assessment using toxicity tests. The July 1976 Red Book criteria (5) were generally chronic exposure (life-time or critical lifestage) safe concentrations which were not to be exceeded at the edge of a "mixing zone." Rather than defining physical dimensions of mixing zones however, the US EPA indicated that the criteria were to be translated into effluent limitations by determining the maximum load of the chemicals that could be discharged at the 7Q10 (7-day, 10-year low flow) without resulting in an exceeding of the criteria values at the point at which physical mixing was achieved at the 7Q10 flow. The US EPA further recommended that the 96-hr LC50 (concentration that results in the death of 50% of the test organisms in a 96-hr exposure - standard acute toxicity test) not be exceeded in the receiving water; a number of states adopted this recommendation.

On July 29, 1985, the US EPA established recommended revised criteria values designed to protect aquatic life against toxicity associated with nine chemicals for which criteria had previously been developed (6). The changes in the criteria reflected updated information on the toxicity of the chemicals to aquatic life, but more importantly, this publication specified the frequency of allowable exceedence of both the acute and chronic exposure criteria and the period of time over which measurements are to be made to determine compliance. For each chemical, two criteria values (equations) were presented: a one-hour average concentration that is not to be exceeded more than once in three years, and a four-day average concentration not to be exceed more than once in three years. The former is designed to protect against acute toxicity (generally lethality after short-term exposure durations), and the latter, against chronic toxicity (generally sublethal effects resulting from life-time or critical lifestage exposure). Another important change that has occurred is that the US EPA has indicated that the 7Q10-approach to definition of allowable loads is only applicable to so-called "stressed" receiving waters, and that a 1Q10 is more appropriate (more stringent) for the protection of aquatic life in "stressed" systems from the toxicity of these chemicals (2). However, for so-called "unstressed" systems, the US EPA (2) recommended that the allowable load be computed based on the 7Q5, or the 1Q5 flow in the receiving water; use of 5-year low flows rather than 10-year low flows would allow greater loads of chemicals to be discharged to the receiving waters.

According to the US EPA (2), the frequency at which the criteria concentrations can be exceeded, and the one-hour and four-day average criteria values are justified based on the normal variabilities that exist in wastewater flows and in contaminant concentrations. In the opinion of the authors, this justification needs further investigation for its appropriateness with particular reference to whether exceeding the four-day average criterion value more than once in three years or exceeding the one-hour maximum more than once in three years will indeed significantly impair the aquatic life-related beneficial uses of the receiving waters. It is possible that the one allowable event in three years in which the criterion value is greatly exceeded would do significant harm to the aquatic life-related beneficial uses of the receiving waters. Similarly, it is quite likely that numerous small-level excursions above the 4-day average (chronic) value can occur without significantly impairing the aquatic life-related beneficial uses of the receiving waters. Further, the application of these criteria with their associated allowed frequency of exceedence would be inappropriate for situations such as the open water dumping of dredged sediment or sewage sludge. The statistical characteristics and the approach for assessing the probability of acute or chronic toxicity events associated with dumping will certainly be significantly different from those associated with a typical municipal or industrial wastewater discharge.

Discussed below are some of the implications of the current chemical-specific approaches to setting discharge limits for particular contaminants of concern in domestic and some industrial discharges.

Ammonia Discharge Limits

One of the most important chemicals in municipal and some industrial wastewaters that can cause toxicity in receiving waters is ammonia. Ammonia is also of interest because there is considerable information available on factors that influence its toxicity to aquatic life as well as on its environmental behavior. Ammonia is present in aquatic systems in the un-ionized (NH_3) form as well as the ionized (NH_4^+) form; the distribution of the total ammonia concentration between these two forms is controlled by pH, temperature, and salinity. This distribution is of significance in that, in large part, it is the un-ionized form which is toxic to aquatic life.

The July 1976 criterion (chronic exposure safe concentration) for un-ionized ammonia was 0.02 mg NH_3/L. It has also generally been found that the 96-hr LC50 for ammonia for a number of forms of aquatic life is on the order of about 0.2 to 0.7 mg NH_3/L. The July 1985 revised criteria (6) specify one-hour average and four-day average criteria concentrations in freshwater for un-ionized ammonia as a function of pH and temperature. At 20 to 25°C with slightly alkaline pH conditions, the one-hour average criterion is about 0.2 mg NH_3/L; no criteria values for the pH and temperature ranges provided were greater than 0.26 mg NH_3/L. At 20 to 25°C with slightly alkaline pH conditions, the four-day average criterion is about 0.02 mg NH_3/L; the criteria are as low as 0.0007 mg NH_3/L for lower temperature and lower pH conditions. For freshwater systems at pH 7.5 and 20°C, the 0.02 mg NH_3/L un-ionized ammonia criterion translates to a total ammonia concentration of 1.5 mg NH_3/L. Since un-nitrified domestic wastewaters typically contain between 15 and 35 mg NH_3/L total ammonia, it is evident that domestic wastewater treatment plants will have to nitrify their effluents unless the 7Q10 or 1Q10 flows provide adequate dilution to achieve a maximum four-day average total ammonia in the receiving waters of 1.5 mg NH_3/L.

As discussed by Lee and Jones (7), there are several factors that need to be considered in implementing the US EPA ammonia criteria into effluent limitations. While some states have allowed municipalities to base their ammonia discharges on the pH and temperature of the effluent, it is the pH, temperature, and salinity of the receiving water that should be used to determine the ammonia discharge limitations. Since in small streams where ammonia is likely to be a problem, the pH can easily fluctuate plus or minus 1 unit and temperature can fluctuate plus or minus 2 to 5 C° over a diel (24-hr) cycle, the time of day at which pH and temperature measurements are made for setting an effluent discharge limitation can significantly affect the allowable discharge.

For example, at 25°C and pH 9, the allowable receiving water total ammonia concentration for protection against chronic toxicity is 0.1 mg NH_3/L while at 20°C and pH 7 the concentration is 1.5 mg NH_3/L. Therefore, depending on the time of day when the "characteristics" of the receiving waters are assessed (e.g., 4 pm vs. 4 am) the allowable discharge may be different by more than a factor of 10. This situation is even further complicated by the fact that the maximum diel pH change will not necessarily be at the point of discharge but may well be at some distance downstream where maximum algal growth (stimulated by the nutrients from the effluent and influenced by increased light penetration) occurs.

Another factor that needs to be considered in implementing the ammonia criteria is the fact that the US EPA has chosen to use a highly conservative endpoint for defining what a "chronic" impact is. While un-ionized ammonia at concentrations slightly greater than 0.02 mg NH_3/L will cause "red gills," a histopathological impact, concentrations as high as 0.06 mg NH_3/L un-ionized ammonia have been found to have no effects on growth, reproduction, or other factors of concern from an aquatic life-related beneficial use point of view. This difference, combined with the variability in toxicity with temperature and pH swings in receiving waters, could result in a 50-fold difference in the allowable ammonia load based on worst-case assumptions. The magnitude of this difference would be even greater in estuarine systems where tide stage would influence the salinity of the receiving waters, which would in turn influence the degree of ionization of ammonia; the greater the salinity, the lower the percent of the total ammonia in the un-ionized form.

It could be suggested that average values of pH and temperature of the receiving waters be used to estimate the appropriate ammonia loads applicable to the 7Q10 (or other specified low flow) conditions. First, it should be noted that the pH and temperature conditions that occur during the 7Q10 event are likely to be different from those which are "typically" present during normal flow periods. Second, the 7Q10 flow may not coincide with the season in which ammonia would present the greatest toxicity threat to aquatic life. For example, the low flow conditions may occur in the winter when the pH and temperature conditions do not favor maximum percentage of ammonia in the un-ionized form. It is quite possible that the truly worst case situation for un-ionized ammonia occurs at a time when the flow is greater than the 7Q10 (or other designated low flow).

In oxic surface waters, ammonia is microbially converted to nitrite which exhibits considerable toxicity to aquatic life. Nitrite, in turn, is converted to nitrate which is not toxic to aquatic organisms. While it is generally assumed that the conversion of nitrite to nitrate is faster than the conversion of ammonia to nitrite, and thus that the presence of nitrite is minimal in surface waters, the authors have found that there are situations in which considerable nitrite concentrations are found in surface waters. Generally these situations occur downstream of domestic wastewater treatment plants which discharge unnitrified effluent. They also tend to occur under cold water conditions where there is limited dilution of the effluent with receiving waters. While the toxicity of nitrite to fish and other aquatic life is not well-defined, it appears from the literature that there are some fish sensitive to nitrite at hundredths of a mg N/L. It appears that there are substantial numbers of types of fish and other aquatic organisms which are sensitive to nitrite at tenths of a mg N/L. Therefore, per unit nitrogen, nitrite is more toxic to fish than ammonia under surface water conditions. It is therefore possible that a domestic wastewater treatment plant could discharge levels of ammonia that would result in non-toxic ammonia concentrations in the receiving waters, but when the ammonia is converted to nitrite downstream of the discharge toxicity could occur due to the nitrite. This is an example of why it is not possible to rely exclusively on effluent and receiving water characteristics at the point and time of discharge to assess the potential toxicity of chemical contaminants in an effluent to aquatic life.

It is evident from this discussion that it is very difficult to properly translate the potential for aquatic life toxicity due to ammonia in receiving waters to concentrations (or load) of ammonia in an effluent on anything other than a site-specific basis involving fairly detailed field studies.

Heavy Metal Discharge Limits

The difficulties of translating effluent concentrations to in-stream toxicity discussed above for ammonia are even more pronounced for many of the heavy metals. It has been well-established in the literature that not all forms of the heavy metals are equally toxic to aquatic life. Many precipitated, sorbed, and complexed heavy metals show little or no toxicity to aquatic life. Similarly, heavy metals present in a crystalline matrix such as in silicate minerals, are generally found not to be toxic. The National Academies of Science and Engineering (8) recognized this problem and recommended to the US EPA that heavy metal toxicity be assessed based on the use of a 96-hr toxicity test and an application factor to relate acute lethal toxicity (96-hr LC50) to chronic safe concentrations. The application factor generally recommended was 0.01. The 96-hr LC50 multiplied by 0.01 was generally believed to be a reliable estimate of the chronic safe concentration. The US EPA July 1976 Red Book (5) water quality criteria adopted this approach for many heavy metals and other contaminants. However, few states used this approach to implement heavy metal standards. This was likely due to the fact that the US EPA had not until recently developed standardized toxicity test procedures that could be used for this purpose.

In its November 1980 announcement of criteria documents for the Priority Pollutants (9), the US EPA abandoned the toxicity test approach for establishing heavy metal criteria in favor of single-value, numeric criteria applied to the total concentration of the metal in the receiving waters. These criteria were adjusted only for water hardness; no attempt was made to consider the effect of the presence of precipitated, complexed, or sorbed forms of the metals on the toxicity of the total concentrations of the heavy metals. Few, if any, states adopted the US EPA November 1980 criteria as the state water quality standards. At this time many states still have not adopted numeric criteria which can be used to establish concentration limitations for heavy metals in wastewater effluents. In the opinion of the authors, this is likely to be due to the facts that the toxicity of heavy metals in aquatic systems is controlled by the characteristics of the receiving waters and that the toxic forms of heavy metals are not readily discernible by chemical analytical means. The importance of using a toxicity-test approach for establishing heavy metal standards was demonstrated by unpublished data of R. Morrison at Colorado State University on the toxicity of several heavy metals to several types of fish in Cache la Poudre River water. His work showed that the toxicity of heavy metals in a series of samples of river water collected at about 20-km intervals from the river varied by a factor of more than 50. The differences in toxicity could not be accounted for by the differences in hardness of the water or toxicity associated with discharges that occurred between the sampling locations.

The US EPA July 29, 1985 revised criteria for the heavy metals (6) are still based on the total concentrations of the heavy metals. Associated with the criteria for the heavy metals in this document, the US EPA stated,

> EPA believes that a measurement such as ´acid-soluble´ would provide a more scientifically correct basis upon which to establish criteria for metals. The criteria were developed on this basis. However, at this time, no EPA approved methods for such a measurement are available to implement the criteria through the regulatory programs of the Agency and the States. The Agency is considering development and approval of methods for a measurement such as ´acid-soluble.´ Until available, however, EPA recommends applying the criteria using the total recoverable method.

Because of this statement, the heavy metal criteria are essentially unimplementable into water quality standards. In the opinion of the authors, the development of an "acid-soluble" test for heavy metals will not significantly change this situation because of the problems associated with interpretation of the results of such extractions. On February 7, 1984, the US EPA had first proposed revised criteria for the heavy metals for which criteria were adopted on July 29, 1985. These proposed criteria were based on the concentration in a water after a pH 4, nitric acid extraction. However, as discussed by Lee and Jones (10) the validity of this approach as well as the details of its implementation must be seriously questioned. There is need for a considerable amount of work to evaluate the validity of this approach.

As was the case with ammonia discussed above, non-toxic forms of heavy metals discharged in an effluent, such as metal sulfides or cyanides, may be converted toxic forms in downstream waters. Thus toxicity tests on a fresh effluent may not show the toxicity that may occur downstream of the discharge.

The net result of the current problems in the development of technically valid criteria for heavy metals and approaches for their implementation into appropriate water quality standards is that the design engineer and the wastewater treatment plant operator do not have available to them defensible numeric standards for heavy metals that can be used to determine the degree of treatment necessary to protect aquatic life in the receiving waters. It is also not likely that meaningful information of this type which can be used to directly assess toxicity in receiving waters, will ever be available for most heavy metals. About all that can be said at this time is that treating wastewaters to achieve July 29, 1985 water quality criteria for heavy metals at a designated low flow, such as 7Q10, will in many instances be protective of aquatic life-related beneficial uses in receiving waters. Also in most instances this degree of treatment is likely to be considerably in excess of that needed to protect the aquatic life-related beneficial uses of the receiving waters.

Chlorine Discharge Limits

The disinfection of municipal wastewaters by chlorination typically leads to the formation of chloramines which are highly toxic to aquatic life. The US EPA July 29, 1985 criteria for chlorine (6) limit the four-day average concentration to about 10 ug Cl_2/L. Since the typical municipal wastewater effluent must contain a chlorine concentration greater than about 500 ug Cl_2/L after a 30-minute contact period in order to have achieved adequate disinfection, it is evident that appreciable chlorine toxicity could occur in receiving waters below municipal wastewater treatment plant discharges. In field studies of the impacts of the effluents of five Colorado municipal wastewater treatment plant effluents on receiving water aquatic organism-related beneficial uses, Lee and Jones (11) found that the chlorine concentrations in the receiving waters were sufficient to potentially cause chronic toxicity to aquatic life for 3 km to approximately 40 km downstream of the discharges (i.e., concentrations exceeded chronic safe concentrations in these reaches). Habitat assessment studies on the receiving waters within the regions where the potential for chronic toxicity existed based on chlorine concentrations, showed that the numbers and types of fish in these regions were about the same as those found in regions where chlorine concentrations were below chronic safe concentrations. This situation raises serious questions about the water quality significance of exceeding chronic safe concentrations of chlorine for extended periods of time. Further information on the studies on the impact of domestic wastewater treatment plant effluent chlorine on receiving water quality is presented by Lee and Jones (11), Heinemann et al. (12), and Newbry et al. (13).

It is evident from these and other studies that it is difficult, if not impossible, to interpret the water quality significance of exceeding US EPA criteria or some state water quality standards in a receiving water. This is, in part, because of the problems of translating the results of laboratory data to field situations (14). There is a variety of factors that influence how a particular contaminant present in excess of the chronic exposure safe concentration impacts the numbers, types, and condition of receiving water aquatic organisms. It is the experience of the authors that the US EPA chronic exposure safe concentration criteria can and should only be used to indicate the potential for water quality problems and that as discussed below, where these levels are exceeded, site-specific investigations should be conducted to determine the significance, if any, of the concentrations to aquatic life.

TOXICITY-BASED EFFLUENT LIMITATIONS

The US EPA September 1985 guidance manual for toxics control (2) indicates that in addition to the water quality criteria and standards approach, the control of toxic chemicals in effluents should include a bioassay/toxicity test evaluation in which the effluent is mixed in various proportions with receiving waters, and the toxicity of these mixtures is assessed. The US EPA manuals (3)(4) prescribe the testing procedures that are to be used. The US EPA recommends that toxicity tests of effluent/receiving water mixtures of 100%, 30%, 10%, 3%, and 1% effluent with three different types of organisms to determine whether an effluent is sufficiently toxic to have an adverse effect on beneficial uses of receiving waters. The US EPA indicates that toxicity test results be used to determine the LC50 or NOEL (no observed effect concentration) which are in turn used to define the number of "toxicity units" (TU´s) associated with the waste, where:

$$TU = \frac{100}{LC50 \text{ or } NOEL} \qquad (1)$$

The use of the LC50 in the equation produces a TU_a (acute toxicity unit) while the use of the NOEL produces a TU_c (chronic toxicity unit).

The US EPA also defines a Criterion Maximum Concentration (CMC) for acute toxicity and a Criterion Continuous Concentration (CCC) for chronic toxicity. It suggests that for protection against acute toxicity, the CMC should not exceed 0.3 TU_a for the most sensitive of at least three test species. For protection from chronic toxicity, the US EPA recommends that the CCC not exceed 1 TU_c. If only two species are tested, the criterion values are divided by 10. The allowable effluent toxicity level is then computed based on the criterion and the effluent dilution at the 7Q10. Further information on this approach is provided in the guidance manual.

This approach suffers from some of the same difficulties of the chemical-specific approach discussed above. It assumes that the chemicals which cause the toxicity in toxicity tests of effluent/receiving water mixtures will behave in the same manner in the toxicity tests as they will in the environment downstream of the discharge. While this is likely to be the case at the point at which the effluent is first mixed with the receiving water, it may not be the case with distance-time downstream. The reactions discussed previously which alter the toxicity of contaminants in an effluent, either increasing it or decreasing it, will also affect the ability to interpret the results of whole-effluent toxicity tests in terms of the actual potential for toxicity due to an effluent in receiving waters. It is also important to consider the impact of the projected toxicity based on whole-effluent toxicity tests, on the aquatic life-related beneficial uses of the receiving waters. Situations such as those experienced with chlorine as discussed above, could readily occur with other chemicals as well where treatment approaches implemented based on projections of potential toxicity from chemical concentrations in receiving waters or effluent toxicity tests could result in little or no improvement in receiving water quality after substantial amounts of money have been spent for the control of the chemical or "toxicity."

It is important to always keep in mind the purpose of the control of contaminants-toxicity, in the instances discussed herein: the maintenance and/or improvement in the aquatic life-related beneficial uses of the receiving waters. The issues of contamination of domestic water supplies, or the bioaccumulation of persistent chemicals within the tissues of fish or other aquatic life which renders them unsuitable for use as food for people, are not addressed by the US EPA aquatic life-based toxicity criteria or effluent toxicity testing of the type discussed herein. The significance of the presence of chemical contaminants in drinking water can best be assessed based on the concentrations that occur in the treated water at the point of consumption. With respect to the potential for bioaccumulation of contaminants in aquatic life, it has been well-established that effluent concentrations and dilution in receiving water provide poor predictions of the extent of bioaccumulation of chemicals such as PCB´s and DDT in receiving water organisms. As discussed by Lee and Jones (14) the primary reason for the lack of predictive ability for bioaccumulation is the fact that many of the chemicals that tend to bioaccumulate in aquatic life also tend to strongly sorb onto natural water particulate matter. This sorption is a significant sink (depository) for contaminants and competes with uptake in organism tissue for the chemical. At this time, the only way to judge the potential for bioaccumulation of a contaminant in an effluent is by measuring the concentrations of the chemical within the edible tissue of organisms in the receiving waters. While not generally done, it may be feasible to improve on the predictive capability of effluent concentration/dilution measurements by determining octanol/water partition coefficients for effluent/receiving water mixtures. Such testing must, however, consider the potential for sorption of the contaminants onto particulate matter that may be present in downstream sources but that is not present at the point of mixing of the effluent and receiving waters.

"Quick & Easy" Toxicity Tests

There is considerable interest in developing a surrogate microorganism toxicity test which could determine the potential for aquatic organism toxicity with "simpler and less expensive" testing procedures than are prescribed by the US EPA in its guidance document and testing manuals (2)(3)(4). It has been claimed by some that a bacterial phosphorescence procedure commercially available as "Microtox" is a suitable method for estimating the aquatic life toxicity of wastewater effluents. A critical review of the information available on this topic, however, shows that the Microtox procedure is not reliable for estimating aquatic life toxicity of complex chemical wastewater effluents and may, in fact, give highly erroneous results (15)(16). For example, as discussed above, ammonia is highly toxic to aquatic life. However, when ammonia-containing effluents are tested with the Microtox procedure toxicity is not found, and in fact, stimulatory effects on the test organisms are found (16). It is obvious that such a procedure cannot be used reliably as a screening test for estimating the toxicity associated with an effluent.

RECOMMENDED APPROACH

While the approaches outlined by the US EPA have significant drawbacks as discussed above, they are useful for indicating situations in which there may be aquatic life-related water quality problems associated with a particular wastewater discharge. Because they are in the general sense, based on worst case conditions, and worst case situations rarely occur or persist for substantial periods of time, it is appropriate to use pertinent aspects of these approaches to flag situations of potential concern that need greater investigation to determine the actual potential hazard associated with the particular identified effluent. It is not generally cost-effective or necessary for protection of aquatic life-related beneficial uses to plan for a generic worst case condition, and indeed in some cases the generic worst case conditions may not be the actual worst case for a particular situation.

Before a major effluent-oriented toxicity potential evaluation study is undertaken, the authors recommend first evaluating the toxicity associated with the receiving waters themselves. As noted previously, this is the real goal of effluent toxicity management. The authors believe that the fish embryo-larval toxicity tests (4) would likely be useful for this screening, although the approach has not yet been fully evaluated for this purpose. If receiving waters in the area of mixing of the effluent with the receiving water and for an appropriate distance downstream exhibit no adverse impact in this test when conducted during worst case times of the year, then less emphasis should have to be placed on effluent characterization and toxicity testing. If, however, waters taken downstream of a discharge show appreciable toxicity in the fish embryo-larval test while upstream waters do not, it is clearly justified to require the removal of those contaminants in the effluent leading to the toxicity in the receiving waters. This type of evaluation approach may be far less expensive and more reliable than conducting a large suite of toxicity and chemical tests on the effluent. For proposed discharges, this type of in situ screening could provide insight into what should be allowable discharge loads as well as on the background potential toxicity of the receiving waters. With a properly designed study plan, this approach could be used to identify sources contributing to the toxicity and lead to load reductions from those contributors.

Discussed below is the approach recommended by the authors for evaluating those situations in which potentially significant amounts of aquatic life toxicity as measured by the embryo-larval test are found in the receiving waters downstream of a particular discharge.

Part A - Chemical-Specific

The authors recommend a two-part approach for evaluating the potential impact of wastewater effluents in order to protect designated beneficial uses of receiving waters at the least possible cost. As part of the specific-chemical portion of this bilateral approach, the discharger should monitor the effluent for specific parameters for which there are US EPA criteria and/or state water quality standards. The concentrations found should be compared with the criteria and standards applicable to the designated beneficial uses after low-flow dilution (if appropriate, at the most critical time of the year) has been taken into account. If the criteria (which are typically chronic exposure safe concentrations) are or would be exceeded based on these measurements and computations, a determination should be made of the ease of implementing control programs to prevent the exceedence of these levels. It is the authors' opinion that if control could be easily accomplished with little cost, serious consideration should be given to implementing the control program without more detailed investigation if the criteria-standards truly represented worst case conditions. For example, while the presence of domestic wastewater chlorine at concentrations above the chronic safe level in a receiving water may not significantly harm the overall aquatic life-related beneficial uses of the receiving water, it certainly could stress some systems or some aspects of a system. Because these effluents can be readily dechlorinated for a cost of a few cents per person per day for the population served, the authors believe that dechlorination should be seriously considered for those effluents which cause the receiving water concentration to substantially exceed the chronic safe concentration for considerable distances downstream of the discharge where it is likely that the removal of the chlorine would result in a significant improvement in the aquatic life-related beneficial uses of the receiving water.

If the control program necessary to reduce the concentrations in the effluent to levels which would meet the worst case criteria and standards would be difficult to develop or implement, or would be expensive, a site-specific field study should be carefully developed and conducted to determine the impact of the particular discharge into the particular receiving water on the beneficial uses of that water. This would be a hazard assessment study of the type discussed elsewhere by the authors (11)(17)(18). A key component of the hazard assessment study should be a habitat assessment to determine the numbers and types of aquatic organisms that would be expected to be present in the receiving waters were it not for the discharge. Information on the use of habitat assessment in water pollution control programs is discussed by Lee and Jones (19). If the results of the hazard assessment study show that the impact is significant or that adoption of the treatment would result in a significant improvement in the water quality of the receiving waters for the designated uses, the treatment program should be adopted. If, however, the results of the hazard assessment show that the impact or improvement would be minimal, the discharger may consider petitioning the state and federal agencies for an exemption from the effluent limitations that would be imposed based on their evaluation approaches.

Part B - Toxicity Testing

The second part of the recommended approach which would be undertaken in conjunction with the specific chemical testing discussed above, is toxicity testing of the effluent to detect potential aquatic life-related beneficial use impacts that may result from chemicals which may not be discretely measured and to evaluate the effects of the effluent as a whole. The worst case concern for aquatic organisms is generally chronic toxicity which impacts growth, reproduction, behavior, etc. of aquatic organism after prolonged, usually life-time or critical lifestage, exposure durations. While full-chronic toxicity tests with fish typically take several years to complete and cost on the order of a hundred thousand dollars, there are techniques available which can be used to estimate the chronic safe concentrations of chemicals in about a week of testing. The most promising of these is the fish embryo-larval test developed by Birge and Black (20) and described in the draft manual of Horning and Weber (4).

It is recommended that in conjunction with the chemical testing above, a series of effluent/receiving water mixtures be subjected to embryo-larval tests ranging from 100% effluent to 100% receiving water. The dilutions are of importance not only from the point of view of reducing effluent concentrations but also from the point of view of examining the effects of interactions between components of the effluent with those of the receiving water. In addition, a parallel suite of tests should be conducted using effluent/receiving water mixtures that have been allowed to age for several days. The latter set of tests would, for many toxicants, identify effluent components which may become more toxic with time-distance downstream. If no chronic toxicity is found in any of the fresh or aged effluent/receiving water tests, it is unlikely that toxicity would occur in the receiving waters for the effluent under the conditions evaluated. If toxicity is found in one or more of the dilutions of the fresh or aged effluent/receiving water systems, the results should be compared with the dilution available under worst case conditions at the particular site to determine whether or not the dilutions tested would likely occur in substantial areas at that site. If they would not occur in significant areas at the site, the state and federal agencies could be petitioned for an exemption from or modification of the limitations based on the worst case conditions.

If toxicity is found in embryo-larval toxicity tests with fresh or aged effluent/receiving water dilutions that could occur in the receiving water, the arbitrary effluent limitations still should not necessarily be accepted apriori unless the discharger did not want to continue testing. Site-specific field investigations using techniques such as in-stream toxicity tests with fish (21) or side-stream testing with fish embryos should be conducted to determine the actual toxicity in the receiving water with distance downstream of the effluent, and with the breadth of the receiving water to determine the area of potential impact on aquatic life-related beneficial uses and the potential impact of the toxicity on the beneficial uses. The authors and their associates discuss elsewhere details of the type of investigations that should be conducted to make these determinations (17)(21). The results of this hazard assessment study should then be used to arrive at possibly alternate degrees of treatment of the effluent from those mandated through the generic regulatory agency approach in order to more cost-effectively achieve appropriate protection for aquatic life-related beneficial uses of the receiving water. It is crucial, however, that the hazard assessment studies be properly designed and carried out in order that convincing assurance is provided that the aquatic life-related beneficial uses would indeed be protected under altered contaminant loading conditions.

For discharges for which it is determined that under certain conditions expected to occur at some times, an adverse impact could occur unless the more stringent effluent limitations were achieved, consideration should be given to seeking a variable discharge permit.

Variable Effluent Discharge Limits

While at this time, wastewater treatment plants are designed for the 7Q10, or in the future, 1Q10 receiving water flows, they do not necessarily have to be routinely operated to achieve the design degree of toxic chemical or characteristic control. Since many of the treatment processes used to remove toxic chemicals are operating-cost-intensive, variable effluent discharge permits for toxic chemicals in which the flow, chemical, and other characteristics of the receiving waters are considered, can provide a technically valid, environmentally protective approach which could save the public and/or the industry a substantial amount of money.

It is strongly recommended that where wastewater treatment facilities are required to practice toxicant control to meet criteria applied to 7Q10 or other designated low flow, the cost of achieving this control be assessed. For those situations in which such costs are judged to be significant, the regulatory agencies should be petitioned for a variable effluent discharge limitation (permit) which appropriately considers the variable conditions of the receiving water that alter the toxicity of the effluent in the receiving water. For example, there is little point in treating a domestic wastewater treatment plant effluent to the extent necessary to not exceed a 10 ug Cl_2/L receiving water limitation under 7Q10 or similar designated critical low flow conditions, when the receiving water flow is substantially greater than the critical low flow amount. Under the higher flow conditions that would be found most of the time (in the case of the 7Q10, for all but one week every 10 years), it should be possible for the treatment plant to practice substantially less chlorine removal; this could result in a significant savings in operating expenses without sacrificing protection of aquatic life-related beneficial uses.

The variable permit approach would require that those responsible for the operations of many treatment plants become more aware of the receiving water characteristics than they may typically be today. As discussed by Lee and Jones (22), those responsible for determining the degree of treatment for a wastewater effluent will in many instances find that appropriate monitoring and assessment of receiving water characteristics (e.g., flow, pH, temperature, turbidity, or any other characteristic pertinent to the impact of the particular contaminant(s) on the beneficial uses of the receiving water) will be highly cost-effective in determining the appropriate degree of treatment necessary to protect designated beneficial uses of the receiving water.

Effluent Toxicity vs. Public Health

The US EPA and state regulatory agencies are working toward developing "non-toxic" effluents. While on the surface, this may seem to be appropriate from an aquatic life protection point of view, from a public health protection point of view, this may not be the case in all instances. As discussed by Lee and Jones (23), the removal of all acute toxicity associated with an effluent such as with the dechlorination of treated municipal wastewaters, will permit fish and other aquatic life to reside for long periods of time near the point of discharge. This will likely result in the accumulation of somewhat higher concentrations of chemicals which can be bioaccumulated in their tissues than would occur if the effluent contained sufficient concentrations of a non-persistent chemical (such as chlorine) which was repulsive or acutely toxic to edible fish and other organisms. The complete detoxification of an effluent could increase its public health impact for those who consume the organisms that have resided for long periods of time near the effluent discharge. Obviously it would be very difficult for regulatory agencies to develop approaches which would allow sufficient toxicity in an effluent to prevent edible organisms' spending long periods of time near the discharge while at the same time, not result in a significant chronic toxicity problem downstream.

While in the opinion of the authors the above-mentioned approach is appropriate for establishing the degree of treatment needed for discharges from most domestic wastewater treatment plants and certain industrial waste treatment facilities, a special situation exists with the effluents from hazardous waste treatment facilities. The approach that is recommended for these types of discharges is discussed below.

HAZARDOUS WASTE TREATMENT FACILITIES

The 1984 amendments to the Resource Conservation and Recovery Act (RCRA) require that the land burial of hazardous wastes be significantly curtailed. This means that in the near future many of the hazardous wastes which are currently being disposed of by land burial will have to be treated/detoxified by physical, chemical, and/or biological means before the residues are disposed of by land burial. Many of these processes will generate a wastewater which will have to be treated before discharge to surface waters. From a public perception point of view, the change from land burial of hazardous waste to hazardous waste treatment will represent a change from groundwater pollution to surface water pollution. It is the experience of the authors that the public will vigorously oppose hazardous waste treatment facilities with a similar or even greater degree of fervor than land burial facilities. In the broadest context, hazardous waste treatment facilities could, if not properly regulated, have a greater impact on more of the public than land burial facilities. It is therefore appropriate to consider the degree of wastewater treatment that such facilities should achieve.

It is the experience of the authors that those who propose to operate such facilities will try to minimize the amount of treatment that has to be provided in order to minimize their costs. However, the wastewater effluents from hazardous waste treatment facilities represent some of the most hazardous wastewater effluents that society has produced. The fact that these treatment facilities will operate on a batch basis, sequentially treating a variety of wastes having a variety of different, highly hazardous components and thereby producing a highly variable effluent, means that special consideration should be given to evaluating and limiting the toxicity of the effluents from these facilities.

Each batch of effluent should be tested by chronic (partial chronic) testing procedures before its release to the environment. Since many of the chemicals associated with such effluents have not been evaluated with respect to their aquatic life toxicity it is appropriate to require that chronic (partial chronic) testing be conducted on each effluent before any of the effluent from a batch or group of batches is discharged. The embryo-larval or other testing procedures described by Horning and Weber (4) such as the Ceriodaphnia testing, or the acute toxicity testing with an acute/chronic ratio of 200 should be used.

In the opinion of the authors, because of the fact that the effluents from hazardous waste treatment facilities will contain a significant number of contaminants which may be highly hazardous to man and the environment and for which there are no criteria or standards, it is appropriate that the wastewater effluent from such facilities be processed through two activated carbon columns in series. When the first column shows breakthrough of an organic component of the effluent that poorly sorbs on the carbon, it is taken out of service and the second column becomes the primary column of the series and a new second column is put into service in series after the new first column. Such an approach should significantly increase the environmental and public health protection afforded by the hazardous waste treatment facility beyond that normally required today.

While this approach will likely significantly increase the cost of wastewater treatment for hazardous waste compared to what is normally practiced today, these costs should be passed on to the public. Those responsible for operating such facilities will thus likely determine the price charged for the treatment based on the costs associated with detoxifying the waste material as well as the removal of the toxicity from the wastewater effluent. These costs would in general result in an insignificant increase in the per-item price of the goods whose manufacture generated the hazardous waste.

TOXICITY OF SEDIMENT-ASSOCIATED CONTAMINANTS

US EPA representatives have indicated (24) that they are attempting to develop sediment criteria which could be used to restrict the discharge of toxicants from wastewater effluents.

It has been found (25) that many US waterway sediments in urban/industrial areas exhibit measurable toxicity to aquatic life when mixed with water. For example, in the studies of the authors and their associates, typically from one to six out of 10 grass shrimp die within 96 hours of exposure to 1:4 mixtures of marine or estuarine sediment and water in static acute tests. While the chemicals responsible for the toxicity have not been identified, Jones and Lee (26) have found that there is a relationship between the amount of ammonia released in these mixtures and the percent mortality after 96 hours for tests on a group of New York/New Jersey harbor sediments. It is clear from these studies, however, that most of the contaminants in US waterway sediments are in non-toxic forms. As reported by Lee and Jones (10), a number of investigators have found that the association of contaminants with particulate matter significantly diminishes their toxicity. This raises serious questions about the validity of the US EPA´s current efforts to develop sediment concentration quality criteria. It is the experience of the authors that such an approach has limited utility in terms of assessing the hazards that sediment-associated contaminants represent to beneficial uses of the waters. Rather than trying to use chemical analyses of the sediments, a far-more reliable approach is to use aquatic organism toxicity tests. Such tests directly assess the availability of contaminants present under the test conditions. There is considerable potential for using a modified fish embryo-larval test on sediments to assess potential chronic toxicity of the sediments to aquatic life (26).

CONCLUSIONS

The amount of treatment that will be required for a particular wastewater (domestic, industrial, or hazardous waste) and the assessment of the adequacy of treatment provided are now coming to be controlled by the amounts of contaminants allowable without adversely affecting the aquatic life-related or public health-related beneficial uses of the receiving water. These concerns will in some instances dictate the type of treatment processes that must be used to achieve the desired degree of treatment as well as the siting of the treatment facilities.

The US EPA has recommended an approach that is designed to begin to address the evaluation and testing issues pertinent to this type of effluent evaluation approach. While there are a number of significant problems with the implementation of this approach in providing cost-effective treatment which will still provide adequate environmental and public health protection in receiving waters, it does represent a significant step toward the regulatory agencies' developing technically defensible control programs for toxics. Following the current (September 1985) US EPA guidance may result in a discharger's treating an effluent to a considerably greater degree than necessary to provide protection of aquatic life-related beneficial uses of the receiving water, or inadequate treatment to achieve this protection. However, using this approach as a starting point, and building on it with site-specific hazard assessment studies will likely be highly cost-effective in developing workable effluent restrictions that provide adequate aquatic life-related beneficial use protection and also insure that money spent for toxics control will improve the water quality of the receiving waters.

REFERENCES

1. NJ DEP, Wastewater Discharge Requirements," NJAC 7:9-5.1, New Jersey Department of Environmental Protection, May (1985).

2. US EPA, "Technical Support Document for Water Quality-based Toxics Control," US EPA Office of Water, Washington, D.C., September (1985).

3. US EPA, "Methods for Measuring the Acute Toxicity of Effluents to Freshwater and Marine Organisms - 3rd ed," EPA/600/4-85/013, US EPA Cincinnati (1985).

4. Horning, W. E., and Weber, C. I., "Draft Methods for Estimating the Chronic Toxicity of Effluents and Receiving Waters to Freshwater Organisms," EPA-600/4-85-014 May (1985).

5. US EPA, <u>Quality Criteria for Water</u>, US Gov't Printing Office (1976).

6. US EPA, Water Quality Criteria; Availability of Documents (Notice of Final Ambient Water Quality Criteria Documents)," <u>Federal Register</u> 50(145):30784-30796 July 29 (1985)

7. Lee, G. F., and Jones, R. A., "Guidance on the Application of US EPA's Ammonia Criterion for Site-Specific Water Quality Standards and Point Source Discharge Limitations," Report to US EPA Duluth (1985).

8. National Academy of Science and National Academy of Engineering, <u>Water Quality Criteria - 1972</u>, EPA/R3-73-033 (1973).

9. US EPA, "Water Quality Criteria Documents; Availability," <u>Federal Register</u> 45(231):79318-79379 November 28 (1980).

10. Lee, G. F., and Jones, R. A., "Water Quality Significance of Contaminants Associated with Sediments: An Overview," Proceedings of Pellston Conference, "The Role of Suspended and Settled Sediments in Regulating the Fate and Effects of Chemicals in the Aquatic Environment," held in Florissant, CO (1984) To be published by SETAC.

11. Lee, G. F., and Jones, R. A., "Water Quality Hazard Assessment for Domestic Wastewaters," In: <u>Environmental Hazard Assessment of Effluents</u>, Pergamon Press, NY, pp 228-246 (1986).

12. Heinemann, T. J., Lee, G. F., Jones, R. A., and Newbry, B. W., "Summary of Studies on Modeling Persistence of Domestic Wastewater Chlorine in Colorado Front Range Rivers," In: <u>Water Chlorination-Environmental Impact and Health Effects</u>, Vol. 4, Ann Arbor Science, Ann Arbor, MI, pp 97-112 (1983).

13. Newbry, B. W., Lee, G. F., Jones, R. A., and Heinemann, T. J., "Studies on the Water Quality Hazard of Chlorine in Domestic Wastewater Treatment Plant Effluents," In: <u>Water Chlorination-Environmental Impact and Health Effects</u>, Vol. 4, Ann Arbor Science, Ann Arbor, MI, pp 1423-1436 (1983).

14. Lee, G. F., and Jones, R. A., "Translation of Laboratory Results to Field Conditions: The Role of Aquatic Chemistry in Assessing Toxicity," In: Aquatic Toxicology and Hazard Assessment: 6th Symposium, ASTM STP 802, ASTM, Philadelphia, pp 328-349 (1983).

15. Weber, C. I., Biological Methods Branch US EPA Cincinnati, OH. Personal Communication to G. Fred Lee (1986).

16. Veith, G., US EPA Environmental Research Laboratory, Duluth. Personal Communication to G. Fred Lee (1986).

17. Lee, G. F., Jones, R. A., and Newbry, B. W., "Alternative Approach to Assessing Water Quality Impact of Wastewater Effluents," Journ. Water Pollut. Control Fed. 54:165-174 (1982).

18. Lee, G. F., Jones, R. A., and Newbry, B. W., "Water Quality Standards and Water Quality," Journ. Water Pollut. Control Fed. 54:1131-1138 (1982).

19. Lee, G. F., and Jones, R. A., "An Approach for Evaluating the Potential Significance of Chemical Contaminants in Aquatic Habitat Assessment," Proc. of Symposium, Acquisition and Utilization of Aquatic Habitat Inventory Information, American Fisheries Society, pp.294-302 (1982).

20. Birge, W. J., and Black, J. A., "In Situ Acute/Chronic Toxicological Monitoring of Industrial Effluents for the NPDES Biomonitoring Program Using Fish and Amphibian Embryo/Larval Stages as Test Organisms," OWEP-82-001, Office of Water Enforcement and Permits, US EPA Washington (1981).

21. Newbry, B. W., and Lee, G. F., "A Simple Apparatus for Conducting In-Stream Toxicity Tests," Journ. of Testing and Evaluation, ASTM 12:51-53 (1984).

22. Lee, G. F., and Jones, R. A., "Active vs. Passive Water Quality Monitoring Programs for Wastewater Discharges," Journ. Water Pollut. Control Fed. 55:405-407 (1983).

23. Lee, G. F., and Jones, R. A., "Fishable Waters Everywhere: Is This an Appropriate Goal?" Industrial Water Engineering 20:14-16 (1984).

24. Dickson, K. L., et al. (eds), The Role of Suspended and Settled Sediments in Regulating the Fate and Effects of Chemicals in the Aquatic Environment, To be published by SETAC.

25. Jones, R. A., Mariani, G. M., and Lee, G. F., "Evaluation of the Significance of Sediment-Associated Contaminants to Water Quality," Proc. Am. Water Resources Assoc. Symposium <u>Utilizing Scientific Information in Environmental Quality Planning</u>, AWRA, Minneapolis, MN pp.34-45 (1981).

26. Jones, R. A., and Lee, G. F., "Toxicity of Waterway Sediments with Particular Reference to New York and New Jersey Harbors," Presented at US EPA Symposium on Chemical and Biological Characterization of Municipal Sludges, Sediments, Dredge Spoils and Drilling Muds, May 1986. To be published by American Society for Testing and Materials.

RESEARCH NEEDS WORKSHOP

Chairman	Co-Chairman
Edward H. Bryan	**A.F. Gaudy, Jr.**
Program Director	*Department of Civil Engineering*
Environmental Engineering	*University of Delaware*
National Science Foundation	*Newark*
Washington, D.C.	

Discussion of paper by Robert W. Peters

Chen commented that "UV-peroxide," "ozonation," and "bioenzyme" are technologies worthy of consideration. Peters replied that his formal paper was more inclusive than the verbal presentation and suggested the further possibility of coupling emerging technologies with traditional approaches such as the activated sludge process. Rozich suggested increased support for research on "natural populations" as a step toward process improvement and Peters responded that this was consistent with his position on research needs.

Discussion of paper presented by Olli H. Tuovinen (co-authored by Conley Hansen)

In response to Chen's question regarding use of activated alumina and carbon as adsorbers for heavy metals in biotreatment processes, Tuovinen suggested the biomass produced by treatment of industrial wastes from fermentation industries could be utilized in a process for stripping the waste from another industry of its valuable metal content. In response to Switzenbaum's question regarding removal of mercury by reduction of the divalent form as contrasted to its methylation and removal by adsorption, Tuovinen indicated that methylation of mercury reduces its solubility and increases its volatility, thus making its recovery more difficult. He also indicated that methylation is not likely because their process is aerobic and methylation occurs primarily under anaerobic conditions.

This is an edited transcript of the Proceedings of this session prepared by the Session Chairman. The content of this section does not constitute an official position of the National Science Foundation or any other agency.

In response to a question from Bishop regarding the future potential for use of special enzymes and "binding proteins" for removal of specific contaminants such as cadmium and mercury. Tuovinen stated that his only direct experience in application of this concept has been with mercury, with respect to which he saw a "great future." Hansen also responded to this question by stating the concept has "a lot of potential" as a complementary or alternative process to those used conventionally indicating there has been some initial work done on both arsenic and selenium.

Preliminary Panel Comments

Following presentation of papers by Peters and Tuovinen a panel was formed to address research needs. Each panelist presented a brief introductory statement. Following all presentations, the panelists discussed issues raised by other panelists and session participants.

Remarks by Morgan Kommer, Manager of Water and Wastewater/Environmental Control, Alcoa, Corporation, Pittsburgh, Pennsylvania.

"I have corporate responsibility for our programs that come under the Clean Water Act, the Safe Drinking Water Act, TOSCA, and now indirectly, Superfund. To industry's way of thinking, at least to Alcoa's, when you talk of research you also talk of development. Many promising concepts that have emerged from research have not been studied in more detail at the pilot scale where you can get some meaningful engineering design parameters. There's a gap between the R and the D and I'd like to see more concepts emerging from research developed to solve real-world pollution control problems."

Remarks by Dolloff F. Bishop, Wastewater Research, Office of Research and Development, U.S. Environmental Protection Agency, Cincinnati, Ohio

"My views reflect the agency needs in the wastewater treatment area for the management of toxics both from the industrial and municipal point of view. One of the first needs that we have is in the detection and identification area, i.e. the monitoring of toxics, especially those in complex wastes and relating the monitoring to improvement of treatment plant operation for control of toxics. We find that we're making good progress in using bioassays for measuring toxicity in an integrated form in terms of its impact on the ecosystem. What we need are less expensive and more near real-time tools that would allow us to characterize the toxicity impact and pass-through the at the central treatment plant. An example of the tools that are being considered for that purpose is the "MICROTOX" analysis.

"There is also an emerging field in the monitoring area for estimating the potential for health effects of specific toxics or complex toxics passing through a treatment facility. There are what the agency calls Level One assays such as the Ames test and mammalian cell assay that indicate a potential for health effects with end-points for carcinogenecity or mutagenicity. The problem with these assays is that we don't know how to relate the results of the relatively inexpensive health effects bioassay to risk of health effects. A better link is needed between the relatively simple bioassay and the risk of health effects. The linkage may be extremely difficult to build and probably very expensive to obtain.

"Regarding the fedbatch reactor and its use in indicating inhibition in the aerobic system, we need to reduce that concept to reasonable automation and develop similar concepts for using the anaerobic process to treat complex wastes containing toxics. This would go a long way toward allowing us to adapt bioreactors for the treatment of complex mixtures of hazardous wastes by a quick assessment of the concentrations that can be tolerated in the reactor without significantly damaging its performance.

"We also need to improve our understanding of the fate and impacts of toxics on treatment processes and relate the fate of the specific toxics to their structural activity properties. The generic toxicity aspects of wastewaters need to be related to the operational capability of treatment plants for removal of toxicity. A better understanding is needed of the fundamental factors controlling the biodegradation of toxic organics in the wastewater treatment system. The literature indicates that xenobiotic or toxic organics can be managed in appropriately controlled reactors under starved carbon conditions. With the adjustment of organic nutrients in the system, the concentration of toxics may be reduced to low levels by manipulation of the SRT and the concentration of biogenic co-metabolites which support the biological process. A much better understanding of this process is needed so that we can learn to manipulate the biological reactors to reduce the specific or selected toxics concentrations to very low levels.

"Of further interest is how combinations of anaerobic and aerobic treatment can be used to complement each other. For example, lindane will pass through an aerobic reactor pretty much untouched. However, lindane can be managed, if acclimated in an anaerobic system. Other innovative concepts are needed that allow us to manipulate the wastewater treatment plants to manage toxicity pass-through which can occur into water, air and solids. This may mean that wastewater treatment facilities will become detoxification facilities where special reactors are developed and maintained to manage special wastes.

"Systems engineering concepts are developing with respect to the management of water quality. Our sister division in Cincinnati is looking into the impact of toxic discharges on cost of water treatment by downstream water users. Models have been developed that can be used to predict fairly well the fate of toxics in the discharges. Water utilities are going to be extremely interested in trying to get the upstream dischargers to share in the increased cost for water treatment attributable to those toxics. Systems engineering concepts will permit us to find the least costly mechanisms for managing toxicity and assist in developing credible management tools."

Remarks by D. B. Chan, Environmental Protection Division, Naval Civil Engineering Research Laboratory, Port Hueneme, California.

"The Navy's hazardous wastes program includes both basic and applied research. The main areas of interest are development, testing and evaluation of emerging and innovative technologies. We are especially interested in field testing laboratory evaluated and proven technologies, using pilot studies and field demonstrations under different environmental conditions.

"Currently, we have two major programs for Navy hazardous wastes. One concerns minimization of waste through process changes at industrial facilities and use of better inventory management systems or by on-site volume reduction. Of particular interest are pre-treatment methodologies such a physicochemical processes or biosorption. The second program is called the Installation Restoration (IR) program which is comparable to EPA's Superfund program. It matches specific remedial actions with specific municipal and industrial disposal sites and includes clean-up of hazardous waste disposal sites used in the past, leaking underground storage tanks and spill-site cleanups. Emphasis is on treatment of previously contaminated soils or groundwater rather than of general wastewater.

"The technology for these types of cleanups may be divided into two stages, the first stage of which involves extraction or washoff of the contaminants from the soil particles. The second stage involves degradation, detoxification, or fixation of the contaminants in soil or in the groundwater. An important concern is whether a given treatment technology will further pollute the general area or environment as a result of increased contaminant mobility or transformation to more toxic products.

"In terms of treatment technology, we are most interested in biological treatments, especially in situ treatment technology, mainly because it is cost-effective and also complete in terms of destruction or detoxification of the contaminants. We are also interested in physiochemical processes such as thermal destruction and chemical reduction or oxidation. The types of technologies that we are now investigating exclude conventional incineration because of its cost. Incineration of PCBs to achieve the required 99.999 percent reduction, is very expensive and difficult to achieve. Permit applications require clarity in process description, knowledge of byproducts and their potential toxicity and analytical techniques all of which are essential research needs in development of alternative technologies."

Remarks by Michael Switzenbaum, Dept. of Civil Engineering, University of Massachusetts, Amherst.

"Most of my work has been in anaerobic treatment of easily treated industrial wastewaters such as food processing wastes and municipal wastewater. We know that anaerobic methane fermentation has a role in the detoxification of certain toxic organics. Regarding Bishop's observations on use of anaerobic and aerobic processes in series for the treatment of certain toxic wastewaters, this may be a highly useful means for the biodegradation of several types of persistant chemicals. Virtually all of the fundamental understanding of the biochemistry of methane fermentation has been developed over the last 15 years. Much of the impetus for research on methanogens is not because they make energy or because they can detoxify certain chlorinated hydrocarbons. It is because they are ancient organisms, and exciting to work with because they are thought to represent phylogenic evidence for the evolution of prokaryotic organisms.

"My second point has to do with pre-treatment which I feel is ore of a political problem than a technical issue. The enforcement of pretreatment standards is an important consideration, particularly from a municipal wastewater point of view. I'll be happy if what I've said here serves as a basis for further discussion.

"My third point relates to superbugs. We hear a lot of talk about superbugs, either bioaugmentation or the genetic manipulation of existing organisms to create an organism for a specific purpose such as the detoxification of a specific compound or perhaps just increased treatment efficiency. I personally am rather skeptical about this. Genetic engineering holds much promise for production of pharmaceuticals and gene therapy as well as wastewater treatment. But the big problem right now is how to keep organisms in a reactor or in an environment into which we introduce them. This might be the bottleneck for the implementation of genetic engineering for wastewater treatment.

"In his paper The Microbial Basis of Biodegradation, Grady points out opportunities for enhancing biodegradation such as better design of reactors or their operation to enhance opportunities for biodegradation with reference to optimum anaerobic/aerobic conditions, high solid retention times in biofilm reactors, continuous flow reactors, and continuous seeding. These are related to the ecological aspects of organisms used in treatment processes which have not been given the attention they deserve for their promise in enhancement of biodegradation."

Remarks by Ronald Unterman, Staff Scientist, Biological Sciences Branch, General Electric Corporation, Schenectady, New York.

"One of the two major biotech programs at the Schenectady site is concerned with the bioconversion of waste chemical products. PCB's have been a major environmental problem for GE, and we are actively studying their bioconversion. In addition we are interested in the properties of metal-binding proteins, both those that exist in interesting organisms like the acidophilic bacteria and those produced synthetically.

"We have devoted about 30 man-years of basic research over the past four years on bioconversion of environmental contaminants and are now ready to leave the laboratory to apply what we've learned about the microbiology, biochemistry and genetics of the organisms we've studied. We expect to begin a site test this summer using some of the organisms that we've isolated and feel that's an important direction for our PCB biodegradation research.

"In our metal-binding proteins project, we are interested in trying to understand the rules which govern metal-binding specificity. For example, why does one domain of metallothionein bind cadmium more strongly than zinc? What structural features confer specificity in these proteins? Once you understand the basic science, you can think about manipulating those proteins to your individual needs, whether it's to clean up a waste problem or to isolate metals from dilute solutions.

"Both waste management projects are at very different stages. The metal-binding protein project is at the basic research level and the PCB-biodegradation project is ready for field testing. We are also interested in anaerobic bioconversions of chlorinated aromatics, specifically dechlorination. Conceptually, one could take a highly chlorinated recalcitrant molecule, initially dechlorinate it anaerobically, and then aerobically biodegrade the less chlorinated product. Such a two-step biodegradation may eventually lead to the treatment of what were previously considered non-biodegradable compounds.

"Regarding the use of recombinant DNA organisms in the environment, this is a general problem especially relevant to biological treatment of contaminated sites. Although our planned summer test is with a native organism, what if GE had a recombinant organism to test in situ? Would we be allowed, and more significantly, willing to use it considering the liability climate today? I don't know how the EPA or any goverment agency is going to deal with this problem. If R-DNA bacteria can really do a significant waste cleanup job, shouldn't we be using them? I personally don't believe GE or any large company can face such unknown liability. And that's an important issue. How then do we as a society implement this potentially beneficial new technology? This question is obviously broader than any specific research needs for waste management."

Remarks by Anthony F. Gaudy, Dept. of Civil Engineering,
University of Delaware, Newark.

"Several of the preceding panelists have stressed the need for developmental research rather than basic research. I think there is a need for both. There are plenty of things that we already know, or think we know, but which need practical developmental research in order to transfer them safely into field practice. Many researchers don't participate enough in this kind of research. I'm pleased that so many of the panelists stressed the developmental aspects.

"Regarding the idea of the selective culture principle, we don't have to go to mutant forms or recombinant forms which would probably be difficult to retain in the system. We would have to find conditions that would hold them in the system or their costly addition would be wasted. That is, there is a general recognition that the conditions that would hold them in the system are the very same conditions that will help select for them naturally. So we shouldn't be rushing into the special culture business."

Comments by Edward H. Bryan, Session Chairman

"Before opening discussion on topics raised by our panel members, I'd like to comment briefly on the concepts described by the words "research" and "development." I like to think of the word "research" as describing what we do when we search for knowledge. Research can be more basic than applied or more applied than basic depending on whether we are seeking something new or how to apply something we already know toward solving a problem.

"In contrast, I think of the word "development" as the process by which we take knowledge that emerges from research and determine its practicality - that is, its technical practicability (e.g. by construction and operation of a pilot plant or the equivalent) and its economic, social, and even political feasibilities. Research leads to knowledge of what is possible while development leads to knowledge of what is deliverable within an economic, social, legal and/or political context. When applying the terms "research" and "development" to topics in environmental engineering, basic and applied research often describe what an individual or relatively small group of individuals do in a laboratory on a "bench" or in an office while development involves leaving the laboratory and office for field work at a pilot plant scale.

"Development is usually a much more expensive activity than research and as a consequence, care must be exercised in selection of the concepts to be developed. Researchers are likely to have a "vested interest" in seeing all of their concepts developed and developers may ignore potential solutions to problems using concepts that have emerged from research outside their immediate organization (the latter is sometimes referred to as the NIH or Not Invented Here Syndrome). Careful and objective review is an important part of the selection process for the investment in development."

Panel Discussion

Comments by C. P. L. Grady (Clemson University)

"Amplifying and emphasizing points made by Switzenbaum and others regarding the ecology of wastewater treatment systems, we started out studying the removal of toxic compounds eight or nine years ago working with both pure and mixed cultures. It has become obvious to me that the development of an ability to predict removals of these materials down to very low levels requires a much better understanding of microbial ecology than we have now. For example, we find that while mixed culture systems can remove materials down to the microgram per liter range, their concentration in the effluent varies considerably. This variability is directly related to microbial competition within the systems and interactions among different microbes. In spite of the importance of microbial interactions we know relatively little about them, and generally, microbiologists have not been particularly interested in studying them. I urge those of you with strong microbiological backgrounds to study the ecology of waste treatment systems and how microbial interactions influence removal of toxic constitutents to very low levels.

"In agreement with a point made by Bishop, toxicity is going to become one of our parameters for evaluating wastewater treatment because it isn't realistic to spend the time and the money required to measure concentrations of individual constituents down in the nanogram per liter range. Thus, toxicity will become one of our surrogate parameters. Given this situation, it is important to note that one of the things that we don't know much about is the natural production of toxicity by microbial systems. We've shown that activated sludge produced by a synthetic wastewater containing no toxic constituents can produce toxic effluents. This is because we're dealing with complex ecosystems in which microbes are fighting for habitat, one mechanism for which is through chemotaxis. Consequently, there are materials produced within those system which register as toxics using microbial assays. Use of toxicity assays assessing treatment systems will require establishment of a baseline for their own natural production of toxicity and answer the question of what actually does constitute significant toxicity?

"Regarding prediction of the removal of toxics to low levels in multi-component systems, our efforts over the past several years have not as yet succeeded. If we grow a pure bacterial culture on a single substrate we can predict the effluent concentration of that constituent. But when we start mixing substrates in continuous culture systems under low-growth rate conditions with nanogram per liter levels of constituents present in their environment, we cannot predict their concentrations in the effluent. Consequently, we need more fundamental work concerning what actually controls concentrations of constituents when organisms are growing slowly. Furthermore, when we use mixed cultures with their added level of complexity, all that we can say is that most of the time they will reduce the concentrations of constitutents below some low level, an observation that may be inadequate for either design or regulatory use.

"I guess this all goes back to the question of whether we should be using toxicity assays to evaluate performance. A fundamental socio-political question is whether we base our effluent standards upon public health concerns or upon ecosystem concerns. In other words, if we protect the ecosystem, does that protect the public health, or if we protect public health, does that protect the ecosystem? Which is the more sensitive one that we should be designing treatment systems for?"

Following Grady's comments, panelists responded to points raised O'Shaughnessy who contrasted the usual process of research followed by reactor design for point-discharges of an industrial waste to the research needs relating to scattered hazardous waste containment sites. He asked: "How do we use biological application to remediate something, not just ship it to Alabama or New York but clean it up in situ. How do we do that?"

Gaudy responded by stating that he tried to stress that in his remarks and stated that: "I think our panelists who opened the session were really talking about that very aspect. How do we now take our research information to the field? Who supports that kind of work? Obviously it cannot be NSF. It may or may not be one of the more mission-oriented organizations, such as the Navy, the Army, EPA, or industry. But I haven't seen any of these organizations rushing in to fill the gap in support of developmental research.

"Many fundamental studies have led to the recommending of some modification, new ways to do things, new methods or processes. However, generally the finding is reported in the literature and is left there. In short, the information is only transmitted in part. The problem is how do we get new ideas tried out in the field. That is the true essence of transfer of research information, and here we are having little success. I believe that what is lacking is sound financial support for the necessary developmental research.

"Some people settle with the word "development" in place of "developmental research." In my opinion the two cannot be separated if you are interested in having the research accomplish some change in the practicing field. Developmental research is the research that comes after the conceptual bench-scale research and it is needed in order to bridge the gap in effecting changes in practice. Everything about a new process isn't found out in the laboratory. When you go to develop a process, that piece of investigation which I would call developemental research feeds back to the more basic stuff.

"Now you say: When will we be able to do that kind of work? I think that is going to happen when you get people interested in supporting that kind of effort. And as engineers, that's an effort we should be devoted to carrying forward. That kind of research needs support, because you can put out all sorts of suggestions for modified or innovative processes in the literature, and you then find people who have read the article and would like to use the ideas but they will invariably ask who has tried it in the field? Do you have any large-scale experimental data? Nobody in practice wants to try these things first or pay the tariff for trying them first. When the Federal government tried out demonstration grants some years back, it was found, in many cases, that the demonstration grants pretty much backfired and the idea became tarnished. I think that the newer ideas will be put into place when the people who have the problem are willing to spend the money to develop potentially innovative processes and use them."

In further discussion, Unterman cited a specific example of how laboratory findings are planned to be subjected to field testing. This will involve a field site where soil has been contaminated by Aerochlor 1242. Several hundred liters of laboratory-grown organisms will be applied to the soil and the rate of biodegredation of the PCB will be carefully studied.

An unidentified participant noted that problems at sanitary landfills were generally complex involving leachate of unknown composition and posing a threat to the quality of groundwater. While skillful operation of a landfill to balance nutrient loading for production of methane is needed, he suggested that the better way is to keep as much of the organic waste materials out of landfills as possible and move them toward bioreactors that are designed specifically to manage them.

In response to Hansen's concern about the problem of obtaining support for the field testing of concepts emerging from research laboratories, Gaudy agreed this was a vital point, stating that: "Development does not have to mean the full-scale field try-out. Development should be the larger than lab scale-pilot plant, where you make the adjustments for transition from the laboratory to full-scale." He indicated that chemical engineers became so effective in the design of some unit processes that they could skip the pilot plant stage, going directly to the field and adjusting the prototype. "While they've gotten so skilled in the processes they deal with that they can do it, we are generally dealing with processes that are extremely complicated such as the decay leg of the carbon/oxygen cycle. If we could really understand that, we would have the solution to our environmental problems nailed down.

"We can't make those kinds of jumps, so we can't consider research as something that is done in the lab, and development as something that's done full-scale. There is an area in between that we vitally need because we don't have all the answers. We can't answer everything with our fundamental research. In fact, most of the best fundamental research that has been done or is being done is telling us that we can't shortcut the process. I think we need to expand our concept of what research is in our applied field. It doesn't stop with what we call the fundamental stuff. I think you have to take the findings to the field in that pilot plant stage to prove things out and then feed back the new information to the fundamental investigational effort and feed forward to the prototype. I think we'll get processes that work a lot better, a lot better control of the environment, and a lot less retrofitting and redesign if we take this extra step. It's a little bit safer and surer way to go. It will enhance and encourage obviously needed changes in practice and can prevent mistakes. In the long run, I think it's very, very cost-effective for the public dollar, whether it comes from Federal or local taxation or a higher price for consumer products of industry."

A suggestion was made by Roe for a concept of "life-cycle costing" to provide credit for the environmental advantages of eliminating or replacing the processes that are being used now citing incineration as a "very environmentally destructive" process. He indicated that with the restraints (such as insurance) that have been put on doing business within the free enterprise system, "In dealing with hazardous wastes, it is very difficult to have innovation especially since there is no credit for using a more environmentally compatable biotechnology to solve the problem."

In response to Roe's comments, there was general discussion from unidentifiable sources regarding economic trade-offs between competitive processes and systems and the procedure by which costs are compared in considering solutions. Duffy cited an example of "a site in upstate New York where we have financially actually looked at the difference between what we do now, which is carbon adsorption and a biological treatment process. Our estimate is that we will save about $18 million going to a biological treatment over a ten-year period." In response to Roe's added comment that "This is the cost, not including life cycle cost credits, for environmental damage of one process vs. another," Duffy continued his response by describing the research that led to pilot-scale evaluation of a bioaugmentation concept which is ready for field application. Blair cited a recent announcement of a "Superfund" award for biological treatment of soil in place as a more economical alternative to hauling it away from the contaminated site. He indicated that: "Biological means are coming into their own, and where it works and properly applied, it is about a third the cost of other methods." Roe continued to suggest that for a valid cost comparison, cost factors need to included for environmental effects in the procedure of comparing costs for competitive processes.

Switzenbaum responded to Duffy's request for augmentation of comments on choice of system for anaerobic processing by pointing out that while anaerobic sludge digesters are a little over 100 years old, anaerobic filters are only 20 years old, sludge blankets are 15 years old, and fluidized and expanded beds, less than 10 year old. He pointed out that it has taken about 70 years of research and operating experience to understand the capabilities of the activated sludge process and its different modifications. A similar understanding is developing as more operating experience is becoming available for the various anaerobic processes. He viewed sludge blankets, anaerobic filters and fluidized beds as "somewhat of a continuum" with most experience in full-scale with sludge blankets. He indicated that "the concept has tremendous potential for the treatment of low-strength waste, particularly in tropical climates. Fluidized beds I think have the most potential because they can overcome mass transfer limitations, but perhaps they are the most complicated."

Comments by Henryk Melcer, Wastewater Technology Centre, Burlington, Ontario.

"The Centre has spent the last six years evaluating high rate anaerobic reactor configurations. Our experience has been that the treatment of each wastewater has a unique solution requiring a specific reactor configuration. The same reactor design does not necessarily address a range of wastewaters successfully. At the present time, we would recommend pilot-scale assessment of a range of options. Having established those which offer a technical solution, a selection is made based on economic concerns.."

In response to general discussion stimulated by Alleman's question regarding the possibility that use of technologies would be hindered because of the "proprietary issue", Unterman asked: "...why would the proprietary nature of it have a tendency to inhibit the use of it? It's just a matter of setting the fee to correspond with what somebody is willing to pay." Melcer responded to this discussion by citing experiences in Canada and by pointing out that: "...not one design is a panacea for all the applications, and a potential user who perhaps doesn't have the experience of understanding the difference between these different reactor configurations can be persuaded by one of these proponents that this is what they need. What they should really be doing is employing a process specialist to come in and do the evaluation for them." Switzenbaum responded as follows: "I would say that sometimes being proprietary hinders and sometimes it helps. What Mr. Melcer just said is certainly true...we used to have heated arguments about the different types of anaerobic processes. I haven't seen it as much recently because I think the market is a little sluggish because of low energy costs. This certainly is a temporary situation. Regarding patents, if you're an industry and you need to put in a waste treatment system and you contact a vendor who makes a certain type of reactor, what are you hiring? Not only their reactor but their expertise, and past performance record and hopefully that it will be applied correctly."

McDermott commented on the ecology of treatment systems, fate of toxics and application of credits to industry for removal of contaminants in municipal treatment plants. He also expressed concern about standard-setting in the absence of relevant scientific data. Gaudy responded by commenting on a conference paper by Pirages in which the question was asked as to whether standards are "going to be based on the public health aspects (which is basically what they are based on now) or are they going to be based on whether the material is biodegradable in the environment?" Gaudy continued "... if a compound is biodegradable in the environment, you've got a right to expect that it should be biodegradable in the controlled engineering environment of the treatment plant. How should we set the standards? If the compound is biodegradable in the environment, then maybe we can permit it to be sewered because we can stop it in the engineered environment at the treatment plant. Maybe that's the way to set the standard. But, let's also realize that this requires a massive investigational effort before issuing a permit to discharge."

Meier stated that he was "...very comfortable with the idea that research and development is motivated by economic needs, and I think that is appropriate industrially and nationwide, but I am very uncomfortable with the effect this is having on the university atmosphere. What I see down the road is a shortage of trained people. Research in the university atmosphere is not a separate thing. It is part of the educational program. You cannot teach people without doing some research. Students that don't do any research will come out not as very useful products for the future. By nature, university-based research has to be more in the area of research than development. We are just not capable of doing development work with students who are there for one, two or three years. So I am a little bit worried about what is happening. NSF is really by now the only one that is generating funds for the support of student research activities. Beyond that, we are finding support for activities that become more like doing jobs that are assigned rather than research that is going to lead to the development of new ideas, new concepts, which we need down the road. So I plead for consideration of doing research as part of the educational program."

In response to general discussion regarding the education of environmental engineers, Chan suggested the concept of a "summer camp" for environmental engineers where they would encounter realistic problems involving both process and hardware and Bishop described a procedure utilized by EPA where professors can spend a sabbatical period working in an EPA laboratory.

Noblock expressed three areas of concern from an industrial standpoint. "Having responsibility for 31 different plants within General Motors, I see research needs in monitoring metal finishing toxics: We need some way of monitoring them up front of the process in order to perform some sort of remedial action right at the waste treatment plant to avoid exceeding the limit of the discharge permit.

"Secondly, I'm interested in the biosorption systems for their possible use in separating some of our metal hydroxide sludges that we generate in our various assembly and manufacturing operations. I've buried tons of copper, chrome, nickel and zinc. We're just putting them in the ground...we should be looking at ways to fulfill the spirit of the law of RCRA by recovering these metals and reusing them.

"My third area of concern is kind of a personal one. At our metal fabrication plants, we use a lot of coolants ranging from free oils to formulations of synthetic compounds. Waste treatment technology hasn't kept up with these new kinds of products. Currently we're putting in three fluidized-bed biological systems to combat soluble organics that are components of these synthetic compounds. I'd like to see if there's a way that we can look at a process whereby we don't have to have acid, caustic, and polymer feed systems, and all these different pumps and circulation systems. I'd like to find other technologies that we can use to economically treat our wastes and also to look at a way of producing a waste stream that we can just discharge to an open waterway or to a municipal wastewater treatment plant. But I think that what we need are different, more economical ways of handling our wastewaters. I think in order for us to make cars efficiently, we have to consider all of our processes, not just wastewater treatment but the whole package."

Gaudy responded by applauding Noblock's "long-term look at metal disposal. "I suppose the only reason you're burying it is for the same reason everybody else is - because it's more cost-effective. You do what you're allowed to do, but the days when you can do that will come to an end, so it is appropriate to take a long-term look. If you have metals in aqueous solution or in suspension, it is rather well-known now that microorganisms will take the metals up. In some cases more than 30 percent of the dry weight of the biomass can be heavy metals without detriment to the biochemical efficiency of the biomass. Furthermore, the metals can be recovered simply by acidification of the biomass and precipitation of the metals. Hydrolysis allows one to remove the metals and recover them. The residual food material can be fed back to the system to develop new biomass, thus regenerating metal removal capability. Those things are possible now but their full-scale application in the field requires developmental research by people who have the problem and are willing to take the long view.

"There is a lot of potential process innovation out in the research literature, and it should be picked up and researched now in the real world sense. Industries or municipalities that have these problems can solve them individually or, perhaps, they can band together to be able to share the solution and the cost of gaining the knowledge. There are things that can be done now, and I'm glad to see that a company as large as General Motors may be willing to take the long-range view and try to do something about it now."

Closing Remarks by the Session Chairman

"My closing remarks have to do with this matter of support for research, development and demonstration. The National Science Foundation (NSF) is authorized to support both basic and applied research but not development or demonstration. Its "mission" is to assure the continued health of science and engineering in the United States. The "mission agencies" such as the Department of Defense or the Environmental Protection Agency are authorized to conduct research as is appropriate for their mission responsibility. They are also authorized to conduct development and demonstration.

"Projects that are supported by grants from the National Science Foundation also receive support from other sources. For example, all of the ten Presidential Young Investigators that are currently receiving support for their research from the Environmental Engineering Program are obtaining funds from industry which qualify them for matching support from the Foundation. Companies and other qualified organizations that are providing this support include General Motors Corporation, General Electric Company, Shell Oil Company, Mobil Oil Corporation, Millipore Corporation, Electric Power Research Institute, the Mellon Foundation, Union Camp Corporation, the Dow Chemical Company, Combustion Engineering Inc., Chemical Waste Management Inc., the Alcoa Foundation, Air Products & Chemicals Corporation, Empire State Electric Energy Research Corporation, Atlantic Richfield Company, E. I. du Pont de Nemours & Company, Hewlett Packard Corporation, United States Steel Corporation, and Stafford County of Virginia.

"The National Science Foundation's Engineering Research Centers Program is in its third year of operation. This program requires grantees to obtain substantial commitment of support from industry. Although several proposals have been submitted for establishment of engineering research centers on hazardous waste management, none have as yet competed successfully for support from this Centers program. However, NSF has provided support for the establishment of a hazardous waste institute at the New Jersey Institute of Technology under the Foundation's Industry/University Cooperative Research Centers Program."

INTRODUCTION TO THE CONSORTIUM FOR BIOLOGICAL WASTEWATER TREATMENT RESEARCH AND TECHNOLOGY

Anthony F. Gaudy, Jr().

Department of Civil Engineering
University of Delaware
Newark, DE

INTRODUCTION

    I have been asked to provide a sort of informal general introductory review of the Consortium for Biological Wastewater Treatment Research and Technology - what it is, how it was started, its goals, and something about its progress to date.
    Originally, the Consortium was simply the banding together of Vanderbilt University, the University of Pittsburgh and the University of Delaware in an association which would enable us to provide investigative and technological services, especially those needed by industry, which we might not be able to handle separately. The New Jersey Institute of Technology has since come into the Consortium, and they share this purpose.
    By coming together in this Consortium, we are expressing our belief that a great weight of responsibility for treatment or degradation of waste products can be placed on the biological processes. We are also stating that we are actively seeking to extend the scope of the natural process through gaining better understanding of it and seeking ways to gain better engineering control of it. We believe that this will lead to truly innovative engineering process modifications, and, we feel, better engineering control over toxic components in

aqueous waste streams.

Who is "we"? Certainly, "we" does not mean all the environmental scientists and engineers and persons in related fields at our individual universities, but it includes all at our institutions who share our interests and goals and who wish to participate to the extent that it best suits their individual desires and needs. It is also worth noting that we do not always act or participate in research in unison. We currently are involved only in our individual investigations at our respective universities, but we stand ready to do joint projects and service should the need and opportunity arise.

BACKGROUND

I would like to give you a little background about how the Consortium was originated. To begin with, nearly all universities have, of late, been rightly concerned with expanding the scope of investigational support and they also are seeking better definition of the role of the university in the community. The first aspect is self-explanatory. There has, of couse, been a long-term marriage between the federal government and the universities regarding financial support of research. Research in environmental engineering and science has traditionally been well-supported, especially when such research was closely associated with public health. Both NIH and NSF actively participated in the marriage. The former is no longer a factor and NSF is only fractionally devoted to large-scale environmental research. Federal agencies which have peripheral interests in environmental research also are oriented toward other major missions; that is, they were established to solve specific national problems or, at any rate, to devote their efforts to specific national concerns. Thus, they would not appear to be, nor have they proven to be, very reliable marriage partners in environmental research over the long haul. Hence, we ought not to build graduate research programs which are totally dependent upon them. The question then is - how do we diversify investigative support? Well, it is not immediately apparent, but the second concern, defining the role of the university in the societal picture, bears on the first concern. Universities usually educate by organizing factual information and disseminating it to students; that is, they engage in teaching. Also, they seek new information to disseminate; that is, they do research. Usually, they do this in a cloistered society consisting of professor and students. Are there other roles that the university should play? Should universities reach out to a broader constituency? Ought not they reach out to other segments of society, such as state and local governmental agencies and commercial and industrial firms, by providing new information and services specifically needed by these segments of the public. One way in which some universities have done this, and have expanded their base of investigative support as well, has been the formation of industrial research complexes and establishment of research institutes aimed at specific industrial interests. These modes of university-industry marriage have been successful in several engineering science areas which are, for the most part, allied to new saleable products, improvements in production methods, etc. I have been personally concerned with possibilities for such marriages regarding pollution control, but have been fully cognizant of the fact that the interest of industry in such activities could not compare with its interest in development of new products.

In any event, it seemed that industrial support for environmental engineering research would require that the investigator be willing or desirous of solving problems of direct interest to industry. That is, all this research could not be fundamentally exploratory, could not be the type of research which seeks to delineate basic principles, for which we may find support through NSF. This problem not only arises when seeking industrial support but also is there to some extent when seeking support from the mission-oriented government agencies, for example, Army, Navy or EPA. Nonetheless, we are applied scientists and engineers and we ought to be able to satisfy the needs of industry and mission-oriented agencies. Furthermore, it would seem that participation in these newer kinds of university-industry activities can, in many ways, be as gratifying to professors and as helpful to students as participation in the traditional mode of graduate training and research projects.

These things were on my mind when Wes Eckenfelder and Yeun Wu posed the idea of forming a Consortium for research on "Fixed Film Reactors." Our chance meeting occurred at the Second International Conference on Fixed Film Reactors in Cincinnati. They accepted my suggestion that the concern be broadened to include all biological processes, and we parted with the feeling that we three agreed on the desirability of forming some kind of confederated research team to offer to potential industrial sponsors. For the next two years or so, we continued to talk about how to organize the Consortium and each tended his own business of training and research. We felt that the Consortium should do more than offer research services. It was essential to seek participation of the sponsors in project design and to foster periodic sponsor review of projects. Also, the university-sponsor activities should include dissemination of new information through sponsorship of conferences and workshops; it should also include state-of-the-art reports on topics of interest to members. Our previous experience with industry and municipalities indicated that, while the success of the usual research project sponsored by NSF or other federal agencies was measured by good publications and usually graduate students, these non-conventional sources of research sponsorship looked beyond this end-product to practical solution, in the field, of the problem which engendered the interest in supporting the investigation. Thus, the people who performed the research might be expected to carry it forward to practical application.

How one gets this across to potential sponsors and how one handles the continuing contact are things we are learning about as we go along. Recall that I referred to the Consortium as a confederation; that is, we did not, by forming it, give up any autonomy in the pursuit of our individual research efforts. Rather, we see it as a way to enhance the individual programs at our universities, while providing an opportunity to engage in joint research when opportunities arise.

We convinced our respective university administrations that they should give us their blessings and a little seed money. This took a little time but was readily accomplished, and we obtained seed money to cover costs of brochures, mailing, etc. Since we were dealing with research and services relating to a non-product area, we could not expect industrial support in amounts which various university institutes can command. We selected a yearly membership fee of $5000.00 as one which would be attractive to industry. Members receive individual copies of any news letters, research publications, manuals of practice or state-of-the-art papers issued under the aegis of the Consortium. Membership also includes the availability of Consortium professors for surveys, studies and services, as well as member input into formulation of topics to be researched under direction of the Consortium. It also includes registration fees at conferences such as this one. Memberships with considerably reduced rights are available for $1000.00 per year. These members receive research publications, news letters, etc., but they do not participate in planning the research efforts of the Consortium and there is no committment to provide seminars.

CURRENT ACTIVITIES

Thus far, Alcoa, Westinghouse, DuPont, and Rockwell International have become members. We have also gathered a distinguished group of scientists and engineers with industrial, governmental and academic backgrounds to serve on our scientific advisory board.

As a means of helping to support research on the biological aspects of environmental science, each member university has an agreement in consonance with its own internal rules to return a portion of the project overhead to the Consortium. For example, at the University of Delaware, 15% of overhead monies generated by a project being done under the auspices of the Consortium is returned to the Consortium. The aim is to place all or part of this overhead fund at the disposal of the principal investigator for enhancement of the goals of the project or to fund exploratory research. The source of support may be traditional (NSF, EPA, etc.) or it may be industrial, municipal or joint support. We encourage development of many sources of research sponsorship. For our part, at the University of Delaware, we have been very much interested in developing non-traditional sponsorship of research. We felt that this was particularly suited to the activity of the Consortium, and it has been a personal professional aim to extend the fundamental type of research which elucidates principles to application in the field. One project we have recently completed involved determination of the range of values of the kinetic constants for a high-strength landfill leachate. This project was sponsored by the Delaware Solid Waste Authority. The research led to functional design recommendations. Also it provided an opportunity to work on a wastewater for which there was a high residual COD in the treated effluent. Thus, it provided an input to basic studies on the relationship between influent feed and effluent COD, a topic we have studied using synthetic wastes of known composition. We will get underway this summer a research project for the City of Baltimore in which our goal is to develop a test procedure for grading influent industrial wastes for their effects on the operational efficiency of the Baltimore activated sludge process. This is a highly applied investigation, yet it has many ramifications and helpful inputs to fundamental kinetics of growth on toxic and nontoxic wastewaters and the work will complement, and expand upon, some of our more fundamental research investigations.

One thing we are finding is that to develop this kind of research support one does not write up his proposal, submit it

through the University Business Office and wait for the review
and decision. These sponsors do not rely on outside review.
Hence, they want to know much more about the project and how it
will contribute to solution of their specific problems. Thus,
one confers with the prospective sponsor many times until there
is a meeting of minds and there is usually some technical input
into the design of the project by the sponsor. Finally, there
are the administrative, contractual and general business aspects
of the project in which the principal investigators must participate. For example, these sponsors contract for work differently than NSF. They are not in the business of supporting research.
One must serve as the contact and bring together the appropriate
business office personnel, auditing personnel, etc., for both
the University and the potential sponsor. This requires of the
professor more than his usual non-participation (sometimes to
the extent of abhorrence of such participation) in business
affairs. However, the project does not move unless there is
such participation on the part of the principal investigator.
Both the University and the potential sponsor need such participation.

In closing, I will say that we are learning how to extend
the University's investigative services to a wider public and
range of sponsorship. This surely does not mean a move away
from seeking federal sponsorship. In fact, federal sponsorship
should be enhanced by development of these other sources of
support because such diversification can provide a wide range
of experience for principal investigators. In any event, it is
a new and, thus far, scientifically rewarding experience for
us.

I hope this rather informal accounting of some of our
Consortium-related activities has been of interest and some
help to the conference participants.